Ergodic Theory via Joinings

Mathematical
Surveys
and
Monographs

Volume 101

Ergodic Theory via Joinings

Eli Glasner

American Mathematical Society

2010 *Mathematics Subject Classification.* Primary 37Axx, 28Dxx, 37Bxx, 54H20, 20Cxx.

Library of Congress Cataloging-in-Publication Data

Glasner, Eli, 1945–
Ergodic theory via joinings / Eli Glasner.
p. cm. — (Mathematical surveys and monographs ; v. 101)
Includes bibliographical references and indexes.
ISBN 0-8218-3372-3 (alk. paper)
1. Ergodic theory. 2. Topological dynamics. 3. Measure theory. I. Title. II. Mathematical surveys and monographs ; no. 101.

QA611.5.G53 2003
515′.42—dc21

2002043617

AMS softcover ISBN 978-1-4704-1951-6

∞ The paper used in this book is acid-free and falls within the guidelines
established to ensure permanence and durability.
Visit the AMS home page at `http://www.ams.org/`

10 9 8 7 6 5 4 3 2 1   19 18 17 16 15 14

To Hillel Furstenberg

Contents

Introduction

Ergodic theory and topological dynamics are two branches of the modern theory of dynamical systems. The first, though not in its broadest definition, deals with groups acting on a probability measure space in a measure-preserving way; the second, with the action of groups on compact spaces as groups of homeomorphisms. Both theories originated as abstractions of certain aspects of the classical theory of dynamical systems that evolved during the second half of the seventeenth century and through the eighteenth and nineteenth centuries, from Newton to Poincaré.

The terminology used today by both branches is almost the same. One speaks of transitivity, ergodicity, weak and strong mixing, distality, rigidity, etc. both in ergodic theory and in topological dynamics. Even more surprising is the fact that major theorems in both areas read almost the same. To mention one conspicuous example, compare the statement of H. Furstenberg's theorem, identifying topologically distal dynamical systems as inverse limit of isometric extensions ([**71**]), with R. Zimmer's theorem, characterizing measure distal systems (i.e. systems having a separating sieve) as systems admitting Furstenberg's towers of (measure) isometric extensions ([**267**], [**268**]). Notwithstanding, the methods and nature of proofs in these two parallel fields are very different. Whereas topological dynamics relies mostly on point set topology and variations on the theme of topological groups, ergodic theory is basically a branch of probability theory, with strong connections, via the representation on the various L^p spaces, to functional analysis and spectral theory on the one hand, and to combinatorics, via Ornstein's combinatorial apparatus, on the other.

In the last thirty years or so, ergodic theory has gradually seen a growing tendency towards the use of an important new tool, that of "joinings". In the ring of integers $\mathbb{Z}$ two integers m and n have no common factor if whenever $k|m$ and $k|n$ then $k = \pm 1$. They are disjoint if $m \cdot n$ is the least common multiple of m and n. Of course in $\mathbb{Z}$ these two notions coincide. In his seminal paper of 1967 [**74**], H. Furstenberg introduced the same notions in the context of dynamical systems, both measure-preserving transformations and homeomorphisms of compact spaces, and asked whether in these categories as well the two are equivalent. The notion of a factor in, say the measure category, is the natural one: for an acting group Γ the dynamical system $\mathbf{Y} = (Y, \mathcal{Y}, \nu, \Gamma)$ is a factor of the dynamical system $\mathbf{X} = (X, \mathcal{X}, \mu, \Gamma)$ if there exists a measurable map $\pi : X \to Y$ with $\pi(\mu) = \nu$ that intertwines the actions of Γ on the phase spaces X and Y. A common factor of two systems $\mathbf{X}$ and $\mathbf{Y}$ is thus a third system $\mathbf{Z}$ which is a factor of both. A joining of the two systems $\mathbf{X}$ and $\mathbf{Y}$ is any system $\mathbf{W}$ which admits both as factors and is in turn spanned by them. According to Furstenberg's definition the systems $\mathbf{X}$ and $\mathbf{Y}$ are disjoint if the product system $\mathbf{X} \times \mathbf{Y}$ is the only joining they admit.

In 1979, D. Rudolph, using joining techniques, provided the first counter example in the category of measure preserving transformations [**217**]. In this important work Rudolph laid the foundation of joining theory. He introduced the class of dynamical systems having "minimal self-joinings" (MSJ), and constructed a rank one mixing dynamical system having minimal self-joinings of all orders. He then showed how any dynamical system with MSJ can be used to construct a counter example to Furstenberg's question as well as a wealth of other counter examples to various questions in ergodic theory. In [**137**] del Junco, Rahe and Swanson were able to show that the classical example of Chacón [**40**] has MSJ, answering a question of Rudolph whether a weakly but not strongly mixing system with MSJ exists. In [**94**] E. Glasner and B. Weiss provide a topological counterexample, which also serves as a natural counterexample in the measure category. The example consists of two horocycle flows which have no nontrivial common factor but are nevertheless not disjoint. It is based on deep results of M. Ratner [**210**] which provide a complete description of the self joinings of a horocycle flow. More recently an even more striking example was given in the topological category by E. Lindenstrauss, where two minimal dynamical systems with no nontrivial factor share a common almost 1-1 extension, [**170**]. The negative answer to Furstenberg's question, as well as many other reasons that will become clear to the reader, show that in order to study the relationship between two dynamical systems it is necessary to know all the possible joinings of the two systems and to understand the nature of these joinings.

Beginning with the pioneering works of Furstenberg and Rudolph, the notion of joinings was exploited by many authors; Furstenberg 1977 [**76**], Rudolph 1979 [**217**], Veech 1982 [**254**], Ratner 1983 [**210**], del Junco and Rudolph 1987 [**138**], Host 1991 [**123**], King 1992 [**155**], Glasner, Host and Rudolph 1992 [**91**], Thouvenot 1993 [**246**], Ryzhikov 1994 [**223**], Kammeyer and Rudolph 1995 [**143**], del Junco, Lemańczyk and Mentzen 1995 [**136**], and Lemańczyk, Parreau and Thouvenot 1999 [**167**], to mention a few.

Since about 1980 I was teaching at Tel Aviv University, intermittently, graduate courses in ergodic theory. The course included, at various years, related subjects in harmonic analysis, spectral theory, Gaussian systems, combinatorial number theory and Lie-group ergodic theory. However, although I intended several times to do so, I never gave a detailed proof of Ornstein's isomorphism theorem in such a course, mainly because it would have taken most of a semester to cover even a survey of it.

My approach to teaching Ornstein's theorem changed under the influence of Rudolph's monograph "The fundamentals of measurable dynamics" [**220**], which appeared in 1990. I immediately liked the book, especially its joining approach to Ornstein's isomorphism theorem, but I found it hard to read. My efforts to put the relevant part of his book into the framework of my course resulted in writing the second part of the present book.

Besides Rudolph's book, I leaned as well on the books of Weiss 1972 [**260**], Parry 1981 [**203**], Furstenberg 1981 [**77**], Walters 1982 [**257**], Petersen 1983 [**205**], Shields 1996 [**235**], and Weiss 2000 [**264**]. Many papers, by various authors, served as bases for the relevant chapters or sections of the book. The reader will be able to follow these more specific attributions when they appear in the text and sometimes in the "Notes" sections at the end of each chapter.

The choice of "joinings" as a leitmotif of the book dictated, to a great extent, the choice of its subjects. The emphasis is on the abstract theoretical aspects of ergodic theory and consequently the reader will find that many important classes of examples are only briefly discussed, or even worse, are hardly mentioned at all. Thus for example the fascinating theories of interval exchange transformations, billiards, and substitution systems are not treated. To the first two of these subjects, the book by Cornfeld, Fomin and Sinai [**43**] is a good introduction. For the third, see Queffelec's book [**208**]. The important class of dynamical systems with minimal self-joinings is treated here in part as a special case of the theory of simple systems. For a detailed study of this class and its outstanding example, the Chacón transformation see Rudolph's book [**220**]. Perhaps the most important applications of the Furstenberg-Zimmer structure theorem, which we study in details in Chapter 10, are Furstenberg's proof of Szemerédi's theorem and numerous related applications. It takes a separate book to describe these developments and fortunately such a book exists. Furstenberg's book [**77**] is the most accessible source for learning about the applications of ergodic theory and topological dynamics to combinatorial number theory.

To sum up, the topics I cover in this book are restricted and many important areas of ergodic theory and topological dynamics are not presented. Of course this bias reflects only my personal tastes, interests and knowledge.

The abstract theory of minimal $\mathbb{Z}$-systems generalizes easily to actions of a general group Γ of homeomorphisms (unless we say otherwise, Γ means a countable discrete group). Accordingly monographs dealing with topological dynamical systems usually present the theory in this general setup (e.g. Gottschalk and Hedlund 1955 [**106**], Ellis 1969 [**58**], Glasner 1976 [**85**], Bronstein 1979 [**36**], de Vries 1993 [**256**] and Auslander 1997 [**11**]). In contrast, the introductory books on ergodic theory usually treat the subject in the more restricted traditional setup of a one parameter acting group (mostly $\mathbb{Z}$ actions, and sometimes $\mathbb{R}$ actions). E.g., in addition to the books mentioned above, Halmos 1956 [**113**], Jacobs 1960 [**127**], Denker, Grillenberger and Sigmund 1976 [**46**], Cornfeld, Fomin and Sinai 1982 [**43**], Krengel 1985 [**159**], and Aaronson 1997, [**1**]. Notable exceptions in this respect are the books by Schmidt 1977 [**226**] and Zimmer 1984 [**269**].

I tried to treat the subjects in the first part of the book within the general framework of a discrete, countable but otherwise general acting group Γ on a standard Lebesgue probability space. Even in this part, though, many of the more specialized results, as well as most of the examples, are presented for $\mathbb{Z}$ actions; either for the sake of convenience, or because the more general theory is not available. In the second part of the book, I give up the attempt to work with general Γ altogether, and shift to the setup of $\mathbb{Z}$ actions.

A brief description of the content of the book follows.

Part I:

The *first chapter* gives a concise introduction to the abstract theory of topological dynamics. Topological dynamics today is a vast area of research and our treatment includes only a few aspects of it. As mentioned above there is, especially in the structure theory of minimal systems, a strong parallelism with ergodic theory and since the topological theory is less technical — being free of measurability worries — sometimes results in topological dynamics are more transparent than their ergodic theory counterparts. This chapter therefore gives me the opportunity to hint at some developments to follow in the latter theory. The first part of the chapter is based on a series of lectures I gave at Luminy in 1996 [**89**]. The second part deals with the following more specialized subjects. First the recently developed theory of almost equicontinuous systems (see [**96**], [**9**] and [**10**]) is described and, in turn, leads to the theory of weakly almost periodic systems. The latter will be used later in the book as a vehicle for introducing some of the most basic definitions of ergodic theory. Also in this chapter, we introduce various families of large subsets of the group Γ (e.g. the families of syndetic and thick sets). These families will play some part throughout the first part of the book and I demonstrate their usefulness, together with the force of the enveloping semigroup, in providing the shortest and most elegant proof I know of van der Waerden's theorem on arithmetical progressions.

In the *second chapter*, the reader will find a treatment of the general setup for ergodic theory, namely the theory of measure-preserving Γ actions on standard Lebesgue probability spaces (= dynamical systems) and their factors. Here I freely use the theory of Polish spaces as developed for example in [**152**].

Motivated by the papers, Ellis and Nerurkar [**60**], Bergelson and Rosenblatt [**21**], and the appendix to Glasner and Weiss [**102**], I chose to introduce the basic notions of ergodicity and the various mixing properties as spectral properties. This means that instead of considering the dynamical system itself one considers its Koopman's representation: the natural unitary representation induced on the associated Hilbert space $L^2(\mu)$. Now in this Hilbert space the group Γ, represented as a group of unitary operators, acts on the unit ball $\mathbf{B}$ of the Banach space $\mathfrak{L}(L^2(\mu))$ of all bounded linear operators from $L^2(\mu)$ to itself. With the appropriate topology, the topological action $(\mathbf{B}, \Gamma)$ forms a weakly almost periodic (WAP) system. This rather abstract approach has the advantage of generality, efficiency, and (once one accepts the abstract setup) simplicity. It also emphasizes the fact that these definitions and results are purely spectral. All of this is to be found in the *third chapter*, where I use the abstract theory of weakly almost periodic systems from Chapter 1 to obtain the classical basic results on ergodicity and mixing. Next in this chapter come a proof of Rohlin's skew-product theorem (taken from Aaronson [**1**]), a detailed study of compact group skew-products, a succinct proof of Birkhoff's ergodic theorem for $\mathbb{Z}$ actions (taken from Katok and Hasselblatt [**146**]), and an account of the hierarchy of mixing properties for $\mathbb{Z}$-systems, from weak mixing to the Kolmogorov property. The chapter ends with a brief discussion of the strong link that connects ergodic theory and probability, namely the construction of a stationary process from a dynamical system and a function. In particular Gauss processes are described.

The next chapter, *Chapter 4*, considers the situation where the measure-preserving dynamical system arises via an invariant probability measure on a compact metric space on which Γ acts as a group of homeomorphisms. The important notions of unique ergodicity and generic points are introduced. We reproduce Parry's proof of the unique ergodicity of the Heisenberg nil-flow and study the classical examples of the geodesic and horocycle flows on a compact Riemann surface. Next the versatile notion of an E-system (i.e. a topologically transitive dynamical system for which there exists an invariant probability measure whose support is the whole space) is introduced and studied. As an example of a useful theorem proved for E-systems, let me mention the following result: an E-system (X, Γ) is (topologically) disjoint from the maximal equicontinuous factor of a minimal system (Y, Γ) iff the product system $(X \times Y, \Gamma)$ is topologically transitive, in particular, every E-system is weakly disjoint from every minimal weakly mixing system (i.e. their product is topologically transitive).

Chapter 5 introduces the rudimentary facts of spectral ergodic theory that will be needed in later chapters. *Chapter 6* introduces the machinery of joinings. Besides the definition indicated above, a joining of two systems can be described by the corresponding Markov commuting operator between their Koopman's representations. This view of joinings obviate the fact that the collection of (two-fold) selfjoinings of a dynamical system forms a compact convex semigroup on which Γ acts in a weakly almost periodic way. Group extensions are studied in light of joining theory and the important theorem of Veech, characterizing group extensions in term of joinings, is proved. The relation between the notions of joinings and factors is greatly clarified by the relative independence theorem. Finally it is shown that spectral singularity implies disjointness.

The next chapter, *Chapter 7*, provides further applications of joinings. First the Halmos-von Neumann theorem and then, a joining characterization of k-fold mixing. This characterization is used in proving a theorem of Rudolph on lifting of mixing and in the proof of (the special case $\mathbb{G} = PSL(2, \mathbb{R})$ of) Mozes' theorem. The latter leads to Marcus' theorem on the mixing of all orders of the horocycle flow. Next, using Rudolph's "counterexamples machine" ([**217**]), examples are presented of: two ergodic systems with no common factor which are not disjoint, and of two weakly isomorphic systems (i.e. each is a factor of the other) which are not isomorphic. I follow del Junco and Lemańczyk [**134**] in using "α-weakly mixing", rather than Rudolph's original "minimal self-joinings", in presenting these examples.

By its very nature the notion of a joining is one that connects two systems. Can one detect the "trace" of a joining of two systems on each of them individually? In fact this is possible and this fact leads one to consider "quasifactors" of a dynamical system. These objects live on the space of probability measures on the phase space of the system and include representations of all the factors of the system but usually much more, namely all the "traces" of joinings with other systems. For example two systems $\mathbf{X}$ and $\mathbf{Y}$ are disjoint iff no factor of $\mathbf{Y}$ is a quasifactor of $\mathbf{X}$. The usefulness of quasifactors is in the fact that, unlike joinings, they are completely determined by the system itself. In the *eighth chapter* I define quasifactors and joining quasifactors and investigate some of their properties. The main result of the chapter is to show, using the de-Finetti-Hewitt-Savage theorem, that quasifactors are isomorphic to countable symmetric selfjoinings. In the last section Choquet's theorem is used to prove the ergodic decomposition theorem for measure-preserving systems of a general group Γ. Here the theory of quasifactors is applied to identify the Choquet decomposition with the decomposition of the measure over the σ-algebra of Γ-invariant sets.

The next two chapters (*Chapters 9 and 10*) expound the beautiful theory of distal extensions due to H. Furstenberg and R. Zimmer ([**76**], [**77**], [**267**] and [**268**]). The main result here is a structure theorem that (in its absolute form) says that, in a canonical way, every ergodic Γ-system is a weakly mixing extension of its largest distal factor.

One of the most intriguing questions left open in ergodic theory is Rohlin's question whether mixing (of order two) implies mixing of all orders. B. Host in his paper [**123**] of 1991, shows (again with the use of joinings) that this is the case for $\mathbb{Z}$-systems whose spectral type is singular. In *Chapter 11* I give a detailed report of this exiting result and the machine built in order to prove it.

A similar (and related) open problem is to be found in the subject of simple systems. These systems, whose definition is in terms of the restricted nature of their selfjoinings, were introduced by Veech in [**254**] and were further investigated in del Junco and Rudolph [**138**], King [**155**] and Glasner, Host and Rudolph [**91**]. The open question is whether two-fold simplicity implies simplicity of all orders. J. King has shown that simplicity of order 4 implies simplicity of all orders [**155**], and in Glasner, Host and Rudolph [**91**] this is improved to show that in fact simplicity of order 3 suffices. The theory of simple systems, mainly based on del Junco and Rudolph [**138**] and Glasner, Host and Rudolph [**91**], is described in *Chapter 12*.

In *Chapter 13* I come back to the situation considered in Chapter 4 where the objects of interest are the invariant measures of a topological dynamical system. It turns out that the structure of the compact convex set $M_\Gamma(X)$ of these invariant probability measures reflects certain algebraic properties of the acting group Γ. The main theorem presented here asserts that an infinite group Γ does not have Kazhdan's property T iff the Choquet simplex $M_\Gamma(\Omega)$ of Γ-invariant probability measures on the topological Bernoulli system (Ω, Γ) has the property that its set of extreme points is dense. Some tools used in the proof of this theorem are: the notion of strong ergodicity, the theory of Gaussian dynamical systems and the ubiquitous notion of joinings; we also have here a first glimpse of the method of symbolic representations that will play a major role in the second part of the book.

Part II:

Chapter 14 introduces briefly the basic definitions and theorems of entropy theory, both topological and measure-theoretical. Then the important "ergodic theorem of information theory" — the Shannon-McMillan-Breiman theorem — is proven. The chapter concludes with some basic examples of computations of entropy.

Chapter 15 introduces the two related notions: "symbolic representations" of a dynamical system and "Rohlin towers". These form the main tools of Ornstein's modern ergodic theory. They provide access to combinatorial methods of investigation and to the study of fine properties of ergodic systems with positive entropy. In this chapter these tools are applied to the study of generic points, to investigation of Ornstein's $\bar{d}$ metric, and to a proof (due to B. Weiss) of the astounding Jewett-Krieger theorem, which provides a uniquely ergodic model for every ergodic system. The presentation I give of this proof is a slight elaboration on that given in Weiss' monograph [**264**].

A special case of the useful method of construction of dynamical systems called "cutting and stacking" is briefly described in *Chapter 16*. This is the "rank one construction" and it is used here to show that α-weakly mixing systems — those used in Chapter 7 to produce counter examples — exist (here I mainly follow del Junco and Lemańczyk [**135**]).

The next three chapters bring us back to entropy theory. In *Chapter 17* a proof is given of the celebrated "variational principle" which characterizes the topological entropy of a topological system (X, T) as the supremum of the measure entropies $h_\mu(X, T)$ over all T-invariant probability measures $\mu \in M_T(X)$. Also presented is a recent extension of this principle, a "variational principle for open covers".

Chapter 18 introduces the Pinsker algebra of a dynamical system as the maximal factor of zero entropy. The Rohlin-Sinai theorem, relating the Pinsker algebra to partition theory, is proved and finally systems of zero entropy are studied; in particular it is shown that distal systems have zero entropy.

Chapter 19 deals with "local" notions of entropy, both measure-theoretical and topological. An entropy pair is a pair of points (x, x') in a dynamical system (X, T) which forces positive entropy in two-set open covers (or measurable partitions) which topologically separate x and x'. The theory of such pairs leads to a better understanding of entropy theory and in particular clarifies the relationship between measure-theoretical entropy and topological entropy. (For example, the variational principle for open covers is used here to a great advantage.) It also provides a successful analogue to the basic notions of K-systems and Pinsker factors in topological dynamics (these are called UPE systems and topological Pinsker factors respectively).

My debt to Rudolph's book [**220**] is easy to detect in the previous chapters as well, but in this last chapter, *Chapter 20*, the treatment is essentially a careful reading of chapter 7 of [**220**]. Following Rudolph I present R. Burton's and A. Rothstein's joining approach to the proofs of Krieger's finite generator theorem and Ornstein's isomorphism theorem. This approach, together with the combinatorial machine of symbolic representations, unifies and simplifies the proofs of these two theorems, which can be viewed as the most profound (but still accessible in an introductory course) achievements of ergodic theory.

Mostly in the first part of the book, some material and many examples are presented as exercises. The more difficult of these are followed by an indication of a solution or a convenient reference (within double brackets [[...]], as in [**256**]).

I did not hesitate to use a heavy theorem from other parts of mathematics in order to save efforts in proving a result in ergodic theory. Some examples are the use of Eberlein-Šmulian's theorem in the treatment of WAP theory, the theory of Polish spaces and analytic sets in the treatment of dynamical systems on Lebesgue spaces, Choquet's theory in proving the ergodic decomposition and the use of the martingale convergence theorem in entropy theory. However I always include (sometimes in the appendix to the book) a precise formulation of the required results together with a detailed reference as to where the reader can find an accessible proof. Of course this means that we assume, in addition to good control of topology and measure theory, some knowledge of basic functional analysis.

In the same manner, in order to keep the book in a reasonable size, I sometimes preferred, even within the confines of ergodic theory — when dealing with the classical standard parts of the theory — short and stream-lined proofs to new, perhaps more transparent, but longer ones. For example I have chosen to follow the martingale approach to the proof of the Shannon-McMillan-Breiman theorem rather than the combinatorial, more conceptual proof of Ornstein and Weiss ([**196**]).

Each chapter ends with a short "Notes" section, where I indicate my main sources for the chapter. I sometimes add a few words on the history of the relevant results and comment on related new developments. I have not tried to trace theorems back to their origin; sometimes the references given merely indicate where I found them.

I follow usual mathematical notations; in particular $\mathbb{N} = \{0, 1, 2, \dots\}$ and $\mathbb{Z}, \mathbb{Q}, \mathbb{R}$ and $\mathbb{C}$ stand for the ring of integers, and the fields of rational, real and complex numbers respectively. The real and imaginary parts of a complex number z are denoted by $\Re z$ and $\Im z$ respectively. $\mathbb{T}$ denotes the 1-dimensional torus considered as the quotient group $\mathbb{R}/\mathbb{Z}$, but sometimes I use the same letter to denote the isomorphic circle group or 1-sphere $S^1 = \{z \in \mathbb{C} : |z| = 1\}$. The cardinality of a set F is denoted by $\operatorname{card} F$, or sometimes by $|F|$. Sequences like $\{a_n\}_{n \in \mathbb{N}}$ are often denoted simply by a_n when there is no room for confusion.

The indicator function of a set A is denoted by $\mathbf{1}_A$ and $\mathbf{1}_X$ is usually denoted by $\mathbf{1}$ when X is the whole space. For a subset A of a space X the set $\{x \in X : x \notin A\}$, the *complement* of A is denoted by A^c. The set $A \cap B^c$ is denoted by $A \setminus B$ and the set $(A \setminus B) \cup (B \setminus A)$, the *symmetric difference* of A and B, is denoted by $A \triangle B$. If $f : X \to Y$ is a function and $A \subset X$ then $f \upharpoonright A : A \to Y$ (and sometimes also $f|_A$) is the restriction of f to A, and $\mathrm{id} : X \to X$ is the identity function $\mathrm{id}(x) = x$, $\forall x \in X$. I often use $x \mapsto f(x)$ to describe the image of the map f at a general point x. The notation $B := A$ is used when this equation defines B.

When (X, d) is a metric space, $x \in X$ and $\epsilon > 0$, we write $B_\epsilon(x)$ for the ϵ-ball $\{x' \in X : d(x, x') < \epsilon\}$. A topological space X is a *Polish* space if it is second countable and metrizable with a compatible complete metric (this is a topological property); it is called *Cantor* if it is zero dimensional (i.e. has a basis for its topology consisting of clopen (= open and closed) sets), compact, perfect (i.e. has no isolated points) and separable (all Cantor spaces are homeomorphic). For a topological space X the Banach space of all (real or complex valued) continuous functions is denoted by $C(X)$ and $C_b(X)$ denotes the subspace consisting of the bounded functions. Thus for a compact X we have $C_b(X) = C(X)$. For a compact metric space X Riesz' representation theorem is used to identify the dual Banach space $C(X)^*$ with the Banach space of finite signed Borel measures on X with the total variation norm. Thus for such a measure ν:

$$\nu(f) = \int_X f \, d\nu, \quad (f \in C(X)), \quad \text{and} \quad \|\nu\| = |\nu|(X),$$

where the measure $|\nu|$ is the *total variation* of ν. In particular the space $M(X)$ of probability measures on X is usually considered as a subset of $C(X)^*$ and is endowed with the (compact) weak* topology inherited from $C(X)^*$. When $\phi : X \to Y$ is a continuous map from the compact space X into the compact space Y and μ is a Borel measure on X we write $\nu = \phi_*(\mu)$ (or sometimes just $\phi(\mu)$) for the image of μ under the covariant map $\phi_* : C(X)^* \to C(Y)^*$ induced by ϕ. Thus

$$\nu(f) = \mu(f \circ \phi), \quad (f \in C(Y)).$$

If $(X, \mathcal{X}, \mu)$ is a probability space (i.e. $\mathcal{X}$ is a σ-algebra of subsets of the set X and $\mu : \mathcal{X} \to \mathbb{R}$ is a probability measure) then for $1 \le p \le \infty$, $L^p(X, \mu) = L^p(\mu)$ is the Banach space of all (classes) of measurable functions $f : X \to \mathbb{R}$ (or $f : X \to \mathbb{C}$) with $\|f\|_p = \left(\int |f|^p \, d\mu\right)^{\frac{1}{p}} < \infty$ (or $\|f\|_\infty = \operatorname{esssup}|f| < \infty$ when $p = \infty$). For a countable set Λ equipped with the (infinite) counting measure, the Banach space $\ell^\infty(\Lambda)$ coincides with the Banach space $C_b(\Lambda)$, where Λ is considered as a topological space with the discrete topology. I usually leave it to the reader to decide, according to the context, whether the various function spaces, $C(X)$ or $L^p(X, \mu)$, are meant to comprise real or complex valued functions. On a probability space $(X, \mathcal{X}, \mu)$ the expectation of a measurable function f (i.e. the integral with respect to μ) is denoted by $\mathbb{E}(f)$ and when $\mathcal{F} \subset \mathcal{X}$ is a sub-σ-algebra $\mathbb{E}^{\mathcal{F}}(f)$ is the conditional expectation of f with respect to $\mathcal{F}$. If $(Y, \mathcal{Y}, \nu)$ is another probability space then $(X \times Y, \mathcal{X} \otimes \mathcal{Y}, \mu \times \nu)$ is the product probability space, where $\mathcal{X} \otimes \mathcal{Y}$ is the σ-algebra generated by the collection $\mathcal{X} \times \mathcal{Y}$. An analogous notation is used for infinite products.

On a measurable space $(X, \mathcal{X})$ the relation of absolute continuity of measures is denoted by $\nu \ll \mu$ (ν is absolutely continuous with respect to μ). The notation $\nu \sim \mu$ means $\nu \ll \mu$ and $\mu \ll \nu$; i.e. ν and μ are equivalent. The greatest lower bound of the measures μ and ν is denoted by $\mu \wedge \nu$ and when $\mu \wedge \nu = 0$, i.e. when μ and ν are mutually singular, I write $\mu \perp \nu$. For a probability measure μ the abbreviation μ a.e. (almost everywhere) is used for a property that holds on a set of μ measure one.

For a compact topological group G, m_G denotes its normalized Haar measure, which is the unique left and right invariant Borel probability measure on G. When μ and ν are finite Borel measures on G, their convolution is denoted by $\mu * \nu$.

I was very fortunate all these years to have two great mathematicians as teachers and friends, Hillel Furstenberg and Benjy Weiss. I owe them a great part of what I know about mathematics. I would like to thank B. Weiss in particular for the constant and invaluable help he offered me in writing this book. I am particularly grateful to Mariusz Lemańczyk who read a preliminary version of the book and whose extensive and illuminating criticism contributed much to improve it. Working with friends on a mathematical problem is even more fun than working on it by yourself. I thank all my co-authors and friends for their cooperation.

Thanks are due to the two anonymous readers of the book for many instructive suggestions.

Part 1

General Group Actions

CHAPTER 1

Topological Dynamics

The objects one studies in topological dynamics are *topological systems* (or, when the context is clear, just systems) (X, Γ) where X is a compact, usually metric, space (sometimes called the *phase space*) and Γ a second countable locally compact topological group represented on X as a group of self homeomorphisms (sometimes called the *acting group*). We often say that Γ *acts on* X. For simplicity we usually take Γ to be infinite countable and discrete. Thus unless we say otherwise Γ denotes an infinite countable and discrete group. Also, unless stated explicitly otherwise, our topological systems are assumed to be metrizable. However we shall have to deal with non-metrizable topological systems when we introduce the "universal" topological systems.

In the first three sections of this chapter we introduce the standard definitions of topological transitivity, minimality, weak mixing, equicontinuity and distality. In Section 4 we present the basic theory of the enveloping semigroup $E(X, \Gamma)$ associated with the system (X, Γ). Section 5 deals with the correspondence between pointed dynamical systems and closed Γ-invariant subalgebras of $\ell^\infty(\Gamma)$. Ellis' joint continuity theorem and Furstenberg's distal structure theorem are stated in Sections 6 and 7 (without proofs; though a proof of Ellis' theorem for the metrizable case, using almost equicontinuity, is given in Section 8). Sections 8-10 introduce and study the classes of almost equicontinuous and weakly almost periodic systems. In the last section (Section 11) I could not resist the temptation to present a succinct "enveloping semigroup proof" of van der Waerden's theorem on the existence of arbitrarily long monochromatic arithmetical progressions in any finite coloring of the set of natural numbers.

In the rest of the book, the only place where we shall use the enveloping semigroup theory is in Chapter 3 where we use a rather restricted part of the theory, namely the part pertaining to the enveloping semigroup of a system $(\mathbf{B}, \Gamma)$ where $\mathbf{B}$ is the unit ball in the Banach algebra $\mathcal{L}(\mathfrak{H})$ of bounded linear operators on a Hilbert space $\mathfrak{H}$ equipped with the weak operator topology, and the group Γ is a group of unitary operators acting by multiplication (a weakly almost periodic system). Accordingly the readers can skip the more technical details of Chapter 1 if they are ready to believe — or otherwise directly prove — the relevant assertions for the system $(\mathbf{B}, \Gamma)$.

1. Topological transitivity, minimality

Given a dynamical system (X, Γ) we let γx denote the image of $x \in X$ under the homeomorphism corresponding to the element $\gamma \in \Gamma$. Let Γx or $\mathcal{O}_\Gamma x$ be the Γ *orbit* of x; i.e., the set $\{\gamma x : \gamma \in \Gamma\}$. $\bar{\mathcal{O}}_\Gamma x$ will denote the orbit closure of x. If (X, Γ) is a system and Y a closed Γ-invariant subset, then we say that (Y, Γ), the restricted action, is a *subsystem* of (X, Γ). For topological systems (X, Γ) and (Y, Γ), their *product system* $(X \times Y, \Gamma)$ is defined by the diagonal action: $\gamma(x, y) = (\gamma x, \gamma y)$. Higher order products are defined analogously and we write (X^n, Γ) for the n-fold product system $(X \times X \times \cdots \times X, \Gamma)$.

Often our examples will be of topological systems with acting group $\mathbb{Z}$. In this case we let T represent the homeomorphism corresponding to the generator 1 of $\mathbb{Z}$ and we write (X, T) instead of $(X, \mathbb{Z})$. The system (X, Γ) is called *topologically transitive* or just *transitive* if for every pair of non-empty open sets U, V in X there exists $\gamma \in \Gamma$ with $\gamma U \cap V \neq \emptyset$. It is called *point transitive* if there exists a point $x_0 \in X$ with $\bar{\mathcal{O}}_\Gamma x_0 = X$. Such x_0 is called a *transitive point*. It is easy to see that for metrizable systems these two notions are the same and in fact the collection of transitive points forms a dense G_δ set in X. The topological system (X, Γ) is called *minimal* if $\bar{\mathcal{O}}_\Gamma x = X$ for every $x \in X$. A point x in a system (X, Γ) is called *minimal* (or *almost periodic*) if the subsystem $\bar{\mathcal{O}}_\Gamma x$ is minimal.

It is often convenient, when working with a topologically transitive system (X, Γ), to distinguish a point $x_0 \in X$ with dense orbit. We call such a system (X, x_0, Γ) a *pointed system*. When (X, x_0, Γ) and (Y, y_0, Γ) are pointed systems, their *join* $(X, x_0, \Gamma) \vee (Y, y_0, \Gamma)$ is by definition the subsystem $\bar{\mathcal{O}}_\Gamma(x_0, y_0) \subset X \times Y$. The join construction applies to any number of topologically transitive pointed systems. Denoting $W = (X, x_0, \Gamma) \vee (Y, y_0, \Gamma)$ we clearly have $\pi_X(W) = X$ and $\pi_Y(W) = Y$, where π_X and π_Y denote the projection maps. In general, for any two systems (X, Γ) and (Y, Γ) any closed invariant set $W \subset X \times Y$ with $\pi_X(W) = X$ and $\pi_Y(W) = Y$ is called a (topological) *joining* of the systems (X, Γ) and (Y, Γ). We say that (X, Γ) and (Y, Γ) are *disjoint* if $X \times Y$ is the only joining of these systems. When (X, Γ) and (Y, Γ) are disjoint at least one of them must be minimal: If both are not minimal and $X_1 \subset X$, $Y_1 \subset Y$ are proper subsystems then $X_1 \times Y \cup X \times Y_1$ and $X \times Y$ are two distinct joinings. Two minimal systems are disjoint iff their product is minimal. For any nontrivial system (X, Γ) we have at least the two distinct selfjoinings $X \times X$ and $\Delta_X = \{(x, x) : x \in X\}$, thus the only system which is disjoint from itself is the trivial one point system.

We call two dynamical systems *weakly disjoint* when the product system is transitive. This is indeed a very weak sense of disjointness as there are systems which are weakly disjoint from themselves. We say that a system (X, Γ) is (topologically) *weakly mixing* if the product system $(X \times X, \Gamma)$ is transitive; i.e. (X, Γ) is weakly disjoint from itself. Clearly disjointness implies weak disjointness.

We say that the system (X, Γ) is *pointwise minimal* (or *pointwise almost periodic*) if it is the union of its minimal subsystems. If this union is dense in X (i.e., the minimal points are dense in X) we say that (X, Γ) satisfies the *Bronstein condition*. If in addition the system (X, Γ) is transitive we say that it is an *M-system*. In general the closure of the union of all minimal subsets of (X, Γ) is called the *mincenter* of (X, Γ). A point $x \in X$ is a *periodic point* if $\mathcal{O}_\Gamma x$ is a finite set. For a periodic point x, $\mathcal{O}_\Gamma x = \bar{\mathcal{O}}_\Gamma x$ is a minimal set and we call a transitive system (X, Γ) a *P-system* when the periodic points are dense. Thus every P-system is an M-system.

When (X, Γ) and (Y, Γ) are topological systems and $\pi : X \to Y$ a continuous onto map which intertwines the Γ actions (i.e., $\pi\gamma x = \gamma\pi x$ for all $x \in X$, $\gamma \in \Gamma$), we say that the system (Y, Γ) is a *factor* of the system (X, Γ), or that (X, Γ) is an *extension* of (Y, Γ) and denote this by $(X, \Gamma) \xrightarrow{\pi} (Y, \Gamma)$. The map π is called a *homomorphism* of topological systems or an *extension* or a *factor map*. Such factor map defines an invariant closed equivalence relation on X (an *icer*):

$$R_\pi = \{(x, x') \in X \times X : \pi(x) = \pi(x')\}.$$

Conversely, an icer $R \subset X \times X$ defines a factor system $(Y, \Gamma) = (X/R, \Gamma)$ and a homomorphism $(X, \Gamma) \xrightarrow{\pi} (Y, \Gamma)$ with $\pi(x)$ denoting the equivalence class of x.

Let (X, Γ) be a Γ-system; a self homeomorphism ϕ of X is called an *automorphism* of (X, Γ) if $\phi\gamma x = \gamma\phi x$ for all $x \in X$ and $\gamma \in \Gamma$. We let $\mathrm{Aut}\,(X, \Gamma)$ denote the group of automorphisms of (X, Γ) . With the topology of uniform convergence of homeomorphisms $\mathrm{Aut}\,(X, \Gamma)$ is a Polish topological group.

If K is a compact subgroup of $\mathrm{Aut}\,(X, \Gamma)$ then the map $x \mapsto Kx$ defines a factor map $(X, \Gamma) \xrightarrow{\pi} (Y, \Gamma)$ with $Y = X/K$ and $R_\pi = \{(x, kx) : x \in X,\ k \in K\}$. Such an extension is called a *group extension*.

An extension $(X, \Gamma) \xrightarrow{\pi} (Y, \Gamma)$ is called *isometric extension* if there exists a continuous function $d : R_\pi \to \mathbb{R}$ such that for every $y \in Y$ the function d restricted to $\pi^{-1}(y) \times \pi^{-1}(y)$ is a metric and for every pair $(x, x') \in R_\pi$, and $\gamma \in \Gamma, d(\gamma x, \gamma x') = d(x, x')$. A well known result (which we shall not prove) shows that an extension of minimal systems $(X, \Gamma) \xrightarrow{\pi} (Y, \Gamma)$ is an isometric extension iff there exists a commutative diagram:

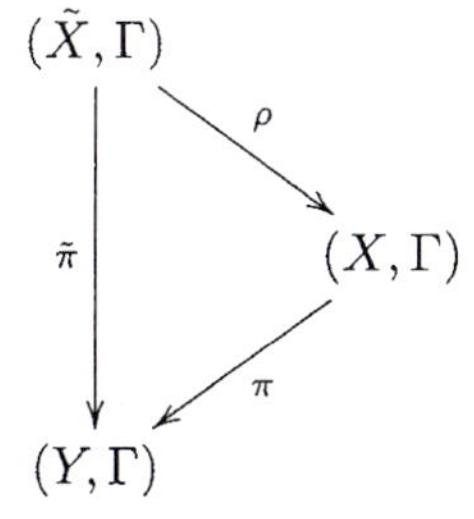

where $(\tilde{X}, \Gamma) \xrightarrow{\tilde{\pi}} (X, \Gamma)$ is a group extension with some compact group K and the map ρ is the quotient map from $\tilde{X}$ onto X defined by a closed subgroup H of K. Thus $Y = \tilde{X}/K$ and $X = \tilde{X}/H$.

For a system (X, Γ) and subsets $A, B \subset X$, we use the following notation

$$N(A, B) = \{\gamma \in \Gamma : \gamma A \cap B \neq \emptyset\}.$$

In particular for $A = \{x\}$ we write

$$N(x, B) = \{\gamma \in \Gamma : \gamma x \in B\}.$$

A subset $S \subset \Gamma$ is (left) *syndetic* if there exists a finite set $F \subset \Gamma$ such that $FS = \Gamma$. $L \subset \Gamma$ is called *thick* if for every finite set $F \subset \Gamma$ there exists $\gamma_0 \in L$ with $F^{-1}\gamma_0 \subset L$; i.e. $L \cap \bigcap_{\gamma \in F} \gamma L \neq \emptyset$.

1.1. Exercises. 1. Show that if (X, Γ) is topologically transitive, there exists in X a dense G_δ subset of points whose orbit is dense. (Recall, the space X is assumed metrizable.)

2. The following are equivalent:
   (a) (X, Γ) is minimal.
   (b) The empty set and X are the only subsystems of (X, Γ).

(c) For every non-empty open set U in X the set $\Gamma U = \bigcup\{\gamma U : \gamma \in \Gamma\}$ is all of X. *(Let U be complement of orbit closure (for $\overline{\gamma_0(c)} = \gamma(a)$))*

(d) For every non-empty open set U in X there exists a finite subset $\{\gamma_1, \ldots, \gamma_k\}$ in Γ with $\bigcup_{j=1}^k \gamma_j U = X$. *(c) ⇒ (d) by compactness*

(e) For every $x \in X$ and neighborhood U of x, the set $N(x, U) = \{\gamma \in \Gamma : \gamma x \in U\}$ is syndetic in Γ.

3. Use Zorn's lemma to show that every system has a minimal subsystem.
4. Prove the latter statement without using Zorn's lemma. [[Let $\{U_n : n \in \mathbb{N}\}$ be a basis for the topology on X. Define inductively a sequence $n_1 < n_2 < \cdots$, as follows. Let n_1 be the first integer n such that $X_1 = X \setminus \Gamma U_n$ is non-empty (if such an n does not exist then (X, Γ) is minimal by Exercise 1.1.2.(d)). Let n_2 be the first integer $n > n_1$ such that $X_2 = X_1 \setminus \Gamma U_n$ is non-empty, etc. Show that $X_\infty = \bigcap_{j=1}^\infty X_j$ has the property that $X_\infty \subset \Gamma U_k$ whenever $X_\infty \cap U_k$ is non-empty.]]
5. The properties "thick" and "syndetic" are *dual* properties: $A \subset \Gamma$ is syndetic iff $A \cap B \neq \emptyset$ for every thick set B. And $B \subset \Gamma$ is thick iff $A \cap B \neq \emptyset$ for every syndetic set A.
6. A subset $B \subset \mathbb{Z}$ is thick iff B contains arbitrarily long intervals.
7. If $B \subset \mathbb{Z}$ is thick then there is an infinite sequence $A = \{a_n\}_{n=1}^\infty$, with $D^+(A) \subset B$, where $D^+(A) = \{a_n - a_m : n > m\}$. [[With no loss of generality assume that $B \cap \mathbb{Z}^+$ is thick. Define inductively a sequence $0 < p_n$, $n \geq 1$ with
$$C = \{p_{n_1} + p_{n_2} + \cdots + p_{n_k} : \forall\, n_1 < n_2 < \cdots < n_k\} \subset B.$$
(Such a set is called an *IP-set*.) Let $p_1 > 0$ be an arbitrary element of B. Since B contains an interval of length $> p_1$ it contains p_2 such that $p_1 + p_2$ also belongs to B. Suppose $p_1, p_2, \cdots, p_n$ have been chosen so that all the partial sums $p_{i_1} + p_{i_2} + \cdots + p_{i_k}$ for $i_1 < i_2 < \cdots < i_k \leq n$ are in B. The set B contains an interval of length $> p_1 + p_2 + \cdots + p_n$. It follows that there exists $p_{n+1} \in B$ with all sums $p_{i_1} + p_{i_2} + \cdots + p_{i_k} + p_{n+1} \in B$ for $i_1 < i_2 < \cdots < i_k \leq n$ are in B. Now set $a_n = p_1 + p_2 + \cdots + p_n$.]]
8. Let (X, Γ) be a minimal system and $f : X \to \mathbb{R}$ a Γ-invariant function with at least one point of continuity (for example this is the case when f is lower or upper semi-continuous or more generally when it is the pointwise limit of a sequence of continuous functions), then f is a constant. [[Let x_0 be a continuity point and x an arbitrary point in X. Since $\{\gamma x : \gamma \in \Gamma\}$ is dense and as the value $f(\gamma x)$ does not depend on γ it follows that $f(x) = f(x_0)$.]]

The following examples of dynamical systems play an important role in the general theory.

*In a group, minimal ⇔ transitive
⇔ no nonconstant invt continuous function*

1.2. EXAMPLES. 1. Let $\alpha = (\alpha_1, \ldots, \alpha_k)$ be an element of $\mathbb{T}^k$, the k-torus, with $\{1, \alpha_1, \alpha_2, \ldots, \alpha_k\}$ independent over $\mathbb{Q}$. Show that the $\mathbb{Z}$-system (X, T), where $X = \mathbb{T}^k$ and $Tx = x + \alpha$, is a minimal system (Kronecker's theorem).

2. Let $\mathbb{A}$ be the dyadic group; i.e., $\mathbb{A} = \{0, 1\}^\mathbb{N}$ and addition is performed modulo 2 with carry to the right. Let $\mathbf{1}$ be the sequence $(1, 0, 0, \ldots)$ and equip $\mathbb{A}$ with the product topology. Then the system (X, T) with $X = \mathbb{A}$ and $Tx = x + \mathbf{1}$ is a minimal system called the *dyadic adding machine*.

More generally for any infinite sequence $\mathbf{a} = (a_0, a_1, a_2, \dots)$ of positive integers greater than one, let

$$\mathbb{A} = \prod_{n \geq 0} \{0, 1, \dots, a_n - 1\}.$$

$\mathbb{A}$ becomes an abelian group — the $\mathbf{a}$-*adic integers* — under the operation of addition mod a_n at the n-th coordinate with carry to the right. Denoting $\mathbf{1} = (1, 0, 0, \dots)$, the transformation $Tx = x + \mathbf{1}$ defines a minimal topological system (X, T) — the $\mathbf{a}$-*adic adding machine*.

3. Let $\mathfrak{L} = \{1, \dots, \ell\}$ (or sometimes $\mathfrak{L} = \{0, \dots, \ell-1\}$) and let $\Omega = \Omega(\ell) = \mathfrak{L}^{\mathbb{Z}}$ be the compact space of bi-infinite sequences on the finite *alphabet* $\mathfrak{L}$. Let (Ω, S) be the $\mathbb{Z}$-system defined on Ω by left shift: $S\omega = \omega'$ where $\omega'_n = \omega_{n-1}$. Show that (Ω, S) is topologically transitive and find sequences in Ω whose orbit closure is infinite and minimal (i.e., non-periodic minimal points). A topological system of the form (X, S), where X is a closed invariant subset of Ω, is called a *subshift*. In general, for a countable group Γ, let (Ω, Γ) be the Γ-system defined on $\Omega = \{1, \dots, \ell\}^{\Gamma}$ by left shift: $(\gamma\omega)_{\gamma'} = \omega_{\gamma^{-1}\gamma'}$. We call the system (Ω, Γ) the *topological ℓ-Bernoulli system* on Γ. Again this is a topologically transitive system.
4. More generally, given any compact metric space X we can form the product space $\Omega(X) = X^{\Gamma}$ and define an action of Γ on $\Omega(X)$ by the formula: $(\gamma\omega)_{\gamma'} = \omega_{\gamma^{-1}\gamma'}$. This is the *topological X-Bernoulli system* which is topologically transitive as well.

1.3. EXERCISES. 1. A subshift $(X, S) \subset \Omega(\ell) = \mathfrak{L}^{\mathbb{Z}}$ is called a *subshift of finite type* (SFT) if there exists a finite list of finite words $\mathcal{W} = \{w_1, \dots, w_k\}$ on the alphabet $\mathfrak{L} = \{1, 2, \dots, \ell\}$ such that $X \subset \Omega$ consists exactly of those sequences $x \in \Omega$ in which these words never appear. The number $m = \max\{|w_j| : 1 \leq j \leq k\}$ (where $|w|$ is the length of the word w) is the *order* of the SFT. Show that every SFT is isomorphic to a SFT (possibly on a different alphabet) of order 2. [[If $(X, S) \subset \Omega(\ell)$ is of order m let the new alphabet be the collection $\mathfrak{L}_m(X)$ consisting of all the blocks of length m that appear in points $x \in X$. Now define the map $\phi : X \to \mathfrak{L}_m(X)^{\mathbb{Z}}$ by $\phi(x)_i = (x_i, x_{i+1}, \dots, x_{i+m-1})$ and set $Y = \phi(X)$. Show that $\phi : (X, S) \to (Y, S)$ is an isomorphism and that (Y, S) is a SFT of order 2.]]
2. With every SFT $(X, S) \subset \Omega(\ell)$ of order 2 we associate the $\ell \times \ell$ matrix $A = (a_{ij})$ of 0's and 1's defined by $a_{ij} = 1$ iff $(i, j) \notin \mathcal{W}$. Show that for every $n \geq 1$ we have $\operatorname{card} \mathfrak{L}_n(X) = \sum_{i,j=1}^{\ell} a_{ij}^{(n)}$ where $A^n = (a_{ij}^{(n)})$.
3. Let $(X, S) \subset \{0, 1\}^{\mathbb{Z}}$ be the SFT of order 2 defined by prohibiting the word 11 (the "golden mean" subshift). Show that $\operatorname{card} \mathfrak{L}_n(X) = f_{n+3}$, where f_n is the n'th Fibonacci number (i.e. $f_0 = 0, f_1 = 1, f_2 = 1$ and $f_{n+1} = f_n + f_{n-1}$ for $n \geq 1$). [[We have $A = \left(\begin{smallmatrix} 1 & 1 \\ 1 & 0 \end{smallmatrix}\right)$ as the associated matrix and $p_A(t) = t^2 - t - 1 = (t - \lambda)(t - \mu)$ is its characteristic polynomial, where $\lambda = \frac{1}{2}(1 + \sqrt{5}), \mu = \frac{1}{2}(1 - \sqrt{5})$. The matrix $P = \left(\begin{smallmatrix} \lambda & \mu \\ 1 & 1 \end{smallmatrix}\right)$ diagonalizes A; i.e. $P^{-1}AP = \left(\begin{smallmatrix} \lambda & 0 \\ 0 & \mu \end{smallmatrix}\right)$, and for every $n \geq 2$ we get

$$A^n = P \begin{pmatrix} \lambda^n & 0 \\ 0 & \mu^n \end{pmatrix} P^{-1} = \begin{pmatrix} f_{n+1} & f_n \\ f_n & f_{n-1} \end{pmatrix}.$$

In particular $f_n = \frac{1}{\sqrt{5}}(\lambda^n - \mu^n)$. Now

$$\begin{aligned}\operatorname{card} \mathfrak{L}_n(X) &= \sum_{i,j=1}^{\ell} a_{ij}^{(n)} = f_{n+1} + f_n + f_n + f_{n-1} \\ &= f_{n+2} + f_{n+1} = f_{n+3},\end{aligned}$$

by the previous exercise.]]

There is an extensive literature on subshifts of finite type investigating their topological as well as their measure-theoretical behaviour. Refer e.g. to the books by Denker, Grillenberger and Sigmund [**46**], Lind and Marcus [**169**], Schmidt [**230**] and Kitchens [**156**]. See also the survey articles by Keane [**149**] and Boyle [**34**].

2. Equicontinuity and distality

The system (X, Γ) is *equicontinuous* if the group Γ acts equicontinuously on X; i.e. for every $\epsilon > 0$ there exists $\delta > 0$ such that $d(x_1, x_2) < \delta$ implies $d(\gamma x_1, \gamma x_2) < \epsilon$, for every $\gamma \in \Gamma$. It is easy to check that, because X is compact, this definition does not depend on the choice of compatible metric and in fact, even when the phase space X is not metrizable we can use the following, more general definition. For every U, a neighborhood of the diagonal $\Delta = \{(x, x) : x \in X\} \subset X \times X$, there exists a neighborhood V of Δ such that for every $\gamma \in \Gamma$, $(\gamma \times \gamma)V \subset U$.

The topological system (X, Γ) is called *distal* if

$$\inf_{\gamma \in \Gamma} d(\gamma x, \gamma x') > 0$$

whenever $x, x' \in X$ are distinct.

1.4. EXERCISES. 1. A system (X, Γ) is equicontinuous iff it is *isometric*; i.e., there exists an equivalent Γ-invariant metric on X. [[Set $D(x, x') = \sup_{\gamma \in \Gamma} d(\gamma x, \gamma x')$.]]
2. Every equicontinuous system is distal.
3. Show that examples 1.2.1 and 1.2.2 are equicontinuous.
4. The definition of distality does not depend on the choice of a compatible metric.

A minimal equicontinuous system will be called *Kronecker*. There is a well known characterization of Kronecker systems (Theorem 1.8 below): they are the systems of the form $(K/H, \Gamma)$ where K is a compact topological group, H a closed subgroup, $\tilde{\Gamma}$ a dense subgroup of K with a homomorphism onto $j : \Gamma \to \tilde{\Gamma}$, and the action of $\gamma \in \Gamma$ on the homogeneous space K/H is by left multiplication with $j(\gamma)$. We always assume that $\bigcap_{k \in K} k^{-1}Hk = \{e\}$, the trivial subgroup consisting of the identity element of K, so that when Γ is abelian (and therefore so is K), we have $H = \{e\}$. See examples 1.2.1 and 1.2.2 above. For a non-abelian example, consider the natural action of a residually finite group Γ on its pro-finite completion. Or take Γ to be a countable dense subgroup of the Lie group $\mathbf{O}(n)$ of all orthogonal $n \times n$ matrices; we get a Kronecker action on a homogeneous space by letting Γ act on the sphere $S^{n-1} = \mathbf{O}(n)/\mathbf{O}(n-1)$.

Every topological system (X, Γ) has a maximal distal and a maximal equicontinuous factor, (X_d, Γ) and (X_{eq}, Γ) respectively. That is, (X_d, Γ) is distal, and every distal factor of (X, Γ) is a factor of (X_d, Γ). (X_{eq}, Γ) has the corresponding property for equicontinuous factors. Thus there are closed Γ-invariant equivalence relations S_d and S_{eq} such that $X/S_d = X_d$ and $X/S_{eq} = X_{eq}$. Since an equicontinuous system is distal, we have $S_d(X) \subset S_{eq}(X)$.

1.5. EXERCISES. Let (X, T) be a minimal $\mathbb{Z}$-system. Call a continuous function $f : X \to \mathbb{C}$ an *eigenfunction* if there exists a $0 \neq \lambda \in \mathbb{C}$ such that $f(Tx) = \lambda f(x)$ for all $x \in X$. We then say that λ is the *eigenvalue* associated with the eigenfunction f. (In the discussion below we use some measure theoretical notions as developed, for example, in Chapter 4.)

1. Show that for an eigenfunction f the corresponding eigenvalue $\lambda = \lambda_f$ has modulus 1 and that $|f| \equiv c$ is a constant. Since f/c is also an eigenfunction with eigenvalue λ we can and shall assume that $|f| \equiv 1$. It now follows that the map $f : X \to \mathbb{T}$ is a homomorphism of the system (X, T) into the system $(\mathbb{T}, \lambda)$ (onto iff λ is not a root of unity).
2. The eigenvalue λ has geometric multiplicity 1. [[If also $g(Tx) = \lambda g(x)$, then $f\bar{g}$ is an invariant function, hence $f\bar{g} \equiv 1$.]]
3. If f_1, f_2 are eigenfunctions corresponding to two distinct eigenvalues then $\|f_1 - f_2\|_\infty \geq \sqrt{2}$. [[Define a map $\phi : X \to \mathbb{T} \times \mathbb{T}, x \mapsto (f_1(x), f_2(x))$. Let $Y = \phi(X)$ and show that $\phi : (X, T) \to (Y, (\lambda_1, \lambda_2))$, where λ_i is the eigenvalue of f_i, $i = 1, 2$, is a homomorphism of dynamical systems. Show that there exists a unique (λ_1, λ_2)-invariant probability measure m on Y. Let μ be a T-invariant probability measure on X such that $\phi_*(\mu) = m$ and show that in $L^2(\mu)$, f_1 and f_2 are orthogonal, whence $\int |f_1 - f_2|^2 \, d\mu = 2$. This clearly implies $\|f_1 - f_2\|_\infty \geq \sqrt{2}$.]]
4. The set $\mathcal{E}$ of (normalized) eigenfunctions is at most countable. [[The fact that the Banach space $C(X)$ is separable now implies that $\mathcal{E}$ is at most countable.]]
5. Define a map $\phi : X \to \prod_{f \in \mathcal{E}} \mathbb{T}_f$, with $\mathbb{T}_f = \mathbb{T}$, by $(\phi(x))(f) = f(x), (x \in X, f \in \mathcal{E})$. Show that ϕ is a homomorphism of the dynamical system (X, T) into the product system (Ω, R) where $\Omega = \prod_{f \in \mathcal{E}} \mathbb{T}_f$ and $(R\omega)(f) = \lambda_f \omega(f)$ (i.e. R acts as multiplication by $\tilde{\lambda} \in \Omega$ with $\tilde{\lambda}(f) = \lambda_f$). Let $Z = \phi(X) \subset \Omega$. Show that (Z, R) is a Kronecker system which is isomorphic to (X_{eq}, T). [[If $\psi : (X, T) \to (K, k_0)$ is any Kronecker factor, with $K = \operatorname{cls}\{k_0^n : n \in \mathbb{Z}\}$ a compact monothetic group, and $\chi : K \to \mathbb{T}$ a character, then $f = \chi \circ \psi$ is in $\mathcal{E}$. Let $\mathcal{E}_0 \subset \mathcal{E}$ be the set of all the elements in $\mathcal{E}$ which are obtained this way. Let $\rho : \Omega \to \Omega_0 = \prod_{f \in \mathcal{E}_0} \mathbb{T}_f$ be the natural projection of Ω onto Ω_0 and denote $\rho(Z) = \tilde{K}$. Now use the fact that the characters separate points on K to show that the systems (K, k_0) and $(\tilde{K}, \rho(\tilde{\lambda}))$ are isomorphic and that the diagram

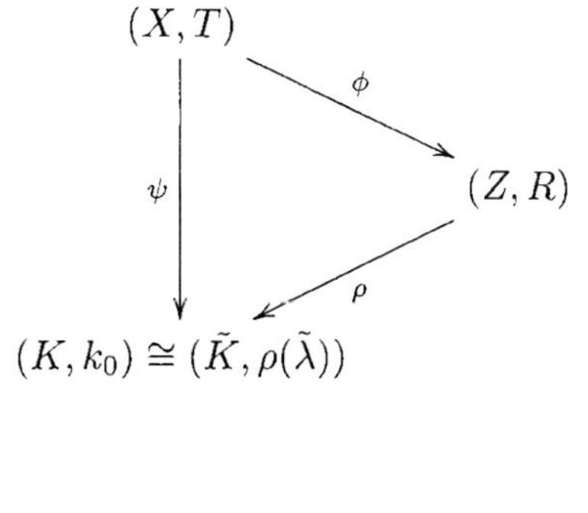

is commutative.

1.6. EXAMPLES. 1. On $X = \mathbb{T}^2$, let $T(x,y) = (x+\alpha, y+x)$ for an irrational α. Show that (X,T) is minimal distal but not equicontinuous. The map $\pi(x,y) = x$ defines a factor map onto the maximal equicontinuous factor of (X,T) which is a group extension. [[The distality follows easily by considering first pairs of points of the form $((x,y),(x,y'))$ and then pairs of the form $((x,y),(x',y'))$ with $x \neq x'$. For minimality show that if M is a *proper* minimal subset of $X \times X$ then $H = \{\beta \in \mathbb{T} : R_\beta M \cap M \neq \emptyset\}$ is the finite subgroup $H = \{0, 1/n, 2/n, \dots, (n-1)/n\}$ for some positive integer n, where $R_\beta(x,y) = (x, y+\beta)$. Show that the set $\{(x, ny) : (x,y) \in M\}$ is a graph of a continuous function $f : \mathbb{T} \to \mathbb{T}$; then use the invariance of M to get $f(x+\alpha) - f(x) = nx$, which is impossible for $n \geq 1$. Finally for non-equicontinuity show that for a sequence $n_i \to \infty$ with $T^{n_i}((0,0)) \to (0,0)$, we have $T^{n_i}((0,0),(1/(2n_i),0)) \to ((0,0),(0,1/2))$.]]

2. Modify the transformation in the previous example to read $T(x,y) = (x+\alpha, y+2x+\alpha)$. Again show that (X,T) is minimal distal. Check that $T^n(0,0) = (n\alpha, n^2\alpha)$ and deduce that the sequence $\{n^2\alpha\}_{n=1}^{\infty}$ is dense in $\mathbb{T}$. (See Furstenberg's book [**77**] for further development of these ideas.)

3. Let G be a topological group and $H \subset G$ a closed syndetic subgroup (i.e. the quotient space G/H is compact). Show that the dynamical system $(G/H, G)$ (where the action of G is given by multiplication on the left: $(g_1, g_2H) \mapsto g_1g_2H, \forall g_1, g_2 \in G)$ is distal iff

$$H = \cap\{HUH : U \text{ a neighborhood of the identity } e \in G\},$$

iff $e \in \operatorname{cls} HgH$ implies $g \in H$.

4. Let G be the *Heisenberg* group; i.e. the nilpotent group

$$G = \left\{ \begin{pmatrix} 1 & x & z \\ 0 & 1 & y \\ 0 & 0 & 1 \end{pmatrix} : x, y, z \in \mathbb{R} \right\}, \qquad H = \left\{ \begin{pmatrix} 1 & a & c \\ 0 & 1 & b \\ 0 & 0 & 1 \end{pmatrix} : a, b, c \in \mathbb{Z} \right\},$$

and

$$T = \begin{pmatrix} 1 & \alpha & \gamma \\ 0 & 1 & \beta \\ 0 & 0 & 1 \end{pmatrix},$$

with $\alpha, \beta, \gamma \in \mathbb{R}$, and α and β independent over $\mathbb{Q}$. Show that the system $(X,T) = (G/H, T)$, where T acts by multiplication on the left, is minimal distal and non equicontinuous. Such a system is called a *nil-system*. [[For the distality use the criterion of the previous exercise to show that even $(G/H, G)$ is distal. For this it suffices to show that $e \in \operatorname{cls} HgH$ implies $g \in H$. Denote

$$Z = \left\{ \begin{pmatrix} 1 & 0 & z \\ 0 & 1 & 0 \\ 0 & 0 & 1 \end{pmatrix} : z \in \mathbb{R} \right\},$$

the center of G, and deduce from $e \in \operatorname{cls} ZHgHZ$ and the fact that the system $(G/ZH, G)$ is equicontinuous, hence distal, that $g \in ZH$. Again use $e \in \operatorname{cls} HgH$ and the fact that the system $(ZH/H, ZH)$ is equicontinuous to deduce that in fact $g \in H$. For the minimality let $M \subset X = G/H$ be a minimal subset for T. Show that the system $(G/ZH, T)$ is isomorphic to the (α, β) rotation on $\mathbb{T}^2$ and in particular is minimal. Deduce that $\pi(M) = Y = G/ZH$, where $\pi : X \to Y$ is the quotient map. Next show that the compact group $K = ZH/H$, which is naturally isomorphic to the circle $\mathbb{R}/\mathbb{Z}$, acts on X by multiplication on the *right* ($R_z : (gH, zH) \mapsto gzH, \forall g \in G, z \in Z$). Set $A = \{zH \in K : R_z M = M\}$ and use the fact that the K and T actions on X commute to show that A is a closed subgroup of K, hence either all of K or a finite subgroup of K. In the first case we have that $X = M$ is minimal. Assume A is finite, let $\sigma : X \to X/A$ be the quotient map and show that the map $\phi : X/A \to Y \times K/A$, defined for $gH \in X$ by $\phi(gH) = (zH, \sigma(gH))$, where $z \in Z$ is such that $R_z(gH) = gzH \in M$, is a homeomorphism. Since the fundamental group $\pi_1(Y) = \mathbb{Z}^2$ it follows that $\pi_1(X/A) \cong \mathbb{Z}^2 \times \mathbb{Z} = \mathbb{Z}^3$. However the homotopy sequence

$$0 = \pi_1(A) \to \pi_1(X) \to \pi_1(X/A) \cong \mathbb{Z}^3,$$

corresponding to the map $\sigma : X \to X/A$ shows that $\pi_1(X)$ is isomorphic to a subgroup of $\mathbb{Z}^3$ contradicting the fact that $\pi_1(X) = H$ is non-abelian. This finishes the proof that (X, T) is minimal. To see that it is not equicontinuous observe that otherwise, being a Kronecker system, $X = G/H$ has the structure of a compact abelian group. Being a manifold it must be a torus and therefore with $\pi_1(X)$ abelian. This system and Example 1.6.1 (due to Furstenberg [**73**]) were the first examples of minimal distal but non equicontinuous $\mathbb{Z}$-systems. See [**12**] for the theory of nil-flows and more generally flows on homogeneous spaces.]]

5. Let G be the nilpotent group

$$G = \left\{ \begin{pmatrix} 1 & q & y \\ 0 & 1 & x \\ 0 & 0 & 1 \end{pmatrix} : q \in \mathbb{Z},\ y, x \in \mathbb{T} \right\}, \qquad H = \left\{ \begin{pmatrix} 1 & q & 0 \\ 0 & 1 & 0 \\ 0 & 0 & 1 \end{pmatrix} : q \in \mathbb{Z} \right\},$$

and

$$T = \begin{pmatrix} 1 & 1 & 0 \\ 0 & 1 & \alpha \\ 0 & 0 & 1 \end{pmatrix}.$$

Show that the system (X, T) from 1.6.1 is isomorphic to the nil-system $(G/H, T)$.

6. Again on $X = \mathbb{T}^2$, let $T_\phi(x, y) = (x + \alpha, y + \phi(x))$, where now $\phi : \mathbb{T} \to \mathbb{T}$ is any continuous function. Any such system is distal and it can be shown that with α irrational and ϕ having a non-vanishing degree (as a map from $\mathbb{T}$ to itself) T_ϕ is minimal. [[See Furstenberg [**72**], Theorem 2.1. and the following remark.]]

Let (X, Γ) be a topological system and $E = E(X, \Gamma)$ the closure of Γ in X^X, the space of all maps from X to itself equipped with the topology of pointwise convergence. The compact space X^X is a semigroup (under composition) and it is easy to see that E forms a subsemigroup. E is called *the enveloping semigroup* of the system (X, Γ) and we shall investigate it in details in Section 4 below. The following theorems follow easily from the enveloping semigroup machinery together with Ellis' joint continuity theorem (Theorem 1.29 below). Their proofs therefore are postponed to a series of exercises in Section 4.

1.7. THEOREM. *Let (X, Γ) be a dynamical system. The following conditions are equivalent.*

1. *(X, Γ) is distal.*
2. *$E(X, \Gamma)$ is a group.*
3. *The product system $(X \times X, \Gamma)$ is pointwise minimal.*

1.8. THEOREM. *Let (X, Γ) be a dynamical system. The following conditions are equivalent.*

1. *(X, Γ) is equicontinuous.*
2. *$E(X, \Gamma)$ is a group of homeomorphisms of X and the topologies of pointwise and uniform convergence coincide on $E(X, \Gamma)$.*
3. *$E(X, \Gamma)$ is a topological group whose action on X is jointly continuous.*
4. *$E(X, \Gamma)$ is a group of homeomorphisms of X.*

When these conditions are satisfied and (X, Γ) is minimal the system (X, Γ) is isomorphic to the homogeneous system $(K/H, \Gamma)$ where K is a compact topological group, H is a closed subgroup of K and Γ is embedded in K as a dense subgroup. If in addition Γ is abelian, $H = \{\mathrm{id}\}$.

3. Proximality and weak mixing

A pair of points $(x_1, x_2) \in X \times X$ in a topological system (X, Γ) is called *proximal* if there exist a sequence $\gamma_i \in \Gamma$ and a point $z \in X$ such that

$$\lim \gamma_i x_1 = \lim \gamma_i x_2 = z.$$

We call the subset P of $X \times X$ consisting of proximal pairs, the *proximal relation* of (X, Γ) thus

$$P = \bigcap \{\Gamma V : V \text{ a neighborhood of the diagonal in } X \times X\}.$$

1.9. EXERCISES.
1. The proximal relation $P \subset X \times X$ is a Γ-invariant symmetric and reflexive relation.
2. P is a G_δ subset of $X \times X$.
3. If P is closed it is already an icer. [[If (x_1, x_2), $(x_2, x_3) \in P$ then the orbit closure of $(x_1, x_2, x_3) \in X \times X \times X$ contains a point of the form (x, x, x') with $(x, x') \in P$ (because P is closed), hence also a diagonal point (y, y, y). Thus $(x_1, x_3) \in P$ and P is an equivalence relation.]]
4. Even for minimal $\mathbb{Z}$-systems P need be neither closed nor an equivalence relation and when P is an equivalence relation, it need not be closed (see Shapiro [**233**]).
5. Clearly (X, Γ) is distal iff $P = \Delta = \{(x, x) : x \in X\}$, the *diagonal* in $X \times X$.

The system (X, Γ) is called *proximal* if every pair is proximal; i.e., $P = X \times X$. For an integer $n \geq 2$ let

$$P^{(n)} = \bigcap \{\Gamma V : V \text{ a neighborhood of the diagonal in } X^n\}.$$

An extension $(X, \Gamma) \xrightarrow{\pi} (Y, \Gamma)$ is a *proximal extension* if $R_\pi \subset P$. An extension $(X, \Gamma) \xrightarrow{\pi} (Y, \Gamma)$ is a *distal extension* when $R_\pi \cap P = \Delta$. A point x in a system (X, Γ) is called *distal* if it is proximal only to itself.

The *regionally proximal relation* on X is defined by

$$Q = \bigcap \{\mathrm{cls}\,(\Gamma V) : V \text{ a neighborhood of the diagonal in } X \times X\}.$$

It is easy to verify that Q is trivial — i.e., equals the diagonal Δ — iff the system is equicontinuous. The relation Q is obviously Γ-invariant, closed, symmetric and reflexive. On a first glance there is no reason to think it is transitive and indeed there are examples of systems where it is not (see Exercise 1.15.2 below). However it turns out that in many important cases $Q = Q(X)$ is an equivalence relation. For example this is the case when Γ is abelian and (X, Γ) is minimal; or more generally, when (X, Γ) is minimal and it admits a Γ-invariant measure (see [**204**], [**143**] or [**11**]). Of course Q is an equivalence relation iff $Q = S_{eq}$.

We have $P \subset S_d$ and $Q \subset S_{eq}$ and the following theorem of Gottschalk and Ellis [**59**] characterizes S_d and S_{eq} in terms of P and Q.

1.10. THEOREM. *Let (X, Γ) be a topological dynamical system, then S_d is the smallest closed Γ-invariant equivalence relation containing P and S_{eq} is the smallest closed Γ-invariant equivalence relation containing Q. The quotient systems X/S_d and X/S_{eq} are respectively the maximal distal and equicontinuous factors of X.*

When $(X, \Gamma) \stackrel{\pi}{\to} (Y, \Gamma)$ is an extension of minimal systems we let:

$$P_\pi = P \cap R_\pi = \bigcap \{\Gamma V \cap R_\pi : V \text{ a neighborhood of the diagonal in } X \times X\}$$

and

$$Q_\pi = \bigcap \{\mathrm{cls}\,(\Gamma V \cap R_\pi) : V \text{ a neighborhood of the diagonal in } X \times X\}.$$

Clearly the extension π is distal iff $P_\pi = \Delta$ and it is not hard to see that π is isometric iff $Q_\pi = \Delta$. Notice that in general we only have $Q_\pi \subset Q \cap R_\pi$.

Recall that the system (X, Γ) is called weakly mixing if the product system $(X \times X, \Gamma)$ is topologically transitive; i.e., for every four non-empty open sets $U_i, i = 1, 2, 3, 4$,

$$N(U_1 \times U_3, U_2 \times U_4) = N(U_1, U_2) \cap N(U_3, U_4) \neq \emptyset$$

Or explicitly, there exists $\gamma \in \Gamma$ with $\gamma U_1 \cap U_2 \neq \emptyset$ and $\gamma U_3 \cap U_4 \neq \emptyset$. Clearly every weakly mixing system is topologically transitive. The system (X, Γ) is (topologically) *mixing* if for every two non-empty open sets $U, V \subset X$, there exists a finite set $F \subset \Gamma$ such that $\gamma U \cap V \neq \emptyset$ for every $\gamma \in \Gamma \setminus F$.

1.11. THEOREM. *For abelian Γ and a system (X, Γ) the following conditions are equivalent*

1. *(X, Γ) is weakly mixing.*
2. *The collection of sets of the form $N(U, V)$, with $U, V \subset X$ non-empty open sets, is a* filter base *of subsets of Γ; i.e. every $N(U, V)$ is non-empty and for every four non-empty open sets U_i, V_i; $i = 1, 2$, there exist non-empty open sets U, V, with $N(U, V) \subset N(U_1, V_1) \cap N(U_2, V_2)$.*
3. *For every $k \in \mathbb{N}$ the k-fold product system (X^k, Γ) is topologically transitive.*
4. *For every pair U, V of non-empty open sets, the set $N(U, V)$ is thick.*
5. *For every pair U, V of non-empty open sets there is an element $\gamma \in \Gamma$ with $\gamma U \cap U \neq \emptyset$ and $\gamma U \cap V \neq \emptyset$.*

PROOF. $1 \Rightarrow 2$. Assume (X,Γ) is weakly mixing and let U_i, V_i; $i = 1,2$, be four non-empty open sets in X. By weak mixing there exists $\gamma \in N(U_1,U_2) \cap N(V_1,V_2)$. Set $U = U_1 \cap \gamma^{-1}U_2$ and $V = V_1 \cap \gamma^{-1}V_2$. If $\gamma' \in N(U,V)$ then

$$\emptyset \neq \gamma' U \cap V \subset (\gamma' U_1 \cap V_1) \cap (\gamma'\gamma^{-1}U_2 \cap \gamma^{-1}V_2).$$

Since Γ is commutative we have $\gamma'\gamma^{-1}U_2 \cap \gamma^{-1}V_2 = \gamma^{-1}(\gamma' U_2 \cap V_2)$, hence $\gamma' \in N(U_1,V_1) \cap N(U_2,V_2)$ and we obtain $N(U,V) \subset N(U_1,V_1) \cap N(U_2,V_2)$ as required.

The implication $2 \Rightarrow 3$ is clear and to get $3 \Rightarrow 4$ we use for $F = \{\gamma_1, \gamma_2, \ldots, \gamma_k\}$, a finite subset of Γ, the topological transitivity of (X^k,Γ) to deduce the fact that

$$\begin{aligned} N(\gamma_1^{-1}U \times \gamma_2^{-1}U \times \cdots \times \gamma_k^{-1}U, V \times V \times \cdots \times V) \\ = \bigcap_{\gamma^{-1}\in F} N(\gamma U, V) = \bigcap_{\gamma \in F} \gamma N(U,V) \neq \emptyset. \end{aligned}$$

$4 \Rightarrow 5$. By assumption $N(U,V)$ is thick, hence non-empty and we choose γ_0 with $W = \gamma_0 U \cap V \neq \emptyset$. Next we use the fact that $N(W,W)$ is thick to pick $\gamma \in N(W,W) \cap \gamma_0^{-1}N(W,W) = N(W,W) \cap N(\gamma_0^{-1}W, W)$. Now

$$\emptyset \neq \gamma W \cap W \subset \gamma\gamma_0 U \cap \gamma_0 U = \gamma_0(\gamma U \cap U),$$

and

$$\emptyset \neq \gamma\gamma_0^{-1} W \cap W \subset \gamma U \cap V.$$

Thus both $\gamma U \cap U \neq \emptyset$ and $\gamma U \cap V \neq \emptyset$.

$5 \Rightarrow 1$. Let U_i, $i = 1, \ldots, 4$, be four non-empty open sets. Using topological transitivity and our assumption we successively find elements $\gamma_i \in \Gamma$, $i = 1,2,3$ so that: $V_1 = \gamma_1 U_1 \cap U_3 \neq \emptyset$, $V_2 = \gamma_2 V_1 \cap \gamma_1 U_2 \neq \emptyset$, and $\gamma_3 V_2 \cap V_2 \neq \emptyset$, $\gamma_3 V_2 \cap U_4 \neq \emptyset$. Then for $\gamma = \gamma_2\gamma_3$ we get

$$\begin{aligned} \gamma_1(\gamma U_1 \cap U_2) &\supset \gamma_1\gamma U_1 \cap \gamma_1 U_2 \cap \gamma U_3 = \gamma V_1 \cap \gamma_1 U_2 \\ &= \gamma_3\gamma_2 V_1 \cap \gamma_1 U_2 \supset \gamma_3 V_2 \cap \gamma_1 U_2 \supset \gamma_3 V_2 \cap V_2 \neq \emptyset, \end{aligned}$$

and

$$\begin{aligned} \gamma\gamma_1 U_1 \cap \gamma U_3 \cap U_4 &= \gamma(\gamma_1 U_1 \cap U_3) \cap U_4 \\ &= \gamma V_1 \cap U_4 \supset \gamma_3 V_2 \cap U_4 \neq \emptyset. \end{aligned}$$

Thus $\gamma U_1 \cap U_2 \neq \emptyset$ and $\gamma U_3 \cap U_4 \neq \emptyset$ as required. □

Let us call a dynamical system (Y,Γ) *syndetically transitive* if $N(U,V)$ is syndetic for every $U, V \subset Y$ non-empty open sets. By Exercise 1.1 every minimal system is syndetically transitive.

1.12. COROLLARY. *For an abelian Γ, a weakly mixing system (X,Γ) has the property that for every syndetically transitive system (Y,Γ), the product system $(X \times Y, \Gamma)$ is topologically transitive.*

PROOF. This follows from the characterization of part 4 of the theorem and Exercise 1.1.5. □

1.13. THEOREM. *If (X,Γ) is minimal and weakly mixing and Γ is abelian then for every $x \in X$ the proximal cell: $P[x] = \{x' : (x,x') \in P\}$, is a dense G_δ subset of X.*

PROOF. Set $V(\gamma, k) = \{(x, x') : d(\gamma x, \gamma x') < 1/k\}$, then clearly $P = \bigcap_{k=1}^{\infty} \bigcup_{\gamma \in \Gamma} V(\gamma, k)$ so that P is a G_δ subset of $X \times X$ (Exercise 1.9). As $P[x] = P \cap (\{x\} \times X)$ it follows that $P[x]$ is a G_δ subset of X and we show next that it is dense. Let $V \subset X$ be a non-empty open set and let U be an open neighborhood of x. By minimality (Exercise 1.1.2), $N(x, U)$ is syndetic, while the weak mixing (Theorem 1.11.4) implies that $N(V, U)$ is thick. By Exercise 1.1.5, there exists $\gamma_0 \in N(x, U) \cap N(V, U)$, so that $\gamma_0 x \in U$ and $\gamma_0 U \cap V$ is not empty. Using this argument repeatedly we now construct inductively two decreasing sequences of open sets U_n and V_n with the following properties. (i) $U_0 = U$ and $V_0 = V$, (ii) for every $n \geq 1$, $x \in U_n$ and $\operatorname{diam}(U_n) < 1/n$, (iii) There exists a sequence $\gamma_n \in \Gamma$ such that

$$\gamma_n x \in U_n, \qquad V_n \subset \bar{V}_n \subset \gamma_{n-1}^{-1} U_{n-1} \cap V_{n-1} \neq \emptyset.$$

By compactness there exists $x' \in \bigcap_{n=1}^{\infty} V_n$, and for every n, $\gamma_n x$ as well as $\gamma_n x'$ are in U_n. Thus $x' \in P[x] \cap V$ and the proof is complete. □

1.14. THEOREM. *If for a minimal dynamical system (X, Γ), the n-th proximal relation $P^{(n)}$ is dense in X^n for every $n \geq 2$, then for every minimal system (Y, Γ), $(X \times Y, \Gamma)$ is topologically transitive. In particular a proximal minimal system is weakly mixing.*

PROOF. Let (Y, Γ) be a minimal system and let $W \subset X \times Y$ be a closed invariant subset with non-empty interior. Let $U \subset X$ and $V \subset Y$ be non-empty open subsets with $U \times V \subset W$. The minimality of (Y, Γ) implies the existence of a finite subset $F = \{\gamma_1, \ldots, \gamma_n\} \subset \Gamma$ such that $Y = \bigcup_{\gamma \in F} \gamma V$.

Let $Q = \gamma_1 U \times \gamma_2 U \times \cdots \times \gamma_n U$, then Q is an open subset of X^n and by our assumption there exists a point $(x_1, x_2, \ldots, x_n) \in Q \cap P^{(n)}$. Since X is minimal, for every $x \in X$, there is a sequence $\zeta_j \in \Gamma$ such that

$$\lim_{j \to \infty} \zeta_j(x_1, x_2, \ldots, x_n) = (x, x, \ldots, x).$$

Fix $y \in Y$. For every j the point $\zeta_j^{-1} y$ belongs to at least one of the sets $\gamma_i V$, $i = 1, 2, \ldots, n$. Thus there exists an i_0, $1 \leq i_0 \leq n$, for which $\zeta_j^{-1} y \in \gamma_{i_0} V$ for infinitely many j. Changing our notation, we can now assume that $\lim_{j \to \infty} \zeta_j(x_1, x_2, \ldots, x_n) = (x, x, \ldots, x)$ and $\zeta_j^{-1} y \in \gamma_{i_0} V$ for all j. Now

$$(x_{i_0}, \zeta_j^{-1} y) \in \gamma_{i_0}(U \times V) \subset \gamma_{i_0} W = W,$$

hence $(\zeta_j x_{i_0}, y) = \zeta_j(x_{i_0}, \zeta_j^{-1} y) \in \zeta_j W = W$, and taking the limit we have $(x, y) \in W$. Since x and y were arbitrary points we conclude that $W = X \times Y$. This proves the topological transitivity of $(X \times Y, \Gamma)$. Finally if (X, Γ) is a proximal minimal system then for every n, $P^{(n)} = X^n$ and from the first part of the theorem we deduce that $(X \times X, \Gamma)$ is topologically transitive; i.e. (X, Γ) is weakly mixing. □

An extension $(X, \Gamma) \xrightarrow{\pi} (Y, \Gamma)$ is called a *weakly mixing extension* if the system (R_π, Γ) is topologically transitive. It is easy to establish that when an extension $(X, \Gamma) \xrightarrow{\pi} (Y, \Gamma)$ is weakly mixing then no nontrivial isometric extension of Y which is a factor of X exists. Unfortunately the converse does not always hold; i.e., the fact that $(X, \Gamma) \xrightarrow{\pi} (Y, \Gamma)$ is not weakly mixing does not always imply the existence of an intermediate nontrivial isometric extension (see, e.g., Example 1.19.3 below).

1.15. EXERCISES. 1. Show that the Bernoulli system is mixing.

2. Let $X = \mathbb{T} = \mathbb{R}/\mathbb{Z}$ and $\alpha \in \mathbb{R}\setminus\mathbb{Q}$. Let a and b be the self homeomorphisms of X defined by:

$$a(t) = t + \alpha \pmod 1), \qquad b(t) = (t - \frac{[4t]}{4})^2 + \frac{[4t]}{4}, \qquad 0 \le t < 1,$$

and let Γ be the group generated by a and b. Show that the system (X, Γ) is minimal and weakly mixing but the relation Q is not an equivalence relation (this example is due to D. McMahon).

1.16. EXAMPLES. 1. Let $\mathbb{G} = SL(2,\mathbb{R})$, Σ a co-compact discrete subgroup and let

$$h_s = \begin{pmatrix} 1 & s \\ 0 & 1 \end{pmatrix}.$$

Then the $\mathbb{R}$-system $(\mathbb{G}/\Sigma, h_t)$—the *horocycle flow*—is a minimal weakly mixing flow (i.e., an $\mathbb{R}$-system). [[The minimality is shown in Chapter 4, Section 6. For the weak mixing use the commutation relation: $g_t h_s g_{t^{-1}} = h_{st^2}$ for $s \in \mathbb{R},\ t > 0$ and

$$g_t = \begin{pmatrix} t & 0 \\ 0 & t^{-1} \end{pmatrix},$$

to show that no nonconstant continuous eigenfunction exists. In fact if f is an eigenfunction with eigencharacter $\chi_\theta(s) = e^{i\theta s}$, $\theta \notin \mathbb{Z}$, $s \in \mathbb{R}$, then $f_s(x) = f(h_s x)$ is an eigenfunction with eigencharacter $\chi_{\theta t^2}(s) = e^{i\theta s t^2}$. As we have already seen (Exercise 1.5.3), whenever f, g are eigenfunctions corresponding to two distinct eigencharacters, $\|f - g\|_\infty \ge \sqrt{2}$ and the separability of the Banach space $C(X)$ is now used to get a contradiction.]]

The minimal system (X, Γ) is called *regular* if every minimal point in $X \times X$ is of the form $(x, \phi(x))$ for $\phi \in \mathrm{Aut}\,(X, \Gamma)$.

An extension $(X,\Gamma) \xrightarrow{\pi} (Y,\Gamma)$ is called an *almost 1-1 extension* if there exists a residual subset $X_0 \subset X$ with $\pi^{-1}(\pi(x)) = \{x\}$ for every $x \in X_0$.

1.17. EXERCISES. 1. For every automorphism ϕ of a system (X, Γ) and a minimal point $x \in X$ the pair $(x, \phi(x))$ is a minimal point. In particular when Γ is abelian and (X, Γ) is minimal, every pair $(x, \gamma x)$ is minimal. Similarly points of the form $(x, \gamma_1 x, \gamma_2 x, \dots, \gamma_n x)$ are minimal points and we conclude that for every $n > 1$ the minimal points are dense in the product system X^n; i.e., X^n satisfies the Bronstein condition (such systems are called *incontractible*).

2. For abelian Γ no nontrivial proximal and minimal system exists.
3. Does there exist a minimal system for which the minimal points are dense in $X \times X$ but not dense in X^n for some $n > 2$ (i.e., X is not incontractible)? [[This is an open problem.]]
4. Proximal as well as Kronecker systems are regular.
5. If $(X,\Gamma) \xrightarrow{\pi} (Y,\Gamma)$ is an extension of topological systems, and $\pi^{-1}(\pi(x_0)) = \{x_0\}$ for a point $x_0 \in X$ with dense orbit, then the set of points $Y_0 \subset Y$ whose inverse image under π is a singleton, forms a dense G_δ subset of Y.

The set $X_0 = \pi^{-1}(Y_0)$ is dense G_δ in X. If moreover (Y, Γ) is minimal, then (X, Γ) is also minimal.

6. For (X, Γ) topologically transitive an extension $(X, \Gamma) \overset{\pi}{\to} (Y, \Gamma)$ is an almost 1-1 extension iff X is the only closed subset of X mapped by π onto Y.
7. An almost 1-1 extension of minimal systems is a proximal extension.

1.18. REMARK. In **[96]** Glasner and Weiss provide examples of extensions $(X, T) \overset{\pi}{\to} (Y, T)$ of minimal $\mathbb{Z}$-systems which are proximal but not almost 1-1.

1.19. EXAMPLES. 1. Let us define an element $\omega \in \Omega = \{0,1\}^{\mathbb{Z}}$ as follows. For a finite block $B \in \{0,1\}^n$ we let $\bar{B}$ be the dual block obtained by replacing each 0 with 1 and vice versa. Then inductively define blocks B_n by:

$$B_0 = 0, \; B_1 = 01, \;\; B_2 = 0110, ..., B_{n+1} = B_n \bar{B}_n.$$

Since for every n, B_n is the initial 2^n subblock of B_{n+1} we get this way an infinite block B_∞ and our ω is now defined by setting $\omega_0\omega_1... = B_\infty$ and $\omega(n) = \omega(-n-1)$ for $n < 0$. We let $X = \bar{\mathcal{O}}_S\omega$. The sequence ω is the famous *Thue Morse sequence* and the system (X, S) is called the *Morse system.* Show that (X, S) is a minimal infinite subshift.

2. The Morse minimal system has the following structure: $X \overset{\rho}{\to} Y \overset{\sigma}{\to} Z$, where ρ is a group extension with group K$=\{e, f\}$ (where $f(\omega) = \bar{\omega}$ and e the identity map), and σ is an almost 1-1 extension of Z, the Kronecker factor of X. The latter coincides with the dyadic adding machine $\mathbb{A}$.
3. Show that for the Morse system the extension $\pi = \sigma \circ \rho : X \to Z$, though not weakly mixing, does not admit an intermediate nontrivial isometric extension; i.e., in every diagram $X \overset{\alpha}{\to} U \overset{\beta}{\to} Z$ with $\pi = \beta \circ \alpha$ and β an isometric extension, β is 1-1.
4. Define inductively blocks B_n by:

$$B_1 = 0010, \;\; B_2 = 0010001010010, ..., B_{n+1} = B_n B_n 1 B_n.$$

Let ω be defined by setting for every n:

$$\omega_{-l_n}\omega_{-l_n+1} \dots \omega_0\omega_1 ... \omega_{l_n-1} = B_n B_n,$$

where l_n is the length of B_n. We let $X = \bar{\mathcal{O}}_S\omega$. The sequence ω is called the *Chacón sequence* and the system (X, T) is the *Chacón system.* Show that (X, S) is a minimal infinite subshift. More difficult to show are the following facts. (X, S) is weakly mixing and regular (in fact every minimal set in $X \times X$ is a graph of some power of S). Moreover (X, S) is a *prime system*; i.e., it admits no nontrivial factors (see [**137**], [**205**]).

5. The last two examples are special cases of a general construction of minimal topological systems called *substitution systems.* For a finite alphabet $\mathfrak{L} = \{0, \dots, \ell - 1\}$ a *substitution* is a function θ which assigns to each symbol of $\mathfrak{L}$ a finite word on $\mathfrak{L}$; i.e., $\theta : \mathfrak{L} \to \mathfrak{L}^* = \cup_{n \geq 1} \mathfrak{L}^n$. When all these words have the same length, say d, we say that θ is of constant length d. Thus the Morse and the Chacón substitutions are defined on $\mathfrak{L} = \{0,1\}$; the Morse substitution is of constant length, while Chacón's is not. Clearly θ induces maps, also denoted θ, from $\mathfrak{L}^*$ into itself and from $\mathfrak{L}^{\mathbb{N}}$ into itself. In

particular iterations of θ make sense. The substitution θ is called *primitive* if

(a) $\theta(0) = 0$.
(b) $|\theta^n(0)| \to \infty$, where $|w|$ denotes the length of the word w.
(c) There exists a positive integer k such that for every $r, t \in \mathfrak{L}$, r appears in $\theta^k(t)$.

It is easy to establish that for a primitive θ the sequence $\theta^n(0)$ converges to a limit $\omega \in \mathfrak{L}^{\mathbb{N}}$ with $\theta(\omega) = \omega$. Define the subshift $X_\theta \subset \mathfrak{L}^{\mathbb{Z}}$ as the collection of all sequences $x \in \mathfrak{L}^{\mathbb{Z}}$ with the property that every finite word in x appears as a subword of ω. Show that the topological system (X_θ, S) is minimal. (See M. Queffelec's book [**208**].)

1.20. EXAMPLES. 1. The subshift $X = \bar{\mathcal{O}}_S\omega$, where $\omega \in \{1,-1\}^{\mathbb{Z}}$ is the sequence

$$\omega(n) = \operatorname{sgn} \cos(2\pi n\alpha)$$

for $\alpha \notin \mathbb{Q}$, is a minimal subshift. Show that its maximal Kronecker factor is the rotation on $(\mathbb{T}, \alpha)$, where the extension $X \to \mathbb{T}$ is an almost 1-1 extension. Show that this extension is *not* a weakly mixing extension.

2. Define a sequence $\omega \in \{0,1\}^{\mathbb{Z}}$ as follows. For every $n \in \mathbb{Z}$, $\omega(2n) = 0$, $\omega(4n+1) = 1$, $\omega(8n+3) = 0$, $\omega(16n+7) = 1$, and in general $\omega(2^k n + 2^{k-1} - 1)$ is zero or one according to whether k is even or odd. Let $X = \bar{\mathcal{O}}_S(\omega)$; show that the subshift (X, S) is a minimal almost 1-1 extension of the dyadic adding machine.
3. In general a sequence $\omega \in \mathfrak{L}^{\mathbb{Z}}$ for an alphabet $\mathfrak{L} = \{1, \dots, \ell\}$ is a *Toeplitz sequence* if for every $n \in \mathbb{Z}$ there exists $p \geq 1$ such that for all $k \in \mathbb{Z}$, $\omega(n + kp) = \omega(n)$. A subshift $X \subset \mathfrak{L}^{\mathbb{Z}}$ is a *Toeplitz system* if X is the orbit closure of a Toeplitz sequence. Show that every Toeplitz system is minimal (see [**128**] and [**266**]).
4. A minimal infinite subshift is Toeplitz iff it is an almost 1-1 extension of an adding machine ([**180**]).

4. The enveloping semigroup

In order to study the asymptotic behavior of a Γ-system (X, Γ) R. Ellis introduced in 1960 the *enveloping semigroup* $E = E(X, \Gamma)$. This is defined as the closure in X^X (with its compact, usually non-metrizable, pointwise convergence topology) of the set Γ considered as a subset of X^X. We identify here the elements of Γ with the corresponding homeomorphism of X. Since in most cases the action of Γ on X is *effective* (i.e., $\gamma x = x$, $\forall x \in X$ implies $\gamma = e$), this abuse of notation will cause no harm.

It follows directly from the definitions that, under composition of maps, E forms a compact semigroup in which the operations

$$p \mapsto pq \qquad \text{and} \qquad p \mapsto \gamma p$$

for $p, q \in E$, $\gamma \in \Gamma$, are continuous. Notice that this makes Γ act on E by left multiplication, so that (E, Γ) is a Γ-system (though usually non-metrizable).

1.21. REMARK. We shall later use the fact that the map $p \mapsto qp$ is continuous whenever $q \in E$ is a continuous map.

The elements of E may behave very badly as maps of X into itself; usually they are not even Borel measurable. However our main interest in E lies in its algebraic structure and its dynamical significance. A key lemma in the study of this algebraic structure is the following:

1.22. LEMMA (Ellis-Namakura). *Let L be a compact Hausdorff semigroup in which all maps $p \mapsto pq$ are continuous. Then L contains an idempotent; i.e., an element v with $v^2 = v$.*

PROOF. By Zorn's lemma, there exists a minimal compact subsemigroup $K \subset L$. For any $v \in K$, Kv is a compact subsemigroup of K whence $Kv = K$ and in particular for some $k \in K$, $kv = v$. Now the set $M = \{l \in K : lv = v\}$ is a non-empty closed subsemigroup of K, and again we deduce that $M = K$. In particular $vv = v$. □

In the next series of exercises we state some useful properties of the enveloping semigroup E. Most of these are easy consequences of the definitions and Lemma 1.22

1.23. EXERCISES.   1. A subset M of E is a minimal left ideal of the semi-group E iff it is a minimal subsystem of (E, Γ). In particular a minimal left ideal is closed. We shall refer to it simply as a *minimal ideal*. Minimal ideals M in E exist and for each such ideal the set of idempotents in M, denoted by $J = J(M)$, is non-empty.
   2. Let M be a minimal ideal and J its set of idempotents then: [$M = \bigoplus_n(v)$ $\forall v \in J$] (by minimality)
      (a) For $v \in J$ and $p \in M$, $pv = p$.
      (b) For each $v \in J$, $vM = \{vp : p \in M\}$ is a subgroup of M with identity element v. For every $w \in J$ the map $p \mapsto wp$ is a group isomorphism of vM onto wM.
      (c) $\{vM : v \in J\}$ is a partition of M. Thus if $p \in M$ then there exists a unique $v \in J$ such that $p \in vM$; we denote by p^{-1} the inverse of p in vM.
   3. Let K, L, and M be minimal ideals of E. Let v be an idempotent in M, then there exists a unique idempotent v' in L such that $vv' = v'$ and $v'v = v$. (We write $v \sim v'$ and say that v' is *equivalent* to v.) If $v'' \in K$ is equivalent to v', then $v'' \sim v$. The map $p \mapsto pv'$ of M to L is an isomorphism of Γ-systems.

[all minimal ideals are canonically isomorphic given an idempotent v]

Often one has to deal with more than one system at a time; e.g., we can be working simultaneously with two systems, their product, subsystems of the product, etc. Or given a topological system (X, Γ) we may have to work with associated systems like the actions induced on the space of probability measures $M(X)$ with its weak* topology, or the space of closed subsets 2^X, with its Hausdorff topology. It is then convenient to have one enveloping semigroup acting on all of the systems simultaneously. This can be easily done by considering the enveloping semigroup of the product of all the systems under consideration. In fact one looses nothing and gain much in convenience as well as in added machinery if one works instead with the "universal" enveloping semigroup.

Let $\beta\Gamma$ denote the Stone-Čech compactification of Γ . We view this compactification as the Gelfand space of the algebra $\ell^\infty(\Gamma)$ (Theorems A.2 and A.3). Thus the elements of $\beta\Gamma$ are the multiplicative functionals $p : \ell^\infty(\Gamma) \to \mathbb{R}$ of norm 1. In particular for every $\gamma \in \Gamma$, $j(\gamma)$, the evaluation at γ, is clearly such a functional on $\ell^\infty(\Gamma)$. The topology on $\beta\Gamma$ is the (compact) weak* topology, the map $j : \Gamma \to \beta\Gamma$ is a homeomorphism (i.e. $j(\Gamma)$ is a discrete subset of $\beta\Gamma$) and $\overline{j(\Gamma)} = \beta\Gamma$. We shall consider Γ as a subset of $\beta\Gamma$. The collection $\{\bar{A} : A \subset \Gamma\}$ is a basis for the topology of $\beta\Gamma$ consisting of clopen sets. Moreover for $\gamma \in \Gamma$, $\gamma \in \bar{A}$ implies $\gamma \in A$. $\beta\Gamma$ is the "universal compactification" of the discrete group Γ. The latter means that every map $f : \Gamma \to X$ where X is a compact space can be uniquely extended to a continuous map $\bar{f} : \beta\Gamma \to X$. For example, each of the maps $L_\gamma : \Gamma \to \beta\Gamma;\ \gamma' \mapsto \gamma\gamma'$ extends to a homeomorphism $\bar{L}_\gamma : \beta\Gamma \to \beta\Gamma$. This construction defines the (non-metrizable) Γ-system $(\beta\Gamma, \Gamma)$, where $\gamma p = \bar{L}_\gamma(p)$.

Using the universal property once more we can now extend, for each $p \in \beta\Gamma$, the map $R_p : \Gamma \to \beta\Gamma;\ \gamma \mapsto \gamma p$, to a continuous map $\bar{R}_p : \beta\Gamma \to \beta\Gamma;\ q \mapsto \bar{R}_p(q) = pq$. This defines a semigroup structure on $\beta\Gamma$ with the same properties as those of the enveloping semigroup discussed above. In fact it is easy to check that the map which sends $p \in \beta\Gamma$ onto the map $q \mapsto pq;\ \beta\Gamma \to \beta\Gamma$, is a topological and algebraic isomorphism of the semigroup $\beta\Gamma$ onto the enveloping semigroup $E(\beta\Gamma, \Gamma)$.

Moreover if (X, Γ) is a system and $x \in X$ then the map $\gamma \mapsto \gamma x$ can be extended uniquely to a map $p \mapsto px$ of $\beta\Gamma$ onto $\bar{\mathcal{O}}_\Gamma x$. This defines a (systems and semigroups) homomorphism of $\beta\Gamma$ onto $E(X, \Gamma)$. In fact, as was mentioned above, it is sometimes convenient to forgo the use of $E(X, \Gamma)$ and use instead the universal enveloping semigroup $\beta\Gamma$.

In $\beta\Gamma$ we fix a minimal ideal $M \subset \beta\Gamma$, we let $J = J(M)$ and fix an idempotent u in J. We say that an idempotent is a *minimal idempotent* if it belongs to a minimal ideal and denote the set of minimal idempotents by $\hat{J}$. We denote the group uM by $\mathfrak{G}$ and use Greek letters for the elements of $\mathfrak{G}$.

Since right multiplication on $\beta\Gamma$ is continuous and since each $v \in J$ is a right identity in M, it follows that for each $\alpha \in \mathfrak{G}$ the map $p \mapsto p\alpha$ is a homeomorphism of M and in fact an automorphism of the (non-metrizable) topological system $(M, \mathfrak{G})$. Thus every element of $\mathfrak{G}$ defines an automorphism of M and it is easy to check that if we identify the elements of $\mathfrak{G}$ with the corresponding automorphisms then actually $\mathfrak{G} = \mathrm{Aut}\,(M, \Gamma)$. Notice that every element p in M can be expressed uniquely as $p = v\alpha$ for $v \in J$ and $\alpha \in \mathfrak{G}$. We then have $p^{-1} = v\alpha^{-1}$.

Emphasizing the importance of the universality properties of $\beta\Gamma$ and M we next state the following theorem.

1.24. THEOREM. 1. *The pointed dynamical system $(\beta\Gamma, e, \Gamma)$ is the universal point transitive Γ-system; i.e. for every pointed Γ-system (X, x_0, Γ) there is a unique homomorphism $\phi : (\beta\Gamma, e, \Gamma) \to (X, x_0, \Gamma)$ with $\phi(e) = x_0$ and $\phi(\beta\Gamma) = \bar{\mathcal{O}}_\Gamma x_0$. Up to a pointed isomorphism $(\beta\Gamma, e, \Gamma)$ is the unique pointed system with this property.*

2. *The dynamical system system (M, Γ) is the universal minimal Γ-system; i.e. for every minimal Γ-system (X, Γ) there is a (usually not unique) homomorphism $\phi : (M, \Gamma) \to (X, \Gamma)$. Every endomorphism of the system*

(M, Γ) is an automorphism and therefore, up to isomorphism, (M, Γ) is the unique minimal Γ-system with this property.

PROOF. Part 1 is already proven. To see 2, suppose (Y, Γ) is another universal minimal Γ-system. Then there are homomorphisms $\phi : M \to Y$ and $\psi : Y \to M$. The composition $j = \psi \circ \phi$ is a homomorphism of M onto itself. We next show that such a homomorphism is necessarily 1-1. Choose an idempotent $u \in M$ and let $j(u) = p$. Then for any $q \in M$ we have

$$(R_{p^{-1}} \circ j)(q) = j(q)p^{-1} = qj(u)p^{-1} = qpp^{-1} = qv = q,$$

where v is the idempotent pp^{-1}. Thus $R_{p^{-1}} \circ j = \mathrm{id}$ and it follows that j and hence also ϕ are isomorphisms. □

Another convenient viewpoint (which we shall not use) is to consider $\beta\Gamma$ as the space of ultrafilters on Γ (see R. Ellis' book [**58**]). Thus for example for a system (X, Γ), if $p \in \beta\Gamma$ and $x \in X$, then px can be interpreted both as the image of x under the image of p in $E(X, \Gamma)$ or as the limit $\lim_p \gamma x$ where we now consider p as an ultrafilter (or a universal net) on Γ. $px = \lim_p \gamma x$, p ultrafilter on Γ

1.25. EXERCISE. The enveloping semigroup of the Bernoulli system (Ω, Γ) is isomorphic (as a Γ-system as well as a compact semigroup) to $\beta\Gamma$. [[Recall that $\{\bar{A} : A \subset \Gamma\}$ is a basis for the topology of $\beta\Gamma$ consisting of clopen sets. Next identify $\Omega = \{0, 1\}^\Gamma$ with the collection of subsets of Γ in the obvious way: $A \longleftrightarrow \mathbf{1}_A$. Now define an "action" of $\beta\Gamma$ on Ω by:

$$p * A = \{\gamma \in \Gamma : \gamma p \in \bar{A}\}.$$

It is easy to see that this action extends the action of Γ on Ω and defines an isomorphism of $\beta\Gamma$ onto $E(\Omega, \Gamma)$.]]

1.26. EXERCISES. Establish the following connections between the dynamical properties of the system (X, Γ) and the algebraic properties of $E(X, \Gamma)$.

1. $\bar{\mathcal{O}}_\Gamma x = Ex$
2. $\bar{\mathcal{O}}_\Gamma x$ is minimal (i.e., x is minimal) iff for every minimal ideal M in E, $\bar{\mathcal{O}}_\Gamma x = Mx$ iff in every minimal ideal there is an idempotent v such that $vx = x$. Thus JX is the set of minimal points of the system (X, Γ). Applying this to the product system we see that $J(X \times X)$ is the set of minimal points in $X \times X$.
3. The pair (x, x') is proximal iff $px = px'$ for some $p \in E$ iff there exists a minimal ideal M in E with $px = px'$ for every $p \in M$.
4. If (X, Γ) is minimal, then
$$P[x] = \{x' \in X : (x, x') \in P\} = \{vx : v \in \hat{J}\}.$$
In particular $x \in X$ is a distal point iff $vx = x$ for every $v \in \hat{J}$.
5. For $v \in \hat{J}$ every pair of distinct points in vX is distal (i.e. not proximal).
6. The relation P is transitive iff E contains a unique minimal ideal.
7. (X, Γ) is distal iff E is a group (Theorem 1.7).
8. A distal system is pointwise minimal.
9. (X, Γ) is distal iff $X \times X$ is pointwise minimal. Now conclude that a factor of a distal system is distal.

10. (X, Γ) is equicontinuous iff E is a group of homeomorphisms of X and the topologies of pointwise and uniform convergence coincide on E (Theorem 1.8). [[Since this condition implies that E is a compact subgroup of the group $H(X)$ of self homeomorphisms of X with the topology of uniform convergence the assertion follows from Arzelà-Ascoli's theorem.]]
11. (X, Γ) is equicontinuous iff E is a topological group whose action on X is jointly continuous. [[By Ellis' joint continuity theorem (Theorem 1.29) it is in fact enough to require that E is a group of continuous maps; see Section 6.]]
12. (X, Γ) is equicontinuous iff for every neighborhood U of the identity $e \in H(X)$, the group of self homeomorphisms of X with the topology of uniform convergence, the set $\{\gamma \in \Gamma : \gamma \in U\}$ is syndetic in Γ.
13. (X, Γ) is minimal and equicontinuous (Kronecker) iff E is a topological group whose action on X is jointly continuous and the system (X, Γ) is isomorphic to the homogeneous system $(K/H, \Gamma)$ where K is a compact topological group, H is a closed subgroup of K and Γ is embedded in K as a dense subgroup.
14. Show that a minimal system (X, Γ) is regular iff it is isomorphic (as a topological system) to some (hence every) minimal ideal in its enveloping semigroup. In particular the (non-metrizable) universal minimal system (M, Γ) is regular.
15. We say that a pointed dynamical system (X, x_0, Γ) is *covering* if for every $x \in X$ there is a homomorphism of pointed systems $\pi_x : (X, x_0, \Gamma) \to (\bar{\mathcal{O}}_\Gamma x, x, \Gamma)$. Show that a transitive dynamical system (X, Γ) is isomorphic as a dynamical system to its enveloping semigroup $E(X, \Gamma)$, iff it is covering.

1.27. PROPOSITION. *Let (X, Γ) be a transitive system and (Y, Γ) a minimal distal system, then (X, Γ) and (Y, Γ) are disjoint iff they are weakly disjoint.*

PROOF. As clearly disjointness implies weak disjointness, we only need to show the converse implication. So assume $W \subset X \times Y$ is a closed invariant set which projects onto both X and Y. Let $(x_0, y_0) \in X \times Y$ be a transitive point. There exists $(x_0, y_1) \in W$ and we shall conclude the proof by showing that (x_0, y_0) is in $\bar{\mathcal{O}}_\Gamma(x_0, y_1)$.

Let $E = E(X \times Y)$ be the enveloping semigroup of the product system and set $L = \{p \in E : px_0 = x_0\}$, then L is a closed subsemigroup. As (x_0, y_0) is a transitive point, there exits $p_0 \in E$ with $p_0(x_0, y_0) = (x_0, y_1)$ and in particular $p_0 \in E$. Now it is easy to check that $Lp_0 \subset L$ is also a closed subsemigroup and the Ellis-Namakura lemma 1.22, implies the existence of an idempotent $v \in Lp_0$; note that $v = qp_0$ for some $q \in L$. Now the enveloping semigroup of the distal system Y is a group (Theorem 1.7) and it follows that v acts on Y as the identity. Therefore $v(x_0, y_0) = (vx_0, vy_0) = (x_0, y_0)$ and we finally have $(x_0, y_0) = v(x_0, y_0) = qp_0(x_0, y_1) = q(x_0, y_1) \in \bar{\mathcal{O}}_\Gamma(x_0, y_0)$. □

5. Pointed systems and their corresponding algebras

For a function $f \in \ell^\infty(\Gamma)$ and $\gamma \in \Gamma$ let ${}_\gamma f(\gamma') = f(\gamma\gamma')$, $(\gamma' \in \Gamma)$. This defines a (anti) representation of Γ as a group of linear isometries (*left translations*) of $\ell^\infty(\Gamma)$.

Let (X, x_0, Γ) be a pointed system. With every $F \in C(X)$ we associate the function $j_{x_0}(F) = f \in \ell^\infty(\Gamma)$ defined by $f(\gamma) = F(\gamma x_0)$. The map $j_{x_0} : C(X) \to \ell^\infty(\Gamma)$ is an isometric isomorphism of the Banach algebra $C(X)$ into $\ell^\infty(\Gamma)$. Let us denote its image by $\mathcal{A}(X, x_0) = j_{x_0}(C(X))$. We have ${}_\gamma f ={}_\gamma (j_{x_0}(F)) = j_{x_0}(F \circ \gamma))$. Therefore $\mathcal{A}(X, x_0)$ is invariant under left translations. The Gelfand space of the algebra $\mathcal{A}(X, x_0)$ (Theorem A.2), which we denote by $|\mathcal{A}(X, x_0)|$, is naturally identified with X and in particular the multiplicative functional $\mathrm{eva}_e : f \mapsto f(e)$, is identified with the point x_0. Moreover the action of Γ on $\mathcal{A}(X, x_0)$ by left translations defines an action of Γ on $|\mathcal{A}(X, x_0)|$ and under this identification the pointed systems (X, x_0, Γ) and $(|\mathcal{A}(X, x_0)|, \mathrm{eva}_e, \Gamma)$ are isomorphic. Thinking of $p \in \beta\Gamma$ as a multiplicative functional on $\ell^\infty(\Gamma)$ we now see that the restriction map $p \mapsto p \restriction \mathcal{A}(X, x_0)$ is just the system homomorphism $p \mapsto px_0$ from the universal pointed system $(\beta\Gamma, e, \Gamma)$ onto (X, x_0, Γ).

Conversely if $\mathcal{A}$ is a Γ-invariant uniformly closed subalgebra of $\ell^\infty(\Gamma)$ containing the constant functions, then its Gelfand space $|\mathcal{A}|$ has a structure of a pointed dynamical system $(|\mathcal{A}|, \mathrm{eva}_e, \Gamma)$. Note that the latter is metrizable iff $\mathcal{A}$ is separable (as a Banach space).

Finally it is easy to check that for any collection $\{(X_\theta, x_\theta, \Gamma) : \theta \in \Theta\}$ of pointed systems we have

$$\mathcal{A}\left(\bigvee\{(X_\theta, x_\theta, \Gamma) : \theta \in \Theta\}\right) = \bigvee\{\mathcal{A}(X_\theta, x_\theta, \Gamma) : \theta \in \Theta\},$$

where the latter is the closed subalgebra of $\ell^\infty(\Gamma)$ generated by the union of the algebras $\mathcal{A}(X_\theta, x_\theta, \Gamma)$.

1.28. EXERCISES. 1. If $\pi : (X, x_0, \Gamma) \to (Y, y_0, \Gamma)$ is a homomorphism of pointed transitive systems then the diagram

$$\begin{array}{ccc} C(Y) & \xrightarrow{\ j_{y_0}\ } & \mathcal{A}(Y, y_0) \\ {\scriptstyle \pi^*}\downarrow & & \downarrow{\scriptstyle i} \\ C(X) & \xrightarrow[\ j_{x_0}\ ]{} & \mathcal{A}(X, x_0) \end{array}$$

commutes. Here $(\pi^* F)(x) = F(\pi x)$, for $F \in C(Y), x \in X$, and i is the inclusion map. Conversely, if $B \subset C(X)$ is a closed Γ-invariant subalgebra containing the constant functions, then the restriction map

$$\pi : (|\mathcal{A}(X, x_0)|, \mathrm{eva}_e, \Gamma), \to (|j_{x_0}(B)|, \mathrm{eva}_e, \Gamma)$$

is a pointed homomorphism of dynamical systems.

2. Show that the pointed system corresponding to $\ell^\infty(\Gamma)$ is the universal point transitive system $(\beta\Gamma, e, \Gamma)$.

6. Ellis' joint continuity theorem

In its full generality Ellis' celebrated theorem, [**56**], is as follows (see also [**11**]).

1.29. THEOREM (Ellis' joint continuity theorem). *Let G be a locally compact space with a group structure such that multiplication is separately continuous. Then*

G is a topological group; i.e. the map $(g, h) \mapsto gh^{-1}$ from $G \times G \to G$ is jointly continuous. Moreover, if X is a locally compact space on which G acts in a separately continuous way then the map $(g, x) \mapsto gx$ from $G \times X \to X$ is jointly continuous.

We shall need the following corollary whose proof (for metrizable X) will be given in Section 8 below.

1.30. COROLLARY. *The dynamical system (X, Γ) is equicontinuous iff $E(X, \Gamma)$ is a group of homeomorphisms of X.*

7. Furstenberg's distal structure theorem

As we have already seen in Examples 1.6, a system which is built as an isometric extension of an isometric system is a distal system. This kind of examples led H. Furstenberg to his remarkable theorem [**73**], describing the structure of a general minimal distal system as a (countable but maybe transfinite) inverse limit of isometric extensions. In this section we precisely state this theorem. For proofs refer to the original paper or to books in topological dynamics (e.g. [**58**], [**36**] or [**11**]).

We say that a (metrizable) minimal system (X, Γ) is an *I system* if there is a (countable) ordinal η and a family of systems $\{(X_\theta, x_\theta)\}_{\theta \le \eta}$ such that (i) X_0 is the trivial system, (ii) for every $\theta < \eta$ there exists an isometric homomorphism $\phi_\theta : X_{\theta+1} \to X_\theta$, (iii) for a limit ordinal $\lambda \le \eta$ the system X_λ is the inverse limit of the systems $\{X_\theta\}_{\theta<\lambda}$ (i.e. $X_\lambda = \bigvee_{\theta<\lambda}(X_\theta, x_\theta)$), and (iv) $X_\eta = X$. The *distal rank* of a system (X, Γ) is the least ordinal η such that there there exists a "tower" of height η realizing it as an I-system.

1.31. THEOREM (Furstenberg's structure theorem). *A minimal system is distal iff it is an I-system.*

1.32. REMARK. In [**17**] Beleznay and Foreman show that, at least for $\mathbb{Z}$-systems, for every countable ordinal η there in fact exists a minimal distal system (X, T) whose distal rank is exactly η.

8. Almost equicontinuity

1.33. DEFINITION. Let (X, Γ) be a topological system. A point $x_0 \in X$ is called an *equicontinuity point* if for every $\epsilon > 0$ there exists a neighborhood U of x_0 such that for every $y \in U$ and every $\gamma \in \Gamma$, $d(\gamma x_0, \gamma y) \le \epsilon$. The topological system (X, Γ) is called *almost equicontinuous* (or is an *AE-system*) if the subset $EQ(X)$ of equicontinuity points is a dense subset of X.

1.34. EXERCISE. A system (X, Γ) is equicontinuous iff every $x \in X$ is an equicontinuity point. [[Given $\epsilon > 0$, let $\mathfrak{U} = \{U_x : x \in X\}$ be a collection of neighborhoods as in the definition of equicontinuity points. Any Lebesgue number δ for the open cover $\mathfrak{U}$ will serve for the equicontinuity condition.]]

1.35. PROPOSITION. *Let (X,Γ) be an almost equicontinuous system, then the set $EQ(X)$ is a dense G_δ subset of X. If (X,Γ) is a transitive almost equicontinuous system then the set $EQ(X)$ of equicontinuity points coincides with the set of transitive points of X. In particular a minimal almost equicontinuous system is equicontinuous.*

PROOF. It is easy to see that in general $EQ(X)$ is a G_δ subset of X. Now assume that (X,Γ) is transitive. Let x_0 be a transitive point and $x \in EQ(X)$. We shall show that also $x_0 \in EQ(X)$. Given $\epsilon > 0$ there exits a $\delta > 0$ such that for all $x', x'' \in B_\delta(x)$, $d(\gamma x', \gamma x'') < \epsilon$ for all $\gamma \in \Gamma$. Since x_0 is a transitive point, there exists $\gamma' \in \Gamma$ and $\eta > 0$ such that $\gamma' B_\eta(x_0) \subset B_\delta(x)$. Thus for every $z \in B_\eta(x_0)$ and every $\gamma \in \Gamma$ we have $d(\gamma\gamma' z, \gamma\gamma' x_0) < \epsilon$ for all $\gamma \in \Gamma$; i.e. $x_0 \in EQ(X)$ and we conclude that the set of transitive points is contained in $EQ(X)$.

Moreover, notations as above, since $\gamma' x_0 \in B_\delta(x)$ we have

$$d((\gamma')^{-1}\gamma' x_0, (\gamma')^{-1}x) = d(x_0, (\gamma')^{-1}x) < \epsilon.$$

Thus $x_0 \in \bar{\mathcal{O}}_\Gamma x$ and we conclude that every equicontinuity point is transitive. The last assertion now follows from Exercise 1.34. □

The next lemma is the key to a variety of "joint continuity" results (see also I. Namioka [**185**]).

1.36. LEMMA. *Let K be a compact (not necessarily metrizable) topological space, let B be a separable Baire space and let (Z,d) be a metric space. Suppose that $\pi : K \times B \to Z$ is continuous in each variable separately, then there is a residual subset $R \subset B$ such that π is jointly continuous at each point of $K \times R$.*

PROOF. **Step 1.** *The metrizable case.* Assume first that K is a metric space with metric ρ. For $m, n \in \mathbb{N}$ set

$$B_{m,n} = \{b \in B : \forall\ k, k' \in K,\ \rho(k,k') < 1/m\ \Rightarrow\ d(\pi(k,b), \pi(k',b)) \le 1/n\}.$$

The continuity of π in b implies that $B_{m,n}$ is closed. The continuity (hence uniform continuity) of π in k implies that for each n, $B = \bigcup_m B_{m,n}$. It therefore follows that the set $R = \bigcap_n \bigcup_m \operatorname{int}(B_{m,n})$ is residual in B.

Now let $(k,b) \in K \times R$. Given n, there exists m such that $b \in \operatorname{int}(B_{m,n})$ and we can find a neighborhood $U \subset \operatorname{int}(B_{m,n})$ of b such that $b' \in U$ implies $d(\pi(k,b), \pi(k,b')) \le 1/n$. Now for k' with $\rho(k,k') < 1/m$ and $b' \in U$ we have

$$d(\pi(k,b), \pi(k',b')) < d(\pi(k,b), \pi(k,b')) + d(\pi(k,b'), \pi(k',b')) < 2/n.$$

This proves the joint continuity at (k,b).

Step 2. *The general case.* Let A be a countable dense subset of B. Clearly the relation $\sim$, defined on K by $k \sim k'$ iff $\pi(k,b) = \pi(k',b)$ for all $b \in B$, is a closed equivalence relation. Therefore the quotient space $\hat{K} = K/\sim$ is a compact Hausdorff space and the map π can be factored through $\hat{K}$ as $\pi = \hat{\pi} \circ (\eta \times \mathrm{id})$, where $\eta : K \to \hat{K}$ is the (continuous) quotient map and $\hat{\pi} : \hat{K} \times B \to Z$ the map induced by π on $\hat{K}$.

Define the map $\phi : K \to Z^A$ by $\phi(k)_a = \pi(k,a)$. Clearly $\eta(k) = \eta(k')$ implies $\phi(k) = \phi(k')$. This defines a continuous induced map $\hat{\phi} : \hat{K} \to Z^A$, $\hat{\phi}(\eta(k)) = \phi(k)$. However, by continuity of π in b and the fact that A is dense it also follows that

$\phi(k) = \phi(k')$ implies $\eta(k) = \eta(k')$. We therefore conclude that the map $\hat{\phi}$ is a homeomorphism of $\hat{K}$ onto the compact metric space $\phi(K) \subset Z^A$.

Now use Step 1 of the proof to deduce that for some residual subset R of B the restriction of $\hat{\pi}$ to $\hat{K} \times R$ is jointly continuous. Since $\pi = \hat{\pi} \circ (\eta \times \mathrm{id})$ it follows that also the restriction of π to $K \times R$ is jointly continuous. □

1.37. THEOREM. *Let (X, Γ) be a metrizable topological system. The following conditions are equivalent.*

1. *(X, Γ) is almost equicontinuous.*
2. *There exists a dense G_δ subset $X_0 \subset X$ such that every member of the enveloping semigroup E is continuous on X_0.*
3. *There exists a residual subset $R \subset X$ such that the evaluation map $\theta : E \times R \to X$, $\theta(p, x) = px$ is continuous.*
4. *There exists a residual subset $R \subset X$ such that the evaluation map $\theta : E \times X \to X$, $\theta(p, x) = px$ is continuous at each point of $E \times R$.*

PROOF. $1 \Rightarrow 2$. Let p be an element of E then at each point x of the G_δ subset $X_0 = EQ(X)$ (Proposition 1.35), given $\epsilon > 0$ there exits a $\delta > 0$ such that for all $x' \in B_\delta(x)$, $d(\gamma x, \gamma x') < \epsilon$ for all $\gamma \in \Gamma$. If $p = \lim \gamma_i$ then for all $x' \in B_\delta(x)$

$$d(px, px') = \lim d(\gamma_i x, \gamma_i x') \le \epsilon,$$

i.e. p is continuous at x.

$2 \Rightarrow 3$. Since a G_δ subset of a metric space is a Baire space this follows from Lemma 1.36.

$3 \Rightarrow 4$. Let $(p, x) \in E \times R$. We show that our assumption that $\theta : E \times R \to X$ is continuous at (p, x) implies in fact that $\theta : E \times X \to X$ is continuous at (p, x). Given $(p, x) \in E \times R$ and $\epsilon > 0$ there exists a $\delta > 0$ and a neighborhood U of p in E such that for $q \in U$ and $x' \in B_\delta(x) \cap R$ we have $d(px, qx') < \epsilon$. In particular this is true for $q = \gamma \in U \cap \Gamma$. However, since γ is continuous and R is dense in X we deduce that $d(px, \gamma z) < \epsilon$ for all $z \in B_\delta(x)$. Finally since $U \cap \Gamma$ is dense in U we also get $d(px, qz) \le \epsilon$ for all $q \in U$ and $z \in B_\delta(x)$.

$4 \Rightarrow 1$. We finish the proof by showing that every point in R as in 4 is an equicontinuity point. Let $x \in R$ and $\epsilon > 0$ be given, then for every $p \in E$ there exists a $\delta_p > 0$ and a neighborhood U_p of p such that $d(px, qx') < \epsilon$ for all $x' \in B_{\delta_p}(x)$ and $q \in U_p$. Since in particular $d(px, qx) < \epsilon$ we have $d(qx, qx') < 2\epsilon$. Using the compactness of E we deduce that there is a $\delta > 0$ such that $x' \in B_\delta(x)$ implies $d(qx, qx') < 2\epsilon$ for all $q \in E$. A fortiori then $d(\gamma x, \gamma x') < 2\epsilon$ for every $\gamma \in \Gamma$ and $x' \in B_\delta(x)$; i.e. x is an equicontinuity point. □

We can now prove Ellis' joint continuity theorem in the metrizable case (see Corollary 1.30).

1.38. THEOREM. *A metrizable dynamical system (X, Γ) is equicontinuous iff $E(X, \Gamma)$ is a group of homeomorphisms of X.*

PROOF. It is easy to check that $E(X, \Gamma)$ is a group of homeomorphisms when (X, Γ) is equicontinuous (see Exercise 1.26.10). Conversely, assume that X is metrizable and that $E = E(X, \Gamma)$ is a group of homeomorphisms. As in the proof of step 2 of Lemma 1.36, choose a countable dense subset $A \subset X$ and define

$\phi : E \to X^A$ by $\phi(p)(a) = pa$, $p \in E, a \in A$. The assumption that each $p \in E$ is continuous implies that ϕ is a homeomorphism of the compact space E into X^A. This implies that E is metrizable. Next apply Theorem 1.37.4, with $X = E$, to deduce the existence of a dense G_δ subset $E_0 \subset E$ such that $\theta : E \times E \to E$, $\theta(p,q) = pq$, is continuous at each point of $E \times E_0$. If q is any point in E choose some $r \in E$ with $qr \in E_0$. Since the map $R_r : s \mapsto sr$ is a self homeomorphism of E, we deduce that θ is continuous also at $(p,q) = (\mathrm{id} \times R_{r^{-1}})(p, qr)$ (for each $p \in E$). Thus $\theta : E \times E \to E$ is continuous everywhere and Arzelà-Ascoli's theorem implies that the action of E on itself by left multiplication is equicontinuous. It follows that the action of E, hence also of Γ, on X is equicontinuous. □

In recent years several authors have tried to formalize the notion of chaos in various ways. One such attempt uses the definition of "sensitive dependence on initial conditions". A chaotic $\mathbb{Z}$-system is defined, according to this school, to be a compact metric space X, together with a continuous self map $T : X \to X$ satisfying the following three properties:

1. Topological transitivity.
2. The T-periodic points are dense in X.
3. Sensitive dependence on initial conditions: there exists an $\epsilon > 0$ such that for all $x \in X$ and all $\delta > 0$ there are $y \in B_\delta(x)$ and $n \in \mathbb{Z}$ with $d(T^n x, T^n y) > \epsilon$.

Now it turns out (see Theorem 1.41 below) that in fact — unless the system is a cyclic permutation of a finite set of points — conditions 1 and 2 (i.e. the condition that (X,T) is a P-system) actually imply condition 3.

In fact the condition that the set of periodic points be dense is much too strong. We replace it in the next theorem with the condition that the system be an M-system and we shall see later that even this can be considerably relaxed (Theorem 4.29).

1.39. DEFINITION. We shall say that a system (X,Γ) is *sensitive* if it satisfies the following condition (sensitive dependence on initial conditions): there exists an $\epsilon > 0$ such that for all $x \in X$ and all $\delta > 0$ there are some $y \in B_\delta(x)$ and $\gamma \in \Gamma$ with $d(\gamma x, \gamma y) > \epsilon$. We say that (X,Γ) is *non-sensitive* otherwise.

We first note the following proposition.

1.40. PROPOSITION. *A transitive dynamical system is almost equicontinuous iff it is non-sensitive.*

PROOF. Clearly an almost equicontinuous system is non-sensitive. Conversely being non-sensitive means that for every $\epsilon > 0$ there exist $x_\epsilon \in X$ and $\delta_\epsilon > 0$ such that for all $y \in B_{\delta_\epsilon}(x_\epsilon)$ and every $\gamma \in \Gamma$, $d(\gamma x_\epsilon, \gamma y) < \epsilon$. For $m \in \mathbb{N}$ set $V_m = B_{\delta_{1/m}}(x_{1/m})$, $U_m = \Gamma V_{1/m}$ and let $R = \bigcap_{m\in\mathbb{N}} U_m$. Suppose $x \in R$ and $\epsilon > 0$. Choose m so that $2/m < \epsilon$, then $x \in U_m$ implies the existence of $\gamma_0 \in \Gamma$ such that $\gamma_0 x \in V_{1/m}$. Put $V = \gamma_0^{-1} V_{1/m}$. We now see that for $y \in V$ and every $\gamma \in \Gamma$

$$d(\gamma x, \gamma y) = d((\gamma\gamma_0^{-1})\gamma_0 x, (\gamma\gamma_0^{-1})\gamma_0 y) < 2/m < \epsilon.$$

Thus the dense G_δ set R consists of equicontinuity points. □

In Proposition 1.35 we have seen that minimality and almost equicontinuity imply equicontinuity. We easily get a stronger result.

1.41. THEOREM. *An almost equicontinuous M-system (X, Γ) is minimal and equicontinuous. Thus an M-system (hence also a P-system) which is not minimal equicontinuous is sensitive.*

PROOF. Let $x_0 \in X$ be an equicontinuity point. Given $\epsilon > 0$ there exists a $0 < \delta < \epsilon$ such that $x \in B_\delta(x_0)$ implies $d(\gamma x_0, \gamma x) < \epsilon$ for every $\gamma \in \Gamma$. Let $x' \in B_\delta(x_0)$ be a minimal point. It then follows that $S = \{\gamma \in \Gamma : \gamma x' \in B_\delta(x_0)\}$ is a syndetic subset of Γ (i.e. $FS = \Gamma$ for some finite subset F of Γ). Collecting these estimations we get, for every $\gamma \in S$,

$$d(\gamma x_0, x_0) \leq d(\gamma x_0, \gamma x') + d(\gamma x', x_0) \leq 2\epsilon.$$

Thus for each $\epsilon > 0$ the set $N(x_0, B_\epsilon(x_0)) = \{\gamma \in \Gamma : d(\gamma x_0, x_0) \leq \epsilon\}$ is syndetic, whence x_0 is minimal. Since, by Proposition 1.35, x_0 is a transitive point it follows that X is minimal, hence also equicontinuous by Proposition 1.35. □

9. Weak almost periodicity

Again we assume in this section that Γ is a countable discrete group. The general case where Γ is a second countable locally compact topological group is treated similarly. One has then to replace the algebra $C_b(\Gamma) = \ell^\infty(\Gamma)$ of (continuous) bounded functions, by its subalgebra $LUC(\Gamma)$ of left uniformly continuous functions.

1.42. DEFINITION. A function $f \in \ell^\infty(\Gamma)$ is called *(weakly) almost periodic* if in the Banach space $C_b(\Gamma) = \ell^\infty(\Gamma)$ the Γ orbit of f, $\{{}_\gamma f : \gamma \in \Gamma\}$, is (weakly) precompact, where ${}_\gamma f(\gamma') = f(\gamma\gamma')$. Denote by $AP(\Gamma)$ and by $WAP(\Gamma)$ the collections of almost periodic and weakly almost periodic functions in $\ell^\infty(\Gamma)$, respectively. If (X, Γ) is a topological system then a function $f \in C(X)$ is called *(weakly) almost periodic* if $\{f \circ \gamma : \gamma \in \Gamma\}$ is (weakly) precompact in the Banach space $C(X)$.

Let $\pi : \Gamma \to \mathfrak{U}(\mathfrak{H})$ be a unitary representation of Γ, i.e. a homomorphism from Γ into $\mathfrak{U}(\mathfrak{H})$ — the space of unitary operators on a separable Hilbert space $\mathfrak{H}$. A *matrix coefficient* of π is any function of the form $f(\gamma) = \langle \pi(\gamma)x, y\rangle$, $x, y \in \mathfrak{H}$. Let B_π be the linear subspace of $\ell^\infty(\Gamma)$ generated by matrix coefficients of π. We denote by $B(\Gamma)$ the subalgebra of $\ell^\infty(\Gamma)$ generated by all the spaces B_π. $B(\Gamma)$ is called the *Fourier-Stieltjes* algebra of Γ.

1.43. THEOREM. 1. *(Grothendieck) For a topological system (X, Γ) (not necessarily metrizable) a function $f \in C(X)$ is WAP iff for every* sequence *$\gamma_n \in \Gamma$ there is a subsequence γ_{n_i} and a function $h \in C(X)$ such that $\lim_{i\to\infty} f \circ \gamma_{n_i}(x) = h(x)$ for every $x \in X$.*

2. *For a WAP function $f \in C(X)$ the weak topology and the topology of pointwise convergence coincide on $D =$ weak-cls$\{f \circ \gamma : \gamma \in \Gamma\}$.*
3. *(Eberlein) $WAP(\Gamma)$ is a closed left and right invariant subalgebra of the algebra $\ell^\infty(\Gamma)$.*
4. *For a WAP function $F \in C(X)$ and $x \in X$ the function $f : \Gamma \to \mathbb{C}$ defined by $f(\gamma) = F(\gamma x)$ is in $WAP(\Gamma)$.*

5.

$$AP(\Gamma) \subset \overline{B(\Gamma)} \subset WAP(\Gamma) \subset \ell^\infty(\Gamma),$$

Also $c_0(\Gamma) \subset WAP(\Gamma)$, $c_0(\Gamma)$ *denoting the algebra of functions vanishing at infinity.*

PROOF. 1. By Riesz' representation theorem, $C(X)^*$ is the space of regular Borel measures on X. By considering the Dirac measures $\{\delta_x : x \in X\}$ and taking into account Eberlein-Šmulian's theorem (Theorem A.12), we see that the condition is necessary.

To get the sufficiency observe that by the other direction of Eberlein-Šmulian's theorem it is enough to show that $\{f \circ \gamma : \gamma \in \Gamma\}$ is sequentially weakly precompact.

Let $f_n = f \circ \gamma_n$ be a sequence of translates of f. By assumption there is a subsequence f_{n_j} and a function $h \in C(X)$ such that $\lim_{j\to\infty} f_{n_j}(x) = h(x)$ for every $x \in X$. By Lebesgue's dominated convergence theorem we now have

$$\lim_{j\to\infty} \int f_{n_j}\, d\mu = \int h\, d\mu$$

for every regular Borel measure μ on X; i.e. $\lim f_{n_j} = h$ weakly and the sequential weak precompactness is proved.

2. Again, considering the Dirac measures $\{\delta_x : x \in X\} \subset C(X)^*$, we see that the identity map id : $(D, \text{weak topology}) \to (D, \text{pointwise convergence topology})$ is continuous. This implies that D is a compact subset with respect to the pointwise convergence topology and, since id is 1-1, it is a homeomorphism.

3. The left and right Γ invariance of $WAP(\Gamma)$ is clear from the definition. Let $\beta\Gamma$ be the Gelfand space of the algebra $\ell^\infty(\Gamma)$; i.e. the Stone-Čech compactification of Γ (Theorem A.3). The identity map id : $\Gamma \to \Gamma$ extends to a topological embedding $j : \Gamma \to \beta\Gamma$ and for each $\gamma \in \Gamma$ the map $L_\gamma : \Gamma \to \Gamma$, defined by $L_\gamma(\gamma') = \gamma\gamma'$, extends to a homeomorphism $L_\gamma : \beta\Gamma \to \beta\Gamma$. The family $\{L_\gamma : \gamma \in \Gamma\}$ defines an action of Γ on $\beta\Gamma$. The embedding $\Gamma \to \beta\Gamma$ defines a canonical isomorphism $C(\beta\Gamma) \cong \ell^\infty(\Gamma)$. Given $f, h \in WAP(\Gamma)$ and a sequence $\gamma_n \in \Gamma$, we can find a subsequence $\gamma_i = \gamma_{n_i}$ such that the weak limits $\lim_{i\to\infty}\ {}_{\gamma_i}f = f'$ and $\lim_{i\to\infty}\ {}_{\gamma_i}h = h'$ exist (of course with $f', h' \in \ell^\infty(\Gamma) \cong C(\beta\Gamma)$). In particular these sequences converge pointwise on Γ as well as on $\beta\Gamma$ and therefore also $\lim_{i\to\infty}\ {}_{\gamma_i}(fh) = f'h'$ pointwise on the compact space $\beta\Gamma$. By part 1, applied to the topological system $(\beta\Gamma, \Gamma)$, we conclude that this convergence is also a weak convergence. Thus also $fh \in WAP(\Gamma)$. The proof that $f + h \in WAP(\Gamma)$ is analogous and we see that $WAP(\Gamma)$ is an algebra. Clearly, f_γ is in $WAP(\Gamma)$ whenever $f \in WAP(\Gamma)$ (where $f_\gamma(h) = f(h\gamma)$) since right and left translations commute. Finally, to see that $WAP(\Gamma)$ is norm closed consider a sequence $f_j \in WAP(\Gamma)$ with $\|f_j - f\| \le \frac{1}{j}$ and a sequence $\gamma_n \in \Gamma$. Applying a diagonal process construct a sub-sequence n_k such that for each j there exists $h_j \in C(\beta\Gamma)$ for which $\lim_{k\to\infty} f_j(\gamma_{n_k}x) = h_j(x)$ for every $x \in \beta\Gamma$. It is easy to check that h_j is a norm Cauchy sequence and that $\lim_{k\to\infty} f(\gamma_{n_k}x) = h(x)$ for every $x \in \beta\Gamma$, where $h \in C(\beta\Gamma)$ satisfies $\lim_{j\to\infty} \|h_j - h\| = 0$. Thus f is also in $WAP(\Gamma)$ and our proof is complete.

4. Define $J_x : C(X) \to \ell^\infty(\Gamma)$ by the formula $J_x(F)(\gamma) = F(\gamma x)$, $\gamma \in \Gamma$. It is easy to see that J_x is a continuous homomorphism of the Banach algebra $C(X)$ into the Banach algebra $\ell^\infty(\Gamma)$ (an isomorphism when $\mathcal{O}_\Gamma x$ is dense in X). This

implies that the family

$$\{f \circ \gamma : \gamma \in \Gamma\} = J_x\{F \circ \gamma : \gamma \in \Gamma\}$$

is weakly precompact in $\ell^\infty(\Gamma)$ as the continuous image of a weakly precompact set in $C(X)$.

5. By Peter-Weyl's theorem (Theorem A.15) $AP(\Gamma)$ is the closed algebra generated by the functions of the form $\gamma \mapsto \langle \pi(\gamma)x, y\rangle$, where π is a *finite dimensional* unitary representation. Hence $AP(\Gamma) \subset \overline{B(\Gamma)}$.

Next consider an arbitrary unitary representation $\pi : \Gamma \to \mathfrak{U}(\mathfrak{H})$. Let $B \subset \mathfrak{H}$ be the unit ball equipped with the weak topology. Then the action $\gamma x = \pi(\gamma)x, x \in B, \gamma \in \Gamma$ defines a compact dynamical system (B, Γ). For a fixed $y \in B$ let $F_y : B \to \mathbb{C}$ be the function $F_y(x) = \langle x, y\rangle,\ x \in B$. We claim that F_y is weakly almost periodic. In fact, if γ_n is a sequence in Γ and we choose a convergent subsequence $\gamma_{n_i}^{-1}y \to y_1 \in B$, then for every $x \in B$,

$$F_y \circ \gamma_{n_i}(x) = \langle \pi(\gamma_{n_i})x, y\rangle = \langle x, \pi(\gamma_{n_i}^{-1})y\rangle \to \langle x, y_1\rangle = F_{y_1}(x).$$

Now any matrix coefficient for π, $f_{x,y}(\gamma) = f(\gamma) = \langle \pi(\gamma)x, y\rangle$, is of the form $f_{x,y} = J_x(F_y)$, therefore the fact that B_π is a subset of $WAP(\Gamma)$ follows from part 4. We conclude that $B(\Gamma) \subset WAP(\Gamma)$, whence also $\overline{B(\Gamma)} \subset WAP(\Gamma)$.

Finally if $f \in \ell^\infty(\Gamma)$ vanishes at infinity and $\gamma_n \in \Gamma$, then either the sequence $\gamma_n f$ is a finite set of functions or it has a subsequence converging weakly to 0. □

1.44. Definition. A (not necessarily metrizable) topological system (X, Γ) is called *weakly almost periodic* (WAP) if every $f \in C(X)$ is WAP.

It is now easy to see that subsystems as well as factors of WAP systems are also WAP systems. We let Z be the Gelfand space of the algebra $WAP(\Gamma)$ (see Theorem A.1 and Section 5). The (left) Γ-invariance of $WAP(\Gamma)$ makes (Z, Γ) a compact point transitive system with $z_0 \in Z$ a transitive point, where $z_0 : WAP(\Gamma) \to \mathbb{C}$ is the evaluation multiplicative homomorphism $z_0(f) = f(e)$. Again it is now easy to check that a point transitive system (X, Γ) with a transitive point x_0 is WAP iff there is a homomorphism $\pi : (Z, z_0) \to (X, x_0)$ which is induced by the embedding $j : C(X) \to WAP(\Gamma)$ defined by $j(F)(\gamma) = F(\gamma x_0),\ F \in C(X), \gamma \in \Gamma$.

1.45. Theorem. *Let (X, Γ) be a topological system. The following conditions are equivalent:*

1. *(X, Γ) is WAP.*
2. *The enveloping semigroup $E(X, \Gamma)$ consists of continuous maps.*

Proof. If $\lim \gamma_i = p \in E$ for a net $\gamma_i \in \Gamma$ then for each $f \in C(X)$, in the topology of pointwise convergence, $\lim f \circ \gamma_i = f \circ p$. Thus when the latter is continuous for every $p \in E$ we deduce that $\{f \circ \gamma : \gamma \in \Gamma\}$ is precompact in the pointwise convergence topology.

Conversely if $\{f \circ \gamma : \gamma \in \Gamma\}$ is precompact in the pointwise convergence topology and $p \in E$, then for some net $\gamma_i \in \Gamma$, $\lim \gamma_i = p$ and without loss of generality the pointwise limit $\lim f \circ \gamma_i = h \in C(X)$ exists. Thus $f \circ p \in C(X)$ for every $f \in C(X)$ and it follows that p is a continuous map. □

1.46. COROLLARY. 1. *A WAP system is almost equicontinuous.*
2. *A WAP minimal system is equicontinuous (Kronecker).*

PROOF. Apply Theorem 1.37. □

1.47. THEOREM. *Let (X,Γ) be a WAP system, $E = E(X,\Gamma)$ its enveloping semigroup.*

1. *E contains a unique minimal left ideal I which is also the unique minimal right ideal in E.*
2. *I is a compact topological group with unit element u.*
3. *$I = uE = Eu$, so that $up = pu$ for every $p \in E$.*
4. *The subset $Y = uX$ is the mincenter of (X,Γ). It is closed and Γ-invariant and as a subsystem it is equicontinuous. The map $u : X \to Y$ is a homomorphism and a retraction.*
5. *If (X,Γ) is point transitive then it has a unique minimal subset Y and $u : X \to Y$ is its maximal Kronecker factor.*

PROOF. Since $E(X,\Gamma)$ is its own enveloping semigroup (acting by left multiplication), we deduce from Theorem 1.45 that the dynamical system (E,Γ) (Γ acting on the left) is WAP as well. If $I \subset E$ is a minimal left ideal then the sub-system (I,Γ) is WAP and minimal hence equicontinuous by Corollary 1.46. It follows that I is a compact topological group.

For every enveloping semigroup the maps $R_q : p \mapsto pq$, $q \in E$, are continuous. Thus the semigroup E acts continuously on itself on the right. In particular this defines a *right* action of the discrete group Γ on E. Now recall that in general for every continuous element q of the enveloping semigroup the map $L_q : p \mapsto qp$ is a continuous map $L_q : E \to E$. Thus for a continuous $q \in E$, if $\lim \gamma_\alpha = p \in E$ for a net $\gamma_\alpha \in \Gamma$ then $\lim q\gamma_\alpha = qp$. By assumption every $q \in E$ is continuous and it follows that the enveloping semigroup for the right action of Γ on E is again E itself, acting by multiplication on the right. Theorem 1.45 implies that the dynamical system E, with respect to the right Γ action, is WAP.

Now the discussion above shows that by symmetry also every minimal *right* ideal is a topological group. Fix a minimal left ideal I with unit element u, and a minimal right ideal R with unit element v, then $vu \in I \cap R$ and it follows that $uv = u = v$, whence also $I = R$. This proves that E contains a unique minimal left ideal I which is also a unique minimal right ideal, that I is a compact topological group with unit element u, and that u is the unique minimal idempotent in E.

For any $p \in E$, pu is an element of I and therefore $upu = pu$, and similarly $upu = up$.

Now part 4 is clear. Part 5 follows since E has a unique minimal ideal and the map $E \to X$, $p \mapsto px_0$, where x_0 is a transitive point in X, is a homomorphism. □

When Γ is abelian we can say much more.

1.48. THEOREM. *Let (X,Γ) be a transitive topological system with Γ abelian, then (X,Γ) is WAP iff the systems (X,Γ) and $(E(X),\Gamma)$ are isomorphic and $E(X)$ is a commutative semitopological semigroup (i.e. multiplication in $E(X)$ is continuous in each variable separately).*

PROOF. Assuming the conditions on $E(X)$ are satisfied we see that every $p \in E(X)$ is a continuous map on X. Theorem 1.45 implies that (X, Γ) is WAP.

Conversely if (X, Γ) is WAP then each $p \in E(X)$ is continuous. Let now $p, q \in E(X)$, say $p = \lim \gamma_i$, $q = \lim \delta_j$, $\gamma_i, \delta_j \in \Gamma$. Then

$$\begin{aligned}
pq &= p(\lim \delta_j) = \lim p\delta_j && \text{(by continuity of } p\text{)} \\
&= \lim_j(\lim_i \gamma_i)\delta_j = \lim_j(\lim_i \gamma_i\delta_j) && \text{(by continuity of right multiplication)} \\
&= \lim_j(\lim_i \delta_j\gamma_i) && \text{(by commutativity of } \Gamma\text{)} \\
&= \lim_j \delta_j(\lim_i \gamma_i) = \lim_j \delta_j p && \text{(by continuity of } \delta_j\text{)} \\
&= qp && \text{(by continuity of right multiplication).}
\end{aligned}$$

This shows that $E(X)$ is a commutative semitopological semigroup. Let $x_0 \in X$ be a transitive point, then the map $p \mapsto px_0$, $E \to X$ is 1-1. In fact if $px_0 = qx_0$ then for every $r \in E(X)$, $prx_0 = rpx_0 = rqx_0$, hence $px = qx$ for every $x \in X$; i.e. $p = q$. Thus $(X, \Gamma) \cong (E(X), \Gamma)$ and the proof is complete. □

10. The unique invariant mean on WAP functions

Recall that for a compact space X we denote by $M(X)$ the compact convex subset of probability measures on X with the weak* topology. When (X, Γ) is a dynamical system $M_\Gamma(X)$ denotes the subset of Γ-invariant measures in $M(X)$.

1.49. LEMMA. *Let (X, Γ) be a WAP dynamical system. Every element $p \in E$ defines an element $p_* \in E(M(X), \Gamma)$ and the map $p \mapsto p_*$ is an isomorphism of $E = E(X, \Gamma)$ onto $E(M(X), \Gamma)$. In particular the dynamical system $(M(X), \Gamma)$ is also WAP.*

PROOF. If $\gamma_i \to p$ is a net of elements of Γ converging to $p \in E = E(X, \Gamma)$, then by Grothendieck's theorem (Theorem 1.43.1), for every $f \in C(X)$, $f \circ \gamma_i \to f \circ p$ weakly in $C(X)$. Therefore, we have for every $\nu \in M(X)$ and $f \in C(X)$:

$$\gamma_i \nu(f) = \nu(f \circ \gamma_i) \to \nu(f \circ p) := p_*\nu(f).$$

It is easy to see that $p \mapsto p_*$ is an isomorphism of flows, whence a semigroup isomorphism. Finally as Γ is dense in both enveloping semigroups, it follows that this isomorphism is onto. □

1.50. LEMMA. *Let (X, Γ) be a WAP dynamical system and let u be the unique idempotent in $E(X, \Gamma)$, then $u_*M(X)$ is the mincenter of the system $(M(X), \Gamma)$. Every measure in $u_*M(X)$ is supported on uX, the mincenter of (X, Γ). The set of Γ-invariant measures $M_\Gamma(X)$ is a subset of $u_*M(X)$. If the system (X, Γ) has a unique minimal set (so in particular when it is point transitive), then (X, Γ) is uniquely ergodic.*

PROOF. By the previous lemma $(M(X), \Gamma)$ is WAP and clearly u_* is the unique idempotent in $E(M(X), \Gamma) \cong E(X, \Gamma)$. Therefore $u_*M(X)$ is the mincenter of $(M(X), \Gamma)$. Now if $f \in C(X)$ and $\mu \in M(X)$ then

$$\int f du_*\mu = \int f \circ u d\mu.$$

Since $f \circ u$ is supported by uX, so is $u_*\mu$.

If $\mu \in M_\Gamma(X)$, then $u_*\mu = \mu$ and thus $M_\Gamma(X) \subset u_*M(X)$. Finally if (X,Γ) has a unique minimal set $Y \subset X$, then the system (Y,Γ) is equicontinuous hence in particular uniquely ergodic. It now follows that also (X,Γ) is uniquely ergodic. □

In the next theorem we consider the universal point transitive WAP system (Z,Γ). Recall that Z is the Gelfand space corresponding to the algebra $WAP(\Gamma)$, that it is isomorphic to its enveloping semigroup $E(Z,\Gamma)$ (Exercise 1.26.15), that it contains a unique minimal ideal $I = Y$ which is in fact a compact group — the Bohr compactification of Γ — and that u, the unit element of I, is a homomorphism and a retraction $u : Z \to Y$. Let $u : WAP(\Gamma) \to WAP(\Gamma)$ also denote the map, defined by $u(f)(z) = f(uz)$, $z \in Z$. Recall as well that a linear functional $m : \mathcal{A} \to \mathbb{C}$ defined on a linear space $\mathcal{A}$ of bounded functions on Γ, closed under conjugation and containing the constant functions, is called a *mean* if

Y = AP(Γ)? cf Loomis (1953) "Abstract harmonic analysis"

$$\overline{m(f)} = m(\overline{f}) \quad \forall\, f \in \mathcal{A},$$
$$m(f) \geq 0 \quad \text{if } f \geq 0, \quad \text{and} \quad m(\mathbf{1}) = 1.$$

cf. Von Neumann (1934) "Almost periodic functions in a group" TAMS

If in addition $\mathcal{A}$ is left (right) Γ-invariant and

$$m({}_\gamma f) = m(f), \quad (m(f_\gamma) = f), \qquad \forall\, \gamma \in \Gamma,$$

then m is called a *left (right) Γ-invariant mean.*

1.51. THEOREM. 1. *The algebra $WAP(\Gamma)$ admits a unique left and right Γ-invariant mean, denoted* $\mathfrak{m}$.

2. *Let $W_0 = W_0(\Gamma) = \ker u = \{f \in WAP(\Gamma) : f \circ u = 0\}$, then*
   (a)
$$W_0 = \{f \in WAP(\Gamma) : \mathfrak{m}(|f|) = 0\}.$$
   (b) *W_0 is a closed two sided invariant subspace of $WAP(\Gamma)$ and it contains the subspace $c_0(\Gamma)$.*
   (c)
$$WAP(\Gamma) = W_0 \oplus AP(\Gamma).$$
   (d) *The constant function $\mathfrak{m}(f)$ is in $Q(f)$, the weakly closed convex hull of the left Γ-orbit of f in $\ell^\infty(\Gamma)$. In fact for every $\epsilon > 0$ there is a convex combination $\psi = \sum_{j=1}^n \lambda_j\, {}_{\alpha_j}f$, $\alpha_j \in \Gamma$ such that $|\psi(\gamma) - \mathfrak{m}(f)| < \epsilon$ for all $\gamma \in \Gamma$.*
   (e) *For $f, g \in WAP(\Gamma)$*
$$|\mathfrak{m}(fg)|^2 \leq \mathfrak{m}(|f|)^2 \cdot \mathfrak{m}(|g|)^2.$$

PROOF. Applying Lemma 1.50 to the universal point transitive WAP system (Z,Γ), we have a unique invariant measure $\mu \in M_\Gamma(Z)$ — Haar measure on $Y \subset Z$. In turn, this implies the existence of a two sided invariant mean on $WAP(\Gamma)$ by defining

$$\mathfrak{m}(f) = \int_Z f\, d\mu = \int_Z uf\, d\mu,$$

where we identify $f \in WAP(\Gamma)$ with its unique extension to Z. This mean $\mathfrak{m}$ is unique and it vanishes on c_0. The decomposition $WAP(\Gamma) = W_0 \oplus AP(\Gamma)$ follows from the definition of W_0 as the kernel of the idempotent operator u, writing $f = (f - uf) + uf$, $(f \in WAP(\Gamma))$. Since μ is supported on $Y = uZ$ it follows that

$\mathfrak{m}(f) = 0$ for every $f \in W_0$. Conversely if $\mathfrak{m}(|f|) = 0$, then $|u(f)| = |f| \restriction Y \equiv 0$, hence $u(f) = 0$.

Next we prove part 2.(d). For $f \in WAP(\Gamma)$ the almost periodic function $u(f)$ is in $Q(f)$, hence we may assume that $f \in AP(\Gamma)$. For such function the weak and the norm closure of the orbit are the same and it follows that Q is a convex norm compact Γ-invariant subset of $\ell^\infty(\Gamma)$ such that the action of Γ on Q is equicontinuous. By Markov-Kakutani's fixed point theorem ([**53**], page 456) such an action admits a fixed point, say $h \in Q$. Now clearly h is a constant function and since $\mathfrak{m}(g) = \mathfrak{m}(f)$ for every $g \in Q$ we conclude that $h \equiv \mathfrak{m}(f)$. This proves the first assertion of part (d). The second now follows from the well known fact that in a Banach space if $x_n \to x$ weakly then some sequence of convex combinations of the points x_n converges to x in norm ([**53**], page 422). Finally for $f, g \in WAP(\Gamma)$ we have

$$|\mathfrak{m}(fg)|^2 = |\int_Z fg\,d\mu|^2 \leq (\int_Z |f|\,d\mu)^2 (\int_Z |g|\,d\mu)^2 = \mathfrak{m}(|f|)^2 \cdot \mathfrak{m}(|g|)^2.$$

This completes the proof of the theorem. □

A function $f \in W_0$ is called a *flight* function. Recall that a subset L of Γ is called *syndetic* if there exists a finite subset S of Γ such that $SL = \Gamma$. It is called *thickly syndetic* if for every finite subset F of Γ the set $\bigcap \{\gamma L : \gamma \in F\}$ is syndetic.

We have the following characterization of flight functions.

1.52. LEMMA. *A function $f \in WAP(\Gamma)$ is a flight function iff for every $\epsilon > 0$ the set $L = \{\gamma \in \Gamma : |f(\gamma)| < \epsilon\}$ is thickly syndetic.*

PROOF. Suppose first that $f \in WAP(\Gamma)$ is a flight function. Let $\epsilon > 0$ be given and observe that for every finite subset $\{\gamma_1, \dots, \gamma_m\} \subset \Gamma$ we have $\mathfrak{m}(\phi) = 0$ where $\phi = \sum_{i=1}^m {}_{\gamma_i}|f|$. Hence, by Theorem 1.51.2.(d), there exists a convex combination $\psi = \sum_{j=1}^n \lambda_j\, {}_{\alpha_j}\phi$, $\alpha_j \in \Gamma$ such that $\psi(\gamma) < \epsilon$ for all $\gamma \in \Gamma$. Since $\epsilon \cdot \mathbf{1}_{\Gamma \setminus L} \leq |f|$, we have

$$\epsilon \cdot \sum_{j=1}^n \lambda_j \sum_{i=1}^m \mathbf{1}_{\Gamma \setminus \gamma_i^{-1} L}(\alpha_j \gamma) < \epsilon.$$

Hence for all $\gamma \in \Gamma$, there exists j such that $\sum_{i=1}^m \mathbf{1}_{\Gamma\setminus\gamma_i^{-1}L}(\alpha_j\gamma) < 1$; i.e. $\alpha_j \gamma \in \gamma_i^{-1} L$ for all $i = 1, \dots, m$. Thus for every $\gamma \in \Gamma$ there exists j such that $\alpha_j \gamma \in \cap_{i=1}^m \gamma_i^{-1} L$, or

$$\Gamma = \bigcup_{j=1}^n \alpha_j^{-1} (\bigcap_{i=1}^m \gamma_i^{-1} L).$$

Thus L is thickly syndetic.

Conversely, assume $f \in WAP(\Gamma)$ satisfies the condition of the lemma and let $\epsilon > 0$ be given. Let $\phi = |f|$ and choose a convex combination $\psi = \sum_{j=1}^n \lambda_j\, {}_{\alpha_j}\phi$ such that $|\psi(\gamma) - \mathfrak{m}(\phi)| < \epsilon$ for all $\gamma \in \Gamma$. Now choose γ in the syndetic set $\cap_{j=1}^n \alpha_j^{-1} L$, where $L = \{\gamma \in \Gamma : |f(\gamma)| < \epsilon\}$; then $\phi(\alpha_j\gamma) < \epsilon$ for $j = 1, \dots, n$ and

$$\begin{aligned}\mathfrak{m}(\phi) &\leq |\mathfrak{m}(\phi) - \sum_{j=1}^{n} \lambda_j \phi(\alpha_j\gamma)| + |\sum_{j=1}^{n} \lambda_j \phi(\alpha_j\gamma)| \\ &\leq \epsilon + \sum_{j=1}^{n} \lambda_j \epsilon = 2\epsilon.\end{aligned}$$

Thus $\mathfrak{m}(\phi) = \mathfrak{m}(|f|) = 0$ and f is a flight function. □

1.53. REMARK. The following observation is a simple corollary of the last lemma. If $f \in WAP(\Gamma)$ is a flight function, then there exists a sequence $\gamma_k \to \infty$ such that $\lim_{k\to\infty} f(\gamma_k) = 0$. However it is easy to deduce this fact directly: given a finite set $K \subset \Gamma$ and $\epsilon > 0$, choose $\phi \in c_0(\Gamma)$ such that $0 \leq \phi \leq 1$ and $\phi \geq \epsilon$ on K. If $|f| \geq \epsilon$ on K^c, then we would have

$$0 = \mathfrak{m}(f) = \mathfrak{m}(|f| + \phi) \geq \inf(|f| + \phi) \geq \epsilon,$$

a contradiction. Thus there exists $\gamma \notin K$ with $|f(\gamma)| < \epsilon$ and we are done.

1.54. EXERCISES. 1. A subset $A \subset \mathbb{Z}$ is thickly syndetic iff for every $k \geq 1$ the set $A_k = \{n \in \mathbb{Z} : [n, n+1, \dots, n+k] \subset A\}$ is a syndetic subset; i.e. there exists $N_k \in \mathbb{N}$ such that every interval of length N_k in $\mathbb{Z}$ contains an element of A_k.

2. Let $\{a_n\}$ be a bounded sequence of rèal numbers. The following conditions are equivalent
   (a) $\lim_{n\to\infty} \frac{1}{n} \sum_{i=0}^{n-1} |a_i| = 0$.
   (b) $\lim_{n\to\infty} \frac{1}{n} \sum_{i=0}^{n-1} |a_i|^2 = 0$.
   (c) For every $\epsilon > 0$,
   $$\lim_{n\to\infty} \frac{1}{n} \text{card}\,(\{i : |a_i| \geq \epsilon\} \cap [0, n-1]) = 0.$$
   (d) There exists a subset $J \subset \mathbb{N}$ of density zero such that
   $$\lim_{\substack{n\to\infty \\ n\notin J}} a_n = 0.$$

   (The subset $J \subset \mathbb{N}$ has density zero when $\lim_{n\to\infty} \frac{1}{n}|J \cap [0, n-1]| = 0$.) [[See [**257**], page 43.]]

3. A function $f \in WAP(\mathbb{Z})$ is a flight function (i.e. $\mathfrak{m}(|f|) = 0$) iff
   $$\limsup_{|I|\to\infty} \frac{1}{|I|} \sum_{i\in I} |f(i)| = 0,$$
   where I is an interval in $\mathbb{Z}$. [[Let $f \in WAP(\mathbb{Z})$. If for some sequence of intervals I_n with $|I_n| \to \infty$,
   $$\lim_{n\to\infty} \frac{1}{|I_n|} \sum_{i\in I} |f(i)| = t$$
   then, using the Hahn-Banach theorem, construct a mean Λ on $\ell^\infty(\mathbb{Z})$ with $\Lambda(|f|) = t$. By the Markov-Kakutani fixed point theorem there exists an invariant mean $\Lambda_0 \in \overline{\text{conv}}(\{\Lambda \circ T^n : n \in \mathbb{Z}\})$ and for Λ_0 as well $\Lambda_0(|f|) = t$.

Since $\mathfrak{m}$ is the unique invariant mean on $WAP(\mathbb{Z})$, we must have $\Lambda_0(|f|) = \mathfrak{m}(|f|) = t$. This clearly implies our assertion.]]

11. Van der Waerden's theorem

In this section we shall use the enveloping semigroup theory to give a short and elegant proof of the celebrated theorem of van der Waerden on the existence of long arithmetical progressions in (at least) one cell of any partition of the integers.

1.55. PROPOSITION. *Let (X, T) be a minimal system. For any $n \geq 1$, let $\tau = T \times T^2 \times \cdots \times T^n$, $\theta = T \times T \times \cdots \times T$, (n times) and denote by $\mathfrak{T}$ the group of homeomorphisms of X^n generated by τ and θ. Let Λ be a θ-minimal subsystem of X^n and set $N = \bar{\mathcal{O}}_\tau(\Lambda) = \mathrm{cls} \bigcup\{\tau^n\Lambda : n \in \mathbb{Z}\}$, then N is $\mathfrak{T}$-minimal.*

PROOF. Let $E = E(N, \mathfrak{T})$ be the enveloping semigroup of $(N, \mathfrak{T})$. Let $\pi_j : N \to X$ be the projection of N on the j-th component, $j = 1, 2, \ldots, n$. We consider the action of the group $\mathfrak{T}$ on the j-th component via the representation $\theta \mapsto T$ and $\tau \mapsto T^j$. With respect to this action of $\mathfrak{T}$ on X the map π_j is a homomorphism $\pi_j : (N, \mathfrak{T}) \to (X, \mathfrak{T})$ and we let $\pi_j^* : E(N, \mathfrak{T}) \to E(X, \mathfrak{T})$ be the corresponding homomorphism of enveloping semigroups. Notice that for this action of $\mathfrak{T}$ on X clearly $E(X, \mathfrak{T}) = E(X, T)$ as subsets of X^X.

Let now $u \in E(N, \theta)$ be any minimal idempotent in the enveloping semigroup of (N, θ). Choose v a minimal idempotent in the closed ideal $E(N, \mathfrak{T})u$; clearly $vu = v$. Set for each j, $u_j = \pi_j^* u$ and $v_j = \pi_j^* v$. We want to show that also $uv = u$ and note that — as an element of $E(N, \mathfrak{T})$ is determined by its projections — it suffices to show that for each j, $u_j v_j = u_j$.

Since for every j the map π_j^* is a semigroup homomorphism, we have $v_j u_j = v_j$. In particular we deduce that v_j is an element of the minimal ideal of $E(X, T)$ which contains u_j. In turn this implies

$$u_j v_j = u_j v_j u_j = u_j,$$

and it follows that indeed $uv = u$. Thus u is an element of the minimal ideal of $E(N, \mathfrak{T})$ which contains v, and therefore u is a minimal idempotent of $E(N, \mathfrak{T})$.

To finish the proof, let x be an arbitrary point in Λ and let $u \in E(N, \theta)$ be a minimal idempotent with $ux = x$. By the above argument, u is also a minimal idempotent of $E(N, \mathfrak{T})$, whence $N = \bar{\mathcal{O}}_{\mathfrak{T}}(x)$ is $\mathfrak{T}$-minimal. □

Using Proposition 1.55 we now prove the following:

1.56. THEOREM (Multiple recurrence theorem). *Let (X, T) be a minimal $\mathbb{Z}$-system, U a non-empty open subset of X, then for every positive integer n there exists a positive integer k with:*

$$U \cap T^k U \cap T^{2k} U \cap \cdots \cap T^{(n-1)k} U \neq \emptyset.$$

PROOF. Set $\Lambda = \Delta_n(X)$ and $N = \bar{\mathcal{O}}_\tau(\Delta_n(X))$, where $\Delta_n(X) = \{(x, \ldots, x) : x \in X\}$ is the diagonal in X^n. Clearly, $\Delta_n(X)$ is a θ-minimal subsystem of (X^n, θ) (canonically isomorphic to (X, T)). Choose any $x \in U$. By Proposition 1.55 the point

$$\hat{x} = (x, x, \ldots, x) \in \Delta_n(X),$$

is an almost periodic point of the system $(N, \mathfrak{T})$. Hence the set

$$N(\hat{x}, \hat{U}) = \{\gamma \in \mathfrak{T} : \gamma\hat{x} \in \hat{U}\},$$

Ex 1.1.2(e)

where $\hat{U} = U \times U \times U \cdots \times U$, is a syndetic subset of $\mathfrak{T}$. In particular, there exist $k > 0$ and $l \in \mathbb{Z}$ such that $\tau^{-k}\theta^l\hat{x} \in \hat{U}$. Thus the point $(T^{l-k}x, \dots, T^{l-k}x)$ is in the set $U \cap T^k U \cap T^{2k} U \cap \cdots \cap T^{(n-1)k} U$. □

Next we use the multiple recurrence theorem to prove:

1.57. THEOREM (Van der Waerden's theorem). *If* $\mathbb{Z} = \bigcup_{j=1}^m A_j$, *then there exists* j *such that* A_j *contains arithmetic progressions of every finite length.*

PROOF. Since

$$\sum_{j=1}^{m} \mathbf{1}_{A_j} \geq \mathbf{1},$$

it follows that there exists j such that the orbit closure $\bar{\mathcal{O}}_T(\mathbf{1}_{Aj})$ in $\{0,1\}^{\mathbb{Z}}$, contains a minimal subset X which is not the trivial set $\{\mathbf{1}_\emptyset\}$. The subset $U = \{x \in X : x(0) = 1\}$ is a non-empty open subset of the minimal system (X, T). Applying the multiple recurrence theorem to U we get $x \in X$ with $x(0) = x(k) = x(2k) = \cdots = x(nk) = 1$. Now the fact that $x \in \bar{\mathcal{O}}_T(\mathbf{1}_{A_j})$ yields an arithmetic progression of length $n + 1$ in A_j. Since there are only finitely many sets A_j, this completes the proof. □

12. Notes

The first few sections of this chapter (1 to 4) are modeled on my Luminy lecture notes [**89**]. General references to topological dynamics, structure theory and Ellis' algebraic theory of minimal systems are [**71**], [**57**], [**85**], [**36**], [**11**] and [**256**].

Proximality and weak mixing: The proof of Theorem 1.13 is due to B. Weiss ([**77**], page 183). Theorem 1.14 is from Glasner [**85**].

The enveloping semigroup: The proof of Proposition 1.27 given here is due to J. Auslander.

Almost Equicontinuity: The main sources for this section are Glasner and Weiss [**97**], and Akin, Auslander and Berg [**9**] and [**10**]. In particular Theorem 1.37 is from [**10**]. See also Banks et al. [**13**].

Weak almost periodicity: The theory of WAP functions on topological groups was developed by W. F. Eberlein, [**55**], A. Grothendieck, [**109**] and I. Glicksberg and K. de Leeuw, [**163**]; see [**107**], [**37**] and [**251**]. Grothendieck shows in [**109**] that a function $f \in \ell^\infty(\Gamma)$ is left WAP iff it is right WAP. The main sources for the two sections on WAP systems are Ellis and Nerurkar [**60**] and Bergelson and Rosenblatt [**21**]. Other sources are [**97**], [**10**], [**56**], [**51**], [**79**] and [**102**].

Van der Waerden theorem: Proposition 1.55 is from [**88**]; the short proof presented here is due to R. Ellis. Theorem 1.56 is from Furstenberg and Weiss [**81**]; see also [**19**] and [**89**].

CHAPTER 2

Dynamical Systems on Lebesgue Spaces

The phase space of a classical dynamical system originating from a set of differential equations is a smooth manifold and the dynamics is represented as a group of diffeomorphisms of this manifold. The conceptual revolution that started with Poincaré and led to modern ergodic theory, abstracted from this richly structured situation only the essential structure of a measure space $(X, \mathcal{X}, \mu)$ and a group Γ of measure-preserving transformations. Furthermore, since events of zero probability (i.e. sets of measure zero) do not really count, we sometimes end up dealing with the mere specter of the phase space X in the form of the measure algebra which is the Boolean algebra $\mathcal{X}$ modulo the ideal of μ-null sets. For example one can start with two dynamical systems $\mathbf{X} = (X, \mathcal{X}, \mu, \Gamma)$ and $\mathbf{Y} = (Y, \mathcal{Y}, \nu, \Gamma)$ described concretely as groups of diffeomorphisms on compact manifolds preserving smooth measures. The systems look very different from each other in that the manifolds and the group actions are obviously non isomorphic in the smooth sense. Nonetheless one discovers that there exists a "joining" of the two systems — i.e. a way of putting them together in a single measure space $(Z, \mathcal{Z}, \lambda, \Gamma)$ with $\mathcal{X}, \mathcal{Y} \subset \mathcal{Z}$, $\lambda \upharpoonright \mathcal{X} = \mu$ and $\lambda \upharpoonright \mathcal{Y} = \nu$ — in such a way that for every $A \in \mathcal{X}$ there is a unique (up to λ-null sets) $B \in \mathcal{Y}$ with $\lambda(A \bigtriangleup B) = 0$ and conversely for every $B \in \mathcal{Y}$ there is a unique $A \in \mathcal{X}$ with $\lambda(A \bigtriangleup B) = 0$. It turns out that the existence of such a joining is sufficient to establish a bi-measurable injection $\phi : X_0 \to Y_0$, where X_0 and Y_0 are full measure subsets of X and Y respectively, which preserves the measures ($\phi(\mu) = \nu$) and intertwines the Γ actions; i.e. an isomorphism of the measure-preserving dynamical systems.

The purpose of this chapter is to make these notions precise. In Section 2, we shall show that given a measure algebra and a group of its automorphisms one can recover a "standard" model $\mathbf{X} = (X, \mathcal{X}, \mu, \Gamma)$ of a dynamical system with X a "nice" phase space. What we mean by nice is clarified in Section 1, where the notions of standard Borel space and Lebesgue measure space are introduced and studied. In section 3 we consider some basic operations on abstract dynamical systems, like forming products and skew-products. In the last section we handle Poincaré's recurrence theorem; an easy yet one of the most fundamental results of ergodic theory.

1. Lebesgue spaces

2.1. Definition. A *measure algebra* is a pair $(\mathcal{A}, \mu)$, consisting of a Boolean σ-algebra $\mathcal{A}$ and a measure: $\mu : \mathcal{A} \to \mathbb{R}^+$, satisfying $\mu(A) = 0$ iff $A = \emptyset$, $\mu(X) < \infty$ and $\mu(\cup_{i=1}^\infty A_i) = \sum_{i=1}^\infty \mu(A_i)$, for every sequence A_i of disjoint members of $\mathcal{A}$. Here X and $\emptyset$ are the unit and zero elements of $\mathcal{A}$ and $\cap$ and $\cup$ are the Boolean operations in $\mathcal{A}$ (thus disjointness means $A_i \cap A_j = \emptyset$ for $i \neq j$).

For a measure algebra $(\mathcal{A}, \mu)$ we let:

$$\rho(A, B) = \mu(A \triangle B).$$

2.2. EXERCISES. 1. $(\mathcal{A}, \rho)$ is a complete metric space and μ is a uniformly continuous function.

2. The topology induced by ρ on $\mathcal{A}$ is the smallest topology relative to which the maps $A \mapsto \nu(A)$ are continuous, for all probability measures ν on the σ-algebra $\mathcal{A}$.
3. The function $A \mapsto A^c$ is an isometry and

$$\rho(A \cup B, C \cup D) + \rho(A \cap B, C \cap D) \leq \rho(A, C) + \rho(B, D),$$

hence the Boolean operations are continuous.

We say that the measure algebra $(\mathcal{A}, \mu)$ is *separable* if the corresponding metric space $(\mathcal{A}, \rho)$ is separable.

2.3. EXAMPLE. Let $(X, \mathfrak{X}, \mu)$ be a measure space (X is a set, $\mathfrak{X}$ a σ-algebra and μ a finite measure on $\mathfrak{X}$). Let $\mathfrak{I} = \{A \in \mathfrak{X} : \mu(A) = 0\}$, be the ideal of zero sets, then the Boolean σ-algebra $\bar{\mathfrak{X}} = \mathfrak{X}/\mathfrak{I}$ obtained as the quotient of $\mathfrak{X}$ by the equivalence relation $A \sim B \iff \mu(A \triangle B) = 0$, together with the measure $\bar{\mu}$ induced by μ on $\bar{\mathfrak{X}}$ is an example of a measure algebra. In fact all of our examples of measure algebras will be of this sort.

2.4. EXERCISE. Show that when $(X, \mathfrak{X}, \mu)$ is a non-atomic measure space the metric space $(\bar{\mathfrak{X}}, \bar{\mu}, \rho)$, is a pathwise connected space. [[First show that for every $B \in \mathfrak{X}$ the range of μ on $B \cap \mathfrak{X}$ is dense in the interval $[0, \mu(B)]$, then show that the range is closed. Use a nesting family of subsets of B, $\{B_t : t \in \mathbb{Q} \cap [0, 1]\}$ with $\mu(B_t) = t\mu(B)$, to define a path connecting B to $\emptyset$.]]

2.5. DEFINITION. A homomorphism $\Phi : (\mathcal{B}, \nu) \to (\mathcal{A}, \mu)$ of measure algebras is a map from $\mathcal{B}$ to $\mathcal{A}$ satisfying

1. $\Phi(B_1 \cup B_2) = \Phi(B_1) \cup \Phi(B_2)$,
2. $\Phi(B^c) = (\Phi(B))^c$,
3. $\mu(\Phi(B)) = \nu(B)$.

Two measure algebras are *isomorphic* if there is a 1-1 homomorphism from one of them onto the other. Such a homomorphism is called an *isomorphism*. An isomorphism $\Phi : (\mathcal{A}, \mu) \to (\mathcal{A}, \mu)$, is called an *automorphism* of the measure algebra $(\mathcal{A}, \mu)$.

2.6. EXERCISES. 1. Every homomorphism is necessarily 1-1 and a σ-homomorphism. [[For the first assertion: $\Phi(B_1) = \Phi(B_2) \Rightarrow \Phi(B_1 \cup B_2) = \Phi(B_1) \cup \Phi(B_2) = \Phi(B_1) \cap \Phi(B_2) = \Phi(B_1 \cap B_2) \Rightarrow \nu(B_1 \cap B_2) = \nu(B_1 \cup B_2) \Rightarrow \nu(B_1 \triangle B_2) = 0$. For the second: if $B = \cup_{n=1}^{\infty} B_n$ then: $\mu(\Phi(B) \setminus \cup_{n=1}^{N} \Phi(B_n)) = \nu(B \setminus \cup_{n=1}^{N} B_n) \to 0 \Rightarrow \mu(\Phi(B)) = \mu(\cup_{n=1}^{\infty} \Phi(B_n)) \Rightarrow \Phi(B) = \cup_{n=1}^{\infty} \Phi(B_n)$.]]

2. Every isomorphism of measure algebras is an isometry.

A *measurable space* $(X, \mathcal{X})$ (i.e. a set X with a σ-algebra $\mathcal{X}$ of subsets of X) is *countably generated* if there exists a sequence $A_n \in \mathcal{X}$ such that $\mathcal{X}$ is the smallest σ-algebra containing the sets A_n. When X is a topological space, the smallest σ-algebra containing the open sets in X is the *Borel* σ-algebra and we denote it by $\mathcal{B}(X)$ or $\mathcal{B}(\mathcal{T})$ if $\mathcal{T}$ is the topology of X.

2.7. PROPOSITION. *Let $(X, \mathcal{X})$ be a measurable space. The following conditions are equivalent:*

1. *There exists a separable metric space Y with $(X, \mathcal{X})$ isomorphic to the measurable space $(Y, \mathcal{B}(Y))$, where $\mathcal{B}(Y)$ is the σ-algebra consisting of Borel subsets of Y.*
2. *$(X, \mathcal{X})$ is isomorphic to the measurable space $(Y, \mathcal{B}(Y))$, where Y is a subset of the Cantor set K.*
3. *$\mathcal{X}$ is countably generated and it separates the points of X (i.e. for every $x \neq x'$ there is an $A \in \mathcal{X}$ with $x \in A$ but $x' \notin A$).*

PROOF. The implications $2 \Rightarrow 1 \Rightarrow 3$ are clear. To prove $3 \Rightarrow 2$ let A_n be a generating sequence for $\mathcal{A}$ and define $\phi : X \to K = \{0,1\}^{\mathbb{N}}$ by: $(\phi(x))_n = \mathbf{1}_{A_n}(x)$. The assumption that $\mathcal{A}$ separates points implies that ϕ is 1-1. If $Y = \phi(X) \subset K$ then, $\phi(A_n) = Y \cap \{y \in K : y_n = 1\}$, hence

$$\phi(x) = y \in [y_n = 1] \iff x = \phi^{-1}(y) \in A_n.$$

From this it is easy to see that both ϕ and ϕ^{-1} are measurable. □

A measurable space $(X, \mathcal{X})$ satisfying the equivalent conditions of the above proposition is called a *Borel space* and we shall denote $\mathcal{X} = \mathcal{B}(X)$ and call the elements of $\mathcal{B}(X)$ the *Borel sets* of X.

A topological space X is called *Polish* if there exists a metric on X such that the metric topology coincides with the original topology and such that with respect to this metric X is complete and second countable. A Borel space $(X, \mathcal{X})$ is called a *standard Borel space* if it is isomorphic to $(Y, \mathcal{B}(Y))$ where Y is a Polish space. For the proofs of the next five theorems refer to Kechris' book [**152**]. (We shall prove some related results in the next section.)

In general when X and Y are standard Borel and $f : X \to Y$ is measurable (i.e. f is a measurable map: $f^{-1}(B) \in \mathcal{B}(X)$ for every $B \in \mathcal{B}(Y)$) then the forward image $f(A) = B$ of a set $A \in \mathcal{B}(X)$ need not be an element of $\mathcal{B}(Y)$. (This fact was discovered by Souslin, correcting a mistake of Lebesgue.) We say that a subset B of a standard Borel space Y is *analytic* if there exist X, f and A as above with $f(A) = B$. We denote by $\mathcal{A}(X)$ the σ-algebra generated by the collection of analytic sets in X. (The complement of an analytic set need not be analytic; in fact $A \subset X$ is Borel iff both A and A^c are analytic.) If X is a standard Borel space and μ a σ-finite Borel measure on X denote by $\mathcal{B}_\mu(X)$ the σ-algebra generated by $\mathcal{B}(X)$ and the collection of μ-null sets (i.e. the completion of $\mathcal{B}(X)$ with respect to μ). The elements of $\mathcal{B}_\mu(X)$ are called *μ-measurable sets*. Every μ-measurable set A has the form $A = B \bigtriangleup N$ with $B \in \mathcal{B}(X)$ and $\mu(N) = 0$. A subset $A \subset X$ is called *universally measurable* if it is μ-measurable for every μ. Thus the collection of universally measurable sets is the σ-algebra $\bigcap_\mu \mathcal{B}_\mu(X)$.

2.8. THEOREM. 1. *Let X be a standard Borel space and $Y \subset X$ a Borel subset, then $(Y, \mathcal{X} \cap Y)$ is a standard Borel space.*

2. *(Souslin) Let X and Y be standard Borel spaces and $f : X \to Y$ a Borel map. If $A \subset X$ is Borel and $f \restriction A$ is 1-1, then $f(A)$ is a Borel subset of Y and f is a Borel isomorphism of A with $f(A)$.*
3. *Let X be a standard Borel space and $\mathcal{A} \subset \mathcal{B}(X)$, a countably generated σ-algebra which separates points on X, then $\mathcal{A} = \mathcal{B}(X)$.*
4. *Let X and Y be standard Borel spaces, then X and Y are Borel isomorphic iff* $\operatorname{card}(X) = \operatorname{card}(Y)$. *In particular any two uncountable standard Borel spaces are isomorphic.*

[[To prove part 3 observe that Theorem 2.7 implies that the measurable space $(X, \mathcal{A})$ is Borel isomorphic to a measurable space $(Y, \mathcal{B}(Y))$, where Y is a separable metric space and $\mathcal{B}(Y)$ is the σ-algebra of Borel subsets of Y. Considering Y as a subset of its completion $\tilde{Y}$ we now have a Borel measurable injection of the standard Borel space $(X, \mathcal{X})$ into the standard Borel space $(\tilde{Y}, \mathcal{B}(\tilde{Y}))$. By part 1 this injection is a Borel isomorphism and it follows that $\mathcal{A} = \mathcal{B}(X)$.]]

2.9. THEOREM (Lusin). *Let X be a standard Borel space, then every set in $\mathcal{A}(X)$, the σ-algebra generated by the analytic sets, is universally measurable.*

see Kechris

2.10. THEOREM (Jankov, von Neumann). *Let X and Y be standard Borel spaces, $f : X \to Y$ a Borel measurable function, then there exists an $\mathcal{A}(X)$-measurable function $g : f(X) \to X$ such that $f(g(y)) = y$ for every $y \in f(X)$.*

2.11. THEOREM. *Let X be a standard Borel space and μ a continuous probability measure on X (i.e. $\mu(\{x\}) = 0$ for every $x \in X$), then there is a Borel isomorphism $\pi : X \to I = [0, 1]$ with $\pi(\mu) = \lambda$ (= Lebesgue measure on I).*

2.12. DEFINITION. We call a measure space $(X, \mathcal{X}, \mu)$ a *standard Lebesgue space* or just a *Lebesgue space* if $(X, \mathcal{X}, \mu)$ is a standard Borel space and μ a regular probability measure on $\mathcal{X}$. Thus, in view of Theorem 2.11, $(X, \mathcal{X}, \mu)$ is isomorphic to the unit interval I with Lebesgue measure λ with the possible addition of a countable number of atoms.

Let $(X, \mathcal{X})$ be a measurable space, $\mathcal{I} \subset \mathcal{X}$ a σ-ideal in $\mathcal{X}$. Then $\mathcal{I}$ defines an equivalence relation on $\mathcal{X}$: $A \sim B$ iff $A \,\triangle\, B \in \mathcal{I}$. The quotient Boolean σ-algebra is denoted by $\mathcal{X}/\mathcal{I}$, and for $A \in \mathcal{X}$ we let $[A]$ be its equivalence class in $\mathcal{X}/\mathcal{I}$. The example we have in mind for a σ-ideal in $\mathcal{X}$ is of course the σ-ideal of null sets for a Borel measure on X.

2.13. THEOREM. 1. *Let $(X, \mathcal{X})$ be a measurable space, $\mathcal{I} \subset \mathcal{X}$ a σ-ideal in $\mathcal{X}$ and Y a non-empty standard Borel space. If $\Phi : \mathcal{B}(Y) \to \mathcal{X}/\mathcal{I}$ is a σ-homomorphism of Boolean algebras, then there is a measurable map $\phi : X \to Y$ such that $\Phi(A) = [\phi^{-1}(A)]$ for every $A \in \mathcal{B}(Y)$. Moreover ϕ is unique mod $\mathcal{I}$ in the following sense: if also $\psi : X \to Y$ is such a map then $\{x \in X : \phi(x) \neq \psi(x)\} \in \mathcal{I}$.*

2. *Let X and Y be standard Borel spaces $\mathcal{I}$ and $\mathcal{J}$ σ-ideals in $\mathcal{X}$ and $\mathcal{Y}$ respectively. Then $\Phi : \mathcal{X}/\mathcal{I} \to \mathcal{Y}/\mathcal{J}$ is an isomorphism of Boolean algebras iff there exist Borel subsets $X_0 \subset X$ and $Y_0 \subset Y$ and a Borel isomorphism*

$\phi : Y_0 \to X_0$ such that $\Phi([A]) = [\phi^{-1}(A \cap X_0)]$ for every Borel subset A of X. Such ϕ is unique mod $\mathcal{Y}$. If both $\mathcal{X}$ and $\mathcal{Y}$ contain uncountable sets, then there exists such ϕ with $X_0 = X$ and $Y_0 = Y$.

2. Dynamical systems and their factors

We can now consider the objects of our study in abstract ergodic theory. Although we prefer to work, and usually will work, with measure spaces and measure-preserving transformations, we shall occasionally have to deal merely with measure algebras and their automorphisms. Fortunately a measure space $(X, \mathcal{X}, \mu)$and a (countable) group Γ of measurable and measure preserving transformations $\gamma : (X, \mathcal{X}, \mu) \to (X, \mathcal{X}, \mu)$; $\gamma \in \Gamma$, can be recovered from a measure algebra and a group of its automorphisms.

2.14. DEFINITION. A *measure-preserving dynamical system*, (or a *dynamical system* or even just a *system* for short) $(\mathcal{X}, \mu, \Gamma)$ is a *separable* measure algebra $(\mathcal{X}, \mu)$ together with a countable group Γ of automorphisms $\gamma^{-1} : \mathcal{X} \to \mathcal{X}$. If $(\mathcal{Y}, \nu, \Gamma)$ is another such system then an *isomorphism* $\Phi : (\mathcal{Y}, \nu, \Gamma) \to (\mathcal{X}, \mu, \Gamma)$ is an isomorphism of the measure algebra $(\mathcal{Y}, \nu)$ into the measure algebra $(\mathcal{X}, \mu)$ that intertwines the Γ actions; i.e. $\Phi \circ \gamma^{-1} = \gamma^{-1} \circ \Phi$ for every $\gamma \in \Gamma$.

2.15. THEOREM. 1. *Let $(\mathcal{X}, \mu, \Gamma)$ and $(\mathcal{Y}, \nu, \Gamma)$ be two measure algebra dynamical systems and let $\Phi : (\mathcal{Y}, \nu, \Gamma) \to (\mathcal{X}, \mu, \Gamma)$ be an isomorphism of the measure algebra system $(\mathcal{Y}, \nu, \Gamma)$ into the measure algebra system $(\mathcal{X}, \mu, \Gamma)$, then there exist Cantor measure-preserving systems (X^*, μ^*, Γ), (Y^*, ν^*, Γ) and a continuous homomorphism $\phi : (X^*, \mu^*, \Gamma) \to (Y^*, \nu^*, \Gamma)$ such that for every $\gamma \in \Gamma$, the following diagram of homomorphisms of measure algebra systems is commutative:*

$$\begin{array}{ccc} (\mathcal{X}^*, \mu^*, \gamma^{-1}) & \xleftarrow{\mathfrak{A}} & (\mathcal{X}, \mu, \gamma^{-1}) \\ \uparrow{\scriptstyle \phi^{-1}} & & \uparrow{\scriptstyle \Phi} \\ (Y^*, \nu^*, \gamma^{-1}) & \xrightarrow[\mathfrak{B}]{} & (\mathcal{Y}, \nu, \gamma^{-1}) \end{array}$$

and the maps $\mathfrak{A}$ and $\mathfrak{B}$ are isomorphisms; i.e. $\mathfrak{A} \circ \Phi = \phi^{-1} \circ \mathfrak{B}$ and for every $\gamma \in \Gamma$, $\mathfrak{A} \circ \gamma^{-1} = \gamma^{-1} \circ \mathfrak{A}$, and $\mathfrak{B} \circ \gamma^{-1} = \gamma^{-1} \circ \mathfrak{B}$.

2. *If $(X, \mathcal{X}, \mu, \Gamma)$, and $(X', \mathcal{X}', \mu', \Gamma)$ are measure-preserving systems on Lebesgue spaces and $\Psi : (\mathcal{X}', \mu', \Gamma) \to (\mathcal{X}, \mu, \Gamma)$ is an isomorphism of the corresponding measure algebra dynamical systems, then there exist Γ-invariant Borel subsets $X_0 \subset X$ and $X'_0 \subset X'$ with $\mu(X_0) = \mu'(X'_0) = 1$, and a Borel isomorphism $\psi : (X_0, \mu, \Gamma) \to (X'_0, \mu', \Gamma)$ satisfying $\psi \circ \gamma = \gamma \circ \psi$; $\gamma \in \Gamma$.*

PROOF. 1. Choose a sequence $\{A_n\}_{n=1}^{\infty}$ of distinct elements of $\mathcal{X}$ with the following properties.

1. The sequence $\{A_n\}_{n=1}^{\infty}$ is dense in the measure algebra $(\mathcal{X}, \mu)$.
2. $\{A_n\}_{n=1}^{\infty}$ is invariant under Γ; i.e for every $\gamma \in \Gamma$ there exists a permutation $\theta_\gamma : \mathbb{N} \to \mathbb{N}$ such that $\gamma^{-1} A_n = A_{\theta_\gamma(n)}$.
3. For some subsequence n_k, $A_{n_k} = \Phi(B_k)$, where the sequence $\{B_k\}_{k=1}^{\infty}$ is dense in the measure algebra $(\mathcal{Y}, \nu)$.

4. For every $\gamma \in \Gamma$ there exists a permutation $\theta'_\gamma : \mathbb{N} \to \mathbb{N}$ such that $\theta_\gamma(n_k) = n_{\theta'_\gamma(k)}$.

Let $\mathcal{A}$ and $\mathcal{B}$ be the countable algebras generated by $\{A_n\}_{n=1}^\infty$ and $\{B_k\}_{k=1}^\infty$ respectively.

Now put $X^* = \{0,1\}^{\mathbb{N}}$ and $A_n^* = \{\xi \in X^* : \xi_n = 1\}$. For every N we let μ_N be the probability measure defined on the finite space $\{0,1\}^N$ by:

$$\mu_N(A_{n_1}^* \cap A_{n_2}^* \cap \cdots \cap A_{n_s}^*) = \mu(A_{n_1} \cap A_{n_2} \cap \cdots \cap A_{n_s}), \qquad n_1 < n_2 < \cdots < n_s \leq N.$$

Let μ_N^* be any extension of μ_N to a probability measure on X^*, and let $\mu_{N_j}^* \to \mu^*$ be a weak* convergent subsequence. The measure μ^* satisfies the equalities above for all finite sequences $n_1 < n_2 \cdots < n_s$ (with μ^* replacing the various μ_N).

The map $\mathfrak{A}(A_n) = A_n^*$ can be extended to an isomorphism $\mathfrak{A} : (\mathcal{A}, \mu) \to (\mathcal{A}^*, \mu^*)$ where $\mathcal{A}^*$ is the Borel algebra of X^*. Moreover for each $\gamma \in \Gamma$, the permutation θ_γ defines a homeomorphism $\gamma : X^* \to X^*$, $\xi \mapsto \gamma\xi$, where

$$(\gamma\xi)_n = \xi_{\theta_\gamma(n)}.$$

Now $\gamma^{-1}(A_n^*) = A_{\theta_\gamma(n)}$ and it follows that $\gamma : X^* \to X^*$ is measure-preserving and that $\mathfrak{A} : (\mathfrak{X}^*, \mu^*, \Gamma) \to (\mathfrak{X}, \mu, \Gamma)$ is an isomorphism of measure algebra systems.

Next define a compact measure-preserving system (Y^*, ν^*, Γ) and an isomorphism $\mathfrak{B}$ by modeling $(\mathcal{Y}, \nu, \Gamma)$ on $Y^* = \{0,1\}^{\mathbb{N}}$ and $B_k^* = \{\eta \in Y^* : \eta_k = 1\}$ as we modeled $(\mathfrak{X}, \mu, \gamma)$ on $\mathfrak{X}^*$ and A_n^*. If we now define $\phi : X^* \to Y^*$ by restriction: $\phi\xi = \eta$ with $\eta(k) = \xi(n_k)$, then it is easy to check that $\phi(\mu^*) = \nu^*$, $\gamma \circ \phi = \phi \circ \gamma$, and that the diagram of homomorphisms of measure algebra systems in the claim of the theorem commutes.

2. Start with $(X, \mathfrak{X}, \mu, \Gamma)$ and, as in the proof of part 1 choose a sequence $\{A_n\}_{n=1}^\infty$ of distinct elements of $\mathfrak{X}$. By adding more sets if necessary, we can further assume that the sets A_n separate points of X and that the sets $\Psi^{-1}(A_n)$ separate points of X'. Also as above we define the Cantor system $(X^*, \mathfrak{X}^*, \mu^*, \Gamma)$ and the homomorphism $\mathfrak{A} : (\mathfrak{X}^*, \mu^*, \gamma^{-1}) \to (\mathfrak{X}, \mu, \gamma^{-1})$. Next define a map $\alpha : X \to X^*$ by:

$$(\alpha(x))_n = \mathbf{1}_{A_n}(x).$$

Since the sequence A_n separates points of X, it is now clear that α is a Borel 1-1 map of X into X^* with $\mathfrak{A} = \alpha^{-1}$ and $\alpha \circ \gamma = \gamma \circ \alpha$. Repeat the same construction with the system $(X', \mathfrak{X}', \mu', \Gamma)$ and the sequence $\Psi^{-1}(A_n)$ to obtain an isomorphism $\alpha' : (X', \mathfrak{X}', \mu', \Gamma) \to (X^*, \mathfrak{X}^*, \mu^*, \Gamma)$. Now put $\alpha(X) = X_1^* \subset X^*$, $\alpha'(X') = X_2^* \subset X^*$ (by Theorem 2.8.2 these are Borel subsets), and $X_0^* = X_1^* \cap X_2^*$. Then X_0^* is a Γ-invariant Borel subset of X^* of full measure and we let $X_0 = \alpha^{-1}(X_0^*)$ and $X_0' = \alpha'^{-1}(X_0^*)$. Finally set $\phi = \alpha^{-1} \circ \alpha' : X_0' \to X_0$. □

Thus, equivalently, by Theorem 2.15, we can talk about a dynamical system as a Lebesgue (or Cantor) space $\mathbf{X} = (X, \mathfrak{X}, \mu)$ (sometimes called the *phase space*) and a group Γ of measurable isomorphisms $\gamma : X \to X$ satisfying $\gamma_*(\mu) = \mu$. The main advantage of using a "concrete" version of a dynamical system on a Lebesgue space, rather than its abstract form defined on a measure algebra, is that along with a Lebesgue space model comes the rich apparatus of analysis; e.g. the full force of Theorems A.6, A.7 and A.8 is available only on such spaces.

We often say that Γ *acts on* $\mathbf{X}$. When dealing with $\mathbb{Z}$ dynamical systems we shall write $(X, \mathfrak{X}, \mu, T)$, with T the automorphism corresponding to $1 \in \mathbb{Z}$, instead of $(X, \mathcal{X}, \mu, \mathbb{Z})$. The letter γ will denote the acting automorphism corresponding to the element $\gamma \in \Gamma$ in all the dynamical systems under consideration. In fact we shall often omit the letter Γ (or T) from the notation of the system and will write $\mathbf{X}$ for $(X, \mathcal{X}, \mu, \Gamma)$, $\mathbf{Y}$ for $(Y, \mathcal{Y}, \nu, \Gamma)$, $\mathbf{Z}$ for $(Z, \mathcal{Z}, \eta, \Gamma)$ etc.

When there is no measure present on a standard Borel space $(X, \mathcal{X})$, an action of Γ on it as a group of Borel isomorphisms will be called a *Borel action.*

I try to be consistent with this notation, so that **throughout the book**, in the measure theoretic context

$$\mathbf{X} = (X, \mathcal{X}, \mu, \Gamma), \quad \text{and} \quad \mathbf{Y} = (Y, \mathcal{Y}, \nu, \Gamma)$$

(or $\mathbf{X} = (X, \mathcal{X}, \mu, T)$ and $\mathbf{Y} = (Y, \mathcal{Y}, \nu, T)$ in the case of $\mathbb{Z}$-systems).

When there is a danger for confusion we may be more explicit and use other letters for our acting automorphism like S, R or $T \times T$, $\gamma \times \gamma$ etc. We shall usually identify two sets when they correspond to the same element in the measure algebra; i.e. when their symmetric difference has measure zero. In particular we shall only distinguish between a σ-algebra and its completion when it is necessary to do so. Similarly, unless we say otherwise, two measurable functions are identified if they are equal almost everywhere.

When dealing with a (Polish) topological space X, $\mathcal{X}$ will denote its Borel σ-algebra.

The morphisms in our category of dynamical systems are the homomorphisms of measure algebras $\Phi : (\mathcal{Y}, \nu) \to (\mathcal{X}, \mu)$ which satisfy $\Phi \circ \gamma = \gamma \circ \Phi$, $\gamma \in \Gamma$, or equivalently measurable, measure-preserving maps $\phi : (X, \mathcal{X}, \mu) \to (Y, \mathcal{Y}, \nu)$, with $\phi \circ \gamma = \gamma \circ \phi$ (so that $\Phi = \phi^{-1}$). We shall write: $\phi : \mathbf{X} \to \mathbf{Y}$, or just $\mathbf{X} \to \mathbf{Y}$ for this homomorphism. We then say that $\mathbf{Y}$ is a *factor* of $\mathbf{X}$ or sometimes that $\mathbf{X}$ is an *extension* of $\mathbf{Y}$. When ϕ is an isomorphism we say that the systems are *isomorphic*. As we have seen, a factor of the dynamical system $\mathbf{X}$ is determined, up to isomorphism, by a Γ-invariant sub-σ-algebra $\mathcal{Y} \subset \mathcal{X}$.

If $\mathbf{X} = (X, \mathcal{X}, \mu, \Gamma)$ is a dynamical system we say that the dynamical system $\hat{\mathbf{X}} = (\hat{X}, \hat{\mathcal{X}}, \hat{\mu}, \Gamma)$ is a *topological (Cantor) model* (or just a model) for $\mathbf{X}$ if $(\hat{X}, \Gamma)$ is a topological (Cantor) system, $\hat{\mu}$ a Γ-invariant Borel probability measure and the systems $\mathbf{X}$ and $\hat{\mathbf{X}}$ are measure theoretically isomorphic. Similarly we say that $\hat{\pi} : \hat{\mathbf{X}} \to \hat{\mathbf{Y}}$ is a *topological model* for $\pi : \mathbf{X} \to \mathbf{Y}$ when $\hat{\pi}$ is a topological factor map and there exist measure theoretical isomorphisms ϕ and ψ such that the diagram

$$\begin{array}{ccc} \mathbf{X} & \xrightarrow{\ \phi\ } & \hat{\mathbf{X}} \\ {\scriptstyle \pi}\downarrow & & \downarrow{\scriptstyle \hat{\pi}} \\ \mathbf{Y} & \xrightarrow[\ \psi\ ]{} & \hat{\mathbf{Y}} \end{array}$$

is commutative.

With a little more effort one can improve Theorem 2.15 to handle countably many factors; we shall not go through the details of this.

As our acting group Γ is assumed to be a discrete countable group, the modeling theorem (Theorem 2.15) is usually an adequate tool. However, as we shall see, a central theme of the structure theory of ergodic systems requires a treatment of "group extensions". Such an extension arises via an action of a compact (second countable) topological group, say K, as a group of automorphisms of the measure algebra. We then have to deal with the action of the locally compact group $\Gamma \times K$

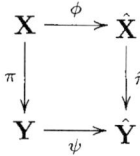

on the measure algebra and it will be necessary to have a modeling theorem for this kind of actions. We next formulate one version of such a theorem and refer to [**247**] for the proof; see also [**176**] page 330, [**209**] page 269, and [**248**] page 13.

2.16. DEFINITION. Let G be a locally compact second countable topological group and $(\mathcal{X}, \mu)$ a measure algebra. We say that $(\mathcal{X}, \mu, G)$ is a *(Boolean) G dynamical system* if G acts on $(\mathcal{X}, \mu)$ (viewed as a complete metric space and a Boolean algebra) as a group of measure-preserving Boolean automorphisms, and for each $A \in \mathcal{X}$ the map $g \mapsto gA$, $G \to \mathcal{X}$ is Borel measurable. Two Boolean dynamical systems are *isomorphic* if there exists a Boolean algebra isomorphism between them that intertwines the G actions and maps one measure to the other. (Recall that such an isomorphism is also an isometry, Exercise 2.6.2.) By a *spatial G dynamical system* we mean an action of G on a standard Lebesgue space $(X, \mathcal{X}, \mu)$ such that the map $(g, x) \mapsto gx$, $G \times X \to X$ is measurable. Two spatial G dynamical systems are isomorphic if there exists a Borel isomorphism between the corresponding standard Borel spaces that intertwines the G actions and maps one measure to the other. Finally we say that a spatial G dynamical system $(X, \mathcal{X}, \mu, G)$ is *topological* if X is a compact space and the map $(g, x) \mapsto gx$, $G \times X \to X$ is continuous.

2.17. EXERCISE. Show that a spatial G-dynamical system induces a Boolean G dynamical system on the associated measure algebra $(\bar{\mathcal{X}}, \bar{\mu})$, where $\bar{\mathcal{X}}$ is $\mathcal{X}$ modulo nullsets, and $\bar{\mu}$ the corresponding measure. [[For every measure ν on $\mathcal{X}$, $A \in \mathcal{X}$ and $g \in G$ we have $\nu(gA) = \int_X \psi(x, g)\, d\nu(x)$, where $\psi(x, g)$ is the indicator function of the subset $\{(h, hx) : h \in G,\ x \in A\}$ of $G \times X$ and ν is considered as a measure on X. Fubini's theorem shows that $g \mapsto \nu(gA)$ is Borel. Now use Exercise 2.2.2 to deduce that $g \mapsto gA$ is Borel.]]

2.18. THEOREM. *Let G be a locally compact second countable topological group.*

1. *Every Boolean G dynamical system is isomorphic to one which is associated to a spatial G-dynamical system.*
2. *Every spatial G-dynamical system is isomorphic to a topological G-dynamical system.*

The second part of the theorem, immediately, yields the following corollary (recall that G is locally compact).

2.19. COROLLARY. *Let $(X, \mathcal{X}, \mu, G)$ be a spatial G dynamical system, then, for every $x \in X$, the stability group $G_x = \{g \in G : gx = x\}$ is a closed subgroup of G and the orbit Gx is a Borel subset of X.*

Let us remark that many interesting examples of dynamical systems in ergodic theory arise as endomorphisms of a Lebesgue space, rather than as automorphisms; i.e. the map $\phi : X \to X$ is not necessarily 1-1 on a conull subset. However, in this course we shall rarely consider such dynamical systems, and our general theory will always be directed towards the invertible case.

3. The automorphism group and some basic constructions

Given an ergodic system $\mathbf{X} = (X, \mathcal{X}, \mu, \Gamma)$, let $\mathrm{End}\,(\mathbf{X})$ be the collection of (classes modulo μ) of *endomorphisms* of $\mathbf{X}$; i.e. measurable measure-preserving maps $\phi : X \to X$ that commute with every element of Γ. Clearly End $(\mathbf{X})$ forms a semigroup and we equip this semigroup with the *weak topology* of convergence in measure. With a Cantor model for $\mathbf{X}$ this topology is given by the metric

$$d(\phi, \psi) = \sum_{n=1}^{\infty} 2^{-n} \mu(\phi^{-1} A_n \bigtriangleup \psi^{-1} A_n),$$

where A_n is a list of all clopen subsets of X. We let Aut $(\mathbf{X}) \subset$ End $(\mathbf{X})$ be the subgroup of *automorphisms* of $\mathbf{X}$; i.e. the invertible (modulo μ) elements of End $(\mathbf{X})$. An equivalent metric for Aut $(\mathbf{X})$ is the metric

$$\hat{d}(\phi, \psi) = \sum_{n=1}^{\infty} 2^{-n} \frac{1}{2} \big(\mu(\phi^{-1} A_n \bigtriangleup \psi^{-1} A_n) + \mu(\phi^{-1} A_n \bigtriangleup \psi^{-1} A_n)\big).$$

With this metric Aut $(\mathbf{X})$ is a complete separable (i.e. Polish) topological group.

2.20. EXERCISES. 1. Show that the weak topology does not depend on the particular Cantor model chosen for $\mathbf{X}$.

2. Prove the assertion that with respect to $\hat{d}$, Aut $(\mathbf{X})$ is a Polish topological group.

Besides taking factors there are several other ways one can construct new dynamical systems from old ones. We next enumerate some important such constructions.

2.21. EXAMPLES.

- **Products:** Given two dynamical systems $\mathbf{X}$ and $\mathbf{Y}$, their *product* system $\mathbf{X} \times \mathbf{Y} = (X \times Y, \mathcal{X} \otimes \mathcal{Y}, \mu \times \nu, \Gamma)$ is defined on the product of the measure spaces by $\gamma(x, y) := \gamma \times \gamma(x, y) = (\gamma x, \gamma y)$. The map $\mathrm{proj}_X : \mathbf{X} \times \mathbf{Y} \to \mathbf{X}$ is clearly a homomorphism with the factor $\mathbf{X}$ arising from the sub σ-algebra $\mathcal{X} \times Y = \{A \times Y : A \in \mathcal{X}\}$. Finite and countable products are defined similarly.
- **Group factors:** Group factors of an ergodic system $\mathbf{X}$ are those factors that arise from compact subgroups of the Polish group Aut $(\mathbf{X})$ = Aut (X, μ, Γ). If $K \subset$ Aut $(\mathbf{X})$ is such a subgroup we set

  $$\mathcal{A}(K) = \{A \in \mathcal{X} : \phi A = A, \ \forall \phi \in K\}.$$

  It is easy to check that $\mathcal{A}(K)$ is a Γ-invariant sub-σ-algebra of $\mathcal{X}$. Thus $\mathcal{A}(K)$ defines a factor system $\mathbf{Y} = (Y, \mathcal{Y}, \nu, \Gamma)$ and a factor map $\pi : \mathbf{X} \to \mathbf{Y}$, with $\pi^{-1}(\mathcal{Y}) = \mathcal{A}(K)$ and $\nu = \pi_*(\mu)$. We call $\mathbf{Y}$ the *group factor* corresponding to the compact group K (or call $\mathbf{X}$ a *group extension* of $\mathbf{Y}$) and denote it by $\mathbf{Y} = \mathbf{X}/K$.
- **Skew products:** Let $\mathbf{Y}$ be a dynamical system and $(U, \mathcal{U}, \rho)$ a standard probability space. Let $\alpha : \Gamma \times Y \to$ Aut (U, ρ) be a measurable cocycle, where Aut (U, ρ) is the Polish group of all invertible measure-preserving transformations of (U, ρ); that is α satisfies the *cocycle equation*

  $$\alpha(\gamma\gamma', y) = \alpha(\gamma, \gamma' y)\alpha(\gamma', y).$$

  We define the *skew-product system* $\mathbf{Y} \underset{\alpha}{\times} (U, \rho)$ to be the system $(Y \times U, \mathcal{Y} \otimes \mathcal{U}, \mu \times \rho, \Gamma)$, where $\gamma(y, u) = (\gamma y, \alpha(\gamma, y)u)$. (Check that this indeed defines an action of Γ on $X = Y \times U$.) When $\Gamma = \mathbb{Z}$ with the measure-preserving invertible map T as the generator of the $\mathbb{Z}$-system, a cocycle α is completely determined by the map $\alpha(y) = \alpha(T, y)$ and we have

$$\alpha(n,y) = \begin{cases} \prod_{i=0}^{n-1} \alpha(T^i y) & \text{for } n \geq 0 \\ (\prod_{i=n}^{-1} \alpha(T^i y))^{-1} & \text{for } n < 0. \end{cases}$$

- **Homogeneous and group skew-products:** An important special case of a skew-product system is when $(U, \rho) = (K/L, \tilde{m})$, with K a compact group, $L \subset K$ a closed subgroup such that $\bigcap_{k \in K} k^{-1} L k = \{e\}$, m the Haar measure on K and $\tilde{m}$ its image on K/L. The cocycle α is a map from $\Gamma \times Y$ into K and $\alpha(\gamma, x) \cdot kL$ is the L coset of the element $\alpha(\gamma, x)k$ of K. We call such an extension a *homogeneous skew-product*. When L is the trivial subgroup $L = \{e\}$, so that $(U, \rho) = (K, m)$, the group K acts on $\mathbf{Y} \underset{\alpha}{\times} (K, m)$ (on the right) by $(y, k')k = (x, k'k)$ and we see that the actions of K and Γ commute. K can then be considered as a compact subgroup of $\mathrm{Aut}\,(\mathbf{Y} \underset{\alpha}{\times} (K, m))$ and the extension $\pi : \mathbf{Y} \underset{\alpha}{\times} (K, m) \to \mathbf{Y}$ is a group extension. Such group extensions we call *group skew-products*. As we shall later see the converse is also true, i.e. every (ergodic) group extension is isomorphic to a group skew-product (Theorem 3.29).
- **Homogeneous and group systems:** When $\mathbf{Y}$ is the trivial one point system, a homogeneous skew-product becomes a homogeneous system: Let K be a compact topological group, m its normalized Haar measure, $L \subset K$ a closed subgroup such that $\bigcap_{k \in K} k^{-1} L k = \{e\}$ and $\alpha : \Gamma \to K$ a homomorphism with dense image. We define a dynamical system $(K/L, \tilde{m}, \Gamma)$, where the phase space is the compact homogeneous space K/L, $\tilde{m}$ is the image of m under the quotient map $\pi : K \to K/L$ and the action of Γ is defined by left multiplication via the homomorphism α; i.e. $\gamma \cdot kL = \alpha(\gamma) kL$, $(\gamma \in \Gamma, k \in K)$. We call such a system a *homogeneous system*. When L is the trivial subgroup $L = \{e\}$ we have a *group* or a *Kronecker* system.
- **Symmetric products:** Given a dynamical system $\mathbf{X}$ and a positive integer r we let S_r be the symmetric group of $r!$ permutations of coordinates on the space X^r. Denote by $\tilde{X}^r$ the open subset of X^r consisting of points $(x_1, x_2, \ldots, x_r)$ with $\{x_1, x_2, \ldots, x_r\}$ a set of r distinct points. The group S_r acts freely on $\tilde{X}^r$ and elements of the quotient space $\tilde{X}^r / S_r$ — which we denote by $X^{\langle r \rangle}$ — are unordered r-tuples $\langle x_1, x_2, \ldots, x_r \rangle$, $x_i \in X$. We let $\mathrm{sym} : \tilde{X}^r \to X^{\langle r \rangle}, (x_1, x_2, \ldots, x_r) \mapsto \langle x_1, x_2, \ldots, x_r \rangle$, be the quotient map. When μ has no atoms, $\mu^r(\tilde{X}^r) = 1$. Clearly elements of S_r are measure-preserving transformations on (X^r, μ^r) which commute with $T \times T \times \cdots \times T$ (r times). We let $\mu^{\langle r \rangle} = \mathrm{sym}_*(\mu^r)$ and let $\mathbf{X}^{\langle r \rangle} = (X^{\langle r \rangle}, \mu^{\langle r \rangle}, T)$ be the quotient system. $\mathbf{X}^{\langle r \rangle}$ is called the *r-symmetric product system* of $\mathbf{X}$.

4. Poincaré's recurrence theorem

One of the earliest results in abstract ergodic theory is the following simple yet surprising fact. The readers may ponder over its "philosophical" implications.

2.22. Exercise (Poincaré's recurrence theorem). 1. If $\mathbf{X}$ is a dynamical system and $A \in \mathcal{X}$ with $\mu(A) > 0$ then the set $N(A, A) = \{\gamma \in \Gamma : \mu(\gamma A \cap$

$A) > 0\}$ meets every set of the form $L^{-1}L\backslash\{e\} = \{\gamma^{-1}\gamma' : \gamma, \gamma' \in L, \gamma \neq \gamma'\}$ with infinite $L \subset \Gamma$. [[The sets $\gamma A, \gamma \in L$, having the same (positive) measure, can not be all disjoint (mod μ). If $\gamma A \cap \gamma' A \neq \emptyset$ (mod μ) then $\mu(\gamma^{-1}\gamma' A \cap A) > 0$ hence $\gamma^{-1}\gamma' \in N(A, A)$.]]

2. In the case of a single measure preserving transformation we prove the following stronger assertion. Let $(X, \mathfrak{X}, \mu)$ be a probability space. Suppose $T : X \to X$ is a measurable map such that $\mu(T^{-1}(E)) = \mu(E)$ for every $E \in \mathfrak{X}$, where $T^{-1}(E) = \{x \in X : Tx \in E\}$. Show that if $B \in \mathfrak{X}$ with $\mu(B) > 0$, then for almost every point $x \in B$ there is a sequence $n_i \nearrow \infty$ such that $T^{n_i}x \in B$. [[Consider the sequence $B, T^{-1}B, T^{-2}B, \dots$. Define for each $N \geq 0$ the set $B_N = \bigcup_{n=N}^{\infty} T^{-n}B$ and let $B_\infty = \limsup T^{-n}B = \bigcap_{N=0}^{\infty} B_N$. Show that for every $j > 0$, $T^{-j}B_\infty = B_\infty$ and that $\mu(B_\infty) \geq \mu(B) > 0$. Now set $A = B \setminus B_\infty \subset B$. If $\mu(A) > 0$ repeat the same argument, with A in place of B, to get a contradiction (as obviously $A_\infty = \emptyset$).]]
3. Show that the situation in 2 applies to the case $X = [0, 1], \lambda =$ Lebesgue measure and $Tx = 2x$ (mod 1).

5. Notes

Lebesgue spaces: My source here is mainly Kechris' book [**152**].

Dynamical systems and their factors: Theorem 2.15 is from the book [**77**] by Furstenberg.

Poincaré's recurrence theorem: The book [**205**] (page 34) by Petersen presents a broad and illuminating discussion of Poincaré's recurrence theorem and more general recurrence properties of dynamical systems.

CHAPTER 3

Ergodicity and Mixing Properties

In a classical mechanical system the phase space X represents the collection of all possible states. Typically these may be described by the spatial and momentum coordinates of a large number of molecules subject to some constraints and to the laws of classical mechanics. The time evolution is then governed by these laws, expressed as a set of differential equations. Starting from a state $x \in X$, the state of the system after one unit of time, Tx, is determined by these equations; $T : X \to X$ being a diffeomorphism of the manifold X. Assuming the conservation of energy, the constant energy levels $X_c = \{x \in X : E(x) = c\}$ are invariant submanifolds and a famous theorem of Liouville tells us that each X_c carries a smooth T-invariant (→ provided they are compact!) probability measure $\mu = \mu_c$ (see e.g. [**43**]). We then interpret $\mu(A)$, for a Borel subset A of X_c, as the likelihood of finding the system at a state $x \in A$. If $f : X_c \to \mathbb{R}$ represents a measurement of some physical magnitude, which we assume is an integrable Borel function, then $\int f\, d\mu$ is the "spatial average" of f. In practice, of course, one can only hope to have some samples of the f measurements, say sequences of the form $f(x), f(Tx), f(T^2x), \ldots, f(T^Nx)$ measuring f at $N+1$ consecutive units of time. It is therefore of great practical as well as theoretical importance to verify the *ergodic hypothesis* which asserts that the "time averages" $\mathbb{A}_N(f,x) = \frac{1}{N+1}\sum_{j=0}^{N} f(T^jx)$ converge (in some sense or another) to the spatial average $\int f\, d\mu$. One can split this hypothesis into two components; first is the question: is there any kind of convergence of the time averages? And, if the answer to this first question is positive, we can further hope that these averages converge to a constant (independent of x) which will then be the spatial average.

An intensive research in the late thirties was started by J. von Neumann, as a consequence of his interest in the mathematical foundation of quantum mechanics. It led him, and then G. D. Birkhoff, to the solution of this old problem: the proofs of what we call today von Neumann's and Birkhoff's (or the mean and pointwise) ergodic theorems. The latter, which is the deeper and more difficult of the two and which was first stated in its modern form by E. Hopf, says that indeed for $f \in L^1(\mu)$, $\lim_{N\to\infty} \mathbb{A}_N(f,x) = f^*(x)$, both μ-a.e. and in $L^1(\mu)$. By taking the union of two unrelated dynamical systems we immediately see that one can not hope for the limit function f^* to be a constant. Those systems for which this is the case (and of necessity then $f^* = \int f\, d\mu$) are called ergodic systems and an important part of the research in ergodic theory following Birkhoff's result (and indeed until today) is concerned with the problem of determining which dynamical systems (many of which have physical origin; e.g various billiard systems) are ergodic systems? See for example Cornfeld, Fomin and Sinai's book [**43**], and the more recent work of Ornstein and Weiss [**198**]. We refer the interested reader to a survey of the history of the ergodic theorems by M. Frank [**69**].

Once ergodicity is established one can ask for more stringent "mixing" properties. In this chapter we deal with weak mixing, mixing, K and Bernoulli systems.

Von Neumann's ergodic theorem (which follows directly from Birkhoff's) asserts that, for $L^2(\mu)$ functions, the convergence $\lim_{N\to\infty} \mathbb{A}_N(f,x) = f^*(x)$ holds in norm in $L^2(\mu)$. It therefore follows that ergodicity is a *spectral property*, i.e. it depends only on the associated unitary operator $U_T : L^2(\mu) \to L^2(\mu)$, defined by $U_T(f) = f \circ T$ (the Koopman representation). Likewise weak mixing and mixing are spectral properties. This is not the case for the K and Bernoulli properties and in fact, as we shall see in the chapter on spectral theory, all K-systems are spectrally isomorphic (i.e. the operators U_T of any two K-systems are unitarily isomorphic).

In Sections 1 and 2 we first introduce our basic definitions in terms of a unitary operator on an abstract Hilbert space, and then apply this theory to dynamical systems via their Koopman representation.

Section 3 deals with a well known, but rarely proved, theorem of Rohlin, which presents an arbitrary factor $\mathbf{X} \to \mathbf{Y}$ of an ergodic dynamical system as a skew-product. In Section 4 we state the basic ergodic decomposition theorem which asserts that every dynamical system is a "direct integral" of its ergodic components; it explains, to a certain extent, the special attention usually devoted to ergodic systems. A proof of this theorem will be given later in Chapter 8. Section 5, is a detailed study of the ergodic decomposition of a compact group skew-product over an ergodic system.

In Section 6 we define amenable groups and prove von Neumann's ergodic theorem for amenable group actions. In Section 7 ergodicity and mixing properties for $\mathbb{Z}$ actions are considered. Section 8 deals with the pointwise ergodic theorem and some of its applications for $\mathbb{Z}$-systems. One such application is a proof of the ergodic decomposition theorem for $\mathbb{Z}$-systems. Section 9 shows how K-systems are related to various mixing properties. In Sections 10 we introduce the probabilistic view point of dynamical systems, that is, the correspondence between Γ-stationary stochastic processes and pairs $(\mathbf{X}, f)$ consisting of a Γ dynamical system $\mathbf{X}$ and a measurable function $f : X \to \mathbb{R}$.

In Section 11 Gaussian systems are defined as those dynamical systems that arise from stationary Gauss processes. In the final section we show that for such systems ergodicity of the system is equivalent to weak mixing of the Koopman representation restricted to the "first chaos".

1. Unitary representations

Given a unitary representation $\pi : \Gamma \to \mathfrak{U}(\mathfrak{H})$, where $\mathfrak{U}(\mathfrak{H})$ is the space of unitary operators on a Hilbert space $\mathfrak{H}$, let

$$B = \{x \in \mathfrak{H} : \|x\| \le 1\} \quad \text{and} \quad \mathbf{B} = \{L \in \mathfrak{L}(\mathfrak{H}) : \|L\| \le 1\},$$

where $\mathfrak{L}(\mathfrak{H})$ is the Banach algebra of bounded linear operators on $\mathfrak{H}$. Unless we say otherwise, $\mathbf{B}$ is equipped with the weak operator topology; i.e. $\lim A_\alpha = A$ for a net A_α and A in $\mathbf{B}$ iff $\lim\langle A_\alpha x, y\rangle = \langle Ax, y\rangle$, for every $x, y \in \mathfrak{H}$. With respect to this topology $\mathbf{B}$ is a compact space and so is B with the relative weak topology induced from $\mathfrak{H}$.

Set

$$\mathbf{E} = \text{cls}\,\{\pi(\gamma) : \gamma \in \Gamma\} \subset \mathbf{B},$$

and observe that $\mathbf{E}$ is the enveloping semigroup of the compact topological system (B, Γ). Let $\mathbf{G} \subset \mathbf{E}$ be the closure in the *strong operator topology* of the group $\{\pi(\gamma) : \gamma \in \Gamma\}$. It is easy to check that in this topology $\mathbf{G}$ is a Polish topological group consisting of unitary operators. For an operator $A \in \mathfrak{L}(\mathfrak{H})$, A^* is the conjugate operator. We sometimes write $\pi(\gamma) = U_\gamma$.

3.1. PROPOSITION. 1. *$\mathbf{B}$ is a compact convex set closed under conjugation.*

2. *The operation of composition makes $\mathbf{B}$ a compact semitopological semigroup, i.e. a semigroup with right and left continuous multiplication; hence the pair $(\mathbf{B}, \Gamma)$, where $\gamma \in \Gamma$ acts by left multiplication by U_γ, is a compact Γ dynamical system and $(\mathbf{E}, \Gamma)$ is a subsystem.*
3. *$\mathbf{E}$ is a closed subsemigroup of $\mathbf{B}$ isomorphic to the enveloping semigroup of all three systems (B, Γ), $(\mathbf{B}, \Gamma)$ and $(\mathbf{E}, \Gamma)$; $\mathbf{E} = E(B, \Gamma) \cong E(\mathbf{B}, \Gamma) \cong E(\mathbf{E}, \Gamma)$.*
4. *The topological systems (B, Γ), $(\mathbf{B}, \Gamma)$ and $(\mathbf{E}, \Gamma)$ are weakly almost periodic. The latter is topologically transitive and therefore contains a unique minimal set, which is also the unique minimal ideal (i.e. minimal left ideal as well as minimal right ideal; see Theorem 1.47) $\mathbf{I}$ of the semigroup $\mathbf{E}$.*
5. *The weak and strong operator topologies coincide on $\mathbf{I}$. The ideal $\mathbf{I}$ is a compact topological group closed under * and $pp^* = p^*p = u$ for every $p \in \mathbf{I}$, where u denotes the unit element of $\mathbf{I}$.*
6. *Let m be the Haar measure on $\mathbf{I}$, then for a continuous function f on $\mathbf{E}$ we have $\mathfrak{m}(f) = \int f(p)\, dm(p)$ where $\mathfrak{m}$ is the unique invariant mean on $WAP(\Gamma)$ and on the left hand side of this equation f is identified with its restriction to $\pi(\Gamma)$.*
7. *When Γ is commutative then $\mathbf{E}$ is a commutative semitopological semigroup.*

PROOF. By definition $\mathbf{E}$ is the enveloping semigroup of the dynamical system (B, Γ) and, as every element of $\mathbf{E}$ is weakly continuous, Theorem 1.45 implies that (B, Γ) is WAP. The isomorphism of the three enveloping semigroups is canonical. The remaining assertions of the proposition follow from the theorems on WAP systems proved in Chapter 1, Sections 9 and 10. □

We let $I = \pi(e)$, the identity operator on $\mathfrak{H}$ and let $\mathfrak{I} = \mathrm{Inv}\,(\mathfrak{H})$ be the closed subspace of $\mathfrak{H}$ consisting of all the Γ-invariant vectors

$$\mathfrak{I} = \{x \in \mathfrak{H} : \pi(\gamma)x = x, \ \forall\, \gamma \in \Gamma\}.$$

$P_{\mathfrak{I}} : \mathfrak{H} \to \mathfrak{I}$ is the corresponding orthogonal projection. Finally $u \in \mathbf{I}$ is the unique minimal idempotent in $\mathbf{E}$. It is also the unit element of the compact group $\mathbf{I}$.

Let B_π be the linear span of the functions $f(\gamma) = \langle U_\gamma x, y\rangle$, $x, y \in \mathfrak{H}$, $\gamma \in \Gamma$ (matrix coefficients of π), then $B_\pi \subset WAP(\Gamma)$.

3.2. DEFINITION. The unitary representation π of Γ on a Hilbert space $\mathfrak{H}$ is called:

1. *Ergodic* if every $f \in B_\pi$ satisfies: $\mathfrak{m}(f) = 0$.
2. It is called *weakly mixing* if $B_\pi \subset W_0(\Gamma)$; i.e. for every $x, y \in \mathfrak{H}$ the function $f(\gamma) = \langle \pi(\gamma)x, y\rangle$ is a flight function (satisfies $\mathfrak{m}(|f|) = 0$).
3. It is *mixing* if $B_\pi \subset c_0(\Gamma)$; i.e. every matrix coefficient of π vanishes at infinity. (Notice that: mixing $\Rightarrow$ weak mixing $\Rightarrow$ ergodicity.)

4. It is *rigid* if the Polish topological group $\mathbf{G}$ (with the strong operator topology) is not a discrete group.
5. It is *isometric* (or *has a discrete spectrum*) if $\mathbf{E} = \mathbf{I} = \mathbf{G}$.

Note that via the characterization of $W_0(\Gamma)$ given in Lemma 1.52 we obtain a combinatorial characterization of weak mixing with no recourse to the invariant mean.

The identity

$$\begin{aligned}\langle x, y\rangle =& \frac{1}{4}\{\langle x+y, x+y\rangle - \langle x-y, x-y\rangle \\ &+ i\langle x+iy, x+iy\rangle - i\langle x-iy, x-iy\rangle\}\end{aligned}$$

implies that the special matrix coefficients of the form $f(\gamma) = \langle \pi(\gamma)x, x\rangle$, $x \in \mathfrak{H}$ already span B_π.

The following lemmas and theorems explain this terminology for those who are familiar with the classical notions of ergodicity and weak mixing.

3.3. LEMMA. 1. *Suppose that π is irreducible and that 0 is the only Γ-invariant vector in $\mathfrak{H}$, then $\mathfrak{m}(f) = 0$ for every $f \in B_\pi$.*

2. *For $f(\gamma) = \langle U_\gamma x, y\rangle$, $x, y \in \mathfrak{H}$, $\gamma \in \Gamma$, we have*

$$\mathfrak{m}(f) = \langle P_{\mathfrak{J}} x, y\rangle.$$

3. *Let $Q = \int_{\mathbf{I}} p\, dm(p)$, where m is Haar measure on $\mathbf{I}$ and the integral is defined in the weak sense; i.e. for every $x, y \in \mathfrak{H}$*

$$\begin{aligned}\langle Qx, y\rangle &= \int_{\mathbf{I}} \langle px, y\rangle\, dm(p) \\ &= \int_{\mathbf{E}} \langle px, y\rangle\, dm(p),\end{aligned}$$

then $Q = P_{\mathfrak{J}}$.

PROOF. 1. It is enough to show that $\mathfrak{m}(f) = 0$, for all the functions of the form $f(\gamma) = \langle \pi(\gamma)x, x\rangle$. Assume now that there exists a vector $x \in \mathfrak{H}$ such that $\mathfrak{m}(f) = a \neq 0$. By the properties of $\mathfrak{m}$ (Theorem 1.51.2.(d)) there exists, for every $\epsilon > 0$, a convex combination $\phi(\gamma) = \sum_{j=1}^n c_j\ {}_{\gamma_j}f(\gamma)$ such that for every $\gamma \in \Gamma$

$$\begin{aligned}|\phi(\gamma) - a| &= |\sum_{j=1}^n c_j \langle \pi(\gamma_j^{-1}\gamma)x, x\rangle - a| \\ &= |\langle x, \pi(\gamma^{-1}) \sum_{j=1}^n c_j \pi(\gamma_j)x\rangle - a| \\ &= |\langle x, \pi(\gamma^{-1})w\rangle - a| < \epsilon,\end{aligned}$$

where $w = w(\epsilon) = \sum_{j=1}^n c_j \pi(\gamma_j)x$. By weak compactness there exists a vector $w_0 \in \mathfrak{H}$ with $\langle \pi(\gamma)x, w_0\rangle = a$ for every $\gamma \in \Gamma$. By irreducibility we see that for every $z \in \mathfrak{H}$, $\langle \pi(\gamma)x, z\rangle$ is independent of γ and in particular $\langle \pi(\gamma)x, x\rangle = \langle x, x\rangle = \|x\|^2$. It follows that $\pi(\gamma)x = x$ for every $\gamma \in \Gamma$. By our assumption $x = 0$, hence $f = 0$, so that $0 = \mathfrak{m}(f) = a$. This contradiction completes the proof of part 1.

2. Again it suffices to work with functions of the form $f(\gamma) = \langle \pi(\gamma)x, x\rangle$. Write $x = x_1 + x_0$ with $P_{\mathfrak{J}}(x) = x_1 \in \mathfrak{J}$, $x_0 \in \mathfrak{J}^\perp$, to get

$$\begin{aligned} \mathfrak{m}(f) &= \mathfrak{m}(\langle U_\gamma x, x\rangle) \\ &= \mathfrak{m}(\langle U_\gamma x_1, x_1\rangle + \langle U_\gamma x_0, x_0\rangle) \\ &= \langle x_1, x_1\rangle + \mathfrak{m}(\langle U_\gamma x_0, x_0\rangle). \end{aligned}$$

By decomposing $\pi \restriction \mathfrak{J}^\perp$ into the direct sum of its irreducible subrepresentations we can apply part 1 to deduce $\mathfrak{m}(f_0) = 0$, where $f_0(\gamma) = \langle U_\gamma x_0, x_0\rangle$. Thus

$$\mathfrak{m}(f) = \langle x_1, x_1\rangle + \mathfrak{m}(f_0) = \langle x_1, x_1\rangle = \langle P_{\mathfrak{J}}x, x\rangle.$$

3. Since m is $*$-invariant ($p^* = p^{-1}$ on $\mathbf{I}$) we have $Q^* = Q$. The invariance of m under multiplication (in $\mathbf{I}$ and hence also in $\mathbf{E}$) implies that $pQ = Qp = Q$ for every $p \in \mathbf{E}$. Thus for every $x, y \in \mathfrak{H}$

$$\begin{aligned} \langle Q^2 x, y\rangle = \langle Qx, Qy\rangle &= \int_{\mathbf{I}} \langle px, Qy\rangle \, dm(p) \\ &= \int_{\mathbf{I}} \langle Qpx, y\rangle \, dm(p) = \int_{\mathbf{E}} \langle Qx, y\rangle \, dm(p) = \langle Qx, y\rangle, \end{aligned}$$

hence $Q^2 = Q$ is a projection. Since in particular $\gamma Q = Q$ for every $\gamma \in \Gamma$, we have $\operatorname{range} Q \subset \operatorname{range} P_{\mathfrak{J}}$. On the other hand, for $x \in \mathfrak{J}$ and $y \in \mathfrak{H}$

$$\langle Qx, y\rangle = \int_{\mathbf{I}} \langle px, y\rangle \, dm(p) = \langle x, y\rangle,$$

hence $Qx = x$ and we also have $\operatorname{range} P_{\mathfrak{J}} \subset \operatorname{range} Q$; i.e. $P_{\mathfrak{J}} = Q$. □

3.4. THEOREM. *The following conditions are equivalent:*

1. *π is ergodic.*
2. *There are no nonzero invariant vectors for π.*
3. *$P_{\mathfrak{J}} = 0$.*

PROOF. Suppose π is ergodic. If $\pi(\gamma)x = x$ for a vector $x \in \mathfrak{H}$ and all $\gamma \in \Gamma$, then $f(\gamma) = \langle \pi(\gamma)x, x\rangle$ is the constant function $\|x\|^2$ and we have $0 = \mathfrak{m}(f) = \|x\|^2$; i.e. $x = 0$.

Conversely, assuming that π admits no non-zero invariant vectors we get $\mathfrak{m}(f) = 0$ for all functions $f \in B_\pi$ from the previous lemma. □

We also have the following characterizations of weak mixing.

3.5. THEOREM. *The following conditions are equivalent:*

1. *π is weakly mixing.*
2. *π contains no non-zero finite dimensional sub-representations.*
3. *$u = P_{\mathfrak{J}} = 0 \in \mathbf{E}$.*
4. *For every unitary representation $\sigma : \Gamma \to \mathfrak{U}(\mathfrak{K})$, on a Hilbert space $\mathfrak{K}$, the unitary representation $\pi \otimes \sigma : \Gamma \to \mathfrak{H} \otimes \mathfrak{K}$ is ergodic.*
5. *The unitary representation $\pi \otimes \pi^J : \Gamma \to \mathfrak{H} \otimes \mathfrak{H}$, is ergodic. Here $\pi^J : \Gamma \to \mathfrak{U}(\mathfrak{H})$ is the unitary representation defined by $\pi^J(\gamma)x = J\pi(\gamma)Jx$ with $J : \mathfrak{H} \to \mathfrak{H}$ a* conjugation; *i.e. a conjugate linear operator satisfying $J^2 = \mathrm{id}$ and $\langle Jx, Jy\rangle = \overline{\langle x, y\rangle}$. (One way to get such a map is by choosing a basis and conjugating the scalars in each coordinate.)*

PROOF. $1 \Rightarrow 2$. Suppose $x \in V \subset \mathfrak{H}$, with V a finite dimensional invariant subspace. Set $f(\gamma) = \langle \pi(\gamma)x, x\rangle$, then $f \in AP(\Gamma)$ and by assumption $f \in W_0(\Gamma)$. By Theorem 1.51.2.(c) $f = 0$, hence $x = 0$. Thus π contains no non-zero finite dimensional sub-representations.

$2 \Rightarrow 3$. By Proposition 3.1.5, for any $x \in \mathfrak{H}$ the vector ux has a strongly compact orbit closure: $\mathrm{cls}\,(\Gamma ux) = \mathrm{cls}\,\{U_\gamma ux : \gamma \in \Gamma\} = \mathbf{E}ux = \mathbf{I}ux$. Therefore the linear span $V = \mathrm{span}(\mathrm{cls}\,(\Gamma ux))$ is a Γ-invariant finite dimensional subspace of $\mathfrak{H}$, which by assumption is $\{0\}$. Thus $ux = 0$ for every $x \in \mathfrak{H}$; i.e. $u = 0$.

$3 \Rightarrow 1$. This follows directly from the definition of the space of flight functions: $W_0 = \ker u = \{f \in WAP(\Gamma) : f \circ u = 0\}$ (see Theorem 1.51).

$1 \Rightarrow 4$. By Theorem 1.51.7 (e) we have for every $x, x' \in \mathfrak{H}$ and $y, y' \in \mathfrak{K}$,

$$|\mathfrak{m}(\langle(\pi \otimes \sigma)(\gamma)x \otimes y, x' \otimes y'\rangle)|^2 = |\mathfrak{m}(\langle\pi(\gamma)x, x'\rangle\langle\sigma(\gamma)y, y'\rangle)|^2$$
$$\leq \mathfrak{m}(|\langle\pi(\gamma)x, x'\rangle|^2) \cdot \mathfrak{m}(|\langle\sigma(\gamma)y, y'\rangle|^2).$$

Now the weak mixing of π implies

$$\mathfrak{m}(|\langle\pi(\gamma)x, x'\rangle|^2) \leq \|f\|_\infty \mathfrak{m}(|f|) = 0,$$

where $f(\gamma) = \langle\pi(\gamma)x, x'\rangle$. Thus $|\mathfrak{m}(\langle(\pi\otimes\sigma)(\gamma)x\otimes y, x'\otimes y'\rangle)| = 0$, for every $x, x' \in \mathfrak{H}$ and $y, y' \in \mathfrak{K}$ and since the vectors of the form $x \otimes y$ generate $\mathfrak{H} \otimes \mathfrak{K}$, this proves the implication.

The implication $4 \Rightarrow 5$ is obvious and finally assuming 5 we have for every $x, x' \in \mathfrak{H}$, and $f(\gamma) = \langle\pi(\gamma)x, x'\rangle$,

$$\mathfrak{m}(|f(\gamma)|^2) = \mathfrak{m}(f(\gamma)\overline{f(\gamma)}) = \mathfrak{m}(\langle\pi(\gamma)x, x'\rangle\langle J\pi(\gamma)x, Jx'\rangle)$$
$$= \mathfrak{m}(\langle(\pi \otimes \pi^J)(\gamma)x \otimes Jx, x' \otimes Jx'\rangle) = 0.$$

The Cauchy-Schwartz inequality

$$\mathfrak{m}(|f|)^2 = |\mathfrak{m}(|f| \cdot \mathbf{1})|^2 \leq \mathfrak{m}(|f|^2) \cdot \mathfrak{m}(\mathbf{1}) = 0,$$

finishes the proof of the implication $5 \Rightarrow 1$. □

3.6. LEMMA. *The restriction of the minimal idempotent u to every finite dimensional invariant subspace of $\mathfrak{H}$ is the identity.*

PROOF. On an invariant finite dimensional space $V \subset \mathfrak{H}$, the weak and strong operator topologies coincide. Thus the restriction of every element $p \in \mathbf{E}$ to V is a unitary operator. In particular the projection u is invertible; i.e. $u \restriction V = \mathrm{id}$. □

3.7. THEOREM. *The following conditions are equivalent:*

1. *π is isometric; i.e. $\mathbf{E} = \mathbf{I} = \mathbf{G}$ is a compact topological group.*
2. *$u = I$.*
3. *The unitary representation π decomposes as a (countable) direct sum of finite dimensional irreducible subrepresentations.*
4. *Every vector $x \in \mathfrak{H}$ is a* compact vector*; i.e. the set $\{U_\gamma x : \gamma \in \Gamma\}$ has a compact closure in the norm topology on $\mathfrak{H}$.*

PROOF. The equivalence of 1 and 2 is clear. Since $\mathbf{I}$ is a compact topological group the implication $1 \Rightarrow 3$ follows from Peter-Weyl's theorem (Theorem A.15). The implication $3 \Rightarrow 2$ follows from Lemma 3.6. Since norm cls $\{U_\gamma x : \gamma \in \Gamma\} = \mathbf{G}x$

we clearly have 1 $\Rightarrow$ 4. Finally, it is easy to see that condition 4 implies that $\{U_\gamma : \gamma \in \Gamma\}$ is precompact in the strong operator topology, whence $\mathbf{G}$ is compact in this topology and therefore $\mathbf{G} = \mathbf{E}$. □

2. The Koopman representation

3.8. DEFINITION. With every dynamical system $\mathbf{X}=(X, \mathfrak{X}, \mu, \Gamma)$ we associate the *Koopman representation* $\pi : \Gamma \to \mathfrak{U}(\mathfrak{H})$ on the Hilbert space $\mathfrak{H} = L^2(\mu)$, defined by $\pi(\gamma) = U_\gamma$, where

$$U_\gamma(f)(x) = f(\gamma^{-1}x) \qquad f \in L^2(\mu),\ g \in \Gamma.$$

(For $\mathbb{Z}$ dynamical systems, though, we shall usually use the traditional definition $U_T(f)(x) = f(Tx)$.) We write $\mathbf{B}(\mathbf{X}), \mathbf{E}(\mathbf{X}), \mathbf{I}(\mathbf{X})$ and $\mathbf{G}(\mathbf{X})$ for the corresponding semigroups of operators as described in the previous section. We let $I = \pi(e)$, the identity operator on $\mathfrak{H}$, $\mathfrak{I} = \mathrm{Inv}\,(\mathfrak{H})$ is the closed subspace of $\mathfrak{H}$ consisting of all the Γ-invariant functions and $\mathfrak{C}$ will denote the one dimensional space of constant functions. $P_{\mathfrak{I}} : \mathfrak{H} \to \mathfrak{I}$ and $P_{\mathfrak{C}} : \mathfrak{H} \to \mathfrak{C}$ are the corresponding orthogonal projections. Finally $u \in \mathbf{I}$ is the unique minimal idempotent in $\mathbf{E}$. It is also the unit element of the compact group $\mathbf{I}$.

3.9. DEFINITION. Let $\mathbf{X}$ be a dynamical system and $\pi : \Gamma \to \mathfrak{U}(L^2(\mu))$ the associated Koopman representation. We let $\pi_0 : \Gamma \to \mathfrak{U}(L_0^2(\mu))$ be the restriction of π to the subspace $L_0^2(\mu)$ of functions with integral zero. We say that $\mathbf{X}$ is

1. *ergodic, weakly mixing, mixing* iff π_0 has the corresponding property as a unitary representation. (Thus: mixing $\Rightarrow$ weak mixing $\Rightarrow$ ergodicity.)
2. *isometric* (or has *discrete spectrum*) if $\mathbf{E}(\mathbf{X}) = \mathbf{I}(\mathbf{X}) = \mathbf{G}(\mathbf{X})$; i.e. $\mathbf{E}(\mathbf{X})$ is a compact topological group.
3. *rigid*, if $\pi(\Gamma) \subsetneq \mathbf{G}(\mathbf{X})$; i.e. the Polish topological group $\mathbf{G}(\mathbf{X})$ is non-discrete.

A straightforward application of the theorems of the previous section to the Koopman representation of a dynamical system yields the following characterizations.

3.10. THEOREM. *The following conditions on a dynamical system* $\mathbf{X}$ *are equivalent:*

1. $\mathbf{X}$ *is ergodic.*
2. *The constant functions are the only invariant functions in* $L^2(\mathbf{X})$.
3. *Every measurable function* f *into a Polish space* P *satisfying* $f \circ \gamma = f$ *a.e. for all* $\gamma \in \Gamma$ *is of the form* $f \equiv p$, *a.e. for a point* $p \in P$.
4. *An invariant subset* $A \in \mathfrak{X}$ *is either* X *or* $\emptyset$ (mod μ).
5. $P_{\mathfrak{I}} = P_{\mathfrak{C}}$. (Projection onto constant functions)

3.11. THEOREM. *The following conditions on a dynamical system* $\mathbf{X}$ *are equivalent:*

1. $\mathbf{X}$ *is weakly mixing.*
2. *The 1-dimensional space of constant functions is the only finite dimensional invariant sub-space of* $L^2(\mathbf{X})$.
3. $u = P_{\mathfrak{I}} = P_{\mathfrak{C}} \in \mathbf{E}$.
4. *For every ergodic system* $\mathbf{Y}$ *the product system* $\mathbf{X} \times \mathbf{Y}$ *is ergodic.*

5. *The product system* $\mathbf{X} \times \mathbf{X}$ *is ergodic.*
6. *An invariant set* $W \subset X \times X$ *is either* $X \times X$ *or* $\emptyset$ $\pmod{\mu \times \mu}$.

A word of warning: for two non-trivial dynamical systems $\mathbf{X}$ and $\mathbf{Y}$ we have $L_0^2(\mu) \otimes L_0^2(\nu) \subsetneq L_0^2(\mu \times \nu)$. This shows why in part 4 we have to require that the system $\mathbf{Y}$ be ergodic while in Theorem 3.5.4 we do not have to require that the representation σ be ergodic; however it also shows that one should be careful in transferring the notions of ergodicity and weak mixing from the abstract representation setup to the corresponding notions for dynamical systems. In part 5 we take J — in the corresponding part of Theorem 3.5 — to be the usual conjugation in $L^2(\mathbf{X})$, so that $JU_\gamma J = U_\gamma$.

3.12. THEOREM. *The following conditions on a dynamical system* $\mathbf{X}$ *are equivalent:*

1. $\mathbf{X}$ *is isometric; i.e.* $\mathbf{E} = \mathbf{I} = \mathbf{G}$ *is a compact topological group.*
2. $u = I$.
3. *The Koopman representation decomposes as a (countable) direct sum of finite dimensional irreducible subrepresentations.*
4. *Every function* $f \in L^2(\mathbf{X})$ *is* compact; *i.e. the set* $\{U_\gamma f : \gamma \in \Gamma\}$ *has a compact closure in the norm topology on* $L^2(\mathbf{X})$.

From Theorems 3.11 and 3.12 we immediately deduce the following result.

3.13. COROLLARY. *A dynamical system which is both weakly mixing and isometric is trivial.*

We also have

3.14. THEOREM. *The dynamical system* $\mathbf{X}$ *is:*

1. *mixing iff* $\mathbf{E} = \{U_\gamma : \gamma \in \Gamma\} \cup \{P_{\mathfrak{C}}\}$
2. *rigid iff* $\mathbf{G}$ *(with the strong operator topology) is a non-discrete group.*

Another convenient characterization of weak mixing is easy to prove.

3.15. THEOREM. *An ergodic system* $\mathbf{X}$ *is weakly mixing iff it admits no nontrivial isometric factor.*

PROOF. Suppose $\mathbf{X}$ is weakly mixing and admits an isometric factor. Now a factor of a weakly mixing system is also weakly mixing and the only system which is both isometric and weakly mixing is the trivial system (Corollary 3.13). Thus a weakly mixing system does not admit a nontrivial isometric factor.

For the other direction, if $\mathbf{X}$ is non-weakly mixing then in the product space $X \times X$ there exists a Γ-invariant measurable subset W such that $0 < (\mu \times \mu)(W) < 1$. For every $x \in X$ let $W_x = \{x' \in X : (x, x') \in W\}$ and let $f_x = \mathbf{1}_{W_x}$, a function in $L^\infty(\mu)$. It is easy to check that $U_\gamma f_x = f_{\gamma^{-1}x}$ so that the map $\pi : X \to L^2(\mu)$ defined by $\pi(x) = f_x, x \in X$ is a Borel factor map. Denote

$$\pi(X) = Y \subset L^2(\mu), \quad \text{and} \quad \nu = \pi_*(\mu),$$

then, since the action of Γ on Y is by isometries (unitary operators), it is easy to deduce that (Y, ν, Γ) is a nontrivial isometric factor. □

3.16. PROPOSITION. *A homogeneous system $(K/L, \tilde{m}, \Gamma)$ is isometric and ergodic.* (Note: Γ is dense in K.)

PROOF. Since the action of the group Γ on K/L is equicontinuous, it follows easily that $\mathbf{G} = \mathbf{E}$ and the system $(K/L, \tilde{m}, \Gamma)$ is isometric. Suppose $f \in L^2(\tilde{m})$ is a Γ-invariant function. We shall think of f as an element of $L^2(K, m)$. By the Peter-Weyl theorem (Theorem A.15) the regular representation π decomposes as a (countable) direct sum of finite dimensional irreducible subrepresentations $\pi = \sum_{n=0}^{\infty} \pi_n$, with $L^2(K, m) = \sum_{n=0}^{\infty} \mathfrak{H}_n$ and $\pi_n = \pi \restriction \mathfrak{H}_n$. Moreover each of the finite dimensional spaces $\mathfrak{H}_n$ consists of continuous functions and π_0 is the restriction of π to the one dimensional space of constant functions $\mathfrak{H}_0$. Let $f = \sum f_n$ be the corresponding decomposition of f, then for every $\gamma \in \Gamma$

$$f(x) = (\pi(\gamma)f)(x) = f(\gamma^{-1}x) = \sum f_n(\gamma^{-1}x) = \sum \pi_n(\gamma)f_n(x) = \sum f_n(x).$$

Hence for every n and γ, $\pi_n(\gamma)f_n(x) = f_n(x)$. If $f_n \neq 0$ then the irreducibility of π_n implies $\pi_n(\gamma) \equiv \mathrm{id}$, where id is the identity operator on the one dimensional space spanned by f_n. Since f_n is a continuous function this implies that f_n is a constant; that is $n = 0$. Thus the isometric system $(K/L, \tilde{m}, \Gamma)$ is ergodic. □

3. Rohlin's skew-product theorem

This surprising and useful theorem asserts that a general extension of an ergodic system is in fact a skew product extension (see Chapter 2, Section 3). In its proof we shall need the following measure theoretic exhaustion lemma. Let $(X, \mathcal{X}, \mu)$ be a probability measure space, we call a collection $\mathcal{H} \subset \mathcal{X}$ *hereditary* if whenever $A \in \mathcal{H}$ and $A \supset B \in \mathcal{X}$ then also $B \in \mathcal{H}$. We say that the hereditary collection $\mathcal{H}$ *saturates* $\mathcal{X}$ if for every $A \in \mathcal{X}$ with $\mu(A) > 0$, there exists $B \in \mathcal{H}$ with $B \subset A$ and $\mu(B) > 0$.

3.17. LEMMA. *If $\mathcal{H}$ is a hereditary collection which saturates $\mathcal{X}$ then there exists a countable measurable partition $\xi = \{A_i : i \in \mathbb{N}\}$ of X, with $A_i \in \mathcal{H}$ for every i.*

PROOF. Let $\epsilon_1 = \sup\{\mu(A) : A \in \mathcal{H}\}$ and choose $A_1 \in \mathcal{H}$ with $\mu(A_1) > \epsilon_1/2$. Inductively define

$$\epsilon_n = \sup\{\mu(A) : A \in \mathcal{H},\ \mu(A \cap (A_1 \cup A_2 \cup \cdots \cup A_{n-1})) = 0\}$$

and choose an element A_n in the above collection with $\mu(A_n) > \epsilon_n/2$. This inductive process will either terminate with $\epsilon_n = 0$ for some $n \in \mathbb{N}$, so that $X = \cup_{j=1}^{n} A_j$, or we shall have a countable disjoint collection $\xi = \{A_i : i \in \mathbb{N}, A_i \in \mathcal{H}\}$. In the latter case we have $\sum_{n=1}^{\infty} \epsilon_n < 2\sum_{n=1}^{\infty} \mu(A_n) < 2$ so that in particular $\epsilon_n \to 0$. If $\cup_{n=1}^{\infty} A_n \neq X$ then $\mu(B) > 0$ for $B = X \setminus \cup_{n=1}^{\infty} A_n$ and since $\mathcal{H}$ saturates $\mathcal{X}$ there is a $C \in \mathcal{H}$ with $\mu(C) > 0$ and $C \subset B$. However from the construction it follows that $\mu(C) \leq \epsilon_n$ for every n, so that $\mu(C) = 0$, a contradiction. □

3.18. THEOREM (Rohlin). *Let $\mathbf{X} \to \mathbf{Y}$ be a factor map of dynamical systems with $\mathbf{X}$ ergodic, then $\mathbf{X}$ is isomorphic to a skew-product over $\mathbf{Y}$. Explicitly, there exist a standard probability space $(U, \mathcal{U}, \rho)$ and a measurable cocycle $\alpha : \Gamma \times Y \to \mathrm{Aut}\,(U, \rho)$ with $\mathbf{X} \cong \mathbf{Y} \underset{\alpha}{\times} (U, \rho) = (Y \times U, \mathcal{Y} \otimes \mathcal{U}, \nu \times \rho, \Gamma)$, where $\gamma(y, u) = (\gamma y, \alpha(\gamma, y)u)$.*

PROOF. Let $\mu = \int \mu_y \, d\nu(y)$ be the disintegration of μ over ν (Theorem A.7). Using a model for $\mathbf{X}$ with $X = [0,1]$ we easily see that the function $p : x \mapsto \mu_{\pi(x)}\{x\} = \lim_{\delta\to 0} \mu_{\pi(x)} B_\delta(x)$ is measurable and since a.e.

$$\mu_{\pi(\gamma x)}\{\gamma x\} = \gamma\mu_{\pi(x)}\{\gamma x\} = \mu_{\pi(x)}\{x\},$$

it is Γ-invariant. By ergodicity of $\mathbf{Y}$ we have $p \equiv c \leq 1$, a constant function. We now have two cases to consider. The first is when $c = 0$ so that a.e. μ_y is a continuous measure. We then let $U = [0,1]$ and ρ Lebesgue measure, and define $\phi : X \to Y \times U$ and $\psi : Y \times U \to X$ by

$$\phi(x) = (\pi(x), \mu_{\pi(x)}([0,x])) \quad \text{and} \quad \psi(y,u) = \min\{x \in X : \mu_y([0,x]) \geq u\},$$

respectively, then $\psi \circ \phi = \mathrm{id}$, whence for $A \in \mathcal{Y}$ and $t \in [0,1]$,

$$\begin{aligned}\mu(\phi^{-1}(A \times [0,t])) &= \int_A \mu_y(\psi(A \times [0,t]))\, d\nu(y)\\ &= \int_A \mu_y(\psi(\{y\} \times [0,t]))\, d\nu(y)\\ &= \int_A \mu_y([0,\psi(y,t)])\, d\nu(y) = t\nu(A).\end{aligned}$$

Thus $\phi(\mu) = \nu \times \rho$.

The second case is when $c > 0$. By ergodicity there exists a positive integer r such that $c = \frac{1}{r}$ and such that a.e. $\mu_y = \frac{1}{r}\sum_{i=1}^r \delta_{x_i}$, with $x_1, x_2, \ldots, x_r$ distinct points of X.

Call a subset $A \in \mathcal{X}$ a *π-section* if $\pi(A) = B \in \mathcal{Y}$ and $\pi \restriction A : A \to B$ is 1-1 and measurable. We claim that there exists a countable measurable partition $\xi = \{A_i : i \in \mathbb{N}, A_i \in \mathcal{X}\}$ of X with A_i a π-section for every i. To see this let $\mathcal{H}$ be the collection of all π-sections in $\mathcal{X}$. Clearly $\mathcal{H}$ is a hereditary collection and we claim that it saturates $\mathcal{X}$. In fact if $\mu(B) > 0$ for a Borel subset $B \in \mathcal{X}$ then $\pi(B) = C$ is an analytic subset of Y. By Jankov-von Neumann's theorem (Theorem 2.10) there is a universally measurable $f : C \to B$ with $\pi \circ f = \mathrm{id}_C$. We have $\nu(C) = \mu(B) > 0$ and we choose a Borel $C' \in \mathcal{Y}$ with $C' \subset C$ and $\nu(C') = \nu(C)$. Set $A = f(C')$, then by Theorem 2.8.2, $A \in \mathcal{X}$. Clearly $A \in \mathcal{H}$ and

$$\mu(A) = \int_{C'} \mu_y(f(y))\, d\nu(y) > 0,$$

since $\nu(C') > 0$ and $\mu_y(f(y)) > 0$, ν-a.e.

Lemma 3.17 yields the partition ξ. By refining ξ we can assume that $\{\pi(A) : A \in \xi\}$ is a partition of Y and it is now possible to define a measurable isomorphism

$$\phi : X \to Y \times \{1, 2, \ldots, r\},$$

with $\phi(\mu) = \nu \times \rho$, where ρ is the uniformly distributed probability measure on $U = \{1, 2, \ldots, r\}$.

In both cases the Γ action on $Y \times U$ and the cocycle α are now defined by the isomorphism ϕ. For $(y,u) = \phi(x)$

$$\gamma(y,u) := (\gamma y, \alpha(\gamma, y)(u)) = \phi(\gamma x).$$

□

We next introduce the important definition of cohomologous cocycles.

3.19. DEFINITION. Let $\mathbf{Y}$ be a dynamical system, $(U, \mathfrak{U}, \rho)$ a standard Lebesgue space. Two measurable cocycles $\alpha, \beta : \Gamma \times Y \to \mathrm{Aut}\,(U, \mathfrak{U}, \rho)$ are said to be *cohomologous* if there exists a measurable map $s : Y \to \mathrm{Aut}\,(U, \mathfrak{U}, \rho)$ such that ν-a.e., for every $\gamma \in \Gamma$, $\beta(\gamma, y) = s(\gamma y)^{-1}\alpha(\gamma, y)s(y)$.

The following lemma shows that a representation of a skew-product with a cohomologous cocycle merely amounts to a change of coordinates; its proof is straightforward.

3.20. LEMMA. *If α and β are cohomologous cocycles as above, then the two skew-product systems $\mathbf{Y} \underset{\alpha}{\times} U$ and $\mathbf{Y} \underset{\beta}{\times} (U, \rho)$ are isomorphic via the map $\phi : (y, u) \to (y, s(y)^{-1}u)$.*

We end up this section with a 'folklore theorem' which shows that the cocycle provided by Rohlin's theorem is in fact unique up to cohomology.

3.21. THEOREM. *Let $\mathbf{X} \to \mathbf{Y}$ and $\mathbf{X}' \to \mathbf{Y}$ be factor maps of dynamical systems with and $\mathbf{X}$ and $\mathbf{X}'$ ergodic. Let $\mathbf{X} = \mathbf{Y} \underset{\alpha}{\times} (U, \rho) = (Y \times U, \mathcal{Y} \otimes \mathfrak{U}, \nu \times \rho, \Gamma)$, and $\mathbf{X}' = \mathbf{Y} \underset{\beta}{\times} (V, \theta) = (Y \times V, \mathcal{Y} \otimes \mathcal{V}, \nu \times \theta, \Gamma)$ be Rohlin representations as in Theorem 3.18. Suppose that there exists an isomorphism $\Phi : \mathbf{X} \to \mathbf{X}'$ of the form $\Phi(y, u) = (y, \phi(y, u))$, μ a.e. Then there exists a measurable map $s : Y \to \mathrm{Iso}\,(\rho, \theta)$ such that ν-a.e., for every $\gamma \in \Gamma$, $\alpha(\gamma, y) = s(\gamma y)^{-1}\beta(\gamma, y)s(y)$. Here* Iso (ρ, θ) *is the Polish space of invertible bi-measurable measure-preserving maps $h : (U, \rho) \to (V, \theta)$.*

PROOF. We have $\mu = \nu \times \rho$, $\mu' = \nu \times \theta$ and $\Phi(\mu) = \mu'$, $\Phi^{-1}(\mu') = \mu$. It follows easily that for ν a.e. y the map $\phi_y : U \to V$ defined by $\phi_y(u) = \phi(y, u)$ is an isomorphism of (U, ρ) onto (V, θ); i.e. an element of Iso (ρ, θ). And, it is not hard to check, that the map $y \mapsto \phi_y$, from Y to Iso (ρ, θ), is measurable.

Now μ a.e., for all $\gamma \in \Gamma$,

$$\gamma\Phi(y, u) = \gamma(y, \phi(y, u)) = (\gamma y, \beta(\gamma, y)\phi(y, u)) =$$
$$\Phi(\gamma(y, u)) = \Phi(\gamma y, \alpha(\gamma, y)u) = (\gamma y, \phi(\gamma y, \alpha(\gamma, y)u)),$$

hence

$$\phi(\gamma y, \alpha(\gamma, y)u) = \beta(\gamma, y)\phi(y, u).$$

Using Fubini's theorem and rewriting this equation in terms of the isomorphisms $s(y) := \phi_y$ we finally get

$$\alpha(\gamma, y) = \phi_{\gamma y}^{-1}\beta(\gamma, y)\phi_y = s(\gamma y)^{-1}\beta(\gamma, y)s(y).$$

□

4. The ergodic decomposition

A great part of the theory of measure-preserving dynamical systems is dedicated to the study of ergodic systems. The main justification for this is the fact that the general dynamical system $\mathbf{X} = (X, \mathcal{X}, \mu, \Gamma)$ can be uniquely "decomposed" into its ergodic components and thereby the general case, in a certain sense, is reduced to the ergodic one. This decomposition is called the *ergodic decomposition* of μ. In this short section we give the precise statement of the ergodic decomposition theorem; its proof will be given in Chapter 8, Section 2, once the appropriate machinery will become available. The special case of $\mathbb{Z}$-systems will be treated in Section 8 of the present chapter (Theorem 3.42).

3.22. THEOREM (The ergodic decomposition). *Let* $\mathbf{X}$ *be a dynamical system* $\mathcal{I} \subset \mathcal{X}$ *the* σ*-algebra of* Γ*-invariant sets. Let* $\pi : \mathbf{X} \to \mathbf{Y}$ *be the factor defined by* $\mathcal{I}$, *so that* $\mathcal{I} = \pi^{-1}(\mathcal{Y})$ *(Theorem 2.15). Let* $\mu = \int_Y \mu_y \, d\nu(y)$ *be the disintegration of* μ *over* ν *and for every* $y \in Y$ *denote* $X_y = \pi^{-1}(y)$ *and* $\mathcal{X}_y = \mathcal{X} \cap X_y$, *then:*

1. Γ *acts as the identity on* Y.
2. *For* ν *a.e.* $y \in Y$, *the system* $(X_y, \mathcal{X}_y, \mu_y, \Gamma)$ *is ergodic.*
3. *This decomposition is unique in the following sense. If* $(Z, \mathcal{Z}, \eta)$ *is a standard probability space and* $\psi : z \mapsto \tilde{\mu}_z$ *a measurable map from* Z *into* $M_\Gamma^{\mathrm{erg}}(X)$ *such that* $\mu = \int_Z \tilde{\mu}_z \, d\eta(y)$, *then there exists a measurable map* $\phi : (Z, \eta) \to (Y, \nu)$ *such that* η*-a.e.* $\mu_{\phi(z)} = \tilde{\mu}_z$.

5. Group and homogeneous skew-products

In this section we analyze group skew-products of the form $\mathbf{X} = \mathbf{Y} \underset{\alpha}{\times} G$, and their associated homogeneous skew-products $\mathbf{Y} \underset{\alpha}{\times} G/H$. Here $\mathbf{Y}$ is an ergodic Γ-system, G a compact second countable topological group, $\alpha : \Gamma \times Y \to G$ a measurable cocycle and $H \subset G$ a closed subgroup. We describe the ergodic decomposition of such systems, establish a convenient criterion for ergodicity, and finally show that any (ergodic) group extension $\mathbf{X} \to \mathbf{Y}$ (Examples 2.21) is in fact a group skew-product (Theorem 3.29). Later on we shall establish an intrinsic characterization of ergodic homogeneous skew-products as "isometric extensions" (Theorem 9.14).

We begin with a lemma about G actions.

3.23. LEMMA. *Let* G *be a second countable compact topological group and* $\mathbf{X} = (X, \mathcal{X}, \mu, G)$ *an ergodic* G *dynamical system. Then there exists a closed subgroup* $H \subset G$ *such that* $\mathbf{X}$ *is isomorphic to the homogeneous* G*-system* $(G/H, \tilde{m}, G)$, *where* $\tilde{m}$ *is the image of the Haar measure* m *on* G *under the quotient map* $\pi : G \to G/H$.

PROOF. By Theorem 2.18 we can assume that $\mathbf{X}$ is a topological G-system; i.e. X is compact and the map $G \times X \to X$, $(g, x) \mapsto gx$, is continuous. Ergodicity implies that $\mu(Gx_0) = 1$ for some point $x_0 \in X$ and we now assume $X = Gx_0$. Let $H = \{g \in G : gx_0 = x_0\}$ be the stability group of x_0. The continuity of the map $\eta : g \mapsto gx_0$ implies that $\eta = \theta \circ \pi$ for a continuous 1-1 onto map $\theta : G/H \to X$. Moreover all three maps η, θ and π are homomorphisms of the corresponding topological G-systems. Since m is the only probability measure on G which is invariant under left translations by elements of G, it follows that $\theta : (G/H, \tilde{m}, G) \to (X, \mu, G)$ is an isomorphism. $\square$

The Mackey range which we now define is an essential tool in the study of cocycles.

3.24. DEFINITION. Let $\mathbf{Y} = (Y, \mathcal{Y}, \nu, \Gamma)$ be an ergodic dynamical system. Let G be a compact second countable topological group with normalized Haar measure $m = m_G$, and $\alpha : \Gamma \times Y \to G$ a measurable cocycle.

Recall that the cocycle α defines a measurable action of Γ on $X = Y \times G$ by $\gamma(y, g) = (\gamma y, \alpha(\gamma, y)g)$ (with $\mathcal{X} = \mathcal{Y} \otimes \mathcal{B}$, where $\mathcal{B}$ denotes the Borel σ-algebra of

G). We denoted this Γ-system by $\mathbf{X}_\alpha = \mathbf{Y} \underset{\alpha}{\times} G$. Note that the compact group G acts as a group of measure preserving maps on both G (with the measure m_G) and X (with the measure $\nu \times m_G$), by right multiplication: $R_g(g') = g'g$ and $R_g(y, g') = (y, g'g)$. The latter action commutes with the Γ action and thus we have a natural action of the product group $\Gamma \times G$ on the standard probability space $(X, \mathcal{X}, \nu \times m_G)$. This action is clearly ergodic. The Γ action on $(X, \mathcal{X}, \nu \times m_G)$ however need not be ergodic and we let $\nu \times m_G = \int_\Omega \omega \, dP(\omega)$ be the corresponding ergodic decomposition of $\nu \times m_G$ (Theorem 3.22). The ergodicity of the $\Gamma \times G$ action on $(X, \mathcal{X}, \nu \times m_G)$ implies that (Ω, P, G) is an ergodic G action and therefore, by Lemma 3.23 (with right instead of left action), has the form of a homogeneous G space $(\Omega, P, G) \cong (H_0 \backslash G, \tilde{m}_G, G)$, where H_0 is a closed subgroup of G and $\tilde{m}_G$ the image of m_G under the quotient map $\phi : G \to H_0 \backslash G = \{H_0 g : g \in G\}$. The ergodic G-system $(H_0 \backslash G, \tilde{m}_G, G)$ is called the *Mackey range* of the cocycle α. Note that the closed subgroup H_0 is defined only up to conjugacy.

We write G_α for the closed subgroup generated by the range of α,

$$G_\alpha = \text{closed group generated by } \{\alpha(\gamma, y) : y \in Y, \gamma \in \Gamma\}.$$

The cocycle α is called *minimal* if there is no cocycle $\beta : \Gamma \times Y \to K$ cohomologous to α with $G_\beta \subsetneq G_\alpha$.

The following theorem is due to Mackey [**175**] and Zimmer [**267**].

3.25. THEOREM. *With notation as above:*

1. *The Mackey range is a cohomology invariant.*
2. *The cocycle α, with Mackey range $(H_0 \backslash G, \tilde{m}_G, G)$, is cohomologous to a cocycle $\beta : \Gamma \times Y \to H$, where H is a closed subgroup of G, iff some conjugate of H_0 is contained in H.*
3. *The cocycle α is minimal with $G_\alpha = G$ iff the Γ-system $\mathbf{X}_\alpha = \mathbf{Y} \underset{\alpha}{\times} G$ is ergodic.*
4. *If α and β are cohomologous minimal cocycles then G_α and G_β are conjugate.*
5. *Any cocycle α is cohomologous to a minimal cocycle $\beta : \Gamma \times Y \to H_0$ for some closed subgroup H_0 of G and*

$$\nu \times m_G = \int_{H_0 \backslash G} R_g(\nu \times m_{H_0}) \, d\tilde{m}_G(H_0 g), \tag{5.1}$$

is the ergodic decomposition of $\nu \times m_G$ with respect to the Γ action defined by β.

PROOF. 1. This follows directly from Lemma 3.20.

2. Suppose first that α is cohomologous to a cocycle $\beta : \Gamma \times Y \to H$, say $\beta(\gamma, y) = u(\gamma y)^{-1} \alpha(\gamma, y) u(y)$, for a measurable map $u : Y \to G$, ν-a.e. for all $\gamma \in \Gamma$ (see Lemma 3.20). It then follows that the map $\phi : (y, g) \to (y, u(y)^{-1} g)$ is an isomorphism of the $\Gamma \times G$ systems $\mathbf{Y} \underset{\alpha}{\times} G$ and $\mathbf{Y} \underset{\beta}{\times} G$. Thus, if $H_0 \backslash G$ and $H_1 \backslash G$ are the Mackey ranges of $\mathbf{X}_\alpha$ and $\mathbf{X}_\beta$ respectively, then H_0 and H_1 are conjugate. Now, if $A \subset G, A \in \mathcal{B}$ is invariant under multiplication on the left by elements of H, then $Y \times A$ is invariant under the Γ action corresponding to β. This means that the map $p : (Y \underset{\beta}{\times} G, \Gamma \times G) \to (H \backslash G, \Gamma \times G)$, defined by $p(y, g) = Hg$,

is a homomorphism (where in the latter system Γ acts trivially). This induces a homomorphism $q : (H_1\backslash G, G) \to (H\backslash G, G)$. It is now easy to deduce that this map is induced by a conjugation of H_1 into H, so that finally also H_0 is conjugate to a subgroup of H.

Conversely, assume that H_0 is conjugate to a subgroup of H. We then assume, with no loss in generality, that $H_0 \subset H$ and we shall show that there exists a cohomologous $\beta : \Gamma \times Y \to H_0$. Let $p : (Y \underset{\alpha}{\times} G, \Gamma \times G) \to (H_0\backslash G, \Gamma \times G)$ be the homomorphism onto the Mackey range. Then, for every $\gamma \in \Gamma$,

$$p(\gamma y, \alpha(\gamma, y)g) = p(\gamma y, e)\alpha(\gamma, y)g = p(y, g) = p(y, e)g, \qquad \nu \times m_G\text{-a.e.},$$

whence $p(y, e) = p(\gamma y, e)\alpha(\gamma, y)$, ν-a.e.

Let $s : H_0\backslash G \to G$ be a Borel section (i.e. $H_0 s(H_0 g) = H_0 g$ for every $g \in G$; such always exists, see for example [**142**] or use Theorem 2.10). Next define $u : Y \to K$ by $u(y) = s(p(y, e))$ and

$$\beta : \Gamma \times Y \to G, \qquad \text{by} \qquad \beta(\gamma, y) = u(y)\alpha(\gamma, y)u(\gamma y)^{-1}.$$

To finish the proof we need to show that ν-a.e. $\beta(\gamma, y) \in H_0$. Now

$$\begin{aligned} H_0 u(y)\alpha(\gamma, y) &= H_0 s(p(y, e))\alpha(\gamma, y) \\ &= p(y, e)\alpha(\gamma, y) \\ &= p(\gamma y, e) \\ &= H_0 s(p(\gamma y, e)) \\ &= H_0 u(\gamma y). \end{aligned}$$

Hence $H_0 u(y)\alpha(\gamma, y)u(\gamma y)^{-1} = H_0$, i.e. $\beta(\gamma, y) \in H_0$.

3. Let α be minimal with $G_\alpha = G$ and let $H_0\backslash G$ be the Mackey range of $\mathbf{X}_\alpha$. If $H_0 \neq G$ then, by the first part of the theorem, there exists a cohomologous $\beta : \Gamma \times Y \to H_0$. Since $G_\alpha = G$ this contradicts the minimality of α. Thus $H_0 = G$, the Mackey range is trivial, and $\mathbf{X}_\alpha$ is an ergodic Γ-system.

Conversely, the ergodicity of the Γ-system $\mathbf{X}_\alpha$ clearly implies $G_\alpha = G$. Now we see that if $\beta : \Gamma \times Y \to H \subset G$ is a cohomologous cocycle then, by part 2, $H \supset G$, hence $H = G$ and it follows that α is minimal.

4. Follows directly from part 2.

5. By part 2, α is cohomologous to some $\beta : \Gamma \times Y \to H_0$, where $H_0\backslash G$ is the Mackey range of α. For such β, with H_0 instead of G, it follows that β is minimal. The rest follows from the definition of the Mackey range. ☐

3.26. THEOREM. *Let $\pi : \mathbf{X} \to \mathbf{Y}$ be a factor of the ergodic system $\mathbf{X} = (X, \mathcal{X}, \mu, \Gamma)$. Suppose $X = Y \times G/H$, with G a second countable compact topological group, $H \subset G$ a closed subgroup with $\bigcap_{g \in G} g^{-1}Hg = \{e\}$, and assume further that the action of Γ on X is given by a cocycle $\alpha : \Gamma \times Y \to G$*

$$\gamma(y, gH) = (\gamma y, \alpha(\gamma, y)gH), \quad \gamma \in \Gamma.$$

There exist then a closed subgroup $K \subset G$, a cohomologous minimal cocycle $\beta : \Gamma \times Y \to K$ with $K = K_\beta$, and a closed subgroup $L \subset K$ so that $\mathbf{X} \cong \mathbf{Y} \underset{\beta}{\times} K/L$, where the latter system is the homogeneous skew-product defined by β; that is, the

Γ-invariant measure on $Y \times K/L$ is $\nu \times \tilde{m}_K$, where m_K is Haar measure on K and $\tilde{m}_K$ its image on K/L.

PROOF. The cocycle α defines a Borel action of Γ on $\hat{X} = Y \times G$ by $\gamma(y, g) = (\gamma y, \alpha(\gamma, y)g)$ and the map

$$\hat{\sigma} : \hat{X} = Y \times G \to Y \times G/H, \qquad \hat{\sigma}(y, g) = (y, gH)$$

intertwines the Γ actions. Recall that the compact group G acts as a group of measure preserving bijections, on both G (with the measure m_G) and $\hat{X}$ (with the measure $\nu \times m_G$), by right multiplication: $R_g(y, g') = (y, g'g)$.

The formula

$$f \mapsto \int_X \int_H f(y, gh)\, dm_H(h)\, d\mu(y, gH),$$

for bounded measurable functions f on $\hat{X}$, defines a Γ-invariant measure $\mu * m_H$ on $\hat{X}$ which projects down to μ. Thus the set Q of Γ-invariant probability measures on $\hat{X}$ projecting onto μ is a non-empty compact convex set. Since μ is ergodic, an extreme point of Q is also an extreme point of $M_\Gamma(\hat{X})$, i.e. a Γ-ergodic measure. Let $\hat{\mu}$ be such ergodic measure; then $\hat{\mathbf{X}} = (Y \times G, \hat{\mu})$ is an ergodic lift of μ; that is, $\hat{\mu}$ is a Γ-invariant ergodic probability measure on $\hat{X}$ and $\hat{\sigma}\hat{\mu} = \mu$

(5.2)

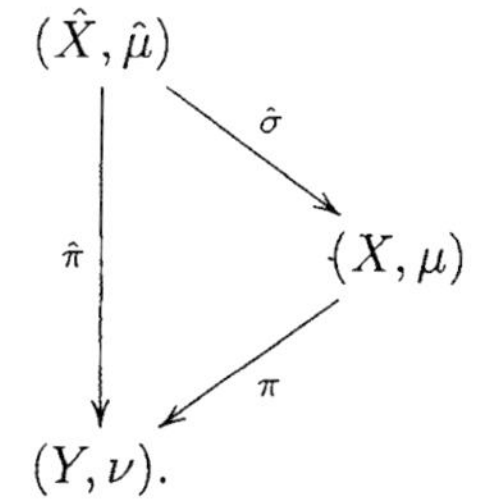

The measures $R_g(\hat{\mu})$, $g \in G$ are all Γ-invariant and ergodic. Set

$$K = \{g \in G : R_g(\hat{\mu}) = \hat{\mu}\},$$

a closed subgroup of G.

Next set

$$\lambda = \int_{K\backslash G} R_g(\hat{\mu})\, d\tilde{m}_G(Kg), \tag{5.3}$$

where $\tilde{m}_G$ is the image of m_G under the quotient map $G \to K\backslash G$. Clearly the projections of λ on $\mathbf{Y}$ and G are the measures ν and m_G respectively and the ergodicity of $\mathbf{Y}$ implies that $\lambda = \nu \times m_G$ (this is easy to vrify; e.g. use Theorem 6.26 below). We now see that (5.3) is the ergodic decomposition of $\nu \times m_G$ with respect to the Γ action corresponding to α. In other words, $(K\backslash G, G)$ is the Mackey range of α.

By Theorem 3.25 there exists a cohomologous minimal cocycle $\beta : \Gamma \times Y \to K$, say, $\beta(\gamma, y) = s(\gamma y)^{-1}\alpha(\gamma, y)s(y)$, with $(Y \underset{\beta}{\times} K, \nu \times m_K, \Gamma)$ ergodic. Moreover, the ergodic decomposition of $(Y \underset{\beta}{\times} G, \nu \times m_G, \Gamma)$ is given by (5.1)

$$\nu \times m_G = \int_{K\backslash G} R_g(\nu \times m_K)\, dm_G(Kg).$$

Denoting the isomorphism given by Lemma 3.20 by

$$\Phi : (Y \underset{\beta}{\times} G, \nu \times m_G, \Gamma) \to (Y \underset{\alpha}{\times} G, \nu \times m_G, \Gamma)$$
$$\Phi(y, g) = (y, s(y)g), \qquad (y \in Y, g \in G).$$

we see, by the uniqueness of the ergodic decomposition (Theorem 3.22) that modulo m_G, $\{R_g(\hat{\mu}) : g \in G\} = \{\Phi R_g(\nu \times m_K) : g \in G\}$. It follows that for some g we have $\Phi R_g(\nu \times m_K) = \hat{\mu}$. With no loss of generality we assume that $\Phi(\nu \times m_K) = \hat{\mu}$.

We now come back to the system $\mathbf{X}$. Define $L = H \cap K$ — a closed subgroup of K — and observe that, denoting $\tilde{K} = \{kH : k \in K\}$, the map

$$\theta : \tilde{K} \to K/L, \qquad \theta : kH \mapsto k(H \cap K)$$

is 1-1 and onto. This implies that the map $\tilde{\phi} : \mathbf{X} \to \mathbf{Y} \underset{\beta}{\times} K/L$, defined on a μ typical point of X by $\tilde{\phi}(y, s(y)kH) = (y, kL)$, is well defined, 1-1, measurable, and it makes the diagram

$$\begin{array}{ccc} (\hat{X}, \hat{\mu}) & \xrightarrow{\ \phi\ } & \mathbf{Y} \underset{\beta}{\times} K \\ \Big\downarrow{\scriptstyle \hat{\sigma}} & & \Big\downarrow{\scriptstyle \sigma} \\ \mathbf{X} & \xrightarrow[\ \tilde{\phi}\ ]{} & \mathbf{Y} \underset{\beta}{\times} K/L \end{array}$$

commutative. Thus $\tilde{\phi} : \mathbf{X} \to \mathbf{Y} \underset{\beta}{\times} K/L$ is the required isomorphism. □

The proof of the previous theorem also provides a proof for the next corollary. We assume that K is a compact second countable topological group and $L \subset K$ a closed subgroup with $\bigcap_{k \in K} k^{-1}Lk = \{e\}$. Normalized Haar measure on K is denoted by m_K and $\tilde{m}$ is the projection of m_K on K/L.

3.27. COROLLARY. *Given an ergodic homogeneous skew-product,*

$$\mathbf{X} = \mathbf{Y} \underset{\beta}{\times} K/L = (Y \times K/L, \nu \times \tilde{m}),$$

we can always assume that β is a minimal cocycle with $K = K_\beta$. Moreover, there exists an ergodic group skew-product $\hat{\mathbf{X}}$ such that the diagram

$$\begin{array}{ccc} \hat{\mathbf{X}} = \mathbf{Y} \underset{\beta}{\times} K & & \\ \Big\downarrow{\scriptstyle \hat{\pi}} & \searrow{\scriptstyle \sigma} & \\ \mathbf{Y} & \xleftarrow[\ \pi\ ]{} & \mathbf{X} = \mathbf{Y} \underset{\beta}{\times} K/L \end{array}$$

commutes. Here $\hat{\mathbf{X}} = \mathbf{Y} \underset{\beta}{\times} K$ is the group skew-product defined by the cocycle β, i.e. $\hat{\mu} = \nu \times m$ where m is Haar measure on K, and $\hat{\mu}$ is ergodic.

Our next goal is to show that for an ergodic system $\mathbf{X}$, being a group extension of a factor $\mathbf{Y}$ is the same as being a group skew-product over $\mathbf{Y}$. For this we shall need the spatial modeling theorem (Theorem 2.18). First we show that, when $\mathbf{Y}$ is ergodic, in such a spatial action the action of the compact group K is essentially free.

3.28. LEMMA. *Let* $\mathbf{X}$ *be an ergodic dynamical system and* $K \subset \mathrm{Aut}(\mathbf{X})$ *a compact subgroup. We assume (as we may by Theorem 2.18) that* $\mathbf{X}$ *is a topological* $\Gamma \times K$ *system; i.e.* X *is compact and the map* $(\gamma, k, x) \to \gamma x k$, $\Gamma \times K \times X \to X$ *is continuous. For every* $x \in X$ *denote by* K_x *the stability group at* x, $K_x = \{k \in K : xk = x\}$. *Then there exists a conull,* $\Gamma \times K$*-invariant subset* $X_0 \subset X$ *such that* $K_x = \{e\}$ *for all* $x \in X_0$.

PROOF. Let $\mathcal{G}$ be the space of all closed subgroups of K. Equipped with the Hausdorff metric, $\mathcal{G}$ is a Polish space and it is easy to check that the map $x \mapsto K_x$, $X \to \mathcal{G}$, satisfies $\limsup K_{x_i} \subset K_x$ whenever $x_i \to x$ in X. Thus this map is upper semi-continuous and in particular Borel measurable. Clearly $K_{\gamma x} = K_x$ for every $x \in X$ and $\gamma \in \Gamma$ and, by ergodicity (Theorem 3.10.3) we conclude that this map is a constant on a conull Γ-invariant set $X_0 \subset X$, say $K_x = K_0$ for $x \in X_0$. Now $k \in K_0$ implies $k = \mathrm{id}$ μ-a.e. so that $k = \mathrm{id}$ as elements of $\mathrm{Aut}(\mathbf{X})$. Thus $K_0 = \{e\}$. □

We now have the following explicit description of ergodic group extensions.

3.29. THEOREM. *Let* $\mathbf{X}$ *be an ergodic system and* $\pi : \mathbf{X} \to \mathbf{Y}$ *a group extension with compact group* $K \subset \mathrm{Aut}(\mathbf{X})$, *then* $\mathbf{X}$ *is isomorphic to a group skew-product* $\mathbf{X} \cong \mathbf{Y} \underset{\beta}{\times} K$ *with* $\beta : \Gamma \times Y \to K$ *a cocycle.*

PROOF. By Theorem 2.18 we can assume that $\mathbf{X}$ is a topological $\Gamma \times K$-system; i.e. X is compact and the map $\Gamma \times K \times X \to X$, $(\gamma, k, x) \mapsto \gamma x k$, is continuous. Now $\pi : X \to Y = X/K$ is just the topological quotient map: $\pi(x) = \pi(x')$ iff $x' = xk$ for some $k \in K$. By Theorem 2.10 there exists a *measurable section* $s : Y \to X$, i.e. a measurable map with $\pi \circ s = \mathrm{id}_Y$. Define a map $\phi : X_0 \to Y \times K$ by $\phi(x) = (\pi(x), u(x))$, where X_0 is the conull subset of X given by Lemma 3.28, and $u(x) \in K$ is defined by the equation $x = s(\pi(x))u(x)$. Clearly then $u(xk) = u(x)k$ for every $k \in K$ and $x \in X_0$. Now for $x, x' \in X_0$ with $\pi(x) = \pi(x')$ and $\gamma \in \Gamma$ we have $x' = xk$ for some $k \in K$ and $u(\gamma x')u(x')^{-1} = u(\gamma x)kk^{-1}u(x)^{-1} = u(\gamma x)u(x)^{-1}$.

Hence the function $\alpha : \Gamma \times Y \to K$ defined by $\alpha(\gamma, y) = u(\gamma x)u(x)^{-1}$, with $y = \pi(x)$, is well defined. It is now easy to check that α is a measurable cocycle and that $\phi : X_0 \to Y \times K$ satisfies $\phi(\gamma x) = \gamma \cdot \phi(x)$, where

$$\gamma \cdot (y, k) = (\gamma y, \alpha(\gamma, y)k),$$

i.e. ϕ is an isomorphism. Let $\mu^* = \phi_*(\mu)$, a Γ-invariant ergodic measure on $Y \underset{\alpha}{\times} K$.

It now follows from Theorem 3.26 (with $H = \{e\}$) that there exists a closed subgroup $K_0 \subset K$ and a cocycle $\beta : \Gamma \times Y \to K_0$, cohomologous to α, such that

$$(Y \underset{\alpha}{\times} K, \mu^*, \Gamma) \cong \mathbf{Y} \underset{\beta}{\times} K_0 = (Y \underset{\beta}{\times} K_0, \nu \times m_{K_0}, \Gamma).$$

Moreover, if $k \in K \setminus K_0$ then, as one can deduce from Theorem 3.26, $R_k(\nu \times m_{K_0})$ and $\nu \times m_{K_0}$ are mutually singular and therefore $k\mu \neq \mu$, contradicting the fact that $k \in \mathrm{Aut}(\mathbf{X})$. Thus $K = K_0$ and the proof is complete. □

We end this section with the following useful theorem which is due to Furstenberg.

3.30. THEOREM. *Let* $\mathbf{X} = (X, \mathcal{X}, \mu, \Gamma)$ *be an ergodic system which is a homogeneous skew product over* $\mathbf{Y}$, *with* $\pi : \mathbf{X} \to \mathbf{Y} = (Y, \mathcal{Y}, \nu, \Gamma)$. *Then* μ *is the unique* Γ*-invariant probability measure on* X *such that* $\pi(\mu) = \nu$. *In particular this is true when* $\pi : \mathbf{X} \to \mathbf{Y}$ *is a group extension.*

PROOF. Assume first that $\mathbf{X}$ is a group skew-product over $\mathbf{Y}$, $\mathbf{X} = \mathbf{Y} \underset{\beta}{\times} K$ with a cocycle $\beta : \Gamma \times Y \to K$. By assumption $\mu = \nu \times m$ is ergodic, where m denotes Haar measure on K. For convenience we also assume that $\mathbf{X}$ is a topological model. For $k \in K$ let R_k denote right multiplication on K, $R_k(y, k') = (y, k'k), (y \in Y, k, k' \in K)$. Let $\lambda \in M_\Gamma(X)$ be a Γ-invariant probability measure on X such that $\pi(\mu) = \nu$. If $\lambda = \int_\Omega \lambda_\omega \, dP(\omega)$ is the ergodic decomposition of λ then, applying π to this equation we get $\nu = \int_\Omega \pi(\lambda_\omega) \, dP(\omega)$. Since ν is ergodic we deduce that P a.s. $\pi(\lambda_\omega) = \nu$. We can therefore assume that λ is ergodic. Set

$$\hat{\lambda} = \lambda * m = \int_K R_k \lambda \, dm(k). \tag{5.4}$$

If $\lambda = \int_Y \delta_y \times \lambda_y \, d\nu(y)$ is the disintegration of λ over ν, then

$$\begin{aligned} \hat{\lambda} &= \int_K \int_Y \delta_y \times R_k \lambda_y \, d\nu(y) \, dm(k) \\ &= \int_Y \delta_y \times \left(\int_K R_k \lambda_y \, dm(k) \right) d\nu(y) \\ &= \int_Y \delta_y \times m_y \, d\nu(y). \end{aligned}$$

For each $y \in Y$, the measure $m_y = \int_K R_k \lambda_y \, dm(k)$ is a right K-invariant probability measure on K, whence $m_y = m$. Thus $\hat{\lambda} = \int_Y \delta_y \times m_y \, d\nu(y) = \nu \times m = \mu$.

Now equation (5.4) reads

$$\mu = \int_K R_k \lambda \, dm(k),$$

and, since for each $k \in K$ the measure $R_k \lambda$ is also ergodic, the ergodicity of μ and the uniqueness of the ergodic decomposition (Theorem 3.22) imply that $R_k \lambda = \mu$, m-a.e. Thus $\lambda = \mu$ and our assertion is proven in the group skew product case.

When $\mathbf{X}$ is a homogeneous skew product over $\mathbf{Y}$ we have, by corollary 3.27, a representation $\mathbf{X} = \mathbf{Y} \underset{\beta}{\times} K/L = (Y \times K/L, \nu \times \tilde{m})$, where β is a minimal cocycle so that in the associated commutative diagram

$$\begin{array}{ccc} \hat{\mathbf{X}} = \mathbf{Y} \underset{\beta}{\times} K & & \\ \downarrow \hat{\pi} & \searrow \sigma & \\ \mathbf{Y} & \xleftarrow{\ \pi\ } & \mathbf{X} = \mathbf{Y} \underset{\beta}{\times} K/L \end{array}$$

the group skew product $\hat{\mathbf{X}}$ (with $\hat{\mu} = \nu \times m$) is ergodic. Given $\lambda \in M_\Gamma(X)$ with $\pi(\lambda) = \nu$, there exists $\hat{\lambda} \in M_\Gamma(\hat{X})$ with $\sigma(\hat{\lambda}) = \lambda$, hence also $\hat{\pi}(\hat{\lambda}) = \nu$. The first part of the proof applies to $\hat{\pi} : \hat{\mathbf{X}} \to \mathbf{Y}$ and we conclude that $\hat{\lambda} = \hat{\mu}$, whence also $\lambda = \mu$.

Finally, when $\pi : \mathbf{X} \to \mathbf{Y} = (Y, \mathcal{Y}, \nu, \Gamma)$ is a group extension with compact group $K \subset \mathrm{Aut}(\mathbf{X})$, we know by Theorem 3.29 that, in fact, $\mathbf{X}$ is a group skew-product over $\mathbf{Y}$ and the proof is complete. □

6. Amenable groups

The class of amenable (countable discrete) groups is vast and contains all solvable (hence all abelian) groups. Among linear groups we have the Titze alternative: A linear group is either amenable, in which case it is a finite extension of a solvable group (*almost solvable*), or it contains a free subgroup on two generators. In general however there is by now a large stock of examples of amenable groups which are not almost solvable (see e.g. [**115**]).

3.31. DEFINITION. Let Γ be a countable discrete group. Given a finite set $K \subset \Gamma$ and $\epsilon > 0$, we say that the finite set $F \subset \Gamma$ is *(K, ϵ)-invariant* if

$$\mathrm{card}\,(F \bigtriangleup KF) < \epsilon \cdot \mathrm{card}\,(F).$$

The group Γ is called *amenable* if for every finite set $K \subset \Gamma$ and $\epsilon > 0$, there exists a (K, ϵ)-invariant finite set $F \subset \Gamma$. A sequence $F_1, F_2, \ldots$ of finite sets of Γ is a *Følner* sequence if for every finite $K \subset \Gamma$ and $\epsilon > 0$ there exists n_0 such that for all $n \geq n_0$, F_n is (K, ϵ)-invariant.

Note that for $\Gamma = \mathbb{Z}$ the sequence $F_n = [0, n]$ is a canonical Følner sequence. We state without proof the following well known theorem (see e.g. [**107**]).

3.32. THEOREM. *A countable discrete group Γ is amenable iff there exists an invariant mean on $\ell^\infty(\Gamma)$.*

Let Γ be an amenable group and $\mathbf{X} = (X, \mathcal{X}, \mu, \Gamma)$ a dynamical system. Let $\mathfrak{I} \subset L^2(\mu)$ be the closed subspace of Γ-invariant functions and let $P : L^2(\mu) \to \mathfrak{I}$ be the corresponding orthogonal projection. For a subset $F \subset \Gamma$ and $f \in L^2(\mu)$ we denote

$$\mathbb{A}_F f(x) = \frac{1}{|F|} \sum_{\gamma \in F} U_\gamma f(x).$$

Let $F_1, F_2, \ldots$ be a Følner sequence for Γ.

3.33. THEOREM (The von Neumann or Mean ergodic theorem). *In the strong operator topology,*
$\lim_{n \to \infty} \mathbb{A}_{F_n} = P$; *i.e.*

$$\lim_{n \to \infty} \|\mathbb{A}_{F_n} f - Pf\| = 0,$$

for every $f \in L^2(\mu)$. In particular, the dynamical system $\mathbf{X}$ is ergodic iff for every $f \in L^2(\mu)$

$$\lim_{n \to \infty} \mathbb{A}_{F_n} f = \int f\, d\mu$$

in L^2-norm.

PROOF. Let $\mathcal{V}_0$ be the linear space consisting of finite linear combinations of functions of the form $F(x) = f(\gamma x) - f(x)$, where f is in $L^2(\mu)$ and γ is some group element. Clearly the Følner condition says that the sequence $\mathbb{A}_n(F)$ tends to zero in norm. The same is then valid for all the functions in $\mathcal{V}_0$ and then also for those in the norm closure $\mathcal{V} = \mathrm{cls}\,(\mathcal{V}_0)$.

We now observe that $L^2(\mu) = \mathcal{V} \oplus \mathfrak{I}$. In fact if $f \perp \mathcal{V}$ then for every $\gamma \in \Gamma$ and $g \in L^2(\mu)$

$$\begin{aligned}\langle U_\gamma f, g\rangle &= \langle f, U_{\gamma^{-1}} g\rangle \\ &= \langle f, U_{\gamma^{-1}} g - g\rangle + \langle f, g\rangle \\ &= \langle f, g\rangle,\end{aligned}$$

whence $f \in \mathfrak{I}$. Since clearly $\mathbb{A}_{F_n} \equiv \mathrm{id}$ on $\mathfrak{I}$, the first assertion of the theorem follows. By Theorem 3.4, applied to the restriction of the Koopman representation to $L_0^2(\mu)$, we see that $\mathbf{X}$ is ergodic iff $\mathfrak{I}$ consists of the constant functions, i.e. iff $Pf = \int f\, d\mu$ for every $f \in L^2(\mu)$; whence the second assertion. □

7. Ergodicity and weak mixing for $\mathbb{Z}$-systems

In this section we consider the meaning of ergodicity and weak mixing for $\mathbb{Z}$-dynamical systems. Thus, a system in this section is a $\mathbb{Z}$-system. In the next few theorems we formulate properties — usually taken as definitions — that characterize ergodicity, weak mixing and mixing. Their proofs mostly reduce to interpretations of the definitions and theorems (especially Theorem 3.33) of the previous sections applied to $\mathbb{Z}$-systems.

We leave their proofs as an exercise.

3.34. THEOREM. *A dynamical system* $\mathbf{X}$ *is:*

1. *Ergodic iff for every* $A \in \mathfrak{I}$, *the* σ*-algebra of* T*-invariant sets in* $\mathcal{X}$, $\mu(A) = 0$ *or* $\mu(A) = 1$.
2. *Weakly mixing iff the product system* $\mathbf{X} \times \mathbf{X}$ *is ergodic; i.e. iff for every* $T \times T$*-invariant subset* $W \subset X \times X$, $\mu \times \mu(W) = 0$ *or* $\mu \times \mu(W) = 1$.

3.35. THEOREM. *The following conditions on the system* $\mathbf{X}$ *are equivalent:*

1. $\mathbf{X}$ *is ergodic.*
2. *For every* $A, B \in \mathcal{X}$ *of positive measure there exists* $n \in \mathbb{Z}$ *with* $\mu(T^n A \cap B) > 0$.
3. $\mu(A) > 0$ *implies* $\mu\{\bigcup T^n A : n \in \mathbb{Z}\} = 1$.
4. *Every measurable function* f *into a Polish space* P *satisfying* $f \circ T = f$ *a.e. is of the form* $f \equiv p$, *a.e. for a point* $p \in P$.
5. *The same for a real (or complex) valued function* f.
6. *The condition* $f \in L^2(\mu)$ *and* $U_T(f) = f \circ T = f$ *implies* $f \equiv$ *constant; i.e. the geometric multiplicity of the eigenvalue* 1 *of the unitary operator* U_T *is* 1.
7. *For every* $f, g \in L^2(\mu)$ *(or* $L^1(\mu)$*),*

$$\lim_{n\to\infty} \frac{1}{n} \sum_{i=0}^{n-1} \int f \circ T^i \cdot \bar{g}\, d\mu = \int f\, d\mu \int \bar{g}\, d\mu. \tag{7.1}$$

3.36. THEOREM. *The following conditions on the system* $\mathbf{X}$ *are equivalent:*

1. $\mathbf{X}$ *is weakly mixing.*
2. *For every* $A, B, C, D \in \mathcal{X}$ *of positive measure there exists* $n \in \mathbb{Z}$ *with* $\mu(T^n A \cap B) > 0$ *and* $\mu(T^n C \cap D) > 0$.

3. *For every* $f, g \in L^2(\mu)$

$$\lim_{n\to\infty} \frac{1}{n}\sum_{i=0}^{n-1}\left|\int f\circ T^i\cdot \bar{g}\,d\mu - \int f\,d\mu\cdot\int \bar{g}\,d\mu\right| = 0, \tag{7.2}$$

i.e. the function $\phi(n) = \langle f\circ T^n, g\rangle$ *is a flight function (see Exercise 1.54).*
4. *The condition* $f \in L^2(\mu)$ *and* $U_T(f) = f\circ T = \lambda f$, *with* $0 \neq \lambda \in \mathbb{C}$, *implies* $f \equiv$ *constant and* $\lambda = 1$; *i.e.* 1 *is the only eigenvalue of the unitary operator* U_T *and the geometric multiplicity of the eigenvalue* 1 *is* 1.
5. *For every ergodic system* $\mathbf{Y}$ *the product system* $(X\times Y, \mathcal{X}\otimes\mathcal{Y}, \mu\times\nu, T\times T)$ *is ergodic.*

3.37. THEOREM. *The following conditions on the system* $\mathbf{X}$ *are equivalent:*
1. $\mathbf{X}$ *is mixing.*
2. *For every* $A, B \in \mathcal{X}$ *of positive measure there exists* $n_0 \in \mathbb{N}$ *with* $\mu(T^nA\cap B) > 0$ *for* $|n| \geq n_0$.
3. *For every* $f, g \in L^2(\mu)$,

$$\lim_{n\to\infty}\int f\circ T^i\cdot\bar{g}\,d\mu = \int f\,d\mu\int\bar{g}\,d\mu. \tag{7.3}$$

3.38. EXERCISE. The system $\mathbf{X} = (X, \mathcal{X}, \mu, T)$ is rigid iff there exists a sequence $n_k \nearrow \infty$ such that $\lim\mu(T^{n_k}A\cap A) = \mu(A)$ for every measurable subset A of X ([**82**]). We say that $\mathbf{X}$ is $\{n_k\}$*-rigid*.

3.39. EXERCISE. In the theorems of this section one can replace the requirement "for all $A, B \in \mathcal{X}$" by "for all $A, B \in \mathcal{S}$" where $\mathcal{S} \subset \mathcal{X}$ is a sub-algebra that generates the σ-algebra $\mathcal{X}$. Likewise one can replace "for all $f, g \in L^2(\mu)$" by "for all $f, g \in V$" where $V \subset L^2(\mu)$ is a dense subspace.

3.40. EXAMPLES.
1. An id-system (i.e. a system $\mathbf{X}$ with T acting as the identity transformation id) is ergodic iff it is the trivial (one point) system.
2. A finite system (i.e. X is a finite space, μ an equidistributed probability on X and T a permutation) is ergodic iff T is cyclic.
3. **Compact group rotations, Kronecker-Weyl's theorem:** Let K be a separable, compact, abelian, topological group. Let μ be normalized Haar measure on K and $a \in K$. Define $T : K \to K$ by $Tz = az$ for $z \in K$, then the system (K, μ, T) is ergodic iff $\{a^n : n \in \mathbb{Z}\}$ is dense in K. (A topological group K with such an element a is called *monothetic*.) [[If $\operatorname{cls}\{a^n : n \in \mathbb{Z}\} = H \neq K$, then there exists a continuous character $\beta \in \hat{K}$, the dual group of K, such that $\beta(z) = 1$ for all $z \in H$ but $\beta \not\equiv 1$. Now $\beta(az) = \beta(a)\beta(z) = \beta(z)$ for every $z \in K$ and we have a nonconstant invariant function in $L^2(\mu)$. Conversely, if $\operatorname{cls}\{a^n : n \in \mathbb{Z}\} = K$ and $f \in L^2(\mu)$ is invariant, then from the Fourier expansion $f = \sum_{\beta\in\hat{K}} c_\beta\beta$, we deduce:

$$f(z) = f\circ T(z) = f(az) = \sum_{\beta\in\hat{K}}\beta(a)c_\beta\beta = \sum_{\beta\in\hat{K}} c_\beta\beta.$$

The uniqueness of the expansion implies $\beta(a)c_\beta = c_\beta$, hence $(1-\beta(a))c_\beta = 0$, for every $\beta \in \hat{K}$. Hence for $\beta \neq \mathbf{1}$, it follows that $c_\beta = 0$ and $f = c_1$, a constant. Of course this is just a special case of Proposition 3.16.]]

4. **Compact group automorphisms:** For K and μ as above, let $Tz = A(z)$, where $A : K \to K$ is a continuous group automorphism, then T is measure-preserving. It is ergodic iff the equations $\beta \circ A^n = \beta$ for $n \neq 0$, $\beta \in \hat{K}$ admit only $\beta = \mathbf{1}$ as a solution. In particular for $K = \mathbb{T}^n = \mathbb{R}^n/\mathbb{Z}^n$, the n-dimensional torus, every automorphism has the form $T : z \mapsto Az$ where $A \in SL(n, \mathbb{Z})$, an $n \times n$ matrix with $\mathbb{Z}$ entries and determinant 1, and such a T is ergodic iff no eigenvalue of A has modulus 1. [[Refer to Walters [**257**] for the proofs. Schmidt's monograph [**231**] is an in-depth study of dynamical systems arising as groups of automorphisms of a compact group.]]
5. **Subshifts:** For an integer $\ell \geq 2$, let $\mathfrak{L} = \{1, \dots, \ell\}$ (or sometimes when convenient $\mathfrak{L} = \{0, 1, \dots, \ell - 1\}$) and set $\Omega = \Omega(\ell) = \mathfrak{L}^{\mathbb{Z}}$. Let $S : \Omega \to \Omega$ be the shift transformation: $(Sx)_n = x_{n+1}$. For integers $m < n$ and a sequence $(a_0, a_1, \dots, a_{n+1-m})$ with $a_j \in \mathfrak{L}$, we write $\mathbf{a}_m^n = [a_0, a_1, \dots, a_{n+1-m}]_m^n$ for the *cylinder set* $\{x \in \Omega : x_{n+j} = a_j;\ 0 \leq j \leq n+1-m\}$. For example $[001]_0^2 = \{x \in \Omega : x_0 = 0, x_1 = 0, x_2 = 1\}$. An S-invariant Borel probability measure μ on Ω is uniquely defined by the values $\mu(\mathbf{a}_m^n)$ and conversely every assignment $\mathbf{a}_m^n \mapsto \mu(\mathbf{a}_m^n)$ which satisfies:
   (a) $0 \leq \mu(\mathbf{a}_1^n) \leq 1$,
   (b) $\mu(\Omega) = 1$
   (c) $\mu(\mathbf{a}_1^k) = \sum\{\mu(\mathbf{a}_1^{k+1}) : \mathbf{a}_1^{k+1}$ is an extension of $\mathbf{a}_1^k\}$, and
   (d) $\mu(\mathbf{a}_m^n) = \mu(\mathbf{b}_{m+1}^{n+1})$ when $b_{j+1} = a_j,\ j = m, m+1, \dots, n$,

   defines such a probability measure on Ω. The collection of all S-invariant Borel probability measures is denoted by $M_S(\Omega)$. Denoting by $\mathcal{B}$ the Borel σ-algebra on Ω we now have an extremely rich supply of dynamical systems: every $\mu \in M_S(\Omega)$ gives rise to the dynamical system $(\Omega, \mathcal{B}, \mu, S)$. Such dynamical systems are called *subshifts* or symbolic systems. ($\mathfrak{L} = \{1, \dots, \ell\}$ is sometimes called the *state space* of the system.) Next we present some well known examples of such subshifts.
6. **Bernoulli shifts (or schemes):** These are determined by a probability vector $p = (p_1, \dots, p_\ell)$. The assignment $\mu([a_0, a_1, \dots, a_n]_0^n) = p_{a_0}p_{a_1}\cdots p_{a_n}$ defines the *Bernoulli measure* $p^{\mathbb{Z}}$ corresponding to p. That is, μ is the product measure $p^{\mathbb{Z}}$. We denote the Bernoulli system corresponding to p by $B(p) = B(p_1, \dots, p_\ell)$. Every Bernoulli measure is mixing. [[Clearly for every non-empty cylinder sets $A, B \subset \Omega$ and for a sufficiently large n the "events" $T^n A$ and B are independent with respect to a Bernoulli measure (since it is a product measure). Thus $\mu(T^n A \cap B) = \mu(A)\mu(B) > 0$.]]
7. One can form Bernoulli shifts over an infinite state space. If $(Y, \mathcal{Y}, \nu)$ is a probability space let $X = Y^{\mathbb{Z}}$ be the product space, $\mu = \nu^{\mathbb{Z}}$ the product measure and $\mathcal{X}$ the product σ-algebra. Again one defines the shift on X by $(Sx)_n = x_{n+1}$ and then the system $(X, \mathcal{X}, \mu, S)$ is called (for reasons to become clear later) a *Bernoulli system of infinite entropy*. For example one can take Y to be any compact second countable space and ν any probability measure on Y. Another example is obtained by taking $Y = \mathbb{N} = \{0, 1, \dots\}$ and ν any probability measure on $\mathbb{N}$.
8. **Markov systems:** Again we have a probability vector $p = (p_1, \dots, p_\ell)$ and in addition we are given an $\ell \times \ell$ stochastic matrix $P = (p_{ij})_{i,j=1}^{\ell}$ (i.e. $p_{ij} \geq$

0 and $\sum_{j=1}^{\ell} p_{ij} = 1$ for every $1 \le i \le \ell$). The corresponding *Markov measure* $\mu \in M_S(\Omega)$ is defined by: $\mu([a_0, a_1, \ldots, a_n]_0^n) = p_{a_0} p_{a_0 a_1} \cdots p_{a_{n-1} a_n}$. This probability measure is S-invariant iff $pP = p$; i.e. p is a left eigenvector of the matrix P with eigenvalue 1. We denote the Markov shift corresponding to the pair (p, P) (satisfying $pP = p$) by $M(p, P)$. The Markov shift $M(p, P)$ is ergodic iff the stochastic matrix P is *irreducible*; i.e. $\forall i, j\ \exists n > 0$ with $p_{ij}^{(n)} > 0$, where $p_{ij}^{(n)}$ is the (i, j) entry of the matrix P^n. (The symbols $\{1, \ldots, \ell\}$ are the *states* of the Markov system and p and P are the *initial* probability vector and the *transition probabilities* matrix of the system respectively.)

9. For the Markov system $M(p, P)$ the following conditions are equivalent:
   (a) The system is weakly mixing.
   (b) The system is mixing.
   (c) The matrix P is irreducible and aperiodic; i.e. $\exists N > 0$ such that the matrix P^N has no zero entries. [[For the elementary theory of Markov measures one needs the Perron-Frobenius theorem on stochastic matrices. Again refer to Walters [**257**]. See also [**83**] and [**65**].]]
   (d) For all $0 \le i, j \le \ell - 1$, $p_{ij}^{(n)} \to p_j > 0$.

   Thus for mixing Markov systems there is in fact a unique probability vector p with $pP = p$ and it is strictly positive: $p_i > 0$ for every i.

8. The pointwise ergodic theorem

For a dynamical system $\mathbf{X} = (X, \mathcal{X}, \mu, T)$ let $\mathcal{I}$ be the σ-algebra of T-invariant sets, and $\mathfrak{I}$ the closed subspace of T-invariant functions in $L^2(\mu)$. The orthogonal projection $P : L^2(\mu) \to \mathfrak{I}$ coincides with the conditional expectation operator $\mathbb{E}^{\mathcal{I}}$. For a function $f \in L^1(\mu)$ and an integer $n \ge 1$ let

$$\mathbb{A}_n f(x) = \frac{1}{n} \sum_{j=0}^{n-1} f(T^j x).$$

3.41. THEOREM (Pointwise ergodic theorem, Birkhoff). *Let* $\mathbf{X}$ *be a dynamical system,* $f \in L^1(\mu)$*, then the averages* $\mathbb{A}_n(f)(x)$ *converge* μ *a.e. and in* $L^1(\mu)$ *to the function* $f^* = \mathbb{E}^{\mathcal{I}}(f)$*. In particular when* $\mathbf{X}$ *is ergodic,*

$$\lim_{n\to\infty} \mathbb{A}_n f(x) = \lim_{n\to\infty} \frac{1}{n} \sum_{j=0}^{n-1} f(T^j x) = \int f\, d\mu,$$

μ *a.e. and in* $L^1(\mu)$.

PROOF. For $\phi \in L^1(\mu)$ set

$$M_n \phi = \max \left\{ \sum_{j=0}^{k-1} \phi \circ T^j : 1 \le k \le n \right\},$$

and observe that $\mathbb{A}_n \phi \le (1/n) M_n \phi$. Therefore

$$\limsup_{n\to\infty} \mathbb{A}_n \phi(x) \le 0,$$

for every x in the set A^c where

$$A = A(\phi) = \{x \in X : \sup_n M_n\phi(x) = \infty\} \in \mathcal{I}.$$

Observe that $M_n\phi$ is an increasing sequence of functions and that

$$M_{n+1}\phi(x) - M_n\phi(Tx) = \phi(x) - \min\{0, M_n\phi(Tx)\} \geq \phi(x).$$

To see this, note that

$$M_n\phi(Tx) = \max\left\{\sum_{j=1}^{k-1} \phi(T^j x) : 2 \leq k \leq n+1\right\}.$$

Thus $M_n\phi(Tx)$ and $M_{n+1}\phi(x)$ are attained at the same k — in which case they differ by $\phi(x)$ — or

$$M_{n+1}\phi(x) = \phi(x) > \phi(x) + M_n\phi(Tx).$$

The latter holds if and only if $M_n\phi(Tx) < 0$.

Therefore, on the set A the sequence $M_{n+1}\phi(x) - M_n\phi(Tx)$ decreases to ϕ and by Lebesgue's dominated convergence theorem:

$$0 \leq \int_A (M_{n+1}\phi - M_n\phi)\, d\mu = \int_A (M_{n+1}\phi - M_n\phi \circ T)\, d\mu \to \int_A \phi\, d\mu.$$

Next let $f \in L^1(\mu)$ and $\epsilon > 0$ be given, and set $\phi = f - f^* - \epsilon$. Since $A = A(\phi) \in \mathcal{I}$, we have $\int_A f\, d\mu = \int_A f^*\, d\mu$, hence

$$\int_A \phi\, d\mu = \int_A (f - f^* - \epsilon)\, d\mu = -\epsilon\mu(A) \leq 0,$$

hence $\int_A \phi\, d\mu = 0$, hence $\mu(A(\phi)) = 0$. Therefore

$$\limsup_{n\to\infty} \mathbb{A}_n\phi(x) \leq 0,$$

holds for almost every $x \in X$.

Since f^* is T-invariant $\mathbb{A}_n\phi = \mathbb{A}_n f - f^* - \epsilon$ and we conclude that

$$\limsup_{n\to\infty} \mathbb{A}_n f(x) \leq f^* + \epsilon.$$

Finally, the same argument applied to $-f$ gives

$$\liminf_{n\to\infty} \mathbb{A}_n f(x) \geq f^* - \epsilon,$$

and the proof of the pointwise convergence is complete.

To prove L^1 convergence, first note that a uniformly bounded, a.e convergent sequence of L^∞ functions converges also in L^1. The decomposition $f = f^+ - f^-$ for $f \in L^1(\mu)$ shows that we can assume that $f \geq 0$, and we let $f_k = \min\{f, k\}$. Now, with L^1-norm,

$$\begin{aligned} ||\mathbb{A}_n f - f^*|| &\leq ||\mathbb{A}_n(f - f_k)|| + ||\mathbb{A}_n f_k - f_k^*|| + ||f_k^* - f^*|| \\ &\leq ||f - f_k|| + ||\mathbb{A}_n f_k - f_k^*|| + ||f_k - f||, \end{aligned}$$

and therefore $\limsup_{n\to\infty} ||\mathbb{A}_n f - f^*|| \leq 2||f - f_k||$. Since $\lim_{k\to\infty} ||f - f_k|| = 0$, this completes the proof. □

We next use the pointwise ergodic theorem to obtain the ergodic decomposition theorem for a $\mathbb{Z}$-dynamical system. In Chapter 8 we shall prove the ergodic decomposition theorem for an arbitrary dynamical system (Theorem 3.22) with the aid of Choquet's theorem A.13.

3.42. THEOREM (The ergodic decomposition for $\mathbb{Z}$-systems). *Let $\mathbf{X}$ be a dynamical system $\mathcal{I} \subset \mathcal{X}$ the σ-algebra of T-invariant sets. Let $(X, \mathcal{X}, \mu) \xrightarrow{\pi} (Y, \mathcal{Y}, \nu)$ be the factor system defined by $\mathcal{I}$, so that $\mathcal{I} = \pi^{-1}(\mathcal{Y})$ (Theorem 2.15). Let $\mu = \int_Y \mu_y \, d\nu(y)$ be the disintegration of μ over ν and for every $y \in Y$ denote $X_y = \pi^{-1}(y)$ and $\mathcal{X}_y = \mathcal{X} \cap X_y$, then:*

1. *T acts as the identity on Y.*
2. *For ν a.e. $y \in Y$, the system $(X_y, \mathcal{X}_y, \mu_y, T)$ is ergodic.*

PROOF. Part 1 of the theorem is clear, as $TA = A$ for every $A \in \mathcal{I}$. We therefore have that, for ν a.e. y, $TX_y \subset X_y$. Now

$$\mu = \int \mu_y \, d\nu(y) = \int T\mu_y \, d\nu(y) = T\mu,$$

and since $\mu_y(X_y) = 1 = T\mu_y(X_y)$ the uniqueness of disintegration (Theorem A.7) implies $T\mu_y = \mu_y$. Thus for ν a.e. y, $(X_y, \mathcal{X}_y, \mu_y, T)$ is a dynamical system.

By the ergodic theorem we have for every $f \in L^1(\mu)$

$$\lim_{n\to\infty} \mathbb{A}_n f(x) = \lim_{n\to\infty} \frac{1}{n} \sum_{j=0}^{n-1} f(T^j x) = f^*(x) = \mathbb{E}^{\mathcal{I}}(f)(x),$$

for every $x \in X_f \subset X$ with $\mu(X_f) = 1$. With no loss of generality we assume that X is a compact metric space. Let f_n be a dense sequence of functions in $C(X)$ and $X_0 = \bigcap_n X_{f_n}$. Now

$$\int_X \mathbf{1}_{X_0} \, d\mu = \int_Y \int_X \mathbf{1}_{X_0} \, d\mu_y \, d\nu = 1$$

and it follows that $\mu_y(X_0) = \mu_y(X_0 \cap X_y) = 1$ for all $y \in Y_0 \subset \pi(X_0)$ with $\nu(Y_0) = 1$. Since for each $y \in Y$ the sequence f_n is dense in $L^1(\mu_y)$, it is easy to verify that for $y \in Y_0$ the system $(X_y, \mathcal{X}_y, \mu_y, T)$ is ergodic (e.g. use condition 7 in Theorem 3.35). □

Let Γ be an amenable group, m a left Haar measure on Γ (the counting measure when Γ is countable).

3.43. DEFINITION. Call a Følner sequence F_n $(n \in \mathbb{N})$ *tempered* if there exists a constant C such that for all n

$$m\Big(\bigcup_{k\le n} F_k^{-1} F_{n+1}\Big) \le C m(F_{n+1}). \tag{8.1}$$

3.44. LEMMA. *Every Følner sequence F_n; $n \in \mathbb{N}$ has a tempered subsequence. In particular, every amenable group has a tempered Følner sequence.*

PROOF. Let F_n be a Følner sequence. Define inductively a sequence n_i as follows. Let $n_1 = 1$ and assume n_j with $j \le i$ have been defined. Set $D_i = \cup_{j\le i} F_{n_j}$

and choose n_{i+1} sufficiently large so that $F_{n_{i+1}}$ is $(D_i^{-1}, \frac{1}{m(D_i)})$-invariant. A simple calculation shows that

$$m(\bigcup_{j\le i} F_{n_j}^{-1}F_{n_{i+1}}) \le 2m(F_{n_{i+1}}),$$

so that F_{n_i} satisfies (8.1) with $C = 2$. □

For a subset $F \subset \Gamma$ and $f : X \to \mathbb{R}$ denote

$$\mathbb{A}_F f(x) = \frac{1}{m(F)} \int_F f(\gamma x)\, dm(\gamma).$$

The following extension of Birkhoff's theorem to amenable groups is a recent result of E. Lindenstrauss [**172**].

3.45. THEOREM (Pointwise ergodic theorem for amenable groups). *Let Γ be an amenable group and $(X, \mathfrak{X}, \mu, \Gamma)$ a dynamical system. Let F_n; $n \in \mathbb{N}$ be a tempered Følner sequence, then for any $f \in L^1(\mu)$ there is a Γ-invariant function $f^* \in L^1(\mu)$ such that*

$$\lim_{n\to\infty} \mathbb{A}_{F_n} f(x) = f^*(x) \quad \mu \text{ a.e.}$$

In particular if $\mathbf{X}$ is ergodic then

$$\lim_{n\to\infty} \mathbb{A}_{F_n} f(x) = \int f\, d\mu \quad \mu \text{ a.e.}$$

9. Mixing and the Kolmogorov property for $\mathbb{Z}$-systems

3.46. DEFINITION. A dynamical system $\mathbf{X}$ is called:

1. *Mixing* if for every $A, B \in \mathfrak{X}$,
$$\lim_{n\to\infty} \mu(T^n A \cap B) = \mu(A)\mu(B).$$
2. *Mixing of order k or k-fold mixing* ($k \ge 1$) if for every $A_0, A_1, \dots, A_k \in \mathfrak{X}$
$$\lim_{n_1,\dots,n_k\to\infty} \mu(A_0 \cap T^{n_1}A_1 \cap T^{n_1+n_2}A_2 \cap \dots \cap T^{n_1+n_2+\dots+n_k}A_k) = \mu(A_0)\mu(A_1)\cdots\mu(A_k).$$
3. *Uniformly mixing* if for every $A, B_1, \dots, B_k \in \mathfrak{X}$
$$\lim_{n\to\infty} \sup\{|\mu(A\cap C) - \mu(A)\mu(C)| : C \in \mathcal{F}(B_1, \dots, B_k; n)\} = 0, \tag{9.1}$$
where $\mathcal{F}(B_1, \dots, B_k; n)$ is the σ-algebra generated by the sets $\{T^j B_i : j \ge n,\ i = 1, \dots, k\}$.

Thus 1-mixing is just mixing and of course k-fold mixing implies l-fold mixing if $k > l$. We also have the following theorem.

3.47. THEOREM. *A uniformly mixing system is mixing of all orders.*

PROOF. For mixing take $A = A_0$ and $B = B_1$, $k = 1$ in the definition of uniform mixing. To get 2-mixing, let $A_0, A_1, A_2 \in \mathfrak{X}$. Apply mixing to get $n_2^{(0)}$ such that for $n_2 \ge n_2^{(0)}$

$$|\mu(A_1 \cap T^{n_2}A_2) - \mu(A_1)\mu(A_2)| < \epsilon.$$

Next apply uniform mixing to the sets $A = A_0$, $B_1 = A_1$ and $B_2 = A_2$ to get $n_1^{(0)}$ such that for $n_1 \geq n_1^{(0)}$

$$|\mu(A_0 \cap T^{n_1}(A_1 \cap T^{n_2} A_2)) - \mu(A_0)\mu(A_1 \cap T^{n_2} A_2)| < \epsilon.$$

Put together these inequalities yield

$$|\mu(A_0 \cap T^{n_1} A_1 \cap T^{n_1+n_2} A_2) - \mu(A_0)\mu(A_1)\mu(A_2)| < 2\epsilon.$$

Proceed by induction. □

One of the outstanding open questions in ergodic theory is the following problem of Rohlin.

3.48. PROBLEM (Rohlin). For $\mathbb{Z}$-systems does mixing imply mixing of all orders?

3.49. DEFINITION. The dynamical system $\mathbf{X}$ is called a *Kolmogorov system* or a *K-system* if there is a sub σ-algebra $\mathcal{K} \subset \mathcal{X}$ with the following properties:

1. $T\mathcal{K} \subset \mathcal{K}$,
2. $\bigvee_{n=1}^{\infty} T^{-n}\mathcal{K} = \mathcal{X}$,
3. $\cap_{n=1}^{\infty} T^{n}\mathcal{K}$ is the trivial algebra $\mathcal{E}_0 = \{\emptyset, X\}$.

3.50. LEMMA. *Every K-system is ergodic and non-atomic.*

PROOF. Let $A \in \mathcal{X}$ be an invariant set and let $\epsilon > 0$ and $m > 0$ be given, then there exists $n_0 \geq 0$ and a subset $B \in T^{-n_0}\mathcal{K}$ such that $\mu(A \bigtriangleup B) < \epsilon$. We then have $T^{m+n_0}B \in T^m\mathcal{K}$ and $\mu(A \bigtriangleup T^{m+n_0}B) < \epsilon$. Since this holds for every $\epsilon > 0$ we conclude that $A \in T^m\mathcal{K}$. This, in turn, holds for every $m > 0$ and therefore $A \in \cap_{n=1}^{\infty} T^m\mathcal{K} = \mathcal{E}_0$. Thus $\mathbf{X}$ is ergodic and it is clear that it can not be a finite system; hence μ is non-atomic. □

3.51. PROPOSITION. *Every Bernoulli system* $\mathbf{X}$ *is a K-system.*

PROOF. Let $(Y, \mathcal{Y}, \nu)$ be the state space of the Bernoulli system. Thus $\mathbf{X} = (X, \mathcal{X}, \mu, S)$, where $X = Y^{\mathbb{Z}}$, $\mu = \nu^{\mathbb{Z}}$ and $(Sx)_n = x_{n+1}$ (see 3.40). For $A \in \mathcal{Y}$ set $\tilde{A} = \{x \in X : x_0 \in A\}$ and let $\mathcal{G} = \{\tilde{A} : A \in \mathcal{Y}\}$, the *time zero* σ-algebra. Now let $\mathcal{K} = \bigvee_{j=1}^{\infty} S^j\mathcal{G}$, the *past* σ-algebra. Clearly $\mathcal{K}$ satisfies properties 1 and 2 in the definition of a K-system. We only have to show that $\cap_{n=1}^{\infty} S^n\mathcal{K} = \mathcal{E}_0 = \{\emptyset, X\}$. Now for a fixed $A \in \cap_{n=1}^{\infty} S^n\mathcal{K} = \cap_{n=1}^{\infty} S^n \bigvee_{j=1}^{\infty} S^j\mathcal{G}$, every k and every $B \in \bigvee_{j=k}^{\infty} S^{-j}\mathcal{G}$ the sets A and B are independent. Since $\bigvee_{k\in\mathbb{Z}} \bigvee_{j=k}^{\infty} S^{-j}\mathcal{G} = \mathcal{X}$, it follows that A is independent of $\mathcal{X}$; i.e. $A \in \mathcal{E}_0 = \{\emptyset, X\}$. □

3.52. REMARK. In every dynamical system $\mathbf{X}$, given a point $x \in X$ and an "event" $A \in \mathcal{X}$, we have $T^{-1}x \in A$ iff $x \in TA$. It thus makes sense to refer to the event TA as the "past" (one unit back in time) of the event A. This motivation for choosing the past direction in this way is reinforced by the example of subshift systems. Since it is natural to refer to events determined by negative, rather than positive, coordinates as past events, we call the σ-algebra $\mathcal{K} = \bigvee_{j=1}^{\infty} S^j\mathcal{G}$, the *past* σ-algebra. For example $x \in S[2]$ iff $S^{-1}x \in [2]$ iff $(S^{-1}x)_0 = x_{-1} = 2$.

The question whether, for $\mathbb{Z}$-systems, mixing implies higher order mixing is still open. On the other hand there are examples of mixing systems which are not K-systems. (As we shall see in Chapter 18, every K-system has positive entropy and there are many examples of mixing systems with entropy zero; one such system is the horocycle system; see Chapter 4, Section 6.) The long standing question whether every K-system is isomorphic to a Bernoulli system was resolved by Ornstein in 1970 [**190**]. In this paper he constructed a K-system which is not isomorphic to a Bernoulli shift (see also Ornstein and Shields's [**193**]). Smooth examples of K non-Bernoulli systems were constructed by A. Katok in [**144**]. In [**140**] S. Kalikow was able to present a more natural example of such a system. His example of a K but not Bernoulli is the famous T, T^{-1} transformation. This is the skew-product where the Bernoulli system (Ω, ν, T) serves as both the "base" and the "fiber" systems (where $\Omega = \{1, -1\}^{\mathbb{Z}}$, ν is the product measure $(\frac{1}{2}, \frac{1}{2})^{\mathbb{Z}}$ and T the shift). The T, T^{-1} transformation R is defined as follows: set $X = \Omega \times \Omega$, $\mu = \nu \times \nu$ and let

$$R(\omega, \omega') = (T\omega, T^{\omega_0}\omega').$$

Our next theorem places the K property at the top of the hierarchy of mixing properties. We prove here only one implication, the proof of the converse implication will be completed in Chapter 18, Section 2, using the Rohlin-Sinai theorem.

3.53. THEOREM. *A system* $\mathbf{X}$ *is a* K*-system iff it is uniformly mixing.*

PROOF. $K \Rightarrow$ **uniform mixing:** Let $\mathcal{K}$ be a σ-algebra making $\mathbf{X}$ a K-system. Since $\bigvee_{n=1}^{\infty} T^{-n}\mathcal{K} = \mathcal{X}$, it is enough to show that (9.1) holds for every $A \in \mathcal{X}$ and $B_1, \ldots, B_k \in \bigcup_{n=1}^{\infty} T^{-n}\mathcal{K}$.

Assume to the contrary that for some such A and $B_1, \ldots, B_k$ the property (9.1) does not hold. Define a function $\rho : \mathcal{X} \to \mathbb{R}$ by

$$\rho(C) = \mu(A \cap C) - \mu(A)\mu(C).$$

It is easy to see that ρ is a finite signed measure; we set $\rho_n = \rho \upharpoonright \mathcal{F}(B_1, \ldots, B_k; n)$. Let $C_n \in \mathcal{F}(B_1, \ldots, B_k; n)$ be the maximal positive set of ρ_n (i.e. $\rho_n(E) \geq 0$ for every $E \in \mathcal{F}(B_1, \ldots, B_k; n)$ with $E \subset C_n$).

Our assumption implies that there exists a $\delta > 0$ such that $\rho_n(C_n) \geq \delta$ for every $n \geq 1$. Now for every $m \geq n$, $C_m \in \mathcal{F}(B_1, \ldots, B_k; n)$, hence

$$\begin{aligned}
&\rho_n(C_m \cup C_{m-1} \cup \cdots \cup C_n) = \\
&\rho_n(C_m) + \rho_n(C_{m-1} \setminus C_m) + \rho_n(C_{m-2} \setminus (C_{m-1} \cup C_m)) + \cdots + \rho_n(C_n \setminus \cup_{j=n+1}^{m} C_j) = \\
&\rho_m(C_m) + \rho_{m-1}(C_{m-1} \setminus C_m) + \rho_{m-2}(C_{m-2} \setminus (C_{m-1} \cup C_m)) + \cdots + \\
&\rho_n(C_n \setminus \cup_{j=n+1}^{m} C_j) \geq \rho_m(C_m) \geq \delta.
\end{aligned}$$

Denoting $D_n = \bigcup_{m=n}^{\infty} C_m$ and $D = \bigcap D_n$ we then have $\rho_n(D_n) \geq \delta$ for every $n \geq 1$ and hence also

$$\begin{aligned}
\mu(A \cap D) - \mu(A)\mu(D) &= \lim_{n \to \infty} (\mu(A \cap D_n) - \mu(A)\mu(D_n)) \\
&= \lim_{n \to \infty} \rho_n(D_n) \geq \delta.
\end{aligned}$$

It now follows that $0 < \mu(D) < 1$.

On the other hand, since $B_1, \ldots, B_k \in \bigcup_{n=1}^{\infty} T^{-n}\mathcal{K}$, there exists an $m \geq 1$ such that $B_1, \ldots, B_k \in T^{-m}\mathcal{K}$ so that $\mathcal{F}(B_1, \ldots, B_k; n) \subset T^{n-m}\mathcal{K}$ and it follows that

$D \in \bigcap_{n=1}^{\infty} T^{n-m}\mathcal{K}$, which by assumption is the trivial algebra $\mathcal{E}_0 = \{\emptyset, X\}$. This contradiction completes the proof.

Uniform mixing $\Rightarrow K$: Observe first that for every $B_1, \dots, B_k \in \mathcal{X}$, the sequence of σ-algebras $\mathcal{F}(B_1, \dots, B_k; n)$ decreases to $\mathcal{E}_0$, the trivial algebra. In fact if $A \in \bigcap_{n=1}^{\infty} \mathcal{F}(B_1, \dots, B_k; n)$, then for every $n \geq 1$

$$\sup_{C \in \mathcal{F}(B_1, \dots, B_k; n)} |\mu(A \cap C) - \mu(A)\mu(C)| \geq |\mu(A) - \mu(A)^2|,$$

hence, by uniform mixing $|\mu(A) - \mu(A)^2| = 0$, whence $A \in \{X, \emptyset\} = \mathcal{E}_0$. Using the Rohlin-Sinai theorem, 18.9, we shall be able to conclude that this fact implies that $\mathbf{X}$ is indeed a K-system. □

10. Stationary stochastic processes and dynamical systems

The strong link between ergodic theory and probability stems from the following construction. Let $(\Omega, \mathcal{B}, P)$ be a probability space and $\{X^\gamma : \gamma \in \Gamma\}$ a system of real valued square integrable random variables (i.e. elements of $L^2(P)$) indexed by the elements of the group Γ satisfying the following condition. For every finite set $F = \{\gamma_1, \gamma_2, \dots, \gamma_m\} \subset \Gamma$ the probability measure P_F on $\mathbb{R}^m$ defined by:

$$P_F(A) = P\{\omega \in \Omega : (X^{\gamma_1}(\omega), \dots, X^{\gamma_m}(\omega)) \in A\}$$

Law of X

for every Borel subset $A \subset \mathbb{R}^m$, depends on F only up to congruence; that is $P_F = P_{\gamma F}$ for every $\gamma \in \Gamma$. In particular the measure $P_\gamma = P_e$ is independent of γ. Such a sequence of random variables is called a *stationary stochastic process*. It is *centered* if $\mathbb{E}(X^e) = \int_\Omega X^e \, dP = 0$ (we use probabilistic language, call the integral expectation and denote it by $\mathbb{E}$). The function $c(\gamma\delta^{-1}) = \mathbb{E}(X^\gamma X^\delta)$, which depend only on $\gamma\delta^{-1}$, is called the *covariance function* of the process.

With every such stationary process we associate the dynamical system $\mathbf{X}(\{X^\gamma : \gamma \in \Gamma\}) = (X, \mathcal{X}, \mu, \Gamma)$, where $X = \mathbb{R}^\Gamma$, $\mathcal{X}$ is the Borel σ-algebra of X (equipped with the product topology), μ is the probability measure on X defined by the requirement that for every finite $F \subset \Gamma$ the corresponding marginal measure on $\mathbb{R}^F$ equals P_F (μ is uniquely defined according to Kolmogorov's consistency theorem, Theorem A.6), and finally the action of Γ on X is by left translations: $(\gamma x)(\gamma') = x(\gamma^{-1}\gamma')$. Of course the original stochastic process can now be recovered if we consider the functions $\hat{X}^\gamma = \pi_\gamma$, where $\pi_\gamma = \pi_e \circ \gamma : X \to \mathbb{R}$ is the projection on the γ coordinate, as a sequence of random variables defined on the probability space $(X, \mathcal{X}, \mu)$. This version of the stochastic process is called the *Kolmogorov representation*.

The reader can easily check that the subshifts presented in Examples 3.40 are dynamical systems associated with stationary stochastic processes as above, simply by taking $(\Omega, \mathcal{B}, P) = (\Omega(\ell), \mathcal{B}, \mu)$ and $X^n : \Omega(\ell) \to \mathfrak{L} = \{1, \dots, \ell\}$ to be the projection on the n-th coordinate. Of course in this case the associated Kolmogorov representation is the subshift itself. In particular Bernoulli shifts and Markov systems are such examples.

One can go in the other direction as well and, given a dynamical system $\mathbf{X}$ and a function $f \in L^2(X, \mu)$, produce a stationary stochastic processes by taking $(\Omega, \mathcal{B}, P) = (X, \mathcal{X}, \mu)$ and setting $X^\gamma(x) = f(\gamma x)$. As is easy to see, the associated dynamical system $\mathbf{Y} = \mathbf{X}(\{X^\gamma : \gamma \in \Gamma\})$ is a factor of $\mathbf{X}$, with $\phi_f(x) = (f(\gamma x)_{\gamma \in \Gamma})$ as the factor map.

In the special case when f has a finite range (i.e. when f corresponds to a finite measurable partition α of X) we shall call the map $\phi_f = \phi_\alpha$ a symbolic representation of $\mathbf{X}$. As we shall see, mainly in the second part of the book, these symbolic representations will serve as the main tool involved in the deep analysis that is required for the proofs of theorems like the Jewett-Krieger theorem or Ornstein's isomorphism theorem for Bernoulli systems.

11. Gaussian dynamical systems

In this section we briefly consider the rich and important class of dynamical systems that arise from Gauss stationary stochastic processes. This discussion will become handy in Chapter 13 (Kazhdan's property). We refer to [**43**] and [**187**] for more details and for proofs. Except for occasional examples the only place where we shall use the theory of Gaussian processes as a tool is in Chapter 13.

3.54. DEFINITION. A centered stationary stochastic process $\{X^\gamma : \gamma \in \Gamma\}$ is called *Gaussian* if for every $F = \{\gamma_1, \gamma_2, \dots, \gamma_m\} \subset \Gamma$, P_F is an m-dimensional Gauss distribution. (In the non-degenerate case, i.e. when the functions X^{γ_j}, $1 \le j \le m$ are independent over $\mathbb{R}$, this means that for every m Borel subsets $C_j \subset \mathbb{R}$

$$P_F(\bigcap\{X^{\gamma_j} \in C_j : 1 \le j \le m\}) = a \cdot \int_{C_1 \times \cdots \times C_m} \exp(-\frac{1}{2}\langle Mt, t\rangle)\, dt_1 \cdots dt_m,$$

where a is a constant, $t = (t_1, \dots, t_m)$ and M is the inverse of the regular covariance matrix $C = (c(\gamma_i \gamma_j^{-1}))$.)

It follows that such a process is uniquely determined by its covariance function $c(\gamma\delta^{-1}) = \mathbb{E}(X^\gamma X^\delta)$. It is easy to check that the covariance function is symmetric $(c(\gamma) = c(\gamma^{-1}))$ and in fact is a positive definite function.

3.55. DEFINITION. A function $\phi : \Gamma \to \mathbb{C}$ is called *positive definite* if it satisfies the following property: for every finite subset $F = \{\gamma_1, \gamma_2, \dots, \gamma_m\} \subset \Gamma$ and every sequence of complex numbers $\{a_j\}_{j=1}^m$

$$\sum_{j=1}^{m}\sum_{i=1}^{m} a_j \bar{a}_i c(\gamma_j \gamma_i^{-1}) \ge 0.$$

One can check that if ϕ is a positive definite function then ϕ has the following properties (see [**120**]):

1. $\phi(e) \ge 0$,
2. $|\phi(\gamma)| \le \phi(e)$ for all $\gamma \in \Gamma$,
3. $\phi(\gamma) = \overline{\phi(\gamma^{-1})}$ for all $\gamma \in \Gamma$, and
4. $|\phi(\gamma) - \phi(\delta)|^2 \le 2\phi(e)(\phi(e) - \Re\phi(\gamma^{-1}\delta))$ for all $\gamma, \delta \in \Gamma$.

Next we show that conversely, any positive definite function on Γ defines a Γ-stationary Gauss process.

3.56. THEOREM. *There is a 1-1 correspondence between positive definite functions $c : \Gamma \to \mathbb{R}$ and covariance functions of centered Γ-stationary Gaussian processes.*

PROOF. We have already seen that the covariance function of a centered Γ-stationary Gauss process is positive definite. Conversely suppose $c : \Gamma \to \mathbb{R}$ is positive definite. We construct a Γ-stationary Gauss process $\{X^\gamma : \gamma \in \Gamma\}$ as follows. By Kolmogorov's consistency theorem (Theorem A.6), it is enough to construct, for any finite set $F = \{\gamma_1, \gamma_2, \dots, \gamma_m\} \subset \Gamma$, a sequence $X_1, \dots, X_m$ of real valued random variables with joint Gauss distribution and with covariance function $\mathbb{E}(X_j X_i) = c(\gamma_j \gamma_i^{-1})$, and this in such a way that the constructions for the various F are consistent (i.e. if $F = F_1 \cup F_2$ then the F_i marginal distributions of the sequence $\{X_j\}_F$ coincide with the distribution of the sequences $\{X_j\}_{F_i}, i = 1, 2$).

The fact that c is positive definite implies that the matrix C with $c_{ij} = c(\gamma_j \gamma_i^{-1})$ is symmetric and positive definite. Thus there exists an orthogonal matrix O such that $OCO^t = D$, a non-negative diagonal matrix. Let R be the diagonal matrix whose entries are the square roots of the corresponding entries of D and set $A = RO$. We have $C = A^t A$ and taking $Z_1, \dots, Z_m$ to be an independent sequence of $\mathcal{N}(0,1)$ random variables we now set $X = (X_1, \dots, X_m) = (Z_1, \dots, Z_m)A$. With $u = (u_1, \dots, u_m)$ we have, by the independence of the vector Z,

$$\begin{aligned}\mathbb{E}(\exp(i\langle u, X\rangle)) &= \mathbb{E}\big(\exp(i \sum_k Z_k(\sum_j u_j a_{kj}))\big)\\ &= \exp(-\frac{1}{2}(\sum_k(\sum_j u_j a_{kj})^2)) = \exp(-\frac{1}{2}uA^tAu)\\ &= \exp(-\frac{1}{2}uCu^t).\end{aligned}$$

This computation of the characteristic function of the vector X shows that it has in fact a Gaussian joint distribution with covariance matrix C. Since the consistency of the constructions for various $F \subset \Gamma$ is clear this concludes the proof of the theorem. □

3.57. Definition. We call a dynamical system $\mathbf{X}$ *Gaussian* if there exists a function $f \in L_0^2(X, \mu)$ such that the corresponding stochastic process $\{f \circ \gamma : \gamma \in \Gamma\}$ is Gaussian and the smallest σ-algebra with respect to which the functions $\{f \circ \gamma : \gamma \in \Gamma\}$ are measurable is all of $\mathcal{X}$. In other words $\mathbf{X}$ is isomorphic to the system $\mathbf{X}(\{f \circ \gamma : \gamma \in \Gamma\})$ defined by the Kolmogorov representation (defined in Section 10).

When $\mathbf{X}$ is a Gaussian system the closed subspace $H \subset L_0^2(X, \mu)$ spanned by $\{f \circ \gamma : \gamma \in \Gamma\}$ is called the *first chaos* and it can be shown that

$$L^2(X, \mu) = \mathbb{C} \oplus \sum_{n=1}^{\infty} H^{\odot n},$$

where $H^{\odot n}$ is the symmetric tensor product of order n (see [**187**], and [**43**]).

In the next section we show that the three conditions: (i) $\mathbf{X}$ is ergodic, (ii) $\mathbf{X}$ is weakly mixing and (iii) the Koopman representation restricted to the first chaos is weakly mixing, are equivalent for a Gaussian Γ-system $\mathbf{X}$.

12. Weak mixing of Gaussian systems

Let $\{X^\gamma : \gamma \in \Gamma\}$ be a Γ-process of centered Gaussian random variables. Let (Ω, μ) be a probability space on which the X^γ are defined and generate the σ-algebra. As was shown in the discussion preceding Theorem 3.56 the action of Γ

on the X_n^γ's defined by

$$\gamma_0(X_n^\gamma) = X_n^{\gamma\gamma_0}$$

defines a Γ action on (Ω, μ) which preserves the measure μ. The corresponding dynamical system $(\Omega, \mathcal{B}, \mu, \Gamma)$ is a Gauss Γ-dynamical system and the closed subspace H of $L_0^2(\Omega, \mu)$ generated by the Gaussian random variables $\{X^\gamma : \gamma \in \Gamma\}$ is the first chaos.

3.58. LEMMA. *Let X, Y be centered Gaussian variables, $\mathbb{E}(X^2) = \mathbb{E}(Y^2) = 1$. Let $\rho = \mathbb{E}(XY)$, then*

$$\mathbb{E}((X^2 - \mathbf{1})(Y^2 - \mathbf{1})) = 2\rho^2.$$

PROOF. Let Z be a Gaussian variable independent of X such that $Y = \rho X + \sqrt{1-\rho^2}Z$. Now

$$\mathbb{E}(X^2Y^2) = \mathbb{E}(X^2(\rho X + \sqrt{1-\rho^2}Z)^2)$$
$$= \rho^2\mathbb{E}(X^4) + 1 - \rho^2.$$

Hence

$$\mathbb{E}((X^2 - \mathbf{1})(Y^2 - \mathbf{1})) = \rho^2(\mathbb{E}(X^4) - 1).$$

An easy computation gives $\mathbb{E}(X^4) = 3$ and our lemma follows. □

3.59. THEOREM. *Let $(\Omega, \mathcal{B}, \mu, \Gamma)$ be a Gauss Γ-dynamical system corresponding to the Gauss stationary Γ-process $\{X^\gamma : \gamma \in \Gamma\}$, then the following conditions are equivalent:*

1. *The dynamical system $(\Omega, \mathcal{B}, \mu, \Gamma)$ is ergodic.*
2. *The dynamical system $(\Omega, \mathcal{B}, \mu, \Gamma)$ is weakly mixing.*
3. *The Koopman representation restricted to the first chaos is weakly mixing.*

PROOF. 2 $\Rightarrow$ 3. Weak mixing of the system $(\Omega, \mathcal{B}, \mu, \Gamma)$ is by definition the same as weak mixing of the Koopman representation on $L_0^2(\Omega, \mu)$, which clearly implies weak mixing on the first chaos.

3 $\Rightarrow$ 2. Conversely, denote the Koopman representation restricted to the first chaos H by π, and assume that π is a weakly mixing representation. By Theorem 3.5, π has no non-trivial finite dimensional subrepresentations.

Recall that the Koopman representation of Γ on $L_0^2(\Omega, \mu)$ is unitarily equivalent to the representation $\bigoplus_{k=1}^\infty \pi^{\odot k}$ where $\pi^{\odot k}$ is the restriction of the tensor product representation $\pi^{\otimes k} = \pi \otimes \pi \otimes \cdots \otimes \pi$ to the subspace $H^{\odot k}$, the symmetric tensor product of order k; i.e. the subspace of the tensor product $H^{\otimes k} = H \otimes H \otimes \cdots \otimes H$ (k times) consisting of the "symmetric" or $\mathfrak{S}_k$-invariant vectors ([**187**]).

Since π (on H) admits no non-trivial finite dimensional subrepresentation, we conclude that also the Koopman representation of Γ on $L_0^2(\Omega, \mu)$ has no non-trivial finite dimensional subrepresentations. By Theorem 3.11 the system (Ω, μ, Γ) is weakly mixing.

2 $\Rightarrow$ 1. Of course weak mixing on $L_0^2(\Omega, \mu)$ implies ergodicity there.

1 $\Rightarrow$ 3. Assume now that π is ergodic on $L_0^2(\Omega, \mu)$. Since functions of the form $\sum_{j=1}^n c_j X^{\gamma_j}$ are dense in H, it is enough to show that $\mathfrak{m}(|\rho(\gamma)|) = 0$, where $\rho(\gamma) = \mathbb{E}(X^\gamma X^e)$ (Definition 3.2). Set $Z = (X^e)^2 - \mathbf{1}$, then $\mathbb{E}Z = 0$ and by ergodicity $\mathfrak{m}\{\langle \pi(\gamma)Z, Z\rangle\} = 0$. By Lemma 3.58,

$$\langle \pi(\gamma)Z, Z\rangle = \mathbb{E}\{[(X^\gamma)^2 - \mathbf{1}][(X^e)^2 - \mathbf{1}]\} = 2\rho(\gamma)^2,$$

hence $\mathfrak{m}(\rho(\gamma)^2) = 0$ and $\rho(\gamma)^2$ is a flight function . Lemma 1.52 implies that also $|\rho(\gamma)|$ is a flight function and the proof is complete. □

13. Notes

Unitary representations: Sources: Mainly Bergelson and Rosenblatt [**21**] and Ellis and Nerurkar. Lemma 3.3.1 and its corollary Theorem 3.4 are from Glasner and Weiss [**102**].

The Koopman representation: This proof of Theorem 3.15 is due to Y. Katznelson.

Rohlin's skew-product theorem: This section is adopted with only small modifications from Aaronson's book [**1**].

Group and homogeneous skew-products: Sources: Mainly Zimmer's papers [**267**], [**268**] and del-Junco and Rudolph [**138**].

Amenable groups: The theory of dynamical systems of amenable groups achieved a profound and comprehensive treatment in Ornstein and Weiss [**195**] (see also the work of Kammeyer and Rudolph [**143**]). The recent papers of Lindenstrauss [**172**], in the direction of convergence theorems, and Rudolph and Weiss [**221**] in entropy theory, seem to herald a new stage in the study of such dynamical systems (see also Danilenko [**44**] and Danilenko and Park [**45**]). Particularly noteworthy is the recent work of Dooley and Golodets, [**49**], where the results of [**221**] are used to obtain a generalization to amenable groups of the well known theorem of Kolmogorov according to which a K system has countable Lebesgue spectrum (see Theorem 5.13, and Exercise 18.18 below). This generalization was conjectured by J.-P. Thouvenot who, together with D. Sinel'shchikov, obtained earlier partial results along these lines. J. von Neumann's mean ergodic theorem (Theorem 3.33 for $\mathbb{R}$ actions) was the first ergodic theorem to be proved, von Neumann [**255**].

The pointwise ergodic theorem: The proof of the pointwise ergodic theorem (for one parameter group of transformations) went a long way from the original papers of Birkhoff [**23**] and [**24**] to the succinct proof produced in [**146**], which I followed here with only slight changes. The main idea of this proof goes back to Garsia [**84**].

Mixing and the Kolmogorov property for $\mathbb{Z}$-systems: Ledrappier has shown that there is a $\mathbb{Z}^2$-system which is mixing but not 5-mixing, [**162**].

Stationary stochastic processes and dynamical systems: Sources: [**43**] and [**186**].

Gaussian dynamical systems: Sources: [**43**], [**186**], [**187**] and [**35**].

Weak mixing of Gauss dynamical systems: The equivalence for a Gauss $\mathbb{R}$-dynamical system of weak mixing and weak mixing on the first chaos is a classical result which goes back to K. Îto [**126**], U. Grenander [**108**], Fomin and Maruyama (see bibliographic notes in [**243**] page 83). The extension of this result to a general group Γ is due to A. Tempelman [**243**]. The equivalence of these conditions with that of ergodicity of the Gauss dynamical system is from Glasner and Weiss [**102**].

CHAPTER 4

Invariant Measures on Topological Systems

In this chapter we put together our two approaches to dynamical systems, the topological and the measure theoretical. We start with a topological dynamical system (X, Γ) and assume that the compact convex set $M_\Gamma(X)$ of Γ-invariant measures is not empty. This is always the case when the acting group Γ is amenable (so in particular when it is abelian and more specifically when $\Gamma = \mathbb{Z}$). In Section 1 we examine the set $M_\Gamma(X)$ and identify its extreme points set $\mathrm{ext}\,(M_\Gamma(X))$ with the set $M_\Gamma^{\mathrm{erg}}(X)$ of Γ-ergodic measures. Viewed in this light, the ergodic decomposition of a measure in $M_\Gamma(X)$ becomes its Choquet representation (see Theorem 3.22 below and [**206**]). The uniqueness of the ergodic decomposition now implies that this representation is unique; i.e. that $M_\Gamma(X)$ is a *Choquet simplex*. It turns out that, at least for $\mathbb{Z}$ actions, every abstract Choquet simplex can be realized as $M_T(X)$ for some minimal dynamical system (X, T). This result is due to T. Downarowicz who showed that in fact one can find all Choquet simplices already within the class of Toeplitz systems, [**50**]. There are many uniquely ergodic systems; that is systems in which $M_\Gamma(X)$ is a singleton and as we shall see (Chapter 13 and 15.15) the simplex $M_S(\Omega)$ of the topological Bernoulli system $(\Omega(\ell), \mu, S)$ is the Poulsen simplex; i.e. the (unique up to affine isomorphism) metrizable Choquet simplex Q in which $\mathrm{ext}\,(Q)$ is dense in Q. In Chapter 13 we investigate a strong connection between the nature of the group Γ and the geometry of $M_\Gamma(\Omega)$, where $\Omega = \Omega(\Gamma) = \{0,1\}^\Gamma$ is the topological Bernoulli system.

Once we choose an element $\mu \in M_\Gamma(X)$, the system $\mathbf{X} = (X, \mathcal{X}, \mu, \Gamma)$ becomes a measure preserving dynamical system. For $\mathbb{Z}$-systems the pointwise ergodic theorem now implies that, when μ is ergodic, for almost every point $x \in X$ and every continuous function $f \in C(X)$ the ergodic sums $\mathbb{A}_n(f)(x)$ converge to the integral $\int f\, d\mu$ (Theorem 4.4). Such points are called generic points for the measure μ. We introduce generic points and study some examples in Section 2.

Another useful concept that combines topological and measure-theoretical aspects is that of strict ergodicity. The topological system (X, Γ) is called strictly ergodic if $M_\Gamma(X)$ consists of a single element μ, i.e. it is uniquely ergodic, and in addition μ is a full measure (i.e. $\mathrm{supp}\,\mu = X$). For a while it was believed that strict ergodicity — which is known to imply some strong topological consequences (like in the case of $\mathbb{Z}$-systems, the fact that *every* point of X is a generic point and moreover that the convergence of the ergodic sums $\mathbb{A}_n(f)$ to the integral $\int f\, d\mu$, $f \in C(X)$ is *uniform*) — entails some severe restrictions on the measure-theoretical behaviour of the system. For example, it was believed that unique ergodicity implies zero entropy; then, first some examples were produced to show that this need not be the case (Furstenberg in [**74**] and Hahn and Katznelson in [**110**] gave examples of uniquely ergodic systems with positive entropy) and later in 1970 R. I. Jewett surprised everyone with his outstanding result: every weakly mixing measure preserving $\mathbb{Z}$-system has a strictly ergodic model, [**129**]. This was strengthened later by Krieger [**161**] who showed that even the weak mixing assumption is redundant and that the result holds for every ergodic $\mathbb{Z}$-system. In an unpublished paper A. Rosenthal outlines a proof of the Jewett Krieger theorem for a (discrete, countable) amenable group [**214**]. We shall give B. Weiss' proof of the Jewett Krieger theorem (for $\mathbb{Z}$-systems) in Chapter 14. In Section 3 we deal with some of the more elementary properties of unique ergodicity. In Section 4 we list many natural examples of uniquely ergodic systems. We also examine here briefly the close connection between the notions of unique ergodicity and uniformly distributed sequences. In Section 5 we examine the important example of the Heisenberg nil-system and show that it is strictly ergodic.

The geodesic and the horocycle flows on compact Riemann surfaces played a crucial role in the history of the theory of dynamical systems and even today generalizations of these flows are of great interest in connection with various aspects of Lie groups and number theory (see for example [**15**]). In Section 6 we show that the geodesic and horocycle flows are mixing and that the horocycle flow is strictly ergodic. In Chapter 7, Section 3 we shall use joining theory to show that the horocycle systems are mixing of all orders.

The connection between the measure-preserving system and the underlying topological system is even closer when one adds the requirement that μ be a full measure. If this is the case, that is if the topological system (X, Γ) admits a full invariant measure, and it is also topologically transitive then we say that it is an E-system. Theorem 4.27 on E-systems is perhaps the principal key to the mysterious connection between topological dynamics and ergodic theory alluded to in the introduction of this book. It tells us that for a full measure $\mu \in M_\Gamma(X)$, topological transitivity, weak mixing as well as mixing of the topological system (X, Γ) are implied by the corresponding measure-theoretical properties (ergodicity, weak mixing and mixing) of the measure-preserving system $(X, \mathcal{X}, \mu, \Gamma)$.

E-systems turn out to be very useful in other respects too and the best result we obtain concerning these systems is Theorem 4.33, which asserts that two minimal E-systems are weakly disjoint iff their maximal equicontinuous factors are disjoint. The last section (Section 7) is dedicated to the study of E-systems.

1. Invariant probability measures

Recall that for a topological system (X, Γ) we let $M(X)$ denote the compact metrizable space of probability measures on X and by $M_\Gamma(X)$ the compact convex subset of Γ-invariant measures. With a general acting group Γ the set $M_\Gamma(X)$ may very well be empty. Consider for example the minimal topological dynamical system (X, Γ), where $X = \mathbb{R} \cup \{\infty\}$, $\Gamma = SL(2, \mathbb{Z})$ and the matrix $\gamma = \begin{pmatrix} a & b \\ c & d \end{pmatrix} \in \Gamma$ acts by the corresponding Möbius transformation $z \mapsto \frac{az+b}{cz+d}$. It is easy to check that for every $\mu \in M(X)$ there exists a sequence $\gamma_n \in \Gamma$ such that the sequence $\gamma_n \mu$ tends weak* to a point mass.

A topological group Γ has the *fixed point property* if whenever (Q, Γ) is an *affine topological dynamical system* — that is a topological dynamical system such that Q is a compact convex subset of a locally convex topological vector space E

and the action of Γ on Q is by affine homeomorphisms — then there is a Γ-fixed point in Q. It is well known that Γ has the fixed point property iff it is amenable (see e.g. [**107**] and [**85**]). Of course whenever (X,Γ) is a topological system the corresponding system $(M(X),\Gamma)$ is such an affine system and we conclude that for an amenable group Γ the subset $M_\Gamma(X)$ is never empty.

For $\mathbb{Z}$-systems there is an easy direct proof for this fact.

4.1. THEOREM (Krylov-Bogolubov). *For a topological $\mathbb{Z}$-dynamical system there is always an invariant Borel probability measure.*

PROOF. Let $x \in X$ be an arbitrary point and for every $n \in \mathbb{N}$ define the probability measure $\mu_n = \frac{1}{n}\sum_{j=0}^{n-1}\delta_{T^j x}$. By the weak* compactness of $M(X)$ there exists a convergent subsequence, say $\mu_{n_k} \to \mu$, and it is easy to see that in fact any such limit point is a T-invariant measure. □

For a measure $\mu \in M(X)$ let supp(μ) be the complement in X of the open subset

$$\bigcup\{U \subset X : U \text{ is open and non-empty and } \mu(U) = 0\}.$$

The measure μ is called *full* when supp$(\mu) = X$; i.e. when $\mu(U) > 0$ for every non-empty open $U \subset X$.

4.2. THEOREM. *The G_δ subset* ext$(M_\Gamma(X))$ *of extreme points of $M_\Gamma(X)$ coincides with the set of Γ ergodic probability measures on X. We shall write $M_\Gamma^{\mathrm{erg}}(X)$ for this set. If $\mu, \nu \in M_\Gamma^{\mathrm{erg}}(X)$ and $\mu \neq \nu$ then μ and ν are mutually singular.*

PROOF. If the measure $\mu \in M_\Gamma(X)$ is non-ergodic then there exists a Γ-invariant measurable subset $A \subset X$ with $0 < \mu(A) < 1$. Now

$$\mu = \mu(A)\mu_A + (1 - \mu(A))\mu_{A^c},$$

(with, for example, $\mu_A(B) := \mu(A)^{-1}\mu(A \cap B)$) is a nontrivial convex combination of elements of $M_\Gamma(X)$ and we conclude that $\mu \notin$ ext$(M_\Gamma(X))$.

Conversely, assume that $\mu \notin$ ext$(M_\Gamma(X))$, so that a nontrivial representation

$$\mu = \alpha\nu + (1 - \alpha)\eta,$$

exists, then $\nu \ll \mu$ and it is easy to check that the Radon-Nikodým derivative $d\nu/d\mu$ is a nontrivial Γ-invariant function in $L^1(\mu)$.

Finally if $\theta = \mu \wedge \nu > 0$, i.e. μ and ν are not mutually singular, then again $d\theta/d\mu$ is a nontrivial Γ-invariant function in $L^1(\mu)$. □

We remark that in order to deduce the equivalence of the properties of being an ergodic measure and being an extreme point of the convex subset $M_\Gamma(X)$ one does not need any topology on the convex set $M_\Gamma(X)$ so that this equivalence holds for any measurable action of Γ on a Lebesgue space $(X, \mathcal{X}, \mu)$ (see e.g. [**206**] Chapter 10 and [**63**]).

2. Generic points

4.3. DEFINITION. Let (X, T) be a topological $\mathbb{Z}$-system, $\mu \in M_T(X)$. A point $x \in X$ is a *generic point for* μ (or μ*-generic*) if

$$\lim_{n\to\infty} \mathbb{A}_n(f)(x) = \lim_{n\to\infty} \frac{1}{n}\sum_{j=0}^{n-1} f(T^j x) = \int f\, d\mu$$

for every $f \in C(X)$. It is called *quasi-generic* if for some sequence $n_i \nearrow \infty$

$$\lim_{i\to\infty} \mathbb{A}_{n_i}(f)(x) = \int f\, d\mu$$

for every $f \in C(X)$. For $x \in X$ we denote the set of measures $\mu \in M_T(X)$ for which x is quasi-generic by $QG(x)$. Clearly $x \in X$ is a generic point for some measure $\mu \in M_T(X)$ iff $QG(x) = \{\mu\}$.

4.4. THEOREM. *Let (X,T) be a topological system, $\mu \in M_T^{\text{erg}}(X)$, then the set of μ-generic points is a Borel subset of X with μ measure 1.*

PROOF. Let $\{f_i\}$ be a dense sequence in $C(X)$. By the ergodic theorem the set of points $x \in X$ for which $\lim_{n\to\infty} \mathbb{A}_n(f_i)(x) = \int f_i\, d\mu$ for every i is a Borel set of measure 1. It is now easy to check that for such x, $\lim_{n\to\infty} \mathbb{A}_n(f)(x) = \int f\, d\mu$ holds for all $f \in C(X)$. □

4.5. EXERCISE. For every $x \in X$ the subset $QG(x) \subset M_T(X)$ is non-empty, closed and connected. [[The compactness of $M_T(X)$ implies that $QG(x) \neq \emptyset$. The connectedness follows by use of the following criterion: A compact metric space X is connected iff for every $x, y \in X$ and every $\epsilon > 0$ there exists a sequence $\{x_i : i = 1, 2, \ldots, n\}$ with $x_1 = x, x_n = y$ and for every $1 \le i < n-1$, $d(x_i, x_{i+1}) < \epsilon$.]]

4.6. PROPOSITION. *When (X,T) is minimal the set $\{x \in X : QG(x) \supset M_T^{\text{erg}}(X)\}$ is residual.*

PROOF. Given $\mu \in M_T(X)$ set

$$X(\mu) = \{x \in X : x \text{ is quasi-generic for } \mu\}.$$

If $\{f_j\}_{j=1}^{\infty}$ is a dense sequence in $C(X)$ then

$$X(\mu) = \bigcap_{k=1}^{\infty} \bigcap_{m=1}^{\infty} \bigcup_{n=1}^{\infty} \bigcap_{j=1}^{m} \{x \in X : \left| \mathbb{A}_n(f_j)(x) - \int f_j\, d\mu \right| < \frac{1}{k}\}$$

so that $X(\mu)$ is a G_δ subset of X. If μ has some quasi-generic point x then $x \in X(\mu)$ and clearly also every translate $T^t x \in X(\mu)$. Since by assumption (X,T) is minimal it follows that $X(\mu)$ is a dense G_δ subset whenever it is non-empty. Of course this is the case for every ergodic μ where we have $\mu(X(\mu)) = 1$ (Theorem 4.4).

Next choose a dense sequence $\{\mu_s\}_{s=1}^{\infty}$ in $M_T^{\text{erg}}(X)$ and let $X_0 = \bigcap_{s=1}^{\infty} X(\mu_s)$. We now see that for every x in the dense G_δ subset X_0, $QG(x) \supset \{\mu_s\}_{s=1}^{\infty}$. Since $QG(x)$ is closed it follows that $QG(x) \supset M_T^{\text{erg}}(X)$. □

3. Unique ergodicity

4.7. DEFINITION. A topological system (X, Γ) is called *uniquely ergodic* if there is on X a unique Γ-invariant probability measure. It is called *strictly ergodic* if it is uniquely ergodic and the unique invariant probability measure on X is full.

4.8. EXERCISES. 1. A topological system (X, Γ) is strictly ergodic iff it is uniquely ergodic and minimal.
2. Strict ergodicity is inherited by factors.
3. Every ergodic group system, hence every ergodic homogeneous system, is strictly ergodic. In particular an irrational rotation on the circle (or more generally on any torus; see Example 1.2) is strictly ergodic. [[This follows from the uniqueness of Haar measure on a compact group.]]

For $\mathbb{Z}$-systems we have the following characterizations.

4.9. THEOREM. *Let (X, T) be a topological system. The following conditions are equivalent.*
1. *(X, T) is uniquely ergodic.*
2. *$C(X) = \mathbb{R} + \bar{B}$, where $B = \{g - g \circ T : g \in C(X)\}$.*
3. *For every continuous function $f \in C(X)$ the sequence of functions*
$$\mathbb{A}_n f(x) = \frac{1}{n} \sum_{j=0}^{n-1} f(T^j x).$$
converges uniformly to a constant function.
4. *For every continuous function $f \in C(X)$ the sequence of functions $\mathbb{A}_n(f)$ converges pointwise to a constant function f^*.*
5. *For every function $f \in A$, for a collection $A \subset C(X)$ which linearly spans a uniformly dense subspace of $C(X)$, the sequence of functions $\mathbb{A}_n(f)$ converges pointwise to a constant function.*

PROOF. Suppose (X, T) is uniquely ergodic with the unique T-invariant probability measure ν and assume that $\mathbb{R} + \bar{B}$ is a proper subspace of $C(X)$. Define a bounded linear functional ϕ on the closed subspace $\mathbb{R} + \bar{B}$ by requiring $\phi(\mathbf{1}) = 1$ and $\phi(f) = 0$ for every $f \in B$. The Hahn-Banach theorem, [**53**], provides an extension $\psi : C(X) \to \mathbb{R}$ such that ψ is a linear functional with norm 1 and such that $\psi(g) \neq \int_X g\, d\nu$ for some function $g \in C(X) \setminus (\mathbb{R} + \bar{B})$. Riesz' representation theorem implies that $\psi(f) = \int_X f\, d\mu$, where μ is a finite signed measure. Let $\mu = \mu^+ - \mu^-$ with μ^+ and μ^- mutually singular positive measures, be the Jordan decomposition of μ, [**112**]. We have $0 = \phi \circ (\mathrm{Id} - T) = \psi \circ (\mathrm{Id} - T)$ and it follows that μ is T-invariant. Now
$$\begin{aligned}\mu &= \mu^+ - \mu^- \\ &= \mu^+ \circ T^{-1} - \mu^- \circ T^{-1}\end{aligned}$$
and the uniqueness of the Jordan decomposition implies that both μ^+ and μ^- are T-invariant. Our assumption of unique ergodicity implies that μ^+ and μ^- are constant multiples of the unique invariant probability measure ν on X and it follows that $\mu = \mu^+ = \nu$. This contradicts the fact that, by construction, $\psi(g) \neq \int_X g\, d\nu$ and we proved that $1 \Rightarrow 2$. A moment's reflection will show that $1 \Leftrightarrow 2$.

Clearly $\mathbb{A}_n(f) = 0$ for every $f \in B$ and by approximation we have the same property for every f in the uniform closure $\bar{B}$. For a constant function f we have $\mathbb{A}_n(f) = f$. It therefore follows that $2 \Rightarrow 3$. The implications $3 \Rightarrow 4 \Rightarrow 5$ are clear and by approximation it is easy to see that $5 \Rightarrow 4$. We shall finish the proof by showing that $4 \Rightarrow 1$.

Assuming 4 define the functional $\phi : C(X) \to \mathbb{R}$ by $\phi(f) = f^*$. It is easy to check that ϕ is a bounded linear positive functional of norm 1, and by Riesz' representation theorem $\phi(f) = \int_X f\, d\mu$ with μ a probability Borel measure on X. Since clearly $\phi(f \circ T) = \phi(f)$ for every $f \in C(X)$ we have $\mu \in M_T(X)$. If now $\nu \in M_T(X)$ then for every $f \in C(X)$, by Lebesgue's dominated convergence theorem,

$$\int_X f\, d\nu = \int_X \mathbb{A}_n(f)\, d\nu = \phi(f) = \int_X f\, d\mu.$$

Thus $\nu = \mu$ and the system (X, T) is uniquely ergodic. □

Of course the constant to which the sequence $\mathbb{A}_n f(x) = \frac{1}{n}\sum_{j=0}^{n-1} f(T^j x)$ converges in condition 3 of Theorem 4.9 is $\int f d\mu$ where μ is the unique element of $M_T(X)$. In other words, condition 4 asserts that every point of X is a generic point for μ. Can every point of, say, a topologically transitive, system be generic, yet the system not being uniquely ergodic? The surprising answer is yes. This can be seen in Katznelson-Weiss [**147**] as well as in [**133**]. In the latter del-Junco and Keane show that this property holds for the square system $(X \times X, T \times T)$ where (X, T) is the Chacón system. Since (X, T) is weakly mixing the system $(X \times X, T \times T)$ is topologically transitive. However, since $\mu \times \mu$ as well as the measure $\mathrm{gr}\,(\mu, \mathrm{id})$ — the image of μ under the map $x \mapsto (x, x)$ from X to $X \times X$ — are distinct elements of $M_T(X \times X)$ for any $\mu \in M_T(X)$ (in fact the Chacón system (X, T) is strictly ergodic and $M_T(X) = \{\mu\}$), it follows that the square system of a non-trivial system is never uniquely ergodic. However we do have the following theorem.

4.10. THEOREM. *If (X, T) is a minimal $\mathbb{Z}$ dynamical system then the condition "every point of X is a generic point for some measure in $M_T(X)$" is equivalent to unique ergodicity.*

PROOF. Condition 3 in Theorem 4.9 shows that our condition is necessary. Conversely, suppose it holds. By Proposition 4.6 the set $\{x \in X : QG(x) \supset M_T^{\mathrm{erg}}(X)\}$ is residual. Let x be any point in this set. By assumption x is generic for some measure $\mu \in M_T(X)$ and since for a generic point the set $QG(x)$ (of measures for which x is quasi-generic) is the singleton $\{\mu\}$, we now have $\{\mu\} = QG(x) \supset M_T^{\mathrm{erg}}(X)$. Thus $M_T(X) = M_T^{\mathrm{erg}}(X) = \{\mu\}$. [[Here is another proof: Our assumption that every point is generic implies that for $f \in C(X)$ the function $f^*(x) = \lim_{n\to\infty} \mathbb{A}_n(f)(x)$ is everywhere defined. Being the pointwise limit of a sequence of continuous functions, f^* has a residual set of points of continuity. Applying Exercise 1.1.8 we conclude that $f^*(x)$ is a constant and unique ergodicity follows from Theorem 4.9.]] □

4. Examples of strictly ergodic systems

As we shall see next, many of the examples of minimal topological systems studied in Chapter 1 are in fact strictly ergodic systems.

4.11. THEOREM. *The following $\mathbb{Z}$-systems are strictly ergodic:*

1. *The skew-product system on $X = \mathbb{T}^2$ given by $T_\phi(x,y) = (x+\alpha, y+\phi(x))$, with α irrational and $\phi : \mathbb{T} \to \mathbb{T}$ satisfying Lipschitz condition and having a non-vanishing degree.*
2. *Every minimal nil-system; i.e. a $\mathbb{Z}$-system of the form $(\mathbb{G}/H, \mu, T)$ where $\mathbb{G}$ is a connected nilpotent Lie group, $H \subset \mathbb{G}$ a discrete co-compact subgroup, μ is the unique (Haar) $\mathbb{G}$-invariant probability measure on $\mathbb{G}/H$ and $T : \mathbb{G}/H \to \mathbb{G}/H$ is defined by $T(gH) = agH$ with a an element of $\mathbb{G}$.*
3. *There exists a minimal distal non-uniquely ergodic system on the torus $\mathbb{T}^2$.*
4. *All horocycle systems; i.e. systems of the form (X, ω, T) where $X = \mathbb{G}/\Sigma$ with $\mathbb{G} = SL(2,\mathbb{R})$ and $\Sigma \subset \mathbb{G}$ a discrete co-compact subgroup, ω is the $\mathbb{G}$-invariant probability measure induced by Haar measure on $\mathbb{G}$, and $T : \mathbb{G}/\Sigma \to \mathbb{G}/\Sigma$ is defined by $T(g\Sigma) = h_s g\Sigma$ with h_s an element of $\mathbb{G}$ of the form*

$$h_s = \begin{pmatrix} 1 & s \\ 0 & 1 \end{pmatrix}$$

   for some $0 \neq s \in \mathbb{R}$.
5. *All systems arising from primitive substitutions; e.g. the Morse and the Chacón systems.*
6. *Every minimal almost 1-1 extension of a Kronecker system $(X,T) \xrightarrow{\pi} (Y,T)$ with $\lambda\left(\{y \in Y : \operatorname{card} \pi^{-1}(y) > 1\}\right) = 0$, where λ is Haar measure on the compact group Y.*
7. *Some Toeplitz systems are strictly ergodic and some are not.*

For proofs refer to the following works. For claim 1 see Furstenberg [**72**], Theorem 2.1. For claim 2 see L. Green's original proof in the important book "Flows on homogeneous spaces" [**12**] by L. Auslander L. Green and F. Hahn, and W. Parry's subsequent simplification of the proof in [**202**]. In the next section we shall follow Parry's proof for the case of Heisenberg nil-systems. For 3 see Furstenberg [**72**]. For claim 4 see Section 6 below and Furstenberg's original proof in [**75**]. For some generalizations see the papers [**178**] by B. Marcus and [**253**] by W. A. Veech. For claim 5 see M. Queffélec [**208**] or B. Host [**124**]. Claim 6 is clear and finally S. Williams [**266**] is a good reference for claim 7.

As an application of the notion of unique ergodicity let us consider next Furstenberg's proof of the fact that for an irrational number α the sequence $\{n^2\alpha\}_{n=0}^{\infty}$ is *uniformly distributed* modulo 1. This means that for every continuous real valued function f on $\mathbb{T}$ we have

$$\lim_{N\to\infty} \frac{1}{N} \sum_{n=0}^{N-1} f(n^2\alpha) = \int f(t)\, d\lambda(t),$$

where λ is Lebesgue measure on $\mathbb{T}$. In fact the proof will show that the sequence is *well distributed* (which means that for every $f \in C(\mathbb{T})$

$$\lim_{N\to\infty} \frac{1}{N} \sum_{n=0}^{N-1} f(x+n^2\alpha) = \int f(t)\,d\lambda(t),$$

uniformly in x). This is a special case of the following classical theorem of Weyl [**265**].

4.12. THEOREM (Weyl). *If $p(t)$ is a real polynomial with at least one coefficient other than the constant term irrational, then the sequence $\{p(n)\}_{n=1}^{\infty}$ is well distributed.*

We refer to Furstenberg's book [**77**] for a "dynamical" proof of this theorem. Here we shall be content with proving the following theorem on the unique ergodicity of group extensions and then sketching a dynamical proof, based on that theorem, of the well distribution of the sequence $n^2\alpha$ in an exercise.

4.13. THEOREM. *Let (Y, Γ) be a uniquely ergodic system with unique invariant measure ν. Let $\pi : (X, \Gamma) \to (Y, \Gamma)$ be a topological group extension; i.e. for a compact group G of automorphisms of (X, Γ), $\pi(x) = \pi(x')$ iff $x' = gx$ for some $g \in G$. Let m denote Haar measure on the compact group G and suppose that the "Haar lift" of ν, i.e. the Γ-invariant probability measure $\mu = \nu * m$ given for $f \in C(X)$ by*

$$\int_X f\,d\mu = \int_Y \int_G f(gx)\,dm(g)\,d\nu,$$

is Γ-ergodic. Then (X, Γ) is uniquely ergodic.

PROOF. By the unique ergodicity of (Y, Γ) we have $\pi(\lambda) = \nu$ for every $\lambda \in M_\Gamma(X)$. Now apply Theorem 3.30 to conclude the proof. □

4.14. REMARK. Here is Furstenberg's original proof ([**72**]), which treated the case of $\mathbb{Z}$ actions. It demonstrates the use of generic points. For simplicity we assume $X = Y \times G$.

We make the following observations: (i) If (y, g') is a generic point for a T-invariant measure $\theta \in M_T(X)$ then for every $g \in G$ the point $R_g(y, g')$ is generic for the T-invariant measure $R_g\theta$. (ii) Since for $\mu = \nu \times m$ we have $R_g\mu = \mu$ for every $g \in G$ it follows that the set of μ-generic points has the form $A \times G$ for $A \subset Y$ with $\nu(A) = 1$. (iii) For every measure $\theta \in M_T^{\mathrm{erg}}(X)$ its projection on Y, $\pi(\theta)$, is the unique T-invariant measure ν on Y and therefore $\theta(A \times G) = 1$. Thus if $X_\theta \subset X$ is the set of θ-generic points then $\theta(X_\theta) = 1$ and it follows that $X_\theta \cap (A \times G) \neq \emptyset$. This finishes the proof since every point in the latter intersection is simultaneously generic for both θ and μ so that $\theta = \mu$.

4.15. EXERCISES. 1. For an irrational number α define the homeomorphism $T : \mathbb{T}^2 \to \mathbb{T}^2$ by $T(x, y) = (x + \alpha, y + x)$ and let $\mu = \lambda \times \lambda$ with λ Lebesgue measure on $\mathbb{T}$. Show that the system $\mathbf{X} = (X, \mathcal{X}, \mu, T)$ (with $X = \mathbb{T}^2$) is strictly ergodic. [[Use the Fourier expansion method to show that a T-invariant function in $L^2(\mu)$ is necessarily a constant; i.e. that the system $\mathbf{X} = (X, \mathcal{X}, \mu, T)$ is ergodic. Next use Theorem 4.13 to deduce that

(X,T) is uniquely ergodic. By Example 1.6 the system (X,T) is minimal and thus (X,T) is strictly ergodic.]]

2. For an irrational number α the sequence $\{n^2\alpha\}_{n=0}^{\infty}$ is well distributed modulo 1. [[Define $S : \mathbb{T}^2 \to \mathbb{T}^2$ by $S(x,y) = (x+\alpha, y+2x+\alpha)$ and repeat the argument above to deduce that $(\mathbb{T}^2, S)$ is strictly ergodic. Now use the fact that $S^n(0,y) = (n\alpha, y+n^2\alpha)$ (Example 1.6) and apply Theorem 4.9 to the continuous functions $F : \mathbb{T}^2 \to \mathbb{R}$ of the form $F(x,y) = f(y)$ with $f \in C(\mathbb{T})$ to deduce that

$$\lim_{N\to\infty} \frac{1}{N} \sum_{n=0}^{N-1} f(y+n^2\alpha) = \int f(t)\, d\lambda(t),$$

uniformly in y.]]

4.16. Example. There exists a non-metrizable minimal dynamical system with no generic points. Moreover, for the universal point transitive dynamical system $(\beta\mathbb{Z}, T)$ (where $\beta\mathbb{Z}$ is the Stone-Čech compactification of $\mathbb{Z}$ and T is the homeomorphism induced by addition of $1 \in \mathbb{Z}$) no point is generic. We see this as follows:

1. Let (X,T) be a minimal distal non-uniquely ergodic system as in Theorem 4.11.3, then the set $X_0 \subset X$ of points which are not generic (for any invariant measure in $M_T(X)$) is non-empty. [[Follows from Theorem 4.10.]]
2. Let $Y \subset X^X$ be the orbit closure of the point $\xi_0 \in X^X$ given by $\xi_0(x) = x$. Show that Y is a minimal, distal, non-metrizable system and that $E(Y,T)$, the enveloping group of the system (Y,T), is isomorphic to $E(X,T)$ via the map $p \mapsto \hat{p}$ where $(\hat{p}(\xi))(x) = px$.
3. For every point $\xi \in Y$ its range $\{\xi(x) : x \in X\}$ is all of X. [[This is clearly the case for ξ_0 and since every $\xi \in Y$ is of the form $\xi = \hat{p}\xi_0$ it follows that range $\xi = pX = X$.]]
4. No point of Y is generic. [[If $\xi \in Y$ is generic for some $\mu \in M_T(Y)$ then, since range $\xi = X$, there exists $x \in X$ with $x_0 = \xi(x) \in X_0$. The projection map $\pi_x : Y \to X$, the restriction to Y of the projection of X^X on the x-coordinate, is a homomorphism and hence $\pi_x(\xi) = x_0$ should be a generic point of X, but being in X_0 it is not.]]
5. The dynamical system $(\beta\mathbb{Z}, T)$ has no generic points. [[The system $(\beta\mathbb{Z}, T)$ being a universal point-transitive system, there is an epimorphism $\phi : (\beta\mathbb{Z}, T) \to (Y,T)$. If $p \in \beta\mathbb{Z}$ is a generic point so must be $\phi(p)$, but we have shown that no point of (Y,T) is generic.]]

5. Minimal Heisenberg nil-systems are strictly ergodic

We have seen in Examples 1.6 that the Heisenberg system $(X,T) = (\mathbb{G}/H, T)$, where

$$\mathbb{G} = \left\{ \begin{pmatrix} 1 & x & z \\ 0 & 1 & y \\ 0 & 0 & 1 \end{pmatrix} : x,y,z \in \mathbb{R} \right\}, \qquad H = \left\{ \begin{pmatrix} 1 & a & c \\ 0 & 1 & b \\ 0 & 0 & 1 \end{pmatrix} : a,b,c \in \mathbb{Z} \right\},$$

and

$$T = \begin{pmatrix} 1 & \alpha & \gamma \\ 0 & 1 & \beta \\ 0 & 0 & 1 \end{pmatrix},$$

with $\alpha, \beta, \gamma \in \mathbb{R}$, α and β independent over $\mathbb{Q}$, is minimal distal and non equicontinuous. In this section we consider this nil-system as a measure preserving $\mathbb{Z}$-system, with μ the unique (Haar) $\mathbb{G}$-invariant probability measure on $\mathbb{G}/H$. Let

$$Z = \left\{ \begin{pmatrix} 1 & 0 & z \\ 0 & 1 & 0 \\ 0 & 0 & 1 \end{pmatrix} : z \in \mathbb{R} \right\},$$

the center as well as the commutator $[\mathbb{G}, \mathbb{G}]$ of $\mathbb{G}$. We have the natural projection $\pi : X = \mathbb{G}/H \to Y = \mathbb{G}/ZH$ and we observe that the factor system $\mathbf{Y}$ is isomorphic to the irrational rotation $T_{(\alpha,\beta)}$ on the two torus $\mathbb{T}^2$ with Lebesgue measure ν. As in the topological setup we say, given a $\mathbb{Z}$-system $\mathbf{X}$, that a measurable non-zero function $f : X \to \mathbb{C}$, is an *eigenfunction* with eigenvalue λ if a.e. $f(Tx) = \lambda f(x)$. Clearly such an eigenvalue has modulus one, $|\lambda| = 1$. Replacing f by $\tilde{f}$, where $\tilde{f} = f/|f|$ when $f(x) \neq 0$ and 0 otherwise, we can assume that $|f|$ is $\{0, 1\}$-valued.

4.17. THEOREM. 1. *The dynamical system $\mathbf{X} = (X, \mathfrak{X}, \mu, T)$ is ergodic.*
2. *Every eigenfunction of $\mathbf{X}$ is a lift of an eigenfunction on $\mathbf{Y}$.*
3. *The topological $\mathbb{Z}$-system (X, T) is strictly ergodic.*

PROOF. We split the proof into several steps.

1. Let $K = Z/Z \cap H$, then K is a group which is isomorphic to the circle group $\mathbb{R}/\mathbb{Z}$. Since Z is the center of $\mathbb{G}$ we see that K acts on $X = \mathbb{G}/H$ as a compact group of homeomorphisms commuting with the natural $\mathbb{G}$ action on X. In particular K is a compact group of automorphisms of the measure preserving $\mathbb{Z}$-system $\mathbf{X}$, and it follows that the homomorphism $\pi : \mathbf{X} \to \mathbf{Y}$ is the group extension which corresponds to $K \subset \mathrm{Aut}\,(\mathbf{X})$.

2. The action of K on $\mathbf{X}$ defines a unitary representation of K on the Hilbert space $\mathfrak{H} = L^2(\mathbb{G}/H, \mu)$. Under this unitary representation $\mathfrak{H}$ decomposes as a direct sum of eigenspaces

$$\mathfrak{H} = \bigoplus_{n \in \mathbb{Z}} V_n,$$

where

$$V_n = \{f \in \mathfrak{H} : f(zx) = e^{2\pi i n z} f(x) \text{ for all } z = \left(\begin{smallmatrix} 1 & 0 & z \\ 0 & 1 & 0 \\ 0 & 0 & 1 \end{smallmatrix}\right) \in Z\},$$

(see e.g. [**120**], Theorem 27.44, page 29). It is easy to check that each V_n is U_T-invariant.

Let now $f : X \to \mathbb{C}$, with $|f| = 0$ or 1, be an eigenfunction for T with eigenvalue λ; i.e. $f(Tx) = \lambda f(x)$, μ a.e. Then in the decomposition $f = \sum_{n \in \mathbb{Z}} f_n$ with $f_n \in V_n$ we have

$$f(Tx) = \lambda f(x) = \sum_{n \in \mathbb{Z}} f_n(Tx) = \sum_{n \in \mathbb{Z}} \lambda f_n(x),$$

and we conclude that a.e. $f_n(Tx) = \lambda f_n(x)$ for each $n \in \mathbb{Z}$.

3. Suppose $f_n \neq 0$ for some $n \neq 0$ and for convenience assume $|f_n| = 1$ or 0. Fix $g \in \mathbb{G}$ and observe that $gT = Tg\zeta(g)$ for $\zeta(g) \in Z$. Hence, a.e.

$$f_n(gTx) = f_n(Tg\zeta(g)x) = e^{2\pi i n \zeta(g)} f_n(Tgx) = e^{2\pi i n \zeta(g)} \lambda f_n(gx)$$

and therefore the function $h_g(x) = f_n(gx)\overline{f_n(x)}$ is an eigenfunction for T with eigenvalue $e^{2\pi i n\zeta(g)}$. Now h_g is clearly Z-invariant and it follows that it is a lift of an eigenfunction of the transformation $T_{(\alpha,\beta)}$ on $\mathbb{T}^2$. In particular $|\int_X h_g(x)\, d\mu(x)|$ is either 1 or 0, depending on whether h_g is a constant function or not. Let $t \mapsto g(t)$, $0 \le t \le 1$ be a continuous path in $\mathbb{G}$ with $g(0) = I$ and $g(1) = g$, and denote $h_t = h_{g(t)}$. Then, as $h_0 \equiv 1$, we see that the continuous $\{0,1\}$-valued function $|\int_X h_t(x)\, d\mu(x)|$ is the constant 1. In particular $|\int_X h_g(x)\, d\mu(x)| = 1$ and we conclude that $\theta(g) = h_g(x) = f_n(gx)\overline{f_n(x)}$ is a constant, independent of x, of modulus 1. Clearly the map $\theta : \mathbb{G} \to S^1$ is a continuous homomorphism (i.e. a character) which coincides with $z \mapsto e^{2\pi i n z}$ on Z. Since $Z = [\mathbb{G}, \mathbb{G}]$ it follows that $\ker\theta \supset Z$. In particular $\theta(z) = e^{2\pi i n z} \equiv 1$; i.e. $n = 0$. This contradiction (we assumed $n \neq 0$) implies that $f = f_0 \in V_0$. Thus every T-eigenfunction is Z-invariant, hence is a lift of a $T_{(\alpha,\beta)}$-eigenfunction and the proof of claim 2 is complete.

4. In particular then, every T-invariant function $f \in L^2(X,\mu)$ is a lift of a $T_{(\alpha,\beta)}$-invariant function in $L^2(\mathbf{Y})$. As the $T_{(\alpha,\beta)}$ action on $\mathbf{Y}$ is ergodic we conclude that every T-invariant function is a constant; i.e. $\mathbf{X}$ is ergodic and claim 1 is proved.

5. As the homomorphism $\pi : \mathbf{X} \to \mathbf{Y}$ is a group extension, an application of Theorem 3.30 shows that μ is the unique measure in $M_T(X)$ which projects onto ν. However, as $(Y, T_{(\alpha,\beta)})$ is uniquely ergodic, every measure in $M_T(X)$ projects to ν and we see that also (X,T) is uniquely ergodic. Since it is also minimal (Example 1.6) we finally get that (X,T) is strictly ergodic and the proof of claim 3 is complete. □

6. The geodesic and horocycle flows

Let $\mathbb{G} = SL(2,\mathbb{R})$ and $\Sigma \subset \mathbb{G}$ a *uniform lattice*; that is, Σ is a discrete subgroup with a compact quotient $X = G/\Sigma$. Our goal in this section is to study the *geodesic* and *horocycle* flows on the compact manifold X. Bekka and Mayer's book [**15**] is an accessible elementary introduction to the subject.

We shall use the following notation and easily verified facts about the simple Lie group $\mathbb{G}$. The letters K, A, N, N^-, P and Q will denote the closed subgroups

$$K = \left\{k_\theta = \begin{pmatrix} \cos\theta & -\sin\theta \\ \sin\theta & \cos\theta \end{pmatrix} : \theta \in \mathbb{T}\right\}, \quad A = \left\{g_a = \begin{pmatrix} a & 0 \\ 0 & a^{-1} \end{pmatrix} : a > 0\right\},$$

$$N = \left\{h_s = \begin{pmatrix} 1 & s \\ 0 & 1 \end{pmatrix} : s \in \mathbb{R}\right\}, \quad N^- = \left\{\begin{pmatrix} 1 & 0 \\ s & 1 \end{pmatrix} : s \in \mathbb{R}\right\}$$

and $P = AN, Q = N^-A$.

We have the *decompositions* of *Iwasawa* $\mathbb{G} = KAN = KP$, *Cartan* $\mathbb{G} = KAK$ and *Bruhat*

$$\mathbb{G} = N^-AN \cup \iota AN = QN \cup \iota P, \qquad \iota = \begin{pmatrix} 0 & 1 \\ -1 & 0 \end{pmatrix},$$

with explicit formulas

$$\begin{pmatrix} a & b \\ c & d \end{pmatrix} = \begin{pmatrix} 1 & 0 \\ c/a & 1 \end{pmatrix} \begin{pmatrix} a & 0 \\ 0 & a^{-1} \end{pmatrix} \begin{pmatrix} 1 & b/a \\ 0 & 1 \end{pmatrix}, \quad \text{if } a \neq 0$$

$$\begin{pmatrix} a & b \\ c & d \end{pmatrix} = \begin{pmatrix} 0 & -1 \\ 1 & 0 \end{pmatrix} \begin{pmatrix} c & 0 \\ 0 & c \end{pmatrix} \begin{pmatrix} 1 & d/c \\ 0 & 1 \end{pmatrix}, \quad \text{if } a = 0$$

for the Bruhat decomposition. It is also easy to verify that $\mathbb{G}$ is the smallest closed group containing N and N^-.

Let λ denote the left Haar measure on $\mathbb{G}$ normalized in such a way that $\lambda(D) = 1$ for a Σ fundamental domain $D \subset \mathbb{G}$ (with $\bar{D}$ compact). We let ω be the corresponding $\mathbb{G}$-invariant measure on $X = G/\Sigma$, which can be identified with D. In the sequel we shall refer to an $\mathbb{R}$-system (or an $\mathbb{R}^*$-system, with $\mathbb{R}^* = \{t \in \mathbb{R} : t > 0\}$, the multiplicative group) as a *flow*. The classical examples of the *geodesic* and *horocycle* flows are the $\mathbb{R}$-systems $(X, \omega, A) = (X, \omega, \{g_t\}_{t>0})$ and $(X, \omega, N) = (X, \omega, \{h_s\}_{s\in\mathbb{R}})$, given by

$$g_t(g\Sigma) = g_t g\Sigma \quad \text{and} \quad h_s(g\Sigma) = h_s g\Sigma \quad (g \in \mathbb{G},\ s \in \mathbb{R},\ t \in R^*),$$

respectively.

The following lemma will play a crucial role in the proofs of the two main results of this section.

4.18. LEMMA. *Let $\{a_n\}$ be a sequence of positive real numbers with $a_n \to 0$.*

1. *For any matrix $g = \left(\begin{smallmatrix} a & s \\ 0 & a^{-1} \end{smallmatrix}\right) \in P$ we have $g_{a_n} g g_{a_n}^{-1} \to \left(\begin{smallmatrix} a & 0 \\ 0 & a^{-1} \end{smallmatrix}\right)$.*
2. *In particular for every $s \in \mathbb{R}$, $g_{a_n} h_s g_{a_n}^{-1} \to e$, where $e = \left(\begin{smallmatrix} 1 & 0 \\ 0 & 1 \end{smallmatrix}\right)$ is the identity element of $\mathbb{G}$.*
3. *When $a_n \to \infty$ we get $g_{a_n} g g_{a_n}^{-1} \to \left(\begin{smallmatrix} a & 0 \\ 0 & a^{-1} \end{smallmatrix}\right)$ for every matrix $g = \left(\begin{smallmatrix} a & 0 \\ s & a^{-1} \end{smallmatrix}\right) \in Q$.*

PROOF.

$$\begin{pmatrix} a_n & 0 \\ 0 & a_n^{-1} \end{pmatrix} \begin{pmatrix} a & s \\ 0 & a^{-1} \end{pmatrix} \begin{pmatrix} a_n^{-1} & 0 \\ 0 & a_n \end{pmatrix} = \begin{pmatrix} a & sa_n^2 \\ 0 & a^{-1} \end{pmatrix}.$$

□

4.19. COROLLARY (Mautner Lemma). *Let $\pi : \mathbb{G} \to \mathfrak{U}(\mathfrak{H})$ be a (strongly continuous) unitary representation; then an A-invariant vector v is also $\mathbb{G}$-invariant.*

PROOF. Fix $0 < a < 1$ then for $s \in \mathbb{R}$

$$\begin{aligned} \|\pi(h_s)v - v\| &= \|\pi(g_a^n)\pi(h_t)v - \pi(g_a^n)v\| \\ &= \|\pi(g_a^n)\pi(h_s)\pi(g_a^{-n})v - v\| \to 0. \end{aligned}$$

Thus $h_s v = v (s \in \mathbb{R})$.

Taking adjoints in Lemma 4.18, we see that $g_a^{-n} \left(\begin{smallmatrix} 1 & 0 \\ s & 1 \end{smallmatrix}\right) g_a^n \to e$ and as above we conclude that $\pi(g)v = v$ for every $g \in N \cup N^-$. Since $N \cup N^-$ topologically generates $\mathbb{G}$ it follows that $\pi(g)v = v$ for every $g \in \mathbb{G}$. □

4.20. THEOREM. *The geodesic flow $(X, \mathcal{X}, \omega, A)$ is mixing.*

PROOF. Let $\mathfrak{H} = L^2(X, \omega)$ and $\mathfrak{C} \subset \mathfrak{H}$ the one dimensional subspace of constant functions. Let $\pi : \mathbb{G} \to \mathfrak{U}(\mathfrak{H})$ be the associated (strongly continuous unitary) Koopman representation. Denoting by $\mathbf{E} = \mathbf{E}(\mathbf{X})$ the closure of $\pi(A) = \{\pi(a) : a \in A\}$ in $\mathfrak{U}(\mathfrak{H})$ with respect to the weak operator topology, what we have to show is that $\mathbf{E} = \pi(A) \cup \{P_{\mathfrak{C}}\}$, where $P_{\mathfrak{C}}$ is the projection on $\mathfrak{C}$; i.e. $P_{\mathfrak{C}}(f) = \int f\, d\omega$ (see Theorem 3.14).

We first observe that $(X, \mathfrak{X}, \omega, A)$ is ergodic. In fact if $f \in \mathfrak{H}$ is A-invariant then by Corollary 4.19, f is $\mathbb{G}$-invariant and as $X = \mathbb{G}/\Sigma$ is a homogeneous space any $\mathbb{G}$-invariant function is in $\mathfrak{C}$.

Next recall that in the compact semigroup $\mathbf{E}$ multiplication on both left and right by a fixed element is continuous. It is also easy to check that in $\mathbf{E}$, if $U_n \to U$ and $P_n \to P$ with U_n, U *unitary* then also $U_n P_n \to UP$ and $P_n U_n \to PU$. Now any element $p \in \mathbf{E} \setminus \pi(A)$ is a limit of a sequence of the form $\pi(g_{a_n})$ for $g_{a_n} \in A$ with $|\log a_n| \to \infty$. By Lemma 4.18, a_n will satisfy $g_{a_n}^{-1} h g_{a_n} \to e$ either for every $h \in N$ or for every $h \in N^-$. We show that when $a_n \to 0$, $\pi(h^-)p = p$ for every $h^- \in N^-$. In fact,

$$\pi(h^-)p = \lim_{n\to\infty} \pi(h^-)\pi(g_{a_n}) = \lim_{n\to\infty} \pi(g_{a_n})(\pi(g_{a_n})^{-1}\pi(h^-)\pi(g_{a_n})) = p\pi(e) = p.$$

In the other case; i.e. when $a_n \to \infty$, we have similarly $\pi(h^+)p = p$ for every $h^+ \in N$. In particular note that p is non-invertible, for otherwise, e.g. $\pi(N)p = p$ implies $\pi(N) = \pi(N)pp^{-1} = \{I\}$, which is clearly false.

Now set $q = p^*p$, then, as $\mathbf{E}$ is a semigroup, $q = q^* \in \mathbf{E}$. Since clearly q as well is non-invertible it follows that also $q = \lim \pi(g_{a_n'})$ with $|\log a_n'| \to \infty$. Since $q^* = q$ we have $q = \lim \pi(g_{a_n'})^* = \lim \pi(g_{a_n'}^{-1})$ and it follows that $\pi(N)q = \pi(N^-)q = q$.

Again, since $N \cup N^-$ topologically generates $\mathbb{G}$, it follows that $\pi(g)q = q$ for every $g \in \mathbb{G}$. Now for every $f \in L^2(X, \omega)$, qf is a $\mathbb{G}$-invariant vector, and by ergodicity $qf \in \mathfrak{C}$. Finally, for every $f \in L_0^2(X, \omega)$, $0 = \langle qf, f\rangle = \langle pf, pf\rangle = \|pf\|^2$ and it follows that $p = P_{\mathfrak{C}}$. □

4.21. EXERCISES. 1. Use Cartan's decomposition to show that, in fact, $\lim \pi(g_n) = P_{\mathfrak{C}}$ for any sequence $g_n \to \infty$ in $\mathbb{G}$ (i.e. for every compact $F \subset \mathbb{G}$ eventually $g_n \notin F$).

2. Deduce the fact that the horocycle flow (X, ω, N) is mixing.
3. Observe that the proof of Theorem 4.20 and part 1 of the present exercise are valid for any Koopman representation of $\mathbb{G}$ corresponding to an ergodic measure preserving $\mathbb{G}$-system.

Our next goal is to prove the strict ergodicity of the horocycle flow. We first need to develop few technical tools. Let λ denote the usual Lebesgue measure on $\mathbb{R}$. Identifying $\mathbb{R}$ with N via the map $s \mapsto h_s$ we shall also write λ for the Haar measure on N. Fix a number $a > 1$ and set $\phi = g_a = \begin{pmatrix} a & 0 \\ 0 & a^{-1} \end{pmatrix}$. Given $M > 0$ define the operator

$$L = L_M : C(X) \to C(X), \qquad (Lf)(x) = \frac{1}{2M}\int_{-M}^{M} f(h_s x)\, d\lambda(s).$$

Clearly L is a bounded linear operator on the Banach space $C(X)$ with norm 1 and it is easy to check that it is also a bounded operator of norm 1 when we consider $C(X)$ as a subspace of the Hilbert space $L^2(X, \omega)$. It therefore follows that there is a unique extension (also denoted by L) $L : L^2(X, \omega) \to L^2(X, \omega)$. Finally let

$$U : L^2(X, \omega) \to L^2(X, \omega)$$

be the Koopman operator corresponding to the action of $\phi = g_a$; i.e.

$$(Uf)(x) = f(\phi x).$$

4.22. LEMMA. *For each $f \in C(X)$ the family of functions $\{LU^n f : n \in \mathbb{N}\}$ is equicontinuous.*

PROOF. Given $g = \left(\begin{smallmatrix} x & y \\ z & w \end{smallmatrix}\right) \in \mathbb{G}$ and $s \in \mathbb{R}$ let

$$h_s g h_{-s} = \begin{pmatrix} x+sz & 0 \\ z & (x+sz)^{-1} \end{pmatrix} \begin{pmatrix} 1 & r \\ 0 & 1 \end{pmatrix} = \alpha(s,g)\cdot\beta(s,g) \in QN,$$

be the Bruhat decomposition of $h_s g h_{-s}$. Explicitly

$$\alpha(s,g) = \begin{pmatrix} x+sz & 0 \\ z & (x+sz)^{-1} \end{pmatrix} \quad \text{and} \quad \beta(s,g) = \begin{pmatrix} 1 & r \\ 0 & 1 \end{pmatrix},$$

with $r = \frac{y+s(w-x)-s^2z}{x+sz}$. Since $r+s = \frac{y+sw}{x+sz}$ we also have

$$\psi(h_s,g) := \beta(s,g)h_s = \begin{pmatrix} 1 & \frac{y+sw}{x+sz} \\ 0 & 1 \end{pmatrix} = h_{\frac{y+sw}{x+sz}}.$$

We fix a neighborhood O of the identity e in $\mathbb{G}$ such that $\psi : J \times O \to N$ is well defined and such that, for each $g \in O$ the map $\psi_g : J \to N$ defined by $\psi_g(h_s) := \psi(h_s, g) = h_{\frac{y+sw}{x+sz}}$, is a diffeomorphism. Note that $\psi_e : J \to N$ is the inclusion map. Set $\nu_g = \psi_*(\lambda \restriction J) = \lambda \circ \psi_g^{-1}$; we then have

$$\lim_{g \to e} \lambda(J \,\triangle\, \psi_g(J)) = 0,$$

and, uniformly on J,

$$\lim_{g \to e} |1 - d\nu_g/d\lambda| = 0.$$

What we want to show is that, given $f \in C(X)$ with $\|f\|_\infty \le 1$, $x \in X$ and $\epsilon > 0$, there exists a neighborhood O_1 of e in $\mathbb{G}$ such that for every $g \in O_1$ and every $n \in \mathbb{N}$

$$|LU^n f(x) - LU^n f(gx)| = \frac{1}{2M} \left| \int_J f(\phi^n h_s x)\, d\lambda(s) - \int_J f(\phi^n h_s g x)\, d\lambda(s) \right| < \epsilon.$$

Our choice of O_1 will be subjected to the following conditions:

(i) $g \mapsto gx$ is a diffeomorphism of O_1 onto an open neighborhood of x in X,
(ii) $O_1 \subset O$,
(iii) $g' = \phi^n \alpha(s,g)\phi^{-n}$ satisfies the condition $|f(g'x) - f(x)| < \epsilon/2$ for every $s \in J$, $n \in \mathbb{N}$ and $g \in O_1$,
(iv) $\lambda(J \,\triangle\, \psi_g(J)) \cdot \sup_J (1 + |\frac{d\nu_g}{d\lambda}|) < \epsilon/4$,
(v) $|1 - d\nu_g/d\lambda| < \epsilon/4$ for every $g \in O_1$

(we can achieve (iii) by use of Lemma 4.18.3).

Next observe that

$$\begin{aligned} f(\phi^n h_s g x) &= f(\phi^n \alpha(s,g)\beta(s,g)h_s x) \\ &= f((\phi^n \alpha(s,g)\phi^{-n})\phi^n \beta(s,g)h_s x), \end{aligned}$$

so that for every $g \in O_1, n \in \mathbb{N}$ and $h_s \in J$, by (iii),

$$|f(\phi^n h_s g x) - f(\phi^n \beta(s,g) h_s x)| < \epsilon/2.$$

In view of the above we now have

$$|LU^n f(x) - LU^n f(gx)| \le \frac{1}{2M}\left|\int_J f(\phi^n h_s gx)\,d\lambda(s) - \int_J f(\phi^n \beta(s,g)h_s x)\,d\lambda(s)\right|$$
$$+\frac{1}{2M}\left|\int_J f(\phi^n \beta(s,g)h_s x)\,d\lambda(s) - \int_J f(\phi^n h_s x)\,d\lambda(s)\right|$$
$$\le \frac{1}{2M}\left|\int_J f(\phi^n h_s x)\,d\lambda(s) - \int_J f(\phi^n \beta(s,g)h_s x)\,d\lambda(s)\right| + \frac{\epsilon}{2}.$$

In turn we get

$$\frac{1}{2M}\left|\int_J f(\phi^n h_s x)\,d\lambda(s) - \int_J f(\phi^n \beta(s,g)h_s x)\,d\lambda(s)\right|$$
$$= \frac{1}{2M}\left|\int_J f(\phi^n h_s x)\,d\lambda(s) - \int_J f(\phi^n \psi_g(h_s)x)\,d\lambda(s)\right|$$
$$\le \frac{1}{2M}\left(\int_J \left|1 - \frac{d\nu_g}{d\lambda}\right| |f|\,d\lambda(s) + \int_{J\triangle\psi_g(J)} \left(1 + \left|\frac{d\nu_g}{d\lambda}\right|\right) |f|\,d\lambda(s)\right)$$
$$\le \epsilon/4 + \epsilon/4 = \epsilon/2.$$

Put together these estimations finally yield $|LU^n f(x) - LU^n f(gx)| \le \epsilon$ as required. □

4.23. THEOREM. *The horocycle flow (X, N) is strictly ergodic.*

PROOF. Let $\mu \in M(X)$ be an N-invariant probability measure. Let $f \in C(X)$ and fix an element $\phi = g_a = \begin{pmatrix} a & 0 \\ 0 & a^{-1} \end{pmatrix} \in A$ with $a > 1$. According to Theorem 4.20 the sequence $U^n f = f \circ \phi^n$ tends weakly to $c = \int f\,d\omega$. Since left multiplication is continuous in the semigroup $\mathbf{B}$ (of linear operators of norm ≤ 1), we also have $LU^n f \to c$ weakly in $\mathfrak{H}$. By Lemma 4.22.2 the family of functions $\{LU^n f : n \in \mathbb{N}\}$ is an equicontinuous family and therefore has the property that any subsequence has a further subsequence which converges uniformly. It follows that $LU^{n_k} f \to c$ uniformly on X. In particular the sequence $(U^{-n}LU^n f)(x) = (LU^n f)(\phi^{-n}x)$ tends pointwise to c and by Lebesgue's dominated convergence theorem

$$\int_X (U^{-n}LU^n f)\,d\mu \to c.$$

On the other hand, by Fubini's theorem, we get

$$\int_X (U^{-n}LU^n f)\,d\mu = \frac{1}{2M}\int_X \int_J f(\phi^n h_s \phi^{-n} x)\,d\lambda(s)\,d\mu(x)$$
$$= \frac{1}{2M}\int_J \int_X f(\phi^n h_s \phi^{-n} x)\,d\mu(x)\,d\lambda(s).$$

Since for every n, $\phi^n h_s \phi^{-n} \in N$ it follows by the N-invariance of μ that the last integral is the constant $d = \int f\,d\mu$, independent of n. Thus

$$d = \int f\,d\mu = \int f\,d\omega = c$$

and, as $f \in C(X)$ is arbitrary, we have $\mu = \omega$. This proves unique ergodicity. Since ω is a full measure (X, N) is a minimal system hence also strictly ergodic. □

4.24. EXERCISES. 1. Let $(X, \{T_t\}_{t\in\mathbb{R}})$ be a minimal flow. Show that the set

$$\{t \in \mathbb{R} : (X, T_t) \text{ is not a minimal } \mathbb{Z}\text{-system}\}$$

is at most countable. [[Fix $0 < s \in \mathbb{R}$, let $Y \subset X$ be a minimal subset of the $\mathbb{Z}$-system (X, T_s). Define $H_s = H = \{t \in \mathbb{R} : T_t Y = Y\}$ and show that, either $H = \mathbb{R}$ or $H = s_0\mathbb{Z}$ for some $s_0 > 0$. Next observe that $\{T_t Y : 0 \le t < s_0\}$ is a partition of X and that the associated function $f : X \to \mathbb{R}/H$ is a continuous eigenfunction of the system $(X, \{T_t\}_{t\in\mathbb{R}})$; i.e. $f(T_t x) = t + f(x)$. As in Chapter 1, Exercise 1.5 show that there are at most countably many eigenfunctions and therefore $H_s = \mathbb{R}$ for all but a countable set of $s \in \mathbb{R}$.]]

2. Let $(X, \{T_t\}_{t\in\mathbb{R}})$ be a strictly ergodic flow. Show that the set

$$\{t \in \mathbb{R} : (X, T_t) \text{ is not a strictly ergodic } \mathbb{Z}\text{-system}\}$$

is at most countable. [[A similar argument, with measurable eigenfunctions, works for unique ergodicity. Let μ denote the unique invariant measure of the flow $(X, \{T_t\}_{t\in\mathbb{R}})$. Fix $s > 0$ and for a T_s-invariant measure ν set $H_s = H = \{t \in \mathbb{R} : T_t\nu = \nu\}$. If $H \ne \mathbb{R}$ then $H = s_0\mathbb{Z}$ for some $s_0 > 0$ and it follows that $\mu = \frac{1}{s_0}\int_{[0,s_0)} T_t\nu\, dt$ is the ergodic decomposition of μ with respect to the $\mathbb{Z}$ action defined by T_{s_0}. This decomposition yields a measurable eigenfunction for the flow — an element of $L^2(\mu)$. Since there are at most countably many such eigenfunctions this argument completes the proof.]]

3. Use the fact that $g_t h_s g_{t^{-1}} = h_{st^2}$ for $s, t > 0$ to deduce that all the horocycle $\mathbb{Z}$-systems (X, h_s), $s > 0$ are conjugate to each other and that they are strictly ergodic.

7. E-systems

Recall that a topologically transitive system is a P-system if the periodic points are dense and an M-system if the union of the minimal sub-systems is dense (Chapter 1, Section 1).

4.25. DEFINITION. A topologically transitive dynamical system (X, Γ) is called an *E-system* if there exists a full measure $\mu \in M_\Gamma(X)$ (i.e. $\operatorname{supp}(\mu) = X$). It is called (topologically) *ergodic* if there exists an ergodic full Γ-invariant measure μ on X. As we shall see in the following exercise a transitive system is an E-system iff the union of the ergodic sub-systems is dense in X, whence the name E-system.

4.26. EXERCISES. 1. For every $\mu \in M_\Gamma(X)$ the subset $\operatorname{supp}(\mu)$ is closed and Γ-invariant and it follows that when the system is minimal *every* measure in $M_\Gamma(X)$ is full. Thus a minimal system (X, Γ), with $M_\Gamma(X) \ne \emptyset$, is an E-system. In particular for an amenable Γ, every minimal system is an E-system.

2. More generally, for amenable Γ every M-system is an E-system.

3. Following Furstenberg [**71**], call a point $x \in X$ in a $\mathbb{Z}$-system *regular* if it is generic for some invariant measure ν such that $\nu(U) > 0$ for every open neighborhood U of x. Show that a generic point for an ergodic measure is regular.

4. Every minimal $\mathbb{Z}$-system is ergodic. An example of Weiss [**259**], shows that an E-system need not be ergodic. (The example constructed there is actually a P-system.)
5. A transitive $\mathbb{Z}$-system (X, T) is an E-system iff the regular points are dense in X, iff the union of the ergodic sub-systems of X is dense in X. [[Let $\mu = \int \omega \, dP(\omega)$ be the ergodic decomposition of a full invariant measure μ. For each ω let $X_\omega \subset X$ be the set of generic points for ω, then $\cup\{X_\omega : \omega \in \Omega\}$ is a dense set of regular points. Conversely, when the regular points are dense in X, it is easy to construct an invariant probability measure which is positive on non-empty open sets.]]
6. More generally, a transitive system (X, Γ) is an E-system iff the union of the ergodic sub-systems is dense in X. [[Use the ergodic decomposition theorem for a general group Γ (Theorem 3.22).]]

The proof of the following theorem is straightforward; one only has to note that for a full measure μ, $\mu(U) > 0$ whenever U is a non-empty open set.

4.27. THEOREM. *Let (X, Γ) be a topological system, $\mu \in M_\Gamma(X)$ a full measure; then (X, Γ) is topologically transitive, weakly mixing, mixing if the measure-preserving system $(X, \mathcal{X}, \mu, \Gamma)$ is ergodic, weakly mixing, mixing respectively.*

The converse of Theorem 4.27 is false. There exists e.g. a topologically mixing strictly ergodic $\mathbb{Z}$-system (X, T) such that the corresponding measure-preserving system $(X, \mathcal{X}, \mu, T)$, with μ the (necessarily full) unique invariant measure, is measure-theoretically isomorphic to an irrational rotation on the circle and thus is not a (measure) weakly mixing dynamical system, Lehrer [**164**]. An even more spectacular example is due to Weiss. He produces in [**262**] a minimal dynamical system (X, T) such that for every ergodic measure-preserving system $(Y, \mathcal{Y}, \nu, T)$, there exists $\mu \in M_T(X)$ with $(X, \mathcal{X}, \mu, T) \cong (Y, \mathcal{Y}, \nu, T)$.

4.28. THEOREM. *For $\mathbb{Z}$-systems.*

1. *An E-system is syndetically transitive.*
2. *Every E-system (X, T) is weakly disjoint from every weakly mixing system (Y, T); i.e. $(X \times Y, T)$ is topologically transitive.*

PROOF. 1. Given two non-empty open sets U, V in X, we choose $k \in \mathbb{Z}$ with $T^k U \cap V \neq \emptyset$. Next set $U_0 = T^{-k} V \cap U$, and observe that $k + N(U_0, U_0) \subset N(U, V)$. Thus it is enough to show that $N(U, U)$ is syndetic for every non-empty open U. By Exercise 1.1.5 we have to show that $N(U, U)$ meets every thick subset $B \subset \mathbb{Z}$. By Poincaré's recurrence theorem (Exercise 2.22), $N(U, U)$ meets every set of the form $A - A = \{n - m : n, m \in A\}$ with A infinite. By Exercise 1.1.7, every thick set B contains some $D^+(A) = \{a_n - a_m : n > m\}$ for an infinite sequence $A = \{a_n\}$. Thus $\emptyset \neq N(U, U) \cap \pm D^+(A) \subset N(U, U) \cap \pm B$. Since $N(U, U)$ is symmetric, this completes the proof.

2. In view of part 1 this follows from Corollary 1.12 □

The following theorem generalizes Theorem 1.41.

4.29. THEOREM. *For $\mathbb{Z}$-systems, an almost equicontinuous E-system is minimal and equicontinuous. A fortiori AE-systems which are also P-systems or M-systems are minimal and equicontinuous.*

PROOF. Let (X,T) be an almost equicontinuous E-system. Given $\epsilon > 0$ there exists a transitive point x_0 and a neighborhood U of x_0 such that $d(T^n x_0, T^n y) \le \epsilon$ for every n and every $y \in U$ (Proposition 1.34).

Let $A = N(U,U) = \{n \in \mathbb{Z} : T^n U \cap U \neq \emptyset\}$. By Theorem 4.28, A is syndetic. Let $l \in A$ then there exists $z \in U$ with $T^l \in U$. Now $T^l x_0 \in U$ implies $\forall n, d(T^{n+l}z, T^n x_0) \le \epsilon$, and $z \in U$ implies $\forall n, d(T^{n+l}z, T^{n+l}x_0) \le \epsilon$. Hence $\forall n, d(T^{n+l}x_0, T^n x_0) \le 2\epsilon$ and it follows that for all $w \in X, d(T^l w, w) \le \epsilon$. We have now proved that for every $\epsilon > 0$ there exists a syndetic subset A of $\mathbb{Z}$ with

$$n \in A \Longrightarrow \sup_{w \in X} d(T^n w, w) \le 2\epsilon.$$

By Exercise 1.26.12. this is a necessary and sufficient condition for equicontinuity. Finally, transitivity implies that (X,T) is minimal. □

Our next result will lead to a characterization (at least for abelian groups) of weakly mixing minimal systems as those minimal systems which have no non-trivial equicontinuous factors. We shall need the following.

4.30. LEMMA. *Let $\pi : (X,\Gamma) \to (Y,\Gamma)$ be a homomorphism of dynamical systems with (X,Γ) minimal. If $U \subset X$ is a non-empty open subset then $\operatorname{int}\pi(U)$ is non-empty.*

PROOF. Let V be a non-empty open set such that $V \subset \bar{V} \subset U$. By minimality there exists a finite subset $F \subset \Gamma$ with $\cup_{\gamma \in F}\gamma V = X$. This implies $\cup_{\gamma \in F}\gamma\pi(\bar{V}) = Y$ and we conclude that $\operatorname{int}\pi(\bar{V}) \neq \emptyset$. Since $\pi(\bar{V}) \subset \pi(U)$ this concludes the proof. □

4.31. THEOREM. *Let (X,Γ) be an E-system and (Y,Γ) a minimal one. Then (X,Γ) is disjoint from the maximal equicontinuous factor of (Y,Γ) iff (X,Γ) is weakly disjoint from (Y,Γ). In particular if the product system $(X \times Y,\Gamma)$ is not transitive then the system (Y,Γ) admits a non-trivial equicontinuous factor.*

PROOF. Weak disjointness of (X,Γ) and (Y,Γ) implies weak disjointness of (X,Γ) and $(\bar{Y},\Gamma)$, the maximal equicontinuous factor of (Y,Γ). By Proposition 1.27 this implies the disjointness of (X,Γ) and $(\bar{Y},\Gamma)$.

Conversely, assuming that $(X \times Y,\Gamma)$ is not transitive we shall construct an equicontinuous factor (Z,Γ) of (Y,Γ) and a non-trivial joining $K \subset X \times Z$. This will imply the non-disjointness of (X,Γ) and (Z,Γ) and a fortiori the non-disjointness of (X,Γ) and the maximal equicontinuous factor of (Y,Γ).

As (X,Γ) is an E-system, there exists an invariant measure $\mu \in M_\Gamma(X)$ with full support. The assumption that $(X \times Y,\Gamma)$ is not transitive implies that there exists an open Γ-invariant subset U of $X \times Y$, such that $M = \overline{U} \subsetneq X \times Y$. Since both (X,Γ) and (Y,Γ) are transitive and since the projections of M to X and Y contain non-empty open sets, it follows that these projections are onto; that is (M,Γ) is a non-trivial topological joining of (X,Γ) and (Y,Γ). For every $y \in Y$ let $M(y) = \{x \in X : (x,y) \in M\}$, and let $f_y = \mathbf{1}_{M(y)}$ be the indicator function of the set $M(y)$, considered as an element of $L^1(X,\mu)$.

Denote by $\pi : Y \to L^1(X,\mu)$ the map $y \mapsto f_y$. We shall show that π is a continuous homomorphism, where we consider $L^1(X,\mu)$ as a dynamical system with the isometric action of the group $\{U_\gamma : \gamma \in \Gamma\}$. Fix $y_0 \in Y$ and $\epsilon > 0$. There exists an open neighborhood V of $M(y_0)$ with $\mu(V \setminus M(y_0)) < \epsilon$. Since M is closed the set map $y \mapsto M(y), Y \to 2^X$ is upper semi-continuous and we can find a neighborhood W of y_0 such that $M(y) \subset V$ for every $y \in W$. Thus for every $y \in W$ we have $\mu(M(y) \setminus M(y_0)) < \epsilon$. In particular, $\mu(M(y)) \leq \mu(M(y_0)) + \epsilon$ and it follows that the map $y \mapsto \mu(M(y))$ is upper semi-continuous. A simple computation shows that it is Γ-invariant, hence, by minimality of (Y, Γ), a constant (see Exercise 1.1.8).

With y_0, ϵ and V, W as above, for every $y \in W$, $\mu(M(y) \setminus M(y_0)) < \epsilon$ and $\mu(M(y)) = \mu(M(y_0))$, thus $\mu(M(y) \Delta M(y_0)) < 2\epsilon$, i.e., $\|f_y - f_{y_0}\|_1 < 2\epsilon$. This proves the claim that π is continuous.

Let $Z = \pi(Y)$ be the image of Y in $L^1(\mu)$. Since π is continuous, Z is compact. It is easy to see that the Γ-invariance of M implies that for every $\gamma \in \Gamma$ and $y \in Y$, $f_{\gamma^{-1}y} = f_y \circ \gamma$ so that Z is Γ-invariant and $\pi : (Y, \Gamma) \to (Z, \Gamma)$ is a homomorphism. Clearly (Z, Γ) is minimal and equicontinuous (in fact isometric).

Set $p = (\mathrm{id} \times \pi) : (X \times Y) \to (X \times Z)$ and let $K = p(M)$, then clearly p is a homomorphism and K is a topological joining of the systems (X, Γ) and (Z, Γ). We next show that this joining is not trivial.

Given $(x, y) \in p^{-1}(p(U))$ there exists $y' \in Y$ with $(x, y') \in U$ and $\pi(y) = \pi(y')$. This implies that $f_y = f_{y'}$ hence $\mu(M(y) \Delta M(y')) = 0$. Since μ is full, $M(y)$ and $M(y')$ have the same interior. Since x belongs to the interior of $M(y')$ it is also in the interior of $M(y)$ so that $(x, y) \in M$. We have shown that $p^{-1}(p(U)) \subset M$. Since $M \subsetneq X \times Y$, there exist non-empty open sets $V \subset X$ and $W \subset Y$ with $(V \times W) \cap M = \emptyset$, hence $(V \times W) \cap p^{-1}(p(U)) = \emptyset$, hence $p(V \times W) \cap p(U) = \emptyset$. By Lemma 4.30 the open set $O = \mathrm{int}\, \pi(W)$ is non-empty and we have

$$p(V \times W) = V \times \pi(W) \supset V \times O \neq \emptyset \quad \text{and} \quad (V \times O) \cap p(U) = \emptyset.$$

Finally, since $\overline{p(U)} = p(M) = K$, we conclude that $(V \times O) \cap K = \emptyset$ and therefore that $K \subsetneq X \times Z$. □

4.32. Corollary. *Let (X, Γ) be a minimal system admitting an invariant measure, then (X, Γ) is weakly mixing iff it has no non-trivial equicontinuous factor.*

Proof. Since (X, Γ) is minimal and admits an invariant measure it is an E-system and, taking (X, Γ)=(Y, Γ) in Theorem 4.31, we see that if (X, Γ) has no non-trivial equicontinuous factor then $(X \times X, \Gamma)$ is topologically transitive; i.e. (X, Γ) is weakly mixing. Conversely, if $\pi : (X, \Gamma) \to (Z, \Gamma)$ is a non-trivial equicontinuous factor, then $(Z \times Z, \Gamma)$ and therefore also $(X \times X, \Gamma)$ are not topologically transitive. □

4.33. Theorem. *Let (X, Γ) be a minimal E-system and (Y, Γ) a minimal system. Let $(\bar{X}, \Gamma)$ and $(\bar{Y}, \Gamma)$ be the maximal equicontinuous factors of (X, Γ) and (Y, Γ) respectively. Then (X, Γ) and (Y, Γ) are weakly disjoint iff $(\bar{X}, \Gamma)$ and $(\bar{Y}, \Gamma)$ are disjoint.*

Proof. Weak disjointness of (X, Γ) and (Y, Γ) implies weak disjointness of $(\bar{X}, \Gamma)$ and $(\bar{Y}, \Gamma)$ and by Proposition 1.27 this implies the disjointness of $(\bar{X}, \Gamma)$ and $(\bar{Y}, \Gamma)$. Conversely, assuming the disjointness of $(\bar{X}, \Gamma)$ and $(\bar{Y}, \Gamma)$, we conclude by Theorem 4.31 (let the E-system $\bar{Y}$ play the role of X and X the role of Y) that (X, Γ) and $(\bar{Y}, \Gamma)$ are weakly disjoint. Proposition 1.27 implies that (X, Γ) and $(\bar{Y}, \Gamma)$ are disjoint and finally applying Theorem 4.31 again we conclude that (X, Γ) and (Y, Γ) are weakly disjoint. □

8. Notes

Generic points: See Furstenberg's paper [**72**].

Unique ergodicity: Source: mainly [**72**]. Theorem 4.10 is from Katznelson-Weiss [**147**]. The proof given here (as well as the proof of Proposition 4.6) I learned from M. Boshernitzan. The alternative proof is from [**147**].

Minimal Heisenberg Nil-flows are uniquely ergodic: As I mentioned above, here I followed Parry's proof of L. Green's criterion for the ergodicity of a nil-flow (the unique ergodicity part is due to Furstenberg). In fact all the essential ingredients of the proof of the general case are already present in this special case of the Heisenberg group. One needs only to apply a simple induction argument on the lower central series of the nilpotent group $\mathbb{G}$ to obtain the general statement of Theorem 4.11.2.

The geodesic and horocycle flows: Much of what was shown in this section for $SL(2,\mathbb{R})$ can be carried out for any connected, non-compact, semisimple Lie group $\mathbb{G}$. Minimality and mixing of the horocycle flow were proved by Hedlund, [**117**], [**118**], and the unique ergodicity by Furstenberg, [**75**]. Ornstein and Weiss show that the geodesic flow is Bernoulli, [**194**]. Theorem 4.20 and the following exercise are special cases of the Howe-Moore theorem, [**125**]. Mixing of all orders of the horocycle flow is shown in [**179**]; see Chapter 7, Section 3 below. My exposition is based on the paper of Ellis and Perizo [**61**] (and also Adams [**3**]), which in turn, is based on the method introduced by Marcus [**178**]. See also the books [**188**] by Nicholls and [**15**] by Bekka and Mayer.

***E*-systems:** The term E-system was introduced in Glasner and Weiss [**97**], which is the main source for the first part of this section. The result of Corollary 4.32 is due to Keynes and Robertson [**153**] who developed an idea of Furstenberg, [**73**]; and independently to K. Petersen [**204**] who utilized a previous work of W. A. Veech, [**250**]. Theorem 4.31 is an elaboration of a result of McMahon [**181**] which is due to Blanchard, Host and Maass, [**29**]; I followed their proof.

CHAPTER 5

Spectral Theory

Although spectral theory is available for general group actions we shall be content in this brief introductory chapter (hence also throughout the book) to deal with the spectral theory of $\mathbb{Z}$-systems. In Section 1 we review the spectral theorem for a unitary operator on a Hilbert space. In Section 2 this theorem is applied to a dynamical system $\mathbf{X} = (X, \mathcal{X}, \mu, T)$ via the Koopman representation $T \mapsto U_T$. In Section 3 we prove a theorem of Kolmogorov identifying the spectral type of a K-system as Lebesgue spectrum of infinite multiplicity. As we shall see later (Exercise 18.18), this same proof yields a much more general result: for every dynamical system the spectral type in the orthogonal complement of the L^2 subspace corresponding to the Pinsker algebra is Lebesgue spectrum of infinite multiplicity. This in essence reduces the study of spectrum in dynamical systems to systems with zero entropy. In the last section (Section 4) we consider the question: can the Koopman representation be irreducible?

We refer to Goodson's paper [**105**] for a survey of modern spectral theory of dynamical systems and to Nadkarni's book [**184**] for a systematic treatment of the subject. The books [**203**], [**208**] and [**165**] provide a short proof of the spectral theorem for unitary operators. In [**203**] there is also a proof of Herglotz' theorem.

1. The spectral theorem for a unitary operator

We have already considered positive definite functions on an arbitrary group Γ as covariance functions for Gaussian processes (Definition 3.55). Here they serve as the means for associating a measure on the circle μ_x (the spectral measure of x) with a pair U, x, where U is a unitary operator on a Hilbert space $\mathfrak{H}$ and x is an element of $\mathfrak{H}$. For convenience we repeat the definition for $\Gamma = \mathbb{Z}$.

5.1. Definition. A function $\phi : \mathbb{Z} \to \mathbb{C}$ is called *positive definite* if for every finite sequence of complex numbers $\{a_n\}_{n=0}^N$

$$\sum_{m=0}^{N} \sum_{n=0}^{N} a_m \bar{a}_n \phi(m-n) \geq 0.$$

5.2. Exercises. 1. Prove the properties 1-3 in Definition 3.55.

2. If $U : \mathfrak{H} \to \mathfrak{H}$ is a unitary operator on a Hilbert space $\mathfrak{H}$, then for every $x \in \mathfrak{H}$ the function

$$\phi(n) = \langle U^n x, x \rangle$$

is positive definite.

3. If μ is a finite measure on $\mathbb{T} = \{z \in \mathbb{C} : |z| = 1\}$, then its Fourier transform, the function

$$\hat{\mu}(n) = \int_{\mathbb{T}} z^n \, d\mu(z),$$

is a positive definite function. [[Let $\mathfrak{H} = L^2(\mu)$ and define $U : \mathfrak{H} \to \mathfrak{H}$ by $Uf(z) = zf(z)$, then apply 2.]]

The converse of the latter statement is the following crucial theorem:

5.3. THEOREM (Herglotz). *Given a positive definite function $\phi : \mathbb{Z} \to \mathbb{C}$ there exists a unique measure μ on $\mathbb{T}$ such that $\phi(n) = \hat{\mu}(n)$.*

5.4. DEFINITION. When $U : \mathfrak{H} \to \mathfrak{H}$ is a unitary operator on a Hilbert space $\mathfrak{H}$ and $x \in \mathfrak{H}$ we denote by $Z(x)$ the *cyclic subspace* generated by x; i.e. the smallest closed subspace of $\mathfrak{H}$ containing the vectors $\{U^n x : n \in \mathbb{Z}\}$. Using Herglotz's theorem and Exercise 5.2 we see that every $x \in \mathfrak{H}$ defines a unique (positive) measure σ_x on $\mathbb{T}$, with $\hat{\sigma}_x(n) = \langle U^n x, x\rangle$. In particular

$$\sigma_x(\mathbb{T}) = \int_{\mathbb{T}} \mathbf{1}\, d\sigma_x = \langle x, x\rangle = \|x\|^2 = \hat{\sigma}_x(0).$$

We call σ_x the *spectral measure* of x (with respect to U).

5.5. EXERCISE. For $x, y \in \mathfrak{H}$ set $a_n = \langle U^n x, y\rangle$ and $b_n = \langle U^n ix, y\rangle = ia_n$. The identities : $\langle U^n x, y\rangle = \overline{\langle U^{-n} y, x\rangle}$ and

$$\langle U^n(x+y), x+y\rangle - \langle U^n x, x\rangle - \langle U^n y, y\rangle = \langle U^n x, y\rangle + \langle U^n y, x\rangle$$

show that the sequence $a_n + \overline{a_{-n}}$ is the Fourier transform of a signed measure on $\mathbb{T}$. Similarly show that this holds for the sequence $b_n + \overline{b_{-n}}$ as well and deduce that there exists a unique complex measure $\sigma_{x,y}$ such that

$$\hat{\sigma}_{x,y}(n) = a_n = \langle U^n x, y\rangle.$$

We then have

$$\sigma_{x+y} = \sigma_x + \sigma_y + \sigma_{x,y} + \sigma_{y,x}.$$

Show that the map $(x, y) \mapsto \sigma_{x,y}$ from $\mathfrak{H} \times \mathfrak{H} \to C(\mathbb{T})^*$ is bilinear and continuous and use Cauchy-Schwarz inequality to show that

$$(|\sigma_{x,y}|(B))^2 \le \sigma_x(B) \cdot \sigma_y(B)$$

for every Borel subset $B \subset \mathbb{T}$. In particular $|\sigma_{x+y}| \ll \sigma_x$. [[See **[208]**.]]

5.6. PROPOSITION. *With notation as above and with $\mathfrak{H} = Z(x)$ define*

$$V : L^2(\sigma_x, \mathbb{T}) \to L^2(\sigma_x, \mathbb{T}) \quad by \quad Vf(z) = zf(z) \quad (z \in \mathbb{T}),$$

then V is a unitary operator and the operators U and V are unitarily equivalent via the isometry $W : Z(x) \to L^2(\sigma_x, \mathbb{T})$ which is the unique (linear isometric) extension of the map $U^n x \mapsto z^n \in L^2(\sigma_x, \mathbb{T})$; i.e. $W^{-1}VW = U$.

PROOF. If $p(z) = \sum_{j=-m}^{m} a_j z^j$ and $q(z) = \sum_{l=-n}^{n} b_l z^l$ are trigonometric polynomial then

$$\begin{aligned}\langle p(U)x, q(U)x\rangle_{Z(x)} &= \sum_{j=-m}^{m}\sum_{l=-n}^{n} a_j \bar{b}_l \langle U^{j-l}x, x\rangle_{Z(x)} = \sum_{j=-m}^{m}\sum_{l=-n}^{n} a_j \bar{b}_l \hat{\sigma}_x(j-l) \\ &= \sum_{j=-m}^{m}\sum_{l=-n}^{n} a_j \bar{b}_l \int_{\mathbb{T}} z^{j-l}\, d\sigma_x = \langle p(z), q(z)\rangle_{L^2(\sigma_x, \mathbb{T})}.\end{aligned}$$

Since elements of the form $p(U)x$ are dense in $Z(x)$, while the trigonometric polynomials $p(z)$ are dense in $L^2(\sigma_x, \mathbb{T})$, this completes the proof. □

5.7. PROPOSITION. *Let $U_i : \mathfrak{H}_i \to \mathfrak{H}_i$, $i = 1, 2$ be unitary operators and $x_i \in \mathfrak{H}_i$, then $U_1 \upharpoonright Z(x_1)$ and $U_2 \upharpoonright Z(x_2)$ are unitarily equivalent iff the spectral measures σ_{x_1} and σ_{x_2} are equivalent ($\sigma_{x_1} \sim \sigma_{x_2}$).*

PROOF. In view of the previous proposition it suffices to show this for the associated operators $V_i : L^2(\sigma_{x_i}) \to L^2(\sigma_{x_i})$. If $WV_1 = V_2 W$ for an intertwining (surjective isometry) $W : L^2(\sigma_{x_1}) \to L^2(\sigma_{x_2})$, then denoting $W(\mathbf{1}) = f$ we have for every $n \in \mathbb{Z}$

$$W(z^n) = WV_1^n \mathbf{1} = V_2^n W \mathbf{1} = V_2^n f = z^n f$$

and we deduce that W is the multiplication operator $W(g) = fg$. Now for a Borel subset $B \subset \mathbb{T}$ we have

$$\begin{aligned}\sigma_{x_1}(B) &= \int \mathbf{1}_B \, d\sigma_{x_1} = \langle \mathbf{1}_B, \mathbf{1}_B \rangle_{\sigma_{x_1}} = \langle f\mathbf{1}_B, f\mathbf{1}_B \rangle_{\sigma_{x_2}} \\ &= \int f\mathbf{1}_B \, \bar{f}\mathbf{1}_B \, d\sigma_{x_2} = \int_B |f|^2 \, d\sigma_{x_2}.\end{aligned}$$

Thus $\sigma_{x_1} \ll \sigma_{x_2}$ (with $\frac{d\sigma_{x_1}}{d\sigma_{x_2}} = |f|^2$) and by symmetry $\sigma_{x_1} \sim \sigma_{x_2}$.

Conversely, if $\sigma_{x_1} \sim \sigma_{x_2}$ with $\frac{d\sigma_{x_1}}{d\sigma_{x_2}} = |f|^2$, $f \geq 0$, then $f > 0$ σ_{x_2} a.e. and defining $W : L^2(\sigma_{x_1}, \mathbb{T}) \to L^2(\sigma_{x_2}, \mathbb{T})$ by $W(g) = fg$ we have

$$\|W(g)\|^2 = \int_{\mathbb{T}} |W(g)|^2 \, d\sigma_{x_2} = \int_{\mathbb{T}} |f|^2 |g|^2 \, d\sigma_{x_2} = \int_{\mathbb{T}} |g|^2 \, d\sigma_{x_1} = \|g\|^2.$$

□

We can now state the spectral theorem for unitary operators. For a proof refer to [**203**], [**208**], or [**165**].

5.8. THEOREM (Spectral theorem for unitary operators). *Let $\mathfrak{H}$ be a separable Hilbert space, $U : \mathfrak{H} \to \mathfrak{H}$ a unitary operator, then there exists a sequence $\{x_n\}_{n=1}^{\infty}$ of vectors in $\mathfrak{H}$ such that*

1. *$\mathfrak{H} = \oplus Z(x_n)$,*
2. *$\sigma_{x_1} \gg \sigma_{x_2} \gg \cdots$.*

Every sequence $\{x'_n\}_{n=1}^{\infty}$ with these properties satisfies $\sigma_{x'_n} \sim \sigma_{x_n}, \forall n$.

Combining this theorem with Propositions 5.6 and 5.7 we see that, up to unitary equivalence, the operator U is determined by the measure class σ_{x_1} and the sequence of Borel sets $A_2 \supset A_3 \supset \cdots$, where $A_n = \{x \in \mathbb{T} : \frac{d\sigma_{x_n}}{d\sigma_{x_1}} > 0\}$. In fact, given a measure $\sigma \in M(\mathbb{T})$ with $\sigma \sim \sigma_{x_1}$ and the sequence $\{A_n\}_{n \geq 2}$, we recover a unitary operator U', unitarily equivalent to U, as follows: Set $\sigma_n = \mathbf{1}_{A_n}\sigma \sim \sigma_{x_n}$, $n = 1, 2, \ldots$ (with $A_1 = \mathbb{T}$) and let $\mathfrak{H}' = \oplus_{n=1}^{\infty} \mathfrak{H}_n$, with $\mathfrak{H}_n = L^2(\sigma_n, \mathbb{T})$. Now define $U' : \mathfrak{H}' \to \mathfrak{H}'$ by letting $U'f(z) = zf(z)$ for $f \in \mathfrak{H}_n$.

The measure σ_{x_1}, or rather its equivalence class, is called the *spectral type* of U and we denote $\sigma_{x_1} = \sigma_U$. The (equivalence class of the) sequence $\sigma_{x_1} \gg \sigma_{x_2} \gg \cdots$

is the *spectral sequence* of U. A slightly more condensed form can be given to the invariants $\{\sigma_U; A_2 \supset A_3 \supset \cdots\}$ by defining the *multiplicity function*

$$M_U = \sum_{n=1}^{\infty} \mathbf{1}_{A_n}, \qquad M_U : \mathbb{T} \to \{1, 2, \dots\} \cup \{\infty\}.$$

We now see that the pair $\{\sigma_U, M_U\}$ completely determines the unitary class of U. The number $\operatorname{esssup} M_U$ is called the *spectral multiplicity* of U. When $M_U \equiv n$ we say that U has *homogeneous spectrum of multiplicity* n (we then have $\sigma_{x_1} \sim \sigma_{x_2} \sim \cdots \sim \sigma_{x_n}$ and $\sigma_{x_k} = 0$ for $k > n$, or $\sigma_{x_1} \sim \sigma_{x_2} \sim \cdots$ when $M_U \equiv \infty$). In particular when $M_U \equiv 1$ we say that U has a *simple spectrum.*

We say that U has *discrete, singular, absolutely continuous or Lebesgue spectrum* when its spectral type σ_U has that same property. Thus for example U has homogeneous Lebesgue spectrum of infinite multiplicity (or infinite Lebesgue spectrum) if its spectral sequence is the constant sequence $\sigma_{x_n} = \lambda$ where λ is Lebesgue measure on $\mathbb{T}$.

2. The spectral invariants of a dynamical system

Given a dynamical system $\mathbf{X} = (X, \mathfrak{X}, \mu, T)$, the spectral invariants $\{\sigma_U, M_U\}$ of its Koopman operator U_T on $\mathfrak{H} = L^2(\mu)$ (or sometimes $L^2_0(\mu)$) are called the *spectral invariants of the dynamical system* $\mathbf{X}$ and we refer to the spectral properties of U_T as the spectral properties of the system.

When $\phi(n) = \langle U_T^n f, f\rangle$, with $f \in L^2(\mu)$, is the positive definite function obtained via the Koopman representation we can avoid the use of Herglotz' theorem and obtain the spectral measure σ_f more directly as follows.

5.9. PROPOSITION. *Let ν_N be the absolutely continuous measure on $\mathbb{T}$ (with respect to Lebesgue measure λ) whose density is given by the function*

$$h_N(z) = \frac{1}{N} \int_X \Big|\sum_{j=1}^{N} z^n f(T^n x)\Big|^2 \, d\mu(x),$$

i.e. $d\nu_N = h_N d\lambda$. Then $\nu_N \to \nu = \sigma_f$ in the weak topology.*

PROOF. A calculation yields,

$$\begin{aligned}
\hat{\nu}_N(k) &= \frac{1}{N} \sum_{n=1}^{N} \sum_{m=1}^{N} \langle U_T^n f, U_T^m f\rangle \int_{\mathbb{T}} z^{n-m-k} \, d\lambda(z) \\
&= \frac{1}{N} \sum_{n=1}^{N} \sum_{m=1}^{N} \langle U_T^{n-m} f, f\rangle \int_{\mathbb{T}} z^{n-m-k} \, d\lambda(z) \\
&= \frac{1}{N} \sum_{n=1}^{N-k} \langle U_T^k f, f\rangle \\
&= (1 - \frac{k}{N}) \langle U_T^k f, f\rangle.
\end{aligned}$$

Thus $\hat{\nu}_N(k) \to \hat{\nu}(k) = \hat{\sigma}_f(k)$. □

5.10. EXERCISES. 1. Let $\mathbf{X} = (X, \mathfrak{X}, \mu, T)$ be a dynamical system. Then $f \in L^2(\mu)$ is an eigenfunction of U_T with eigenvalue $\alpha \in \mathbb{T}$ iff $\sigma_f = \|f\|^2 \delta_\alpha$, where δ_α is the point mass at α. Deduce that $\mathbf{X}$ is ergodic iff $f \equiv c$ (a constant) is the only function whose spectral measure $\sigma_f = \|f\|^2 \delta_1$, and that $\mathbf{X}$ is weakly mixing iff σ_f is a continuous (non-atomic) measure for every $f \in L_0^2(\mu)$.

2. Let $\mathbf{X} = (X, \mathfrak{X}, \mu, T)$ be an ergodic system. For an integer $n \neq 0$ the system $(X, \mathfrak{X}, \mu, T^n)$ is not ergodic iff U_T admits an n-th root of unity as an eigenvalue.
3. The dynamical system $\mathbf{X} = (X, \mathfrak{X}, \mu, T)$ is called *totally ergodic* if the system $(X, \mathfrak{X}, \mu, T^k)$ is ergodic for every $0 \neq k \in \mathbb{Z}$. If $(X, \mathfrak{X}, \mu, T^n)$ is not totally ergodic then either $\mathbf{X}$ admits a nontrivial adding machine as a factor or $\mathbf{X}$ admits a finite factor $\mathbf{X} \to \{0, 1, \dots, d-1\}$ such that $\mathbf{X}$ is isomorphic to $\mathbf{X}_0 \times \{0, 1, \dots, d-1\}$, where $\mathbf{X}_0$ is a totally ergodic T^d-system and for $x \in X_0, 0 \le j < d-1, T(x, j) = (x, j+1)$, while $T(x, d-1) = (T^d x, 0)$.
4. The spectrum of a Kronecker system is simple and discrete. Find a cyclic vector for $\mathfrak{H} = L^2(\mathbb{T}, \lambda)$ when λ is Lebesgue's measure and $U : \mathfrak{H} \to \mathfrak{H}$ is the Koopman operator corresponding to $T_\alpha : \mathbb{T} \to \mathbb{T}$, an irrational rotation of the circle.
5. A measure $\sigma \in M(\mathbb{T})$ is called *Rajchman* if $\lim_{|n| \to \infty} \hat{\sigma}(n) = 0$. Show that the system $\mathbf{X}$ is mixing iff σ_f is Rajchman for every $f \in L_0^2(\mu)$. Use the Riemann-Lebesgue lemma to deduce that $\mathbf{X}$ is mixing when its spectral type σ_{U_T} (on $L_0^2(\mu)$) is absolutely continuous.
6. The system $\mathbf{X}$ has a homogeneous Lebesgue spectrum of infinite multiplicity iff there exists an orthonormal sequence $\{f_n\}_{n=1}^\infty \subset L^2(X, \mu)$ such that $\{U_T^m f_n : m \in \mathbb{Z},\ n = 1, 2, \dots\}$ is an orthonormal basis for $L^2(X, \mu)$. [[We have $U_T = W^{-1} V W$, where $V : \oplus_{n=1}^\infty \mathfrak{H}_n \to \oplus_{n=1}^\infty \mathfrak{H}_n$, $\mathfrak{H}_n = L^2(\mathbb{T}, \lambda)$, $Vf(z) = zf(z)$ for $f \in \mathfrak{H}_n$, and $W : L^2(X, \mu) \to \oplus_{n=1}^\infty \mathfrak{H}_n$ is the surjective isometry defined by $Wf_n = \mathbf{1}_n$, where $\mathbf{1}_n = \mathbf{1}_\mathbb{T} \in \mathfrak{H}_n$.]]
7. The spectrum of a Bernoulli system is homogeneous Lebesgue spectrum of infinite multiplicity. [[Let $(X, \mathfrak{X}, \mu, S)$ with $X = Y^\mathbb{Z}$, $\mu = \nu^\mathbb{Z}$ and $(Sx)_n = x_{n+1}$ be the Bernoulli system with state space $(Y, \mathcal{Y}, \nu)$. For each $n \in \mathbb{Z}$ let $\{f_j^{(n)}\}_{j=1}^\infty$ be a basis for $L^2(Y, \nu)$. Show that the functions of the form $f_{j_1, j_2, \dots, j_k}^{n_1, n_2, \dots, n_k} = f_{j_1}^{(n_1)} f_{j_2}^{(n_2)} \cdots f_{j_k}^{(n_k)}$, for $n_1 < n_2 < \cdots < n_k$, form a basis for $L^2(X, \mu)$ as required in the previous exercise.]]
8. See Exercise 10.17 below.

5.11. THEOREM. *The spectral type of a rigid system is singular with respect to Lebesgue measure.*

PROOF. Let $\sigma = \sigma_f$ be the spectral measure of any bounded function f in $L^2(\mu)$ with $\|f\| = 1$, where $\mathbf{X} = (X, \mathfrak{X}, \mu, T)$ is a rigid system with respect to the sequence $n_k \nearrow \infty$. With $U = U_T$ we have

$$\begin{aligned}\int_\mathbb{T} |z^{n_k} - \mathbf{1}|^2 \, d\sigma &= \int_\mathbb{T} (z^{n_k} - \mathbf{1})(z^{-n_k} - \mathbf{1}) \, d\sigma = 2(1 - \Re \int_\mathbb{T} z^{n_k} \, d\sigma) \\ &= 2(1 - \Re \hat{\sigma}(n_k)) = 2(1 - \Re \langle U^{n_k} f, f \rangle) \to 0.\end{aligned}$$

Hence, on a subset $A \subset \mathbb{T}$ of σ measure 1, a subsequence of the sequence of functions z^{n_k} tends to the function $\mathbf{1} = \mathbf{1}_{\mathbb{T}}$. Denoting Lebesgue measure on $\mathbb{T}$ by λ, the Riemann-Lebesgue lemma implies $\lambda(A) = \int_{\mathbb{T}} \mathbf{1}_A \, d\lambda = \lim \int_{\mathbb{T}} \mathbf{1}_A z^{n'_k} \, d\lambda = 0$ and the measures σ and λ are singular. □

We close this section with the following intriguing open question which is attributed to Banach.

5.12. PROBLEM (Banach). Does there exist an ergodic dynamical system $\mathbf{X} = (X, \mathcal{X}, \mu, T)$ with simple Lebesgue spectrum?

3. The spectral type of a K-system

5.13. THEOREM (Kolmogorov). *A K-system has an infinite Lebesgue spectral type.*

PROOF. Let $\mathbf{X}$ be a K-system. By definition there exists a sub σ-algebra $\mathcal{K} \subset \mathcal{X}$ with the following properties

1. $T\mathcal{K} \subset \mathcal{K}$,
2. $\bigvee_{n=1}^{\infty} T^{-n}\mathcal{K} = \mathcal{X}$,
3. $\bigcap_{n=1}^{\infty} T^{n}\mathcal{K}$ is the trivial algebra $\mathcal{E}_0 = \{\emptyset, X\}$.

We have already seen (Lemma 3.50) that $\mathbf{X}$ is ergodic and that μ is non-atomic. Let $K = L^2(\mathcal{K}, \mu) \subset L^2(\mu)$ be the subspace of all the $\mathcal{K}$-measurable elements of $L^2(\mu)$. If we set $Uf(x) = f(Tx)$ then properties 1 and 2 imply $U^{-1}K \subsetneq K$ and we let $V = K \ominus U^{-1}K$. Now for every $m, n > 0$ we have

$$\begin{aligned} U^n K &= U^n(V \oplus U^{-1}K) = U^n V \oplus U^{n-1}(V \oplus U^{-1}K) \\ &= U^n V \oplus U^{n-1} V \oplus U^{n-2} K = \cdots \\ &= U^n V \oplus U^{n-1} V \oplus \cdots \oplus U^{-m} V \oplus U^{-m-1} K. \end{aligned}$$

If $f \in \bigcap_{m=0}^{\infty} U^{-m}K$ then f is $T^m\mathcal{K}$-measurable for every $m > 0$ and by property 3 we see that f is a constant. On the other hand every f that is $T^{-n}\mathcal{K}$-measurable is an element of the subspace $U^n K$ and property 2 implies that

$$L^2(\mu) = \mathbb{C} \oplus \bigoplus_{-\infty}^{\infty} U^n V.$$

In view of Exercise 5.10 the proof will be complete when we show that as a Hilbert space $\dim V = \infty$.

Set $T\mathcal{K} = \mathcal{L}$. Let A be a set in $\mathcal{K} \setminus \mathcal{L}$ and let $f_0 = \mathbf{1}_A - \mathbb{E}^{\mathcal{L}} \mathbf{1}_A$. Let $f = \mathbb{E}^{\mathcal{L}}(|f_0|^2)$, $C = \{x \in X : f > 0\}$ and finally

$$g = \begin{cases} \frac{f_0 \cdot \mathbf{1}_C}{\sqrt{f}} & x \in C \\ 0 & x \in C^c. \end{cases}$$

Then clearly $C \in \mathcal{L}$, $\mu(C) > 0$, $\mathbb{E}^{\mathcal{L}}(g) = 0$ and we have

$$\mathbb{E}^{\mathcal{L}}(|g|^2) = \frac{\mathbf{1}_C}{f} \mathbb{E}^{\mathcal{L}}(|f_0|^2) = \mathbf{1}_C.$$

Therefore

$$\int |g|^2 \, d\mu = \mathbb{E}(|g|^2) = \mathbb{E}(\mathbb{E}^{\mathcal{L}}(|g|^2)) = \mathbb{E}(\mathbf{1}_C) = \mu(C).$$

Since μ is non atomic, the Borel subset C is a standard Borel space isomorphic to the unit interval and $W = \mathbf{1}_C U^{-1} K$ is the subspace consisting of all the $\mathcal{L}$-measurable functions in $L^2(\mu)$ which are supported on C, clearly a Hilbert space of infinite dimension. Let $\{e_1, e_2, \dots\}$ be an orthonormal basis for W consisting of *bounded* functions. Set $g_k = g \cdot e_k$. We shall show next that for every k, $g_k \in V$ and that $\{g_k\}$ is an orthonormal sequence; thus proving that $\dim V = \infty$.

Clearly g_k is $\mathcal{X}$-measurable and as e_k is bounded it is an element of K. However, for any $h \in U^{-1}K$ we have

$$\langle g_k, h\rangle = \mathbb{E}(g_k \bar{h}) = \mathbb{E}(\mathbb{E}^{\mathcal{L}}(g_k \bar{h})) = \mathbb{E}(\bar{h} e_k \mathbb{E}^{\mathcal{L}}(g)) = 0,$$

and therefore $g_k \in V = K \ominus U^{-1}K$. Finally

$$\begin{aligned}\langle g_k, g_l\rangle &= \mathbb{E}(g_k \bar{g}_l) = \mathbb{E}(g e_k \bar{g} \bar{e}_l)\\ &= \mathbb{E}(|g|^2 e_k \bar{e}_l) = \mathbb{E}(\mathbb{E}^{\mathcal{L}}(|g|^2 e_k \bar{e}_l))\\ &= \mathbb{E}(e_k \bar{e}_l \mathbb{E}^{\mathcal{L}}(|g|^2)) = \mathbb{E}(e_k \bar{e}_l \mathbf{1}_C)\\ &= \mathbb{E}(e_k \bar{e}_l) = \langle e_k, e_l\rangle = \delta_{kl}.\end{aligned}$$

□

4. Irreducible Koopman representations

The Koopman representation, $n \mapsto U_{T^n}$ of the group of integers $\mathbb{Z}$ corresponding to a nonatomic $\mathbb{Z}$ dynamical system $(X, \mathcal{X}, T, \mu)$ (whether ergodic or not), restricted to the subspace $L_0^2(\mu)$, is never irreducible. In fact if $Z(h) \subset L_0^2(\mu)$, with $0 \neq h \in L_0^2(\mu)$, is any cyclic subspace then, by Proposition 5.6, $U_T \restriction Z(h)$ is unitarilly equivalent to

$$V : L^2(\sigma_h, \mathbb{T}) \to L^2(\sigma_h, \mathbb{T}) \quad \text{defined by} \quad Vf(z) = zf(z) \quad (z \in \mathbb{T}).$$

Now to every Borel subset $B \subset \mathbb{T}$ corresponds the V-invariant subspace $\mathbf{1}_B L^2(\sigma_h, \mathbb{T})$ and thus $U_T \restriction Z(h)$ is evidently reducible. It can be shown similarly that no Koopman representation corresponding to a nonatomic Γ-system $(X, \mathcal{X}, \Gamma, \mu)$ is irreducible on $L_0^2(\mu)$, whenever Γ is abelian. Can any Koopman representation corresponding to nonatomic ergodic system $(X, \mathcal{X}, \Gamma, \mu)$ be irreducible? The answer is yes and we shall show next that this can be achieved with Γ a countable amenable group.

Let $\mathbb{Z}_2 = \{0, 1\}$ and $X = \mathbb{Z}_2^{\mathbb{N}}$, the compact topological group with respect to product topology and coordinatewise addition modulo 2. Let $Y = \bigoplus_{n=0}^{\infty} \mathbb{Z}_2 \subset X$ be the countable subgroup

$$Y = \{x \in X : \text{there is some } n \geq 0 \text{ with } x_j = 0 \text{ for } j \geq n\}.$$

For $y = (y_0, y_1, \dots, y_n, 0, 0, 0, \dots) \in Y$ let the function $\chi_y : X \to \mathbb{Z}_2$ be defined by

$$\chi_y(x) = (-1)^{\sum_{j=0}^{\infty} y_j x_j}.$$

It is easy to verify that for each $y \in Y$, χ_y is a continuous character of the compact abelian group X — i.e. an element of the dual group $\hat{X}$ — and that the map $\phi : y \mapsto \chi_y$ is an isomorphism onto $\phi : Y \to \hat{X}$. We ignore the topology Y inherits from X, and consider it as a discrete group thus identifying Y with $\hat{X}$.

Denoting the normalized Haar measure on X by m we let $\mathfrak{H} = L_0^2(X,m)$ be the subspace of the Hilbert space $L^2(X,m)$ consisting of functions with zero integral. Let $G = \operatorname{Aut}(X,m)$ be the Polish group of automorphisms of the probability measure space (X,m), i.e. the group of invertible bi-measurable maps $T : X \to X$ such that $m \circ T^{-1} = m$, equipped with the topology of convergence in measure or equivalently the weak operator topology on G obtained by identifying $T \in G$ with the corresponding unitary operator $U_T : f \mapsto f \circ T^{-1}$ on $\mathfrak{H}$. We call this unitary representation $\pi : G \to \mathfrak{U}(\mathfrak{H}), \pi(T) = U_T$, the Koopman representation.

5.14. THEOREM. *The Koopman representation* $\pi : G \to \mathfrak{U}(\mathfrak{H})$ *is irreducible.*

PROOF. 1. Consider two elements $y, z \in Y$ both different from zero. We shall show that there is $T \in G$ with $\chi_y \circ T = \chi_z$. In fact, denoting $\ker \chi_y = A$, $\ker \chi_z = B$, we observe that A and B are clopen subgroups of X with index 2 hence with $m(A) = m(B) = 1/2$. It now suffices to choose any $T \in G$ such that $TA = B$.

2. For a function $f \in \mathfrak{H}$ let $Z(f)$ denote the cyclic subspace of $\mathfrak{H}$ corresponding to f, i.e. the smallest closed G-invariant subspace of $\mathfrak{H}$ containing f. We have to show that $Z(f) = \mathfrak{H}$ whenever $f \neq 0$. Let

$$f = \sum_{y \in Y \setminus \{0\}} c_y \chi_y$$

be the Fourier expansion of f. The fact that $f \neq 0$ is equivalent to the statement that there exists $0 \neq y_0 \in Y$ with $c_{y_0} \neq 0$. Now, as we have seen, for every $0 \neq y \in Y$ there is $T_y \in G$ such that $\chi_{y_0} = \chi_y \circ T_y$, whence

$$\langle f \circ T_y^{-1}, \chi_y \rangle = \langle f, \chi_y \circ T_y \rangle = \langle c_{y_0} \chi_{y_0}, \chi_{y_0} \rangle = c_{y_0} \neq 0.$$

3. Denote $\mathfrak{K} = Z(f)^{\perp}$ and let $h \in \mathfrak{K}$. Set, for $y_0 \in Y$ and $z \in X$,

$$h_{y_0}(z) = \int_X h(z+x) \chi_{y_0}(x)\, dm(x),$$

and observe that $h_{y_0} \in \mathfrak{K}$. If $h = \sum_{y \in Y \setminus \{0\}} a_y \chi_y$ is the Fourier expansion of h then

$$\begin{aligned} h_{y_0}(z) &= \int_X h(z+x)\chi_{y_0}(x)\, dm(x) \\ &= \sum_{y \in Y \setminus \{0\}} a_y \int_X \chi_y(z+x)\chi_{y_0}(x)\, dm(x) \\ &= \sum_{y \in Y \setminus \{0\}} a_y \chi_y(z) \int_X \chi_y(x)\chi_{y_0}(x)\, dm(x) \\ &= a_{y_0}\chi_{y_0}(z). \end{aligned}$$

It follows that $\chi_{y_0} \in \mathfrak{K}$ whenever $a_{y_0} \neq 0$. However, by part 2, no χ_{y_0} is in $\mathfrak{K}$, whence $h = 0$. Thus $\mathfrak{K} = Z(f)^{\perp} = \{0\}$, whence $Z(f) = \mathfrak{H}$. $\square$

5.15. REMARKS. 1. Since all non-atomic Lebesgue spaces (Ω, μ) are isomorphic we can take instead of (X,m) any other model and get the same result. For example we can take (Ω, μ) to be the unit interval or the circle with Lebesgue measure.

2. Let $\Gamma \subset G$ be any countable dense subgroup, then the restriction of π to Γ yields an example of a Koopman irreducible representation for Γ. For example if we

take $X = [0,1]$, the unit interval, α_n the dyadic partition $[0,1] = \cup_{j=0}^{2^n-1}[j/2^n, j + 1/2^n]$ and Γ_n the group of permutations of these intervals, then we find that the amenable group $\Gamma = \cup_{n=1}^{\infty}\Gamma_n$ admits an irreducible Koopman representation.

3. Using the method described above, examples of irreducible Koopman representations can be constructed directly for groups Γ (which are not necessarily dense in $G = \mathrm{Aut}\,(X, m)$) with other properties like being solvable, or being finitely generated.

5. Notes

Material for this chapter is drawn from Parry [**203**], Cornfeld, Fomin and Sinai [**43**], Lemańczyk [**165**] and Weiss [**260**]. Proposition 5.9 is from Bourgain's paper [**31**].

The spectral type of a K-system: Theorem 5.13 is due to Kolmogorov [**157**]; the first detailed proof is to be found in Rohlin [**211**]. See the recent work of Dooley and Golodets [**49**] for an extension of this result to amenable groups.

Irreducible Koopman representations: The question whether a Koopman representation can be irreducible was raised in a conversation with P. de la Harpe. The material of this section was developed in several discussions of the problem with H. Furstenberg and B. Weiss. Independently similar examples were produced by B. Bekka and P. de la Harpe.

CHAPTER 6

Joinings

A joining λ of two dynamical systems $\mathbf{X}$ and $\mathbf{Y}$ can be presented as an invariant measure on the product space $X \times Y$ with marginals μ and ν respectively, as a Markov operator $P_\lambda : L^2(X,\mu) \to L^2(Y,\nu)$ that intertwines the Γ actions on $\mathbf{X}$ and $\mathbf{Y}$; or finally as a homomorphism ϕ of one of the systems, say $\mathbf{Y}$, into the space of probability measures on the other $M(X)$, with the property that the barycenter of the image $\phi_*(\nu)$ equals μ. All three approaches are useful and in this chapter we introduce and investigate the two first possibilities (in the first and second sections respectively). The third approach leads to the notion of a quasifactor and will be studied in Chapter 8.

One of the first achievements of the theory of joinings was Veech's theorem characterizing group extensions in terms of joinings. This result is described in Section 3 and we extend it, in Section 4, to a characterization of a general homogeneous skew-products (i.e. isometric extensions, see Theorem 9.14 below). After a short section on finite type joinings (Section 5) we introduce in Section 6 the important notion of disjointness and prove the fundamental "relative independence theorem" which illuminates the relation between joinings and common factors. We conclude the chapter with a short section (Section 7) where it is shown that spectral disjointness is stronger than disjointness.

1. Joinings of two systems

6.1. Definition. Let $\mathbf{X} = (X, \mathcal{X}, \mu, \Gamma)$ and $\mathbf{Y} = (Y, \mathcal{Y}, \nu, \Gamma)$ be ergodic systems. A probability measure λ on $(X \times Y, \mathcal{X} \otimes \mathcal{Y})$ is a *joining* of $\mathbf{X}$ and $\mathbf{Y}$ if it is Γ-invariant, and has μ and ν as marginals; i.e. $\mathrm{proj}_X(\lambda) = \mu$ and $\mathrm{proj}_Y(\lambda) = \nu$. We let $J(\mathbf{X},\mathbf{Y})$ (or $J(\mu,\nu)$) be the space of all joinings of $\mathbf{X}$ and $\mathbf{Y}$; this is never empty as $\mu \times \nu \in J(\mathbf{X},\mathbf{Y})$. We fix topological Cantor models for $\mathbf{X}$ and $\mathbf{Y}$ and then it is clear that $J(\mathbf{X},\mathbf{Y})$ becomes a convex compact subset of $M_\Gamma(X \times Y)$ with its weak* topology. We denote its subset of extreme points by $J_e(\mathbf{X},\mathbf{Y})$. When $\mathbf{Y} = \mathbf{X}$ we refer to such joinings as *self-joinings* (of order 2) and write $J(\mathbf{X},\mathbf{Y}) = J(\mathbf{X}) = J(\mu)$. More generally if $\{\mathbf{X}_i = (X_i, \mathcal{X}_i, \mu_i, \Gamma)\}_{i\in I}$ is a collection of dynamical systems, a probability measure λ on $\Pi_{i\in I}\mathbf{X}_i$ is a *joining* of the systems $\{\mathbf{X}_i\}_{i\in I}$ if λ is Γ-invariant, and has μ_i as marginals; i.e. $\mathrm{proj}_{X_i}(\lambda) = \mu_i$ for every $i \in I$. We let $J(\{\mathbf{X}_i\}_{i\in I})$ be the space of all these joinings. When $I = \{1, 2, \dots, n\}$ we call such joinings *n-fold joinings*; and when in addition $\mathbf{X}_i = \mathbf{X}$ for $i = 1, 2, \dots, n$, *n-fold self-joinings*.

6.2. Theorem. *With respect to this topology on $J(\mathbf{X},\mathbf{Y})$, a sequence $\lambda_n \to \lambda$ iff $\lambda_n(A \times B) \to \lambda(A \times B)$ for every $A \in \mathcal{X}$ and $B \in \mathcal{Y}$. Thus the topology on $J(\mathbf{X},\mathbf{Y})$ does not depend on the Cantor models we choose. The subset $J_e(\mathbf{X},\mathbf{Y})$ coincides with the set of ergodic measures in $J(\mathbf{X},\mathbf{Y})$.*

PROOF. If our condition on sequences is satisfied for all A, B, it is satisfied as well for clopen A, B and it follows that $\lambda_n \to \lambda$ weak*. Conversely, if $A \in \mathcal{X}, B \in \mathcal{Y}$ and $\epsilon > 0$ are given we can choose A', B' clopen with $\mu(A \triangle A') < \epsilon$ and $\nu(B \triangle B') < \epsilon$ and then for every $\kappa \in J(\mathbf{X}, \mathbf{Y})$, $\kappa(A \times B \,\triangle\, A' \times B') < 2\epsilon$. It is now easy to see that the weak* convergence $\lambda_n \to \lambda$ implies $\lambda_n(A \times B) \to \lambda(A \times B)$ for every $A \in \mathcal{X}$ and $B \in \mathcal{Y}$. Finally the ergodic measures in $M_\Gamma(X \times Y)$ are just the extreme points of that set (Theorem 4.2). Thus clearly an ergodic measure in $J(\mathbf{X}, \mathbf{Y})$ is an extreme point of $J(\mathbf{X}, \mathbf{Y})$.

Conversely, if $\lambda \in J(\mathbf{X}, \mathbf{Y})$ is written as $\lambda = \frac{1}{2}(\kappa + \theta)$ with $\kappa, \theta \in M_\Gamma(X \times Y)$, then, projecting onto X, we see by the extremality of μ, that the X-marginal measure of both κ and θ must be μ. Similarly their Y-marginal measure is ν and we conclude that κ and θ are in fact joinings. Thus if λ is not ergodic it is not in $J_e(\mathbf{X}, \mathbf{Y})$. □

Here are some important examples of joinings.

6.3. EXAMPLES. • **Graph joinings:** Let $\phi : \mathbf{X} \to \mathbf{Y}$ be a homomorphism of ergodic systems; we let

$$\mathrm{gr}\,(\mu, \phi) = \int_X (\delta_x \times \delta_{\phi(x)})\, d\mu(x) = (\mathrm{id} \times \phi)_*(\mu),$$

where $(\mathrm{id} \times \phi) : X \to X \times Y$ is the map given by $(\mathrm{id} \times \phi)(x) = (x, \phi(x))$. Equivalently $\mathrm{gr}\,(\mu, \phi) \in J_e(\mathbf{X}, \mathbf{Y})$ is defined by:

$$\mathrm{gr}\,(\mu, \phi)(A \times B) = \mu(A \cap \phi^{-1}B).$$

- **Isomorphism graph joinings:** When $\phi : \mathbf{X} \to \mathbf{Y}$ is an isomorphism of ergodic systems (e.g. an automorphism of $\mathbf{X}$), we say that $\mathrm{gr}\,(\mu, \phi)$ is an *isomorphism graph joining*. In particular, when Γ is abelian we have the *off-diagonal* joinings $\{\mathrm{gr}\,(\mu, \gamma) : \gamma \in \Gamma\} \subset J(\mathbf{X})$. When $\phi : \mathbf{X} \to \mathbf{X}$ is an automorphism we also use the notation $\mathrm{gr}\,(\mu, \phi) = \Delta_\mu^\phi$. In particular we let $\mathrm{gr}\,(\mu, \mathrm{id}) = \Delta_\mu^{\mathrm{id}} = \Delta_\mu$.

  An ergodic system $\mathbf{X}$ is called 2-*simple* if each ergodic self-joining of order 2 is either the graph of an automorphism, or $\mu \times \mu$ (when $\mathbf{X}$ is weakly mixing):

  $$J_e(\mathbf{X}) \setminus \{\mu \times \mu\} = \{\mathrm{gr}\,(\mu, \phi) : \phi \in \mathrm{Aut}\,(\mathbf{X})\}.$$

  We say that $\mathbf{X}$ has *minimal self-joinings* (of order 2) when Γ is abelian and every ergodic self-joining is either an off-diagonal, or $\mu \times \mu$ (when $\mathbf{X}$ is weakly mixing):

  $$J_e(\mathbf{X}) \setminus \{\mu \times \mu\} = \{\mathrm{gr}\,(\mu, \gamma) : \gamma \in \Gamma\}.$$

- **Relatively independent joinings:** If $\pi : \mathbf{X} \to \mathbf{Z}$ and $\sigma : \mathbf{Y} \to \mathbf{Z}$ are homomorphisms of ergodic systems, we let

  $$\mu = \int_Z \mu_z \, d\eta(z), \quad \nu = \int_Z \nu_z \, d\eta(z)$$

  be the disintegrations of μ and ν over η respectively, and put

  $$\lambda_{\pi,\sigma} = \int_Z (\mu_z \times \nu_z)\, d\eta(z),$$

the relatively independent joining of $\mathbf{X}$ *and* $\mathbf{Y}$ *over their common factor* $\mathbf{Z}$. It is easy to check that λ is in fact a joining of $\mathbf{X}$ and $\mathbf{Y}$, though it need not be ergodic. We denote this relatively independent joining by $\lambda = \mu \underset{\eta}{\times} \nu$. Taking $\mathbf{Z}$ to be the trivial system we of course get $\lambda = \mu \times \nu$.

- **Factor joinings:** In the last example, for $\mathbf{X} = \mathbf{Y}$ and $\pi = \sigma : \mathbf{X} \to \mathbf{Z}$, we get the *factor joining*

$$\mu \underset{\eta}{\times} \mu = \lambda_\pi = \int_Z (\mu_z \times \mu_z)\, d\eta(z),$$

with the product measure $\lambda = \mu \times \mu$, when $\mathbf{Z}$ is the trivial factor.

- **Joinings over a common factor:** Again let $\pi : \mathbf{X} \to \mathbf{Z}$ and $\sigma : \mathbf{Y} \to \mathbf{Z}$ be homomorphisms of ergodic systems so that $\mathbf{Z} = (Z, \mathcal{Z}, \eta)$ is a common factor. Every element of $\lambda \in J(\mathbf{X}, \mathbf{Y})$ projects in a natural way to a self-joining $(\pi \times \sigma)_* \lambda \in J(\mathbf{Z})$. We let $J^{\mathbf{Z}}(\mathbf{X}, \mathbf{Y})$ (or $J_\eta(\mathbf{X}, \mathbf{Y})$) be the subset consisting of those $\lambda \in J(\mathbf{X}, \mathbf{Y})$ whose image in $J(\mathbf{Z})$ is $\Delta_\eta = \mathrm{gr}\,(\eta, \mathrm{id})$, the diagonal measure on $\mathbf{Z} \times \mathbf{Z}$. Always, $\mu \underset{\eta}{\times} \nu \in J^{\mathbf{Z}}(\mathbf{X}, \mathbf{Y})$. When $\pi : \mathbf{X} \to \mathbf{Y}$ is a factor map we write $J^{\mathbf{Y}}(\mathbf{X})$ for $J^{\mathbf{Y}}(\mathbf{X}, \mathbf{X})$.
- **Joinings over a fixed joining:** In the setup of the previous example set

$$J^\rho(\mathbf{X}, \mathbf{Y}) = \{\lambda \in J(\mathbf{X}, \mathbf{Y}) : (\pi \times \pi)(\lambda) = \rho\},$$

where ρ is a fixed element of $J(\mathbf{Z})$. In this notation

$$J^{\mathbf{Z}}(\mathbf{X}, \mathbf{Y}) = J^{\mathrm{gr}\,(\eta, \mathrm{id})}(\mathbf{X}, \mathbf{Y}).$$

Another useful example to consider is the space $J^{\eta \times \eta}(\mathbf{X}, \mathbf{Y})$.

- **Joinings induced by a common extension:** Given a dynamical system $\mathbf{X} = (X, \mathcal{X}, \mu, \Gamma)$ and two factors $\mathcal{X}_1, \mathcal{X}_2 \subset \mathcal{X}$, the sub-$\sigma$-algebra $\mathcal{Y} \subset \mathcal{X}$, which is spanned by $\mathcal{X}_1$ and $\mathcal{X}_2$, defines a factor system $\pi : \mathbf{X} \to \mathbf{Y} = (Y, \mathcal{Y}, \mu, \Gamma)$. The latter can be viewed as a joining of the two factor systems $\pi_i : \mathbf{X} \to \mathbf{X}_i = (X_i, \mathcal{X}_i, \mu, \Gamma), i = 1, 2$. In fact, it is easy to see that the system $(X_1 \times X_2, \mathcal{X}_1 \otimes \mathcal{X}_2, \lambda, \Gamma)$, with $\lambda = (\pi_1 \times \pi_2)_*(\mu)$, is isomorphic to $\mathbf{Y}$.
- **Inverse limits:** Let D be a (countable) directed set and for every $\sigma \in D$ let $\mathbf{X}_\sigma = (X_\sigma, \mathcal{X}_\sigma, \mu_\sigma, \Gamma)$ be a dynamical system. Assume further that for every pair $\sigma > \tau$ there exits a factor map $\phi_{\sigma\tau} : \mathbf{X}_\sigma \to \mathbf{X}_\tau$ such that for any triple $\sigma > \tau > \rho$ in D the diagram

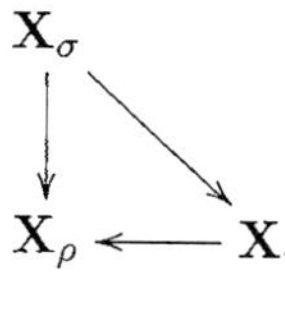

commutes. Set

$$X = \{x \in \prod_{\sigma \in D} X_\sigma : \phi_{\sigma\tau}(x_\sigma) = x_\tau, \text{ for all } \sigma > \tau\},$$

and let $\phi_\sigma : X \to X_\sigma$ be the projection maps. Clearly X is a Borel Γ-invariant subset of the product space and Kolmogorov's consistency theorem (Theorem A.6) yields a unique probability measure $\mu \in M_\Gamma(X)$ with the right marginals. The system $\mathbf{X} = (X, \mathcal{X}, \mu, \Gamma)$ — a joining of the family $\{\mathbf{X}_\sigma : \sigma \in D\}$ — is called the *inverse limit of the directed family* $\{\mathbf{X}_\sigma : \sigma \in D\}$ and we write

$$\mathbf{X} = \text{inv} \lim_D \{\mathbf{X}_\sigma : \sigma \in D\}.$$

We leave to the reader the proof of the following proposition concerning the inverse limit of directed families.

6.4. PROPOSITION. *Let $\{\mathbf{X}_\sigma : \sigma \in D\}$ be a directed family of dynamical systems and $\mathbf{X} = \text{inv} \lim_D \{\mathbf{X}_\sigma : \sigma \in D\}$.*

1. *$\mathbf{X}$ has the following universal property: If $\mathbf{Y}$ is a dynamical system and $\psi_\sigma : \mathbf{Y} \to \mathbf{X}_\sigma$ are factor maps such that for all $\tau > \rho$, the diagrams*

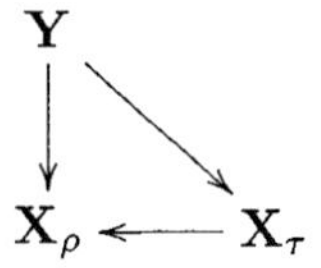

commute, then there exists a unique factor map $\pi : \mathbf{Y} \to \mathbf{X}$ such that all the diagrams

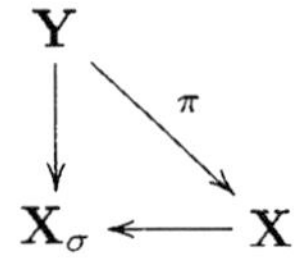

commute.

2. *Up to an isomorphism, $\mathbf{X}$ is the unique system with this universal property.*
3. *If all the systems $\mathbf{X}_\sigma$ are either ergodic, weakly mixing or isometric, then so is $\mathbf{X}$.*

In the next theorem we use the, slightly imprecise, notation $\mathcal{X} \times Y = \{A \times Y : A \in \mathcal{Y}\}$ and a similar definition for $X \times \mathcal{Y}$.

6.5. THEOREM. *Let $\mathbf{X}$ and $\mathbf{Y}$ be ergodic systems, $\lambda \in J_e(\mathbf{X}, \mathbf{Y})$. Then λ is a graph joining iff*

$$\mathcal{X} \times Y \supset X \times \mathcal{Y} \qquad (\text{mod } \lambda).$$

It is an isomorphism graph joining iff

$$\mathcal{X} \times Y = X \times \mathcal{Y} \qquad (\text{mod } \lambda).$$

PROOF. We prove the isomorphism graph joining case. When $\phi : \mathbf{X} \to \mathbf{Y}$ is an isomorphism we have, for $\lambda = \text{gr}(\mu, \phi)$ and every $A \in \mathcal{X}$

$$\begin{aligned}\lambda((A \times Y) \triangle (X \times \phi(A))) &= \lambda(A \times \phi(A)^c \cup A^c \times \phi(A)) \\ &= \mu(A \cap \phi^{-1}\phi(A^c)) + \mu(A^c \cap \phi^{-1}\phi(A)) = 0,\end{aligned}$$

and similarly for every $B \in \mathcal{Y}$, $\lambda((\phi^{-1}(B) \times Y) \triangle (X \times B)) = 0$.

Conversely our assumption $\mathcal{X} \times Y = X \times \mathcal{Y} \quad (\text{mod } \lambda)$ defines maps $\Phi : \mathcal{Y} \to \mathcal{X}$ and $\Psi : \mathcal{X} \to \mathcal{Y}$ such that, $\lambda((A \times Y) \triangle (X \times \Psi(A))) = 0$ and $\lambda((\Phi(B) \times Y) \triangle (X \times B)) = 0$. Clearly these maps are well defined on the measure algebras and we have $\Phi \circ \Psi = \text{id}_\mathcal{X}$ and $\Psi \circ \Phi = \text{id}_\mathcal{Y}$. Now apply Theorem 2.15. □

It is now easy to deduce the following.

6.6. THEOREM. *Let* $\mathbf{X}$ *and* $\mathbf{Y}$ *be two* Γ*-systems and* $\lambda \in J(\mathbf{X},\mathbf{Y})$*, then*

$$\mathcal{A} = \{A \in \mathcal{X} : \exists B \in \mathcal{Y},\ \lambda((A \times Y) \triangle (X \times B)) = 0\}$$

and

$$\mathcal{B} = \{B \in \mathcal{Y} : \exists A \in \mathcal{X},\ \lambda((A \times Y) \triangle (X \times B)) = 0\}$$

are Γ*-invariant sub-*σ*-algebras. We have*

$$\mathcal{A} \times Y = \mathcal{X} \times Y \cap X \times \mathcal{Y} = X \times \mathcal{B} \qquad (\text{mod } \lambda),$$

and the corresponding factors of $\mathbf{X}$ *and* $\mathbf{Y}$ *are isomorphic.*

Using the above isomorphism, we can identify the algebras $\mathcal{A} = \mathcal{B} = \mathcal{Z}_\lambda$, and consider $\mathcal{Z}$ as a common factor. For suitable Cantor models, where this common factor is denoted by $\mathbf{Z}$, we have the commutative diagram

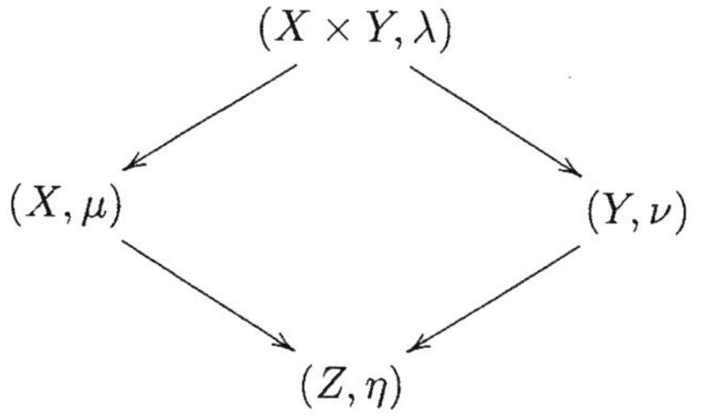

We call $\mathbf{Z} = \mathbf{Z}_\lambda$ the *common factor of* $\mathbf{X}$ *and* $\mathbf{Y}$ *determined by* λ, or just the *factor of* $\mathbf{X}$ *determined by* λ when $\mathbf{Y} = \mathbf{X}$.

6.7. EXERCISES. 1. Let $\pi : \mathbf{X} \to \mathbf{Z}$ and $\sigma : \mathbf{Y} \to \mathbf{Z}$ be two systems with a common factor $\mathbf{Z}$, then $\mathbf{Z}$ is the common factor of $\mathbf{X}$ and $\mathbf{Y}$ determined by $\lambda = \mu \underset{\eta}{\times} \nu$. [[If $\lambda((A \times Y) \triangle (X \times B)) = 0$ then, in $L^2(\lambda)$ and with the usual notation, $\mathbf{1}_A = \mathbb{E}^{\mathcal{X}}(\mathbf{1}_{A\times Y}) = \mathbb{E}^{\mathcal{X}}(\mathbf{1}_{X\times B})$. However, as $\lambda = \int_X \delta_x \times \nu_{\pi(x)}\, d\mu(x)$, it follows that $\mathbb{E}^{\mathcal{X}}(\mathbf{1}_{X\times B})(x) = \nu_{\pi(x)}(B)$, so that $\mathbf{1}_A$ is $\mathcal{Z}$-measurable.]]

2. Let $\pi : \mathbf{X} \to \mathbf{Z}$ and $\sigma : \mathbf{Y} \to \mathbf{Z}$ be two systems with a common factor $\mathbf{Z}$. The joining $\lambda \in J(\mathbf{X},\mathbf{Y})$ is the relative product measure $\mu \underset{\eta}{\times} \nu$ iff $L^2(\mathbf{X}) \ominus L^2(\mathbf{Z}) \perp L^2(\mathbf{Y}) \ominus L^2(\mathbf{Z})$.

2. Composition of joinings and the semigroup of Markov operators

Given a joining $\lambda \in J(\mathbf{X},\mathbf{Y})$ of the systems $\mathbf{X} = (X, \mathcal{X}, \mu, \Gamma)$ and $\mathbf{Y} = (Y, \mathcal{Y}, \nu, \Gamma)$ we define the *adjoint* joining $\lambda^* \in J(\mathbf{Y},\mathbf{X})$ as the dual joining of $\mathbf{Y}$ and $\mathbf{X}$; thus λ is a measure on $X \times Y$ whereas λ^* is defined on $Y \times X$. Let $\lambda = \int \lambda_y \times \delta_y\, d\nu(y)$ be the disintegration of λ over ν. We denote by $P_\lambda : L^2(X,\mu) \to L^2(Y,\nu)$ the conditional expectation operator given by

$$P_\lambda f(y) = \mathbb{E}^{\mathcal{Y}}_\lambda(f) = \int f(x)\, d\lambda_y(x) \quad \text{for} \quad \nu\text{-a.e. } y.$$

Equivalently P_λ is defined by

$$\int P_\lambda f(y) g(y)\, d\nu(y) = \int f(x) g(y)\, d\lambda(x,y)$$

$(f \in L^2(X,\mu), g \in L^2(Y,\nu))$. It is easy to check that $P_{\lambda^*} : L^2(Y,\nu) \to L^2(X,\mu)$ is the adjoint of P_λ. When $\pi : \mathbf{X} \to \mathbf{Y}$ is a factor map we write $P_\mathbf{Y} : L^2(X,\mu) \to L^2(Y,\nu)$ for the orthogonal projection (or the conditional expectation operator) $P_\mathbf{Y}(f) = \mathbb{E}^{\mathcal{Y}}(f) = \int f(x)\, d\lambda_y(x)$. Of course $P_\mathbf{Y} = P_{\lambda_\pi}$, where P_{λ_π} is the factor joining associated with π.

6.8. THEOREM. *A necessary and sufficient condition for a joining $\lambda \in J(\mathbf{X},\mathbf{Y})$, to be of the form $\mu \underset{\eta}{\times} \nu$, where $\mathbf{Z} = (Z, \mathcal{Z}, \eta, \Gamma)$ is the common factor of $\mathbf{X}$ and $\mathbf{Y}$ determined by λ, is that the operator $P_\lambda : L^2(X,\mu) \to L^2(Y,\nu)$ is the projection $P_\mathbf{Z}$ of $L^2(X,\mu)$ onto $L^2(Z,\eta)$. In particular $\lambda = \mu \times \nu$ iff P_λ is the projection on the constant functions; i.e. $P_\lambda(f) = \int f\, d\mu$. A self-joining $\lambda \in J(\mathbf{X})$ is a factor joining iff P_λ is a conditional expectation; i.e. iff $P_\lambda^* = P_\lambda$ and $P_\lambda^2 = P_\lambda$ is an idempotent.*

PROOF. The fact that $P_\lambda = P_\mathbf{Z}$ is satisfied for $\lambda = \mu \underset{\eta}{\times} \nu$ is easily seen. Conversely, under this condition, for every $f \in L^2(\mu)$, the function $P_\lambda(f) = \mathbb{E}_\lambda^{\mathcal{Y}}(f) = \mathbb{E}_\lambda^{\mathcal{Z}}(f)$ is $\mathcal{Z}$-measurable, hence for every $g \in L^2(\nu)$ (with $\mathbb{E} = \mathbb{E}_\lambda$):

$$\begin{aligned}
\mathbb{E}^{\mathcal{Z}}(f \cdot g) &= \mathbb{E}^{\mathcal{Z}}(\mathbb{E}^{\mathcal{Y}}(f \cdot g)) = \mathbb{E}^{\mathcal{Z}}(\mathbb{E}^{\mathcal{Y}}(f) \cdot g) \\
&= \mathbb{E}^{\mathcal{Z}}(P_\lambda(f) \cdot g) = \mathbb{E}^{\mathcal{Z}}(P_\mathbf{Z}(f) \cdot g) \\
&= \mathbb{E}^{\mathcal{Z}}(\mathbb{E}^{\mathcal{Z}}(f) \cdot g) = \mathbb{E}^{\mathcal{Z}}(f)\mathbb{E}^{\mathcal{Z}}(g).
\end{aligned}$$

This clearly implies $\lambda = \mu \underset{\eta}{\times} \nu$. □

Let $\lambda \in J(\mathbf{X},\mathbf{Y})$ and $\tau \in J(\mathbf{Y},\mathbf{Z})$ be given. Note that the relatively independent joining of λ and τ over (Y,ν,Γ), denoted $\lambda \underset{Y}{\times} \tau$, can be viewed as the measure on $X \times Y \times Z$ given by

$$\lambda \underset{Y}{\times} \tau = \int \lambda_y \times \delta_y \times \tau_y^*\, d\nu(y)$$

or equivalently by the equation

$$\int f(x)g(y)h(z)\, d\lambda \underset{Y}{\times} \tau = \int P_\lambda f(y) \cdot g(y) \cdot P_\tau^* h(y)\, d\nu(y)$$

$(f, g, h$ bounded measurable functions on X, Y and Z respectively).

6.9. DEFINITION. The measure $\tau \circ \lambda$ on $X \times Z$ given by

$$\tau \circ \lambda = \int \lambda_y \times \tau_y^*\, d\nu(y) = \mathrm{proj}_{X \times Z}(\lambda \underset{Y}{\times} \tau)$$

is called the *composition* of λ and τ. It is an element of $J(\mathbf{X},\mathbf{Z})$. The joinings $\lambda \in J(\mathbf{X},\mathbf{Y})$ and $\tau^* \in J(\mathbf{Z},\mathbf{Y})$ are said to be *orthogonal over* $\mathbf{Y}$ if $\tau \circ \lambda$ is the product measure $\mu \times \eta$.

6.10. PROPOSITION.

$$P_{\tau \circ \lambda} = P_\tau P_\lambda.$$

PROOF. We work in $L^2(X \times Y \times Z, \lambda \underset{Y}{\times} \tau)$ and expectations are calculated with respect to $\lambda \underset{Y}{\times} \tau$. By Theorem 6.8 we have, for $f \in L^2(X)$, $\mathbb{E}_{\lambda \underset{Y}{\times} \tau}^{\mathcal{Y} \otimes \mathcal{Z}} f = P_\mathbf{Y} f$. Now in

$L^2(X \times Y \times Z, \lambda)$, $P_{\mathbf{Y}} \restriction L^2(X) = P_\lambda$, hence,

$$\begin{aligned} P_{\tau\circ\lambda} f &= \mathbb{E}^{\mathcal{Z}}_{\lambda\circ\tau} f = \mathbb{E}^{\mathcal{Z}}_{\lambda \underset{Y}{\times} \tau} f = \mathbb{E}^{\mathcal{Z}}_{\lambda \underset{Y}{\times} \tau}(\mathbb{E}^{\mathcal{Y}\otimes\mathcal{Z}}_{\lambda \underset{Y}{\times} \tau} f) \\ &= \mathbb{E}^{\mathcal{Z}}_{\lambda \underset{Y}{\times} \tau}(P_{\mathbf{Y}} f) = \mathbb{E}^{\mathcal{Z}}_{\lambda \underset{Y}{\times} \tau}(P_\lambda f) = P_\tau P_\lambda f. \end{aligned}$$

□

6.11. PROPOSITION. *For a self-joining $\lambda \in J(\mathbf{X}, \mathbf{Y})$, $\lambda \circ \lambda^* = \mu \underset{Z}{\times} \mu$ for some factor $\pi : \mathbf{X} \to \mathbf{Z} = (Z, \mathcal{Z}, \eta, \Gamma)$ iff there exists a factor $\sigma : \mathbf{Y} \to \mathbf{Z}$ such that $\lambda = \mu \underset{Z}{\times} \nu$. In particular λ is orthogonal to λ^* over $\mathbf{Y}$ iff it is the independent joining $\lambda = \mu \times \nu$.*

PROOF. Clearly $\lambda = \mu \underset{Z}{\times} \nu$ implies $\lambda \circ \lambda^* = \mu \underset{Z}{\times} \mu$. Conversely assume $\lambda \circ \lambda^* = \mu \underset{Z}{\times} \mu$ and apply $\pi \times \pi$ to get

$$\mathrm{gr}\,(\eta, \mathrm{id}) = (\pi \times \pi)(\mu \underset{Z}{\times} \mu) = \int (\pi \times \pi)(\lambda_y \times \lambda_y)\, d\nu(y),$$

where $\eta = \pi(\mu)$. This means that for ν a.e. y the measure $(\pi \times \pi)(\lambda_y \times \lambda_y)$ is supported on the diagonal $\Delta_Z \subset Z \times Z$. Since $(\pi \times \pi)(\lambda_y \times \lambda_y) = \pi(\lambda_y) \times \pi(\lambda_y)$ is a product measure, it must have the form $\delta_z \times \delta_z$ where $z = \sigma(y)$ is a point in Z. This yields a factor map $\sigma : \mathbf{Y} \to \mathbf{Z}$.

Next disintegrate ν over η as $\nu = \int \nu_z\, d\eta(z)$ so that

$$\begin{aligned} \mu \underset{Z}{\times} \mu &= \int \mu_z \times \mu_z\, d\eta(z) \\ &= \int \lambda_y \times \lambda_y\, d\nu(y) \\ &= \int\!\!\int \lambda_y \times \lambda_y\, d\nu_z(y)\, d\eta(z). \end{aligned}$$

Since the measure $\int (\lambda_y \times \lambda_y)\, d\nu_z(y)$ is supported on the set $\pi^{-1}(z) \times \pi^{-1}(z)$, the uniqueness of disintegration implies $\int \lambda_y \times \lambda_y\, d\nu_z(y) = \mu_z \times \mu_z$, η a.e.

Now for every bounded $f \in L^2(\mu_z)$, denoting by F the function on Y given by

$$F(y) = \int_X f(x)\, d\lambda_y(x),$$

we get

$$\begin{aligned} \int_Y \int_{X\times X} f \otimes f\, d(\lambda_y \times \lambda_y)\, d\nu_z(y) &= \int_Y \left(\int_X f\, d\lambda_y\right)^2 d\nu_z(y) \\ &= \int_Y (F(y))^2\, d\nu_z(y), \end{aligned}$$

while

$$\int_{X\times X} f\otimes f\, d(\mu_z\times\mu_z) = \left(\int_X f(x)\, d\mu_z(x)\right)^2$$
$$= \left(\int_Y\int_X f(x)\, d\lambda_y(x)\, d\nu_z(y)\right)^2$$
$$= \left(\int_Y F(y)\, d\nu_z(y)\right)^2 .$$

Equality in Cauchy-Schwartz's inequality, implies that $F \equiv \int F\, d\nu_z$ is a constant ν_z a.e. Now functions of the form F above, considered as functions on the space $M(X)$ of probability measures on X, distinguish between the points of $M(X)$. We therefore conclude that λ_y is a constant measure ν_z a.e.. Since $\int \lambda_y\, d\nu_z(y) = \mu_z$, we have $\lambda_y = \mu_z$, ν_z a.e. and this for η almost every z. Hence

$$\lambda = \int \lambda_y\times\delta_y\, d\nu(y) = \int\int \mu_z\times\delta_y\, d\nu_z(y)\, d\eta(z)$$
$$= \int \mu_z\times\nu_z\, d\eta(z) = \mu \underset{Z}{\times} \nu.$$

□

6.12. Definition. Let $\mathbf{X}=(X,\mathcal{X},\mu)$ and $\mathbf{Y}=(Y,\mathcal{Y},\nu)$ be ergodic systems. An operator $P\in\mathfrak{L}(L^2(\mu),L^2(\nu))$ is called a *commuting Markov operator* if it satisfies the following conditions:

1. $P\geq 0$ and $P^*\geq 0$ ($P\geq 0$ means $Pf\geq 0$ for $f\geq 0$).
2. $P\mathbf{1}_X=\mathbf{1}_Y$ and $P^*\mathbf{1}_Y=\mathbf{1}_X$.
3. $PU_\gamma = U_\gamma P$ for every $\gamma\in\Gamma$.

Let $\mathbf{M}(\mathbf{X},\mathbf{Y})$ (or $\mathbf{M}(\mu,\nu)$) be the space of all such Markov operators and equip it with the weak operator topology. Write $\mathbf{M}(\mathbf{X})$ (or $\mathbf{M}(\mu)$) when $\mathbf{Y}=\mathbf{X}$. $\mathbf{M}(\mathbf{X})$ is a compact semitopological semigroup with $\mathbf{M}(\mathbf{X})\subset\mathbf{B}(\mathbf{X})$; however $\mathbf{E}(\mathbf{X})\subset\mathbf{M}(\mathbf{X})$ iff Γ is abelian. When this is the case then $(\mathbf{M}(\mathbf{X}),\Gamma)$ is a subsystem of the topological WAP system $(\mathbf{B}(\mathbf{X}),\Gamma)$ with group action $\gamma P = U_\gamma P$, $\gamma\in\Gamma, P\in\mathbf{M}(\mathbf{X})$. (See Chapter 3, Section 2 for the definitions of $\mathbf{B}(\mathbf{X})$ and $\mathbf{E}(\mathbf{X})$.)

We leave the straightforward proof of the following theorem as an exercise.

6.13. Theorem. 1. *The map $\phi:\lambda\mapsto P_\lambda$ from $J(\mathbf{X},\mathbf{Y})$ to $\mathbf{M}(\mathbf{X},\mathbf{Y})$ is a surjective affine homeomorphism. For $P\in\mathbf{M}(\mathbf{X},\mathbf{Y})$ the joining $\lambda=\phi^{-1}P$ is given by*

$$\lambda(A\times B) = \int_Y P(\mathbf{1}_A)\mathbf{1}_B\, d\nu.$$

2. *The selfadjoint idempotents of the semigroup $\mathbf{M}(\mathbf{X})$ correspond in a 1-1 way to factors of the system $\mathbf{X}$.*

When Γ is abelian we have

3 *The map $\phi: J(\mathbf{X})\to\mathbf{M}(\mathbf{X})$ satisfies*

$$\phi(\mathrm{gr}\,(\mu,\gamma)\circ\lambda) = U_\gamma P_\lambda, \qquad \gamma\in\Gamma.$$

Thus $(J(\mathbf{X}),\Gamma)$ is a WAP topological system isomorphic to $(\mathbf{M}(\mathbf{X}),\Gamma)$, where the action of Γ is given by $\gamma\lambda = \mathrm{gr}\,(\mu,\gamma)\circ\lambda$.

The next proposition justifies our terminology for orthogonal joinings.

6.14. PROPOSITION. *Let* $\mathbf{X} = (X, \mathcal{X}, \mu, \Gamma)$, $\mathbf{Y} = (Y, \mathcal{Y}, \nu, \Gamma)$, $\mathbf{Z} = (Z, \mathcal{Z}, \eta, \Gamma)$ *be dynamical systems. The joinings* $\lambda \in J(\mathbf{X}, \mathbf{Y})$ *and* $\tau^* \in J(\mathbf{Z}, \mathbf{Y})$ *are orthogonal over* $\mathbf{Y}$ *(i.e.* $\tau \circ \lambda = \mu \times \eta$*) iff the subspaces* $P_\lambda L_0^2(\mu)$ *and* $P_{\tau^*} L_0^2(\eta)$ *of* $L_0^2(\nu)$ *are orthogonal.*

PROOF. Suppose first that $\tau \circ \lambda = \mu \times \eta$, then given $f \in L_0^2(\mu), g \in L_0^2(\eta)$ we have

$$\begin{aligned} 0 &= \int_{X \times Z} f \otimes g \, d\mu \times \eta = \int_{X \times Z} f \otimes g \, d\tau \circ \lambda \\ &= \int_{X \times Y \times Z} f \otimes \mathbf{1}_Y \otimes g \, d\lambda \underset{Y}{\times} \tau \\ &= \int_Y \left(\int_X f(x) \, d\lambda_y(x) \int_Z g(z) \, d\tau_y^*(z) \right) d\nu(y) \\ &= \int P_\lambda f(y) P_{\tau*} g(y) \, d\nu(y) \\ &= \langle P_\lambda f, P_{\tau*} g \rangle. \end{aligned}$$

Conversely, if $P_\lambda L_0^2(\mu) \perp P_{\tau^*} L_0^2(\eta)$ then, given $f \in L^2(\mu), g \in L^2(\eta)$ with $\int f = a, f = f_0 + a, \int g = b, g = g_0 + b$, we have

$$\begin{aligned} \int_{X \times Z} f \otimes g \, d\tau \circ \lambda &= \int_{X \times Y \times Z} f \otimes \mathbf{1}_Y \otimes g \, d\lambda \underset{Y}{\times} \tau \\ &= \int_{X \times Y \times Z} (f_0 + a) \otimes \mathbf{1}_Y \otimes (g_0 + b) \, d\lambda \underset{Y}{\times} \tau \\ &= \int_{X \times Y \times Z} a \otimes \mathbf{1}_Y \otimes b \, d\lambda \underset{Y}{\times} \tau = a \cdot b \\ &= \int_{X \times Z} f \otimes g \, d\mu \times \eta. \end{aligned}$$

□

3. Group extensions and Veech's theorem

In this section we study group factors of an ergodic system $\mathbf{X}$. We recall (see Chapter 2, Section 3) that these factors arise from compact subgroups of the Polish group $\mathrm{Aut}\,(\mathbf{X}) = \mathrm{Aut}\,(X, \mu, \Gamma)$. If $K \subset \mathrm{Aut}\,(\mathbf{X})$ is such a subgroup we let

$$\mathcal{A}(K) = \{A \in \mathcal{X} : \phi A = A, \ \forall \phi \in K\}.$$

It is easy to check that $\mathcal{A}(K)$ is a Γ-invariant sub-σ-algebra of $\mathcal{X}$; i.e. $\mathcal{A}(K)$ defines a factor system $\mathbf{Y} = (Y, \mathcal{Y}, \nu, \Gamma)$ and a factor map $\pi : \mathbf{X} \to \mathbf{Y}$, with $\pi^{-1}(\mathcal{Y}) = \mathcal{A}(K)$ and $\nu = \pi_*(\mu)$. We call $\mathbf{Y}$ the *group factor* corresponding to the compact group K (or call $\mathbf{X}$ a *group extension* of $\mathbf{Y}$) and denote it by $\mathbf{Y} = \mathbf{X}/K$.

Let m be Haar measure on K and define a Markov commuting operator $P_K \in \mathbf{M}(\mathbf{X})$ by the formula

$$P_K(f)(x) = \int_K f(\phi x) \, dm(\phi).$$

It is easy to check that P_K is indeed a commuting Markov operator and moreover that $P_K^* = P_K$ and $P_K^2 = P_K$. Thus by Theorem 6.8, $P_K = P_{\mathbf{Z}}$ is a projection operator (a conditional expectation): $P_{\mathbf{Z}} : L^2(X) \to L^2(Z)$, for a factor map $\sigma : \mathbf{X} \to \mathbf{Z}$. Since $P_K \circ \phi = P_K$ for every $\phi \in K$, it follows that $\phi A = A$ for every $\phi \in K$ and $A \in \mathcal{Z}$, hence $\mathcal{Z} \subset \mathcal{Y}$. On the other hand $P_K(\mathbf{1}_A) = \mathbf{1}_A$ for every $A \in \mathcal{Y}$, so that $\mathcal{Z} = \mathcal{Y}$ and $\mathbf{Z} = \mathbf{Y} = \mathbf{X}/K$.

It now follows that μ a.e. the measure $\mu_{\pi(x)} = \int_K \delta_{\phi(x)}\, dm(\phi)$ depends only on $\pi(x)$ with $P_K(f)(x) = \int_X f\, d\mu_{\pi(x)}$ and we have $\mu = \int \mu_y\, d\nu(y)$ for the disintegration of μ over $\mathbf{Y} = \mathbf{X}/K$. Finally, the corresponding relatively independent product $\lambda = \mu \underset{Y}{\times} \mu$ has the following decomposition, which by the uniqueness, must be its ergodic decomposition:

$$
\begin{aligned}
\lambda &= \int_Y \mu_y \times \mu_y \, d\nu(y) \\
&= \int_Y \int_X \delta_x \times \left(\int_K \delta_{\phi(x)}\, dm(\phi) \right) d\mu_y(x)\, d\nu(y) \\
&= \int_K \int_X \int_Y (\delta_x \times \delta_{\phi(x)})\, dm(\phi)\, d\mu_y(x)\, d\nu(y) \qquad (3.1) \\
&= \int_K \int_X (\delta_x \times \delta_{\phi(x)})\, d\mu(x)\, dm(\phi) \\
&= \int_K \mathrm{gr}\,(\mu, \phi)\, dm(\phi).
\end{aligned}
$$

This discussion proves the first part of the following proposition.

6.15. PROPOSITION. *Let* $\mathbf{X}$ *be an ergodic system,* $K \subset \mathrm{Aut}\,(\mathbf{X})$ *a compact subgroup and* $\mathbf{Y} = \mathbf{X}/K$ *the corresponding group factor. Let* $\lambda = \int_Y \mu_y \times \mu_y\, d\nu(y)$ *be the relatively independent joining over* $\mathbf{Y}$.

1. $\lambda = \int_K \mathrm{gr}\,(\mu, \phi)\, dm(\phi)$ *is the ergodic decomposition of* λ.
2. *If* $\tau \in J_e^Y(\mathbf{X})$ *is an ergodic self-joining which projects onto* $\mathrm{gr}\,(\nu, \mathrm{id}) \in J(\mathbf{Y})$ *then* $\tau = \mathrm{gr}\,(\mu, \phi)$ *for some* $\phi \in K$.

PROOF. The first part follows from (3.1). In order to prove the second we let $\zeta = \int (\phi \times \mathrm{id})\tau\, dm(\phi)$, where m is Haar measure on K. Writing $\tau = \int \tau_x \times \delta_x\, d\mu(x)$ we observe that for μ-a.e. x, $\pi(\tau_x) = \delta_{\pi(x)}$ and

$$
\begin{aligned}
\int_K \phi(\tau_x)\, dm(\phi) &= \int_K \int_X \delta_{\phi x'}\, d\tau_x(x')\, dm(\phi) \\
&= \int_X \int_K \delta_{\phi x'}\, dm(\phi)\, d\tau_x(x') \\
&= \int_X \mu_{\pi(x)}\, d\tau_x(x') = \mu_{\pi(x)}.
\end{aligned}
$$

Therefore

$$\begin{aligned}\zeta &= \int_K (\phi \times \mathrm{id})\tau \, dm(\phi) = \int_K (\phi \times \mathrm{id}) \int (\tau_x \times \delta_x)\, d\mu(x)\, dm(\phi) \\ &= \int_X \int_K (\phi\tau_x \times \delta_x)\, dm(\phi)\, d\mu(x) = \int_X (\mu_{\pi(x)} \times \delta_x)\, d\mu(x) \\ &= \int_Y \int_X (\mu_y \times \delta_{x'})\, d\mu_y(x')\, d\nu(y) = \int_Y (\mu_y \times \mu_y)\, d\nu(y) \\ &= \mu \underset{Y}{\times} \mu = \lambda.\end{aligned}$$

Now the uniqueness of the ergodic decomposition shows that m-a.e., $(\phi \times \mathrm{id})\tau = \mathrm{gr}\,(\mu, \psi)$ for some $\psi \in K$, hence $\tau = \mathrm{gr}\,(\mu, \phi^{-1} \circ \psi)$. □

With every factor $\pi : \mathbf{X} \to \mathbf{Y}$ we associate the subgroup $\mathrm{Aut}^{Y}(\mathbf{X}) = H(\mathbf{Y}) = H(\mathcal{Y})$:

$$\begin{aligned}H(\mathcal{Y}) &= \{\phi \in \mathrm{Aut}\,(\mathbf{X}) : \phi A = A,\ \forall A \in \mathcal{Y}\} \\ &= \{\phi \in \mathrm{Aut}\,(\mathbf{X}) : \pi \circ \phi = \pi\}.\end{aligned}$$

Using Proposition 6.15 we now have the following "Galois" relation.

6.16. PROPOSITION. *For a compact group $K \subset \mathrm{Aut}\,(\mathbf{X})$ and $\mathbf{Y} = \mathbf{X}/K$ we have*

$$H(\mathcal{Y}) = H(\mathcal{A}(K)) = K, \quad \textit{hence} \quad \mathcal{A}(H(\mathcal{Y})) = \mathcal{Y}.$$

PROOF. The inclusion $K \subset H(\mathcal{A}(K))$ is trivial. Suppose $\phi_0 \in H(\mathcal{A}(K))$, then since by definition $\mathcal{Y} = \mathcal{A}(K)$, ϕ_0 acts as the identity on $\mathcal{Y}$. This and $\phi_0\mu = \mu$, imply $\phi_0\mu_y = \mu_y$, ν a.e., hence $(\mathrm{id} \times \phi_0)\lambda = \lambda$. Now

$$\begin{aligned}\lambda = (\mathrm{id} \times \phi_0)\lambda &= \int_K (\mathrm{id} \times \phi_0)\mathrm{gr}\,(\mu, \phi)\, dm(\phi) \\ &= \int_K \mathrm{gr}\,(\mu, \phi_0\phi)\, dm(\phi) = \int_K \mathrm{gr}\,(\mu, \phi)\, dm(\phi).\end{aligned}$$

The uniqueness of the ergodic decomposition implies that m a.e. $\phi_0\phi \in K$, hence $\phi_0 \in K$. Finally $\mathcal{A}(H(\mathcal{Y})) = \mathcal{A}(H(\mathcal{A}(K))) = \mathcal{A}(K) = \mathcal{Y}$. □

6.17. EXERCISE. Let $\hat{\pi} : \hat{\mathbf{X}} \to \mathbf{Y} = \hat{\mathbf{X}}/K$ be an ergodic group extension; show that every intermediate factor

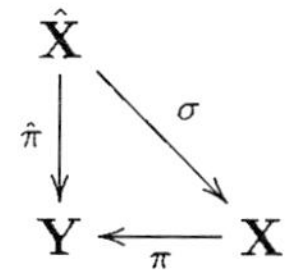

is of the form $\mathbf{X} = \hat{\mathbf{X}}/L$ for a closed subgroup $L \subset K$. [[Define $L = H(\mathcal{X})$ and use Proposition 6.16.]]

The following is a very useful converse to Proposition 6.15.

6.18. THEOREM (Veech). *Let* $\pi : \mathbf{X} \to \mathbf{Y}$ *be a factor of the ergodic system* $\mathbf{X}$. *Let*

$$\mu = \int_Y \mu_y \, d\nu(y) \qquad \text{and} \qquad \lambda = \mu \underset{\nu}{\times} \mu = \int_Y \mu_y \times \mu_y \, d\nu(y)$$

be the disintegrations of μ *and the relatively independent joining* $\lambda = \mu \underset{\nu}{\times} \mu$ *over* ν. *Let*

$$\lambda = \int_\Omega \omega \, dm(\omega),$$

be the ergodic decomposition of λ. *Suppose that in this decomposition* m *a.e.* ω *is of the form* $\omega = \mathrm{gr}\,(\mu, \phi)$ *for* $\phi = \phi_\omega \in \mathrm{Aut}\,(\mathbf{X})$, *then there is a compact subgroup* $K \subset \mathrm{Aut}\,(\mathbf{X})$ *such that* $\mathbf{Y} = \mathbf{X}/K$, *moreover* $K = \mathrm{Aut}^{\mathbf{Y}}(\mathbf{X})$ *and we can take* $K = \Omega$ *and* m *to be Haar measure on* K.

PROOF. Let

$$\lambda = \int_\Omega \omega \, dm(\omega), \tag{3.2}$$

where m a.e. ω is a graph joining $\omega = \mathrm{gr}\,(\mu, \phi_\omega)$ for some $\phi = \phi_\omega \in \mathrm{Aut}\,(\mathbf{X})$. Since clearly, for a.e. ω, $(\pi \times \pi)\omega = \mathrm{gr}\,(\nu, \mathrm{id})$, it follows that ϕ_ω fixes every element of the sub-σ-algebra $\mathcal{Y}$ and therefore $\pi \circ \phi_\omega = \pi$, m a.e.

Since $H(\mathbf{Y}) = \mathrm{Aut}^{Y}(\mathbf{X})$ is the closed subgroup of those elements $\phi \in \mathrm{Aut}\,(\mathbf{X})$ that satisfy $\pi \circ \phi_\omega = \pi$, we can identify Ω as a subset K of $\mathrm{Aut}^{\mathbf{Y}}(\mathbf{X})$ and m as a probability measure on this Polish group.

Now for $\phi \in \mathrm{Aut}^{\mathbf{Y}}(\mathbf{X})$ the equation $\phi\mu = \mu$ and the uniqueness of disintegration imply that $\phi\mu_y = \mu_y$, hence also $(\mathrm{id} \times \phi)\lambda = \lambda$. Applying $\mathrm{id} \times \phi$ to the ergodic decomposition (3.2), we see that for m a.e. ω, $(\mathrm{id} \times \phi)\mathrm{gr}\,(\mu, \phi_\omega) = \mathrm{gr}\,(\mu, \phi\phi_\omega) \in \Omega = K$ and that $\phi m = m$. Thus $\phi m = m$ for every $\phi \in K$ and we can therefore apply Mackey-Weil's theorem (Theorem 11.3, see also [**112**]) to deduce that $K \subset \mathrm{Aut}^{Y}(\mathbf{X})$ is a compact group and m its Haar measure.

If we let $Z = X/K$ and $\mathbf{Z} = \mathbf{X}/K = (Z, \mathcal{Z}, \eta)$ be the corresponding group factor then $\mathcal{Z} = \mathcal{A}(K) = \{A \in \mathcal{X} : \phi A = A, \ \forall \phi \in K\}$, hence $\mathcal{Y} \subset \mathcal{Z}$. Denoting the corresponding factor maps by $\rho : \mathbf{X} \to \mathbf{Z}$ and $\sigma : \mathbf{Z} \to \mathbf{Y}$ (so that $\pi = \sigma \circ \rho$), we have $\rho \circ \phi = \rho$ for every $\phi \in K$, hence

$$\begin{aligned} \eta \underset{\nu}{\times} \eta = (\rho \times \rho)\lambda &= \int_K (\rho \times \rho)\omega \, dm(\omega) \\ &= \int_K \mathrm{gr}\,(\eta, \mathrm{id}) \, dm(\omega) = \mathrm{gr}\,(\eta, \mathrm{id}). \end{aligned}$$

Thus σ is 1-1, $\mathbf{Y} = \mathbf{Z} = \mathbf{X}/K$ and $K = \mathrm{Aut}^{\mathbf{Y}}(\mathbf{X})$. □

4. A joining characterization of homogeneous skew-products

Let $\mathbf{X}$ be an ergodic system and $\pi : \mathbf{X} \to \mathbf{Y}$ a factor. Let $\mu = \int_Y \mu_y \, d\nu(y)$ be the disintegration of μ over ν and

$$\lambda = \int_Y \mu_y \times \mu_y \, d\nu(y),$$

the relatively independent product over $\mathbf{Y}$. We form the following infinite self-joinings

$$\lambda_\infty = \int_Y (\cdots \times \mu_y \times \mu_y \times \cdots)\, d\nu(y)$$

$$\text{and} \quad \tilde{\lambda}_\infty = \lambda_\infty \underset{\nu}{\times} \mu = \int_Y (\cdots \times \mu_y \times \mu_y \times \cdots) \times \mu_y \, d\nu(y)$$

on $X^{\mathbb{Z}}$ and $X^{\mathbb{Z}} \times X$ respectively. Denote by $p : X^{\mathbb{Z}} \times X \to X^{\mathbb{Z}}$ and $q : X^{\mathbb{Z}} \times X \to X$ the corresponding projections, so that $p_*(\tilde{\lambda}_\infty) = \lambda_\infty$ and $q_*(\tilde{\lambda}_\infty) = \mu$.

6.19. THEOREM. *The extension $\pi : \mathbf{X} \to \mathbf{Y}$ is a homogeneous skew-product iff almost every ergodic component $\tilde{\zeta}$ of $\tilde{\lambda}_\infty$ is a graph joining corresponding to a factor map $(X^{\mathbb{Z}}, \zeta) \to (X, \mu)$, with $\zeta = p_*(\tilde{\zeta})$; i.e. with respect to $\tilde{\zeta}$, in $X^{\mathbb{Z}} \times X$, the X coordinate is a.e. determined by the $X^{\mathbb{Z}}$ coordinates. Moreover, when this condition is satisfied then for almost every ergodic component ζ of λ_∞ we have the following commutative diagram*

$$\begin{array}{ccc} (X^{\mathbb{Z}}, \zeta) \cong \mathbf{Y} \underset{\alpha}{\times} K & & \\ \downarrow \hat{\pi} & \searrow^{\sigma} & \\ \mathbf{Y} & \xleftarrow[\pi]{} & \mathbf{X} \cong \mathbf{Y} \underset{\alpha}{\times} K/L \end{array}$$

In particular for almost every ζ the system $(X^{\mathbb{Z}}, \zeta)$ is the group extension that is associated with the homogeneous skew-product extension π.

PROOF. **Part I:** We assume that $\mathbf{X}$ is an ergodic system and that $\pi : \mathbf{X} \to \mathbf{Y}$ is a homogeneous skew-product; that is, for some compact group K, a closed subgroup $L \subset K$ and a cocycle $\alpha : \Gamma \times Y \to K$, $\mathbf{X} = (Y \times K/L, \tilde{m}, \Gamma)$, where m is Haar measure on K, $\mu = \tilde{m}$ its image on K/L, and for $\gamma \in \Gamma$ and $(y, kL) \in Y \times K/L$, $\gamma(y, kL) = (\gamma y, \alpha(\gamma, y)kL)$. We shall assume, as we well may with no loss in generality, that $\bigcap_{k \in K} kLk^{-1} = \{e\}$.

By Corollary 3.27 we can assume that the corresponding group extension $\hat{\mathbf{X}} = \mathbf{Y} \underset{\alpha}{\times} K$, with the invariant measure $\hat{\mu} = \nu \times m$, is ergodic. On $\hat{X}^{\mathbb{Z}}$ define

$$\xi_\infty = \int_Y (\cdots \times \hat{\mu}_y \times \hat{\mu}_y \times \cdots)\, d\nu(y),$$

where $\hat{\mu} = \int \hat{\mu}_y \, d\nu(y)$, $\hat{\mu}_y = \delta_y \times m$. If we let Δ_K be the subgroup

$$\Delta_K = \{(\ldots, k, k, \ldots) \in K^{\mathbb{Z}} : k \in K\},$$

then it is easy to check that the ergodic components of ξ_∞ are supported on the invariant subsystems $\mathbf{Y} \underset{\alpha}{\times} \Delta_K(\ldots, k_{-1}, k_0, k_1, \ldots)$ with $(\ldots, k_{-1}, k_0, k_1, \ldots) \in K^{\mathbb{Z}}$, each of them canonically isomorphic to $\hat{\mathbf{X}}$ via

$$q_0 : \mathbf{Y} \underset{\alpha}{\times} \Delta_K(\ldots, k_{-1}, k_0, k_1, \ldots) \to \hat{\mathbf{X}},$$

$$q_0(y, (\ldots, kk_{-1}, kk_0, kk_1, \ldots)) = (y, kk_0),$$

the projection on the zero coordinate.

The map $r : \hat{X}^{\mathbb{Z}} \to X^{\mathbb{Z}}$ defined by

$$r(y, (\ldots, k_{-1}, k_0, k_1, \ldots)) = (y, (\ldots, k_{-1}L, k_0L, k_1L, \ldots))$$

sends ξ_∞ onto λ_∞ and it follows that for a.e. ergodic component ζ of λ_∞ the system $(X^{\mathbb{Z}}, \zeta)$ is the image, under r, of an ergodic component of ξ_∞. Whence ζ is supported on a set of the form

$$r(Y \times \Delta_K(\dots, k_{-1}, k_0, k_1, \dots)) = \{(y, (\dots, kk_{-1}L, kk_0L, kk_1L, \dots)) : y \in Y, k \in K\},$$

for some $(\dots, k_{-1}, k_0, k_1, \dots) \in K^{\mathbb{Z}}$.

Set $f = r \circ q_0^{-1} : \mathbf{Y} \underset{\alpha}{\times} K \to X^{\mathbb{Z}}$,

$$\begin{aligned}(y, k) &\overset{q_0^{-1}}{\mapsto} (y, (\dots, kk_0^{-1}k_{-1}, k, kk_0^{-1}k_1, \dots)) \\ &\overset{r}{\mapsto} (y, (\dots, kk_0^{-1}k_{-1}L, kL, kk_0^{-1}k_1L, \dots)),\end{aligned}$$

and let

$$M(\zeta) = \bigcap_{i \in \mathbb{Z}} k_i L k_i^{-1}.$$

Then note that for $k, k' \in K$ we have $f(y, k) = f(y, k')$ iff $k^{-1}k' \in k_0^{-1}M(\zeta)k_0$. Thus the map $(y, kk_0^{-1}M(\zeta)k_0) \mapsto (y, (\dots, kk_{-1}L, kk_0L, kk_1L, \dots))$ is an isomorphism of the systems $\mathbf{Y} \underset{\alpha}{\times} K/k_0^{-1}M(\zeta)k_0$ and $(Y \underset{\alpha}{\times} (K/L)^{\mathbb{Z}}, \zeta)$.

It only remains to show that for almost every ζ the subgroup $M = M(\zeta) = \{e\}$. For this we first observe that if the set of elements in the sequence $\{k_i\}_{i \in \mathbb{Z}}$ forms a dense subset in K then indeed $M = \bigcap_{i \in \mathbb{Z}} k_i L k_i^{-1} = \{e\}$.

Since clearly the points $(\dots, k_{-1}, k_0, k_1, \dots) \in K^{\mathbb{Z}}$ for which $\{k_i\}_{i \in \mathbb{Z}}$ is dense in K form a set of $m^{\mathbb{Z}}$ measure one, it follows that for a.e. ergodic component θ of ξ_∞ the corresponding sequence $\{k_i\}_{i \in \mathbb{Z}}$ is dense in K. Projecting onto $Y \times (K/L)^{\mathbb{Z}}$ we conclude that indeed for a.e. component ζ of λ_∞, $M(\zeta) = \{e\}$.

Thus we established the fact that a.e. ergodic component ζ of λ_∞ is isomorphic to $\hat{\mathbf{X}}$ and (since the measures λ_∞ and $\tilde{\lambda}_\infty$ are basically one and the same) this clearly implies that almost every ergodic component $\tilde{\zeta}$ of $\tilde{\lambda}_\infty$ is a graph joining defined by the projection on the last coordinate as required. This completes the proof of the first part of the theorem.

Part II: We now assume that the condition of the theorem is satisfied. Our strategy will be to show that for a typical ergodic component ζ of λ_∞, almost every ergodic component of $\zeta \underset{\nu}{\times} \zeta$ is a graph joining and then apply Veech's theorem 6.18 to deduce that $\hat{\pi} : (X^{\mathbb{Z}}, \zeta) \to \mathbf{Y}$ is a group extension.

We therefore have to deduce this fact from the information given to us about the ergodic components $\tilde{\zeta}$ of $\tilde{\lambda}_\infty$.

Let

$$\tilde{\lambda}_\infty = \lambda_\infty \underset{\nu}{\times} \mu = \int \tilde{\zeta}\, d\tilde{P}(\tilde{\zeta}), \qquad \lambda_\infty = \int \zeta\, dP(\zeta)$$

$$\text{and} \quad \tilde{P} = \int P_\zeta\, dP(\zeta)$$

be the ergodic decompositions of $\tilde{\lambda}_\infty$, λ_∞ and the disintegration of $\tilde{P}$ over P respectively. By Fubini's theorem we have

$$\lambda_\infty \underset{\nu}{\times} \mu = \left(\int \zeta\, dP(\zeta) \right) \underset{\nu}{\times} \mu = \int \zeta \underset{\nu}{\times} \mu\, dP(\zeta).$$

For every ζ let $\zeta \underset{\nu}{\times} \mu = \int \hat{\zeta}\, dQ_\zeta(\hat{\zeta})$ be the ergodic decomposition of $\zeta \underset{\nu}{\times} \mu$, then

$$\tilde{\lambda}_\infty = \int \tilde{\zeta}\, d\tilde{P}(\tilde{\zeta}) = \int \int \tilde{\zeta}\, dP_\zeta(\tilde{\zeta})\, dP(\zeta) = \int \int \hat{\zeta}\, dQ_\zeta(\hat{\zeta})\, dP(\zeta).$$

The latter equality yields two expressions for the ergodic decomposition of the measure $\tilde{\lambda}_\infty$ and by the uniqueness of the ergodic decomposition we conclude that

$$\tilde{P} = \int P_\zeta\, dP(\zeta) = \int Q_\zeta\, dP(\zeta).$$

In turn, by the uniqueness of disintegration (both P_ζ and Q_ζ are concentrated on the set of measures $\tilde{\zeta}$ which project onto ζ), this implies that $P_\zeta = Q_\zeta$ for almost every ζ, whence

$$\zeta \underset{\nu}{\times} \mu = \int \tilde{\zeta}\, dP_\zeta(\tilde{\zeta})$$

is the ergodic decomposition of $\zeta \underset{\nu}{\times} \mu$.

Now we can use our assumption and deduce the fact that for a typical ergodic component ζ of λ_∞, almost every ergodic component $\tilde{\zeta}$ (with $p_*(\tilde{\zeta}) = \zeta$) of the measure $\zeta \underset{\nu}{\times} \mu$ is a graph joining; i.e. it has the property that with respect to it almost everywhere the last coordinate in $X^{\mathbb{Z}} \underset{Y}{\times} X$ is determined by the first $X^{\mathbb{Z}}$ coordinates.

Fixing $i \in \mathbb{Z}$ and considering the projection $q_i : X^{\mathbb{Z}} \underset{Y}{\times} X^{\mathbb{Z}} \to X$ defined by

$$q_i(y, (\dots, x_{-1}, x_0, x_1, \dots), (\dots, x'_{-1}, x'_0, x'_1, \dots)) = (y, (\dots, x_{-1}, x_0, x_1, \dots), x'_i),$$

and recalling that for almost every ergodic component ζ of λ_∞ we have

$$(q_i)_*(\zeta \underset{\nu}{\times} \zeta) = \zeta \underset{\nu}{\times} \mu,$$

we now conclude that for almost every ergodic component of $\zeta \underset{\nu}{\times} \zeta$, almost everywhere all the coordinates in the second $X^{\mathbb{Z}}$ factor are determined by the first $X^{\mathbb{Z}}$ factor. This means that every such joining is a graph joining; in fact, by symmetry, a graph of an isomorphism. Now apply Veech's theorem 6.18 to deduce that $\hat{\pi} : (X^{\mathbb{Z}}, \zeta) \to \mathbf{Y}$ is indeed a group extension. The proof of the theorem is now complete. □

5. Finite type joinings

Let $\lambda \in J_e(\mathbf{X}, \mathbf{Y})$ be an ergodic joining of the ergodic systems $\mathbf{X}$ and $\mathbf{Y}$ and $\lambda = \int \lambda_y \times \delta_y\, d\nu(y)$ the disintegration of λ over ν, then as is easy to see, either for ν-a.e. y, λ_y is a continuous measure or there exists a positive integer r such that for ν-a.e. y, λ_y is an atomic measure equidistributed on a finite subset of X of cardinality r. In the latter case $(X \times Y, \lambda, \Gamma)$ is an r to one extension of $\mathbf{Y}$ and we say that λ is of *finite type* (or more precisely of *type* r) over $\mathbf{Y}$.

6.20. LEMMA. *Let λ and τ be two ergodic joinings of the ergodic systems $\mathbf{X}$ and $\mathbf{Y}$ and let $\Delta \subset X \times X$ be the diagonal, then $\tau^* \circ \lambda(\Delta) > 0$ iff $\tau = \lambda$ and λ is of finite type. In this case $\tau^* \circ \lambda(\Delta) = \frac{1}{r}$ where r is the type of λ.*

PROOF. Since

$$\tau^* \circ \lambda(\Delta) = \int \lambda_y \times \tau_y(\Delta)\, d\nu(y) = \int \sum \lambda_y\{x\}\tau_y\{x\}\, d\nu(y),$$

it follows that when $\tau = \lambda$ and this joining is of finite type r, then $\tau^* \circ \lambda(\Delta) = r\frac{1}{r^2} = r$. Conversely suppose $\tau^* \circ \lambda(\Delta) > 0$, then the above formula shows that both λ and τ are of finite type, say s and t respectively, and that

$$\tau^* \circ \lambda(\Delta) = \frac{1}{t \cdot s} \int \operatorname{card}(A_y \cap B_y)\, d\nu(y)\ ,$$

where A_y and B_y are the finite supports of λ_y and τ_y respectively. Since for every $\gamma \in \Gamma$ $\gamma A_y = A_{\gamma y}$ and $\gamma B_y = B_{\gamma y}$ ν-a.e. it follows that the subset $E = \{(x, y) : x \in A_y \cap B_y\}$ of $X \times Y$ is Γ-invariant. Since moreover

$$\begin{aligned}\lambda(E) &= \int \lambda_y \times \delta_y(E)\, d\nu(y) = \frac{1}{s} \int \operatorname{card}(A_y \cap B_y)\, d\nu(y) \\ &= t(\tau^* \circ \lambda)(\Delta) > 0,\end{aligned}$$

the ergodicity of λ implies $\lambda(E) = 1$. Thus card $(A_y \cap B_y) = s$, whence $A_y \subset B_y$ for ν-a.e. y. By symmetry also $B_y \subset A_y$ and we conclude that $\lambda = \tau$ and that λ is of type $r = s = t$. □

6. Disjointness and the relative independence theorem

In this section we introduce Furstenberg's notion of disjointness and prove the relative independence theorem. This theorem unveils the mystery behind Furstenberg's question on disjointness versus "no common factor" (see the introduction, Chapter 0). It reveals how a general joining of two systems arises from a relatively independent joining over a common factor of one of the systems with a symmetric infinite self-joining of the other. As an application of the relative independence theorem we deduce the disjointness of isometric and weakly mixing systems. We shall come back to these questions in Chapter 8 and consider them from another viewpoint, namely that of quasifactors.

6.21. DEFINITION. Two systems $\mathbf{X}$ and $\mathbf{Y}$ are *disjoint* if $\mu \times \nu$ is the unique joining of the two systems. More generally if $\mathbf{Z}$ is a common factor:

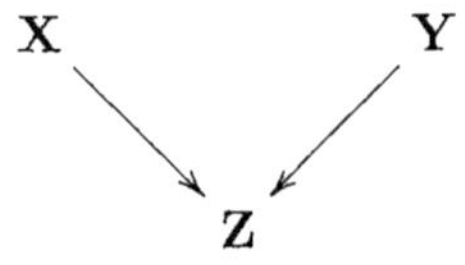

we say that $\mathbf{X}$ and $\mathbf{Y}$ are *disjoint over* $\mathbf{Z}$ if $\mu \underset{\eta}{\times} \nu$ is the unique element of $J_\eta(\mu, \nu)$.

6.22. EXERCISES. 1. Show that the systems $\mathbf{X}$ and $\mathbf{Y}$ are disjoint iff whenever $\mathbf{Z} \xrightarrow{\alpha} \mathbf{X}$ and $\mathbf{Z} \xrightarrow{\beta} \mathbf{Y}$ is a common extension then there exits a factor map $\mathbf{Z} \xrightarrow{\theta} \mathbf{X} \times \mathbf{Y} = (X \times Y, \mathcal{X} \otimes \mathcal{Y}, \mu \times \nu)$ such that the following diagram

commutes

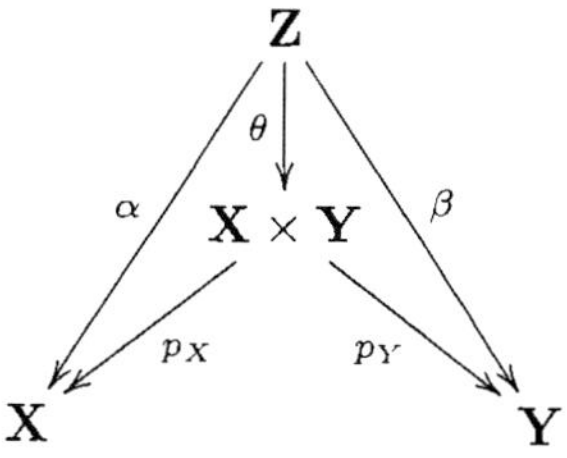

2. If Γ is amenable, (X, Γ) and (Y, Γ) are strictly ergodic topological systems with unique invariant measures μ and ν respectively, and the corresponding measure-preserving systems are disjoint, then (X, Γ) and (Y, Γ) are topologically disjoint.

In general, even for strictly ergodic $\mathbb{Z}$-systems, topological disjointness does not imply measure disjointness. This is not easy to show; one needs sophisticated constructions in order to produce a counter example. One such example is as follows. Use a theorem of Lehrer **[164]** to produce a topologically mixing, strictly ergodic topological system (X, μ, T) such that there is a *measure theoretical* isomorphism $\phi : \mathbf{X} = (X, \mathcal{X}, \mu, T) \to \mathbf{Y}$, where $\mathbf{Y} = (Y, \mathcal{Y}, \nu, R_\alpha)$ is an irrational rotation on the circle Y. It then follows that the topological systems (X, T) and (Y, R_α) are disjoint but on $X \times Y$ there are at least two $T \times R_\alpha$-invariant measures, namely $\mu \times \nu$ and $\mathrm{gr}\,(\mu, \phi)$. Evidently both these measures are joinings of μ and ν, so that the measure-preserving systems $\mathbf{X}$ and $\mathbf{Y}$ are not disjoint.

However we have the following:

6.23. THEOREM. *If (Z, η, Γ) is a strictly ergodic system and $(Z, \Gamma) \xrightarrow{\alpha} (X, \Gamma)$ and $(Z, \Gamma) \xrightarrow{\beta} (Y, \Gamma)$ are topological factors such that $\alpha^{-1}(U) \cap \beta^{-1}(V) \neq \emptyset$ whenever $U \subset X$ and $V \subset Y$ are nonempty open sets, then the measure-preserving systems $\mathbf{X} = (X, \mathcal{X}, \mu, \Gamma)$ and $\mathbf{Y} = (Y, \mathcal{Y}, \nu, \Gamma)$ are measure-theoretically disjoint. In particular this is the case if the systems (X, Γ) and (Y, Γ) are topologically disjoint.*

PROOF. It suffices to show that the map $\alpha \times \beta : Z \to X \times Y$ is onto since this will imply that the topological system $(X \times Y, \Gamma)$ is strictly ergodic. We establish this by showing that the measure $\lambda = (\alpha \times \beta)_*(\eta)$ (a joining of μ and ν) is full; i.e. that it assigns positive measure to every set of the form $U \times V$ with U and V as in the statement of the theorem. In fact, since by assumption η is full we have

$$\lambda(U \times V) = \eta((\alpha \times \beta)^{-1}(U \times V)) = \eta(\alpha^{-1}(U) \cap \beta^{-1}(V)) > 0.$$

This completes the proof of the first assertion. The second follows since topological disjointness of (X, Γ) and (Y, Γ) implies that $\alpha \times \beta : Z \to X \times Y$ is onto. □

We say that $\pi : \mathbf{X} \to \mathbf{Y}$ is an *ergodic extension* if every Γ-invariant $\mathcal{X}$-measurable function is $\mathcal{Y}$-measurable.

6.24. LEMMA. *Let $(Y, \mathcal{Y}, \nu, \Gamma) \xrightarrow{\phi} (Z, \mathcal{Z}, \eta, \mathrm{id})$ be an extension of Γ-systems, where* id *denotes the trivial Γ action. The extension $\mathbf{Y} \to \mathbf{Z}$ is ergodic iff whenever $(X, \mathcal{X}, \mu, \mathrm{id}) \xrightarrow{\pi} (Z, \mathcal{Z}, \eta, \mathrm{id})$, then $(X, \mathcal{X}, \mu, \mathrm{id})$ and $(Y, \mathcal{Y}, \nu, \Gamma)$ are relatively disjoint over $(Z, \mathcal{Z}, \eta, \mathrm{id})$; i.e. $\lambda = \mu \underset{\eta}{\times} \nu$ is the only joining of X and Y over Z. In particular the Γ-system $\mathbf{Y}$ is ergodic iff it is disjoint from every* id*-system $\mathbf{X} = (X, \mathcal{X}, \mu, \mathrm{id})$.*

PROOF. Let λ be any joining of the systems $(X, \mathcal{X}, \mu, \mathrm{id})$ and $(Y, \mathcal{Y}, \nu, \Gamma)$ over their common factor $(Z, \mathcal{Z}, \eta, \mathrm{id})$. Let

$$\mu = \int_Z \mu_z \, d\eta(z) \quad \text{and} \quad \nu = \int_Z \nu_z \, d\eta(z),$$

be the disintegrations of μ and ν over η respectively, and let

$$\lambda = \int_Y \lambda_y \times \delta_y \, d\nu(y).$$

be the disintegration of λ over ν. Then for every $\gamma \in \Gamma$

$$\lambda = (\mathrm{id} \times \gamma)\lambda = \int_Y \lambda_y \times \delta_{\gamma y} \, d\nu(y) = \int_Y \lambda_{\gamma^{-1} y} \times \delta_y \, d\nu(y).$$

By uniqueness of disintegration we have $\lambda_y = \lambda_{\gamma^{-1}y}$ ν-a.e., hence, by ergodicity of the extension $Y \xrightarrow{\phi} Z$, $\lambda_y = \lambda_{\phi(y)} = \mu_{\phi(y)}$ ν-a.e. (the latter equality follows by projecting the disintegration of λ onto the X coordinate). Thus

$$\begin{aligned} \lambda &= \int_Y \mu_{\phi(y)} \times \delta_y \, d\nu(y) \\ &= \int_Z \int_Y \mu_z \times \delta_y \, d\nu_z(y) \, d\eta(z) \\ &= \int_Z \mu_z \times \left(\int_Y \delta_y \, d\nu_z(y) \right) \, d\eta(z) \\ &= \int_Z \mu_z \times \nu_z \, d\eta(z) = \mu \underset{\eta}{\times} \nu. \end{aligned}$$

Conversely, if ϕ is not an ergodic extension we can find a bounded Γ-invariant $\mathcal{Y}$ but not $\mathcal{Z}$-measurable function $0 \le f \le 1$, with $\int f \, d\nu = 1/2$. Then, setting $X = Z \times \{0, 1\}$ with product measure $\mu = \eta \times \frac{1}{2}(\delta_0 + \delta_1)$, we define probability measures λ_i, $i = 0, 1$, by

$$\lambda_0 = 2 \int_Z (\delta_z \times \delta_0 \times f \cdot \nu_z) \, d\eta(z), \qquad \lambda_1 = 2 \int_Z (\delta_z \times \delta_1 \times (1 - f) \cdot \nu_z) \, d\eta(z).$$

Now $\lambda = \frac{1}{2}(\lambda_0 + \lambda_1) \in J^{\mathbf{Z}}(\mathbf{X}, \mathbf{Y})$ and $\lambda \neq \mu \underset{\eta}{\times} \nu$. □

6.25. THEOREM (The relative independence theorem). *Let $(X, \mathcal{X}, \mu)$ and $(Y, \mathcal{Y}, \nu)$ be Lebesgue spaces (not necessarily Γ-systems), λ a joining of μ and ν (i.e. λ projects to μ and ν). Let*

$$\lambda = \int_Y \lambda_y \times \delta_y \, d\nu(y),$$

be the disintegration of λ over ν and define the probability measures λ_∞ on $X^{\mathbb{Z}} \times Y$ and μ_∞ on $X^{\mathbb{Z}}$ by:

$$\lambda_\infty = \int_Y (\cdots \times \lambda_y \times \lambda_y \times \cdots) \times \delta_y \, d\nu(y)$$

$$\text{and} \quad \mu_\infty = \int_Y (\cdots \times \lambda_y \times \lambda_y \times \cdots) \, d\nu(y) \quad \text{respectively.}$$

Let $\mathcal{Z}$ denote the largest σ-algebra common to the algebras $\mathcal{X}^{\mathbb{Z}}$ and $\mathcal{Y}$ mod λ_∞ and let $(Z, \mathcal{Z}, \eta)$ denote the corresponding factor Lebesgue space. Then $\mathcal{X}^{\mathbb{Z}}$ and $\mathcal{Y}$ are *relatively independent over $\mathcal{Z}$ with respect to λ_∞. In particular, if $\mathbf{X}$ and $\mathbf{Y}$ are Γ dynamical systems and $\lambda \in J(\mathbf{X}, \mathbf{Y})$, then the corresponding joining λ_∞ of the systems $\mathbf{X}^{\mathbb{Z}} = (X^{\mathbb{Z}}, \mu_\infty)$ and $\mathbf{Y}$ is the relatively independent joining of these systems over their common factor $\mathbf{Z}$. We then have the following commutative diagram:*

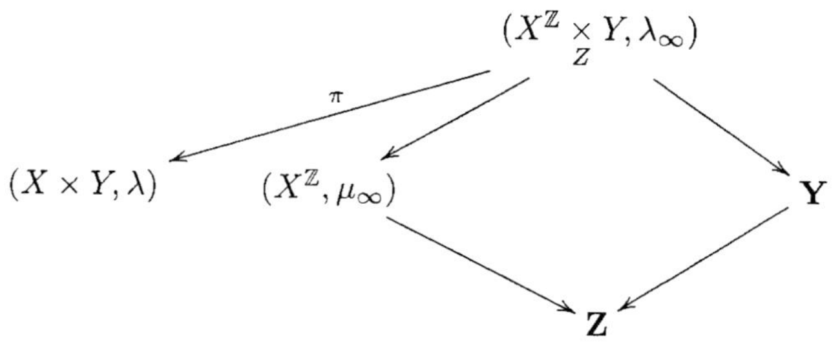

Here π is the map $\pi_0 \times \mathrm{id} : X^{\mathbb{Z}} \times Y \to X \times Y$ with $\pi_0 : X^{\mathbb{Z}} \to X$ the projection on the zero coordinate.

PROOF. Define a transformation $S : X^{\mathbb{Z}} \times Y \to X^{\mathbb{Z}} \times Y$ by $S(x, y) = (\sigma x, y)$ where $x = (\cdots, x_{-1}, x_0, x_1, \cdots) \in X^{\mathbb{Z}}$ and σ is the left shift on $X^{\mathbb{Z}}$. If $f(x, y)$ is an S-invariant function on $X^{Z} \times Y$ then for every y the function $f_y(x) = f(x, y)$ is a σ-invariant function on $(X^{\mathbb{Z}}, \lambda_y^{\mathbb{Z}})$, a Bernoulli $\mathbb{Z}$-system, hence a constant; i.e. $f(x, y) = f(y)$, λ_∞ a.e.

Thus every S-invariant function is $\mathcal{Y}$-measurable and in particular the extension $(X^{\mathbb{Z}}, \mu_\infty, \sigma) \to (Z, \eta, \mathrm{id})$ is an ergodic extension. Now apply Lemma 6.24 to the diagram

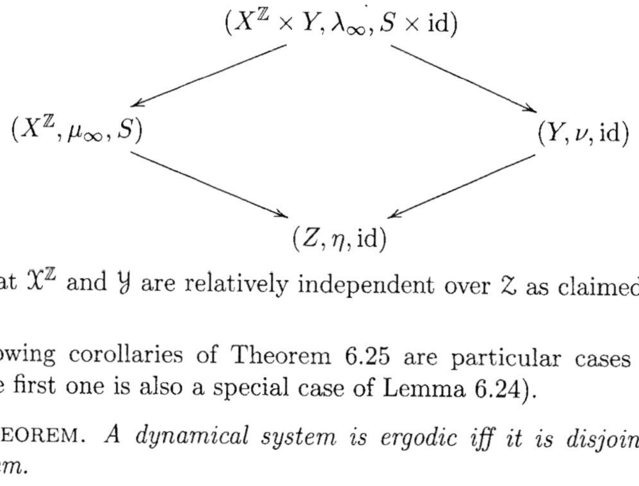

to deduce that $\mathcal{X}^{\mathbb{Z}}$ and $\mathcal{Y}$ are relatively independent over $\mathcal{Z}$ as claimed. □

The following corollaries of Theorem 6.25 are particular cases of a general principle (the first one is also a special case of Lemma 6.24).

6.26. THEOREM. *A dynamical system is ergodic iff it is disjoint from every identity system.*

PROOF. Disjointness from every identity system implies no non-trivial identity factor hence, by Theorem 3.10 ergodicity. For the converse we first observe that the class of identity systems is closed under factors as well as formation of countable joinings. Thus if $(X, \mathcal{X}, \mu, \mathrm{id})$ is an identity system so are the systems $(X^{\mathbb{Z}}, \mathcal{X}^{\mathbb{Z}}, \mu_\infty, \mathrm{id})$ and $(Z, \mathcal{Z}, \eta, \mathrm{id})$. However, if $(Y, \mathcal{Y}, \nu, \Gamma)$ is ergodic so is its factor $(Z, \mathcal{Z}, \eta, \mathrm{id})$ and this clearly implies that Z is a trivial one point space. Thus $\mathcal{Y}$ and $\mathcal{X}^{\mathbb{Z}}$ are independent with respect to the joining λ_∞ and a fortiori $\mathcal{Y}$ and $\mathcal{X}$ are independent with respect to λ; i.e. $\lambda = \mu \times \nu$. □

6.27. THEOREM. *An ergodic system is weakly mixing iff it is disjoint from every ergodic isometric system.*

PROOF. Disjointness from every isometric system implies no nontrivial isometric factor hence, by Theorem 3.11 weak mixing. For the converse we first note that, by Theorem 3.12, any ergodic joining λ of any (countable) collection of ergodic isometric systems $\mathbf{X}_n$, $n \in \mathbb{N}$, is an isometric system. In fact the restriction of u to each subspace $L^2(\mu_n) \subset L^2(\lambda)$ is the identity operator and since the collection of subspaces $L^2(\mu_n)$ generates $L^2(\lambda)$ the operator $u : L^2(\lambda) \to L^2(\lambda)$ is the identity operator. Clearly also any factor of an isometric system is isometric. Next observe that, by Theorem 3.11, a factor of a weakly mixing system is also weakly mixing.

Now let $\mathbf{X}$ be an ergodic, isometric system and $\mathbf{Y}$ a weakly mixing system. Applying Theorem 6.25 we see that the common factor $\mathbf{Z}$ is both isometric and weakly mixing, hence trivial (Theorem 3.13). Thus $\mathcal{Y}$ and $\mathcal{X}^{\mathbb{Z}}$ are independent with respect to the joining λ_∞ and a fortiori $\mathcal{Y}$ and $\mathcal{X}$ are independent with respect to λ; i.e. $\lambda = \mu \times \nu$. □

7. Joinings and spectrum

6.28. THEOREM. *If $\mathbf{X}$ and $\mathbf{Y}$ are two dynamical systems with mutually singular spectral types (on the corresponding L_0^2 spaces), then they are disjoint.*

PROOF. Let λ be a joining of μ and ν and let $\mathfrak{H} = L^2(\lambda)$. Given $f \in L^2(\mu) \subset \mathfrak{H}$ and $g \in L^2(\nu) \subset \mathfrak{H}$, we let $H_0 = Z(f_0)$ be the cyclic space generated by the function $f_0 = f - \int f$, and let P be the orthogonal projection $P : \mathfrak{H} \to H_0$. Now the spectral measure σ_{Pg} is absolutely continuous with respect to both spectral types $\sigma(T, \mu)$ and $\sigma(T, \nu)$, hence must be zero. Thus $Pg = 0$ and we conclude that

$$\begin{aligned}\int f_0 \cdot g \, d\lambda &= \int Pf_0 \cdot g \, d\lambda = \int f_0 Pg \, d\lambda = 0 \\ &= \int f_0 \, d\lambda \int g \, d\lambda,\end{aligned}$$

hence

$$\int fg \, d\lambda = \int f \, d\mu \int g \, d\nu.$$

Thus $\lambda = \mu \times \nu$ and our assertion is proved. □

6.29. EXERCISE. Use Theorem 6.28 to give alternative proofs for Theorems 6.26 and 6.27.

8. Notes

Group extensions and Veech's theorem: My main source here is [**138**]. Theorem 6.19, an extension of Veech's characterization of group extensions to isometric extensions, seems to be new (see also Lemańczyk, Thouvenot and Weiss [**168**] for related results).

Disjointness and the relative independence theorem: Theorem 6.23 is due to B. Weiss, [**261**]. The relative independence theorem is proved in the papers by Lemańczyk, Parreau and Thouvenot [**167**] and by Glasner, Thouvenot and Weiss [**93**]. In [**167**] the authors note that one can deduce from it the disjointness of the following pairs of families:

- id $\perp$ ergodic,
- distal $\perp$ weakly mixing,
- rigid $\perp$ mild mixing,
- zero entropy $\perp$ K-systems.

See also Corollary 8.16 and Theorem 18.16 below and the paper [**103**] by Glasner and Weiss.

Joinings and spectrum: Hahn and Parry [**111**], Thouvenot [**246**].

CHAPTER 7

Some Applications of Joinings

Most of the applications in the present chapter are to $\mathbb{Z}$-systems. In Section 1 the classical Halmos-von Neumann theorem receives a neat joining proof. In section 2 we characterize k-fold mixing in terms of joinings and use this characterization to show that for an abelian acting group mixing lifts through group extensions when the extended system is weakly mixing. Another application of the joining characterization of k-fold mixing is exhibited by Mozes' theorem on the mixing of all orders of $PSL(2,\mathbb{R})$-systems. A corollary of this theorem is the mixing of all orders of the horocycle flows. All of this is the content of Section 3. Section 4 introduces the classes of α-weakly mixing systems. These are weakly mixing systems where some rigidity is present. It turns out that they are sufficiently rigid to force severe restrictions on their spectral measures. This in turn is used in the last section (Section 5) to produce, via α-weakly mixing systems, some of the counter examples originally obtained by Rudolph via systems with minimal self-joinings. Specifically, we present examples of two non-isomorphic ergodic systems which are weakly isomorphic; and an example of two non-disjoint ergodic systems with no non-trivial common factor. The questions whether such examples exist were open for a long time. The first was shown to exist by S. Polit [**207**], the second by D. Rudolph [**217**]. The existence of α-weakly mixing systems will be demonstrated in Chapter 16.

1. The Halmos-von Neumann theorem

7.1. THEOREM. *An ergodic $\mathbb{Z}$ dynamical system $\mathbf{X}$ is isomorphic to a rotation of a compact monothetic group $\mathbf{K}$, iff U_T has discrete spectrum.*

PROOF. Let $\mathbf{X}$ have discrete spectrum; i.e. we assume that $L^2(\mu)$ is spanned by U_T-eigenfunctions. Let $G = spec(\mathbf{X})$ be the spectrum of $\mathbf{X}$; i.e. the set of eigenvalues of U_T. Then G is a countable subgroup of the circle $S^1 = \{z \in \mathbb{C} : |z| = 1\}$. There exists a compact monothetic group K such that $\hat{K}$, the dual group of K, is isomorphic to G (see e.g. [**119**], Corollary 24.32, page 390). Let $\mathbf{K} = (K, \mathcal{K}, m, R_a)$ be the group system corresponding to a generator a. Let $\lambda \in J_e(\mathbf{X}, \mathbf{K})$ be any ergodic joining. We shall show that

$$L^2(X \times \mathcal{K}, \lambda) = L^2(\mathcal{X} \times K, \lambda). \tag{1.1}$$

Then, for every $A \in \mathcal{X}$ there exists $g \in L^2(m)$ such that $\mathbf{1}_A \times \mathbf{1}_K = \mathbf{1}_X \times g$. Since $\mathbf{1}_A^2 = \mathbf{1}_A$, we also have $g^2 = g$ and we conclude that $g = \mathbf{1}_B$ for some $B \in \mathcal{K}$. Similarly for every $B \in \mathcal{K}$ we get an $A \in \mathcal{X}$, so that $X \times B = A \times K \pmod{\lambda}$. Thus $\mathcal{X} \times K = X \times \mathcal{K} \pmod{\lambda}$ and Theorem 6.5 yields the required isomorphism. It remains to show (1.1).

By assumption the set of normalized eigenfunctions $\{f_\gamma : \gamma \in \hat{K}\}$ forms an orthonormal basis for $L^2(X,\mu)$. We also know that the characters $\{\chi_\gamma : \gamma \in \hat{K}\}$ form an orthonormal basis for $L^2(K,m)$. When considered as elements of $L^2(\lambda)$, we see that both f_γ and χ_γ are eigenfunctions corresponding to the same eigenvalue γ, and the ergodicity of λ implies that (mod λ), $f_\gamma = c \cdot \chi_\gamma$, with $|c| = 1$. This clearly implies (1.1) and the proof is complete. $\square$

2. A joining characterization of mixing

In this section we characterize mixing and k-fold mixing in terms of joinings and then apply this characterization to show that when Γ is abelian k-fold mixing lifts to isometric extensions provided the extended systems are weakly mixing (see Theorem 9.14).

When the acting group Γ is not abelian, given a dynamical system $\mathbf{X} = (X, \mathcal{X}, \mu, \Gamma)$ and $\gamma \in \Gamma$, the probability measure $\operatorname{gr}(\mu, \gamma) = (\operatorname{id} \times \gamma)\Delta_\mu$ on the product space $X \times X$ still has the right marginals on the two X coordinates, namely $\pi_i(\operatorname{gr}(\mu, \gamma)) = \mu, i = 1, 2$; however $\operatorname{gr}(\mu, \gamma)$ is no longer Γ-invariant (unless γ is an element of the center).

Let $\mathbf{X}$ be a dynamical system, $k \geq 2$, an integer and consider the compact convex space $\hat{J}^{(k)}(\mathbf{X})$ of probability measures λ on the k-fold cartesian product X^k with marginals $\pi_i(\lambda) = \mu, i = 1, 2, \ldots, k$. The graph measures

$$\operatorname{gr}(\mu, \gamma_1 \times \gamma_2 \times \cdots \times \gamma_k) = (\gamma_1 \times \gamma_2 \times \cdots \times \gamma_k)\Delta_\mu^{(k)}$$

are elements of $\hat{J}^{(k)}(\mathbf{X})$; for convenience we still call them joinings. In section 2 we have defined k-fold mixing for $\mathbb{Z}$-systems. This definition generalizes in an obvious way to any infinite group Γ.

7.2. DEFINITION. *The dynamical system* $\mathbf{X}$ *is said to be* k-fold mixing *if for any* $k+1$ *measurable subsets* $A_0, A_1, \ldots, A_k \in \mathcal{X}$ *and any sequence* $\boldsymbol{\gamma}_n = (\gamma_n^{(0)}, \gamma_n^{(1)}, \ldots, \gamma_n^{(k)}) \in \Gamma^{k+1}$ *such that* $(\gamma_n^{(i)})^{-1}\gamma_n^{(j)} \to \infty$ *for* $i \neq j$, *we have*

$$\lim_{n \to \infty} \mu\left(\bigcap_{j=0}^{k}(\gamma_n^{(j)})^{-1}A_j\right) = \prod_{j=0}^{k}\mu(A_j).$$

7.3. PROPOSITION. *The dynamical system* $\mathbf{X}$ *is*
1. *mixing iff in* $\hat{J}^{(2)}(\mathbf{X})$,

$$\lim_{\gamma \to \infty} \operatorname{gr}(\mu, \gamma) = \mu \times \mu,$$

2. k-*fold mixing iff in* $\hat{J}^{(k+1)}(\mathbf{X})$

$$\lim_{n \to \infty} \operatorname{gr}(\mu, \boldsymbol{\gamma}_n) = \mu^{k+1},$$

for every sequence $\boldsymbol{\gamma}_n = (\gamma_n^{(0)}, \gamma_n^{(1)}, \ldots, \gamma_n^{(k)}) \in \Gamma^{k+1}$ *such that* $(\gamma_n^{(i)})^{-1}\gamma_n^{(j)} \to \infty$ *for* $i \neq j$.

PROOF. By the invariance of μ we can always assume that for all n, $\gamma_n^{(0)} = e$, the identity element of Γ. We then have, for $A_j \in \mathcal{X}$, $0 \leq j \leq k$,

$$\operatorname{gr}(\mu, \boldsymbol{\gamma}_n)(A_0 \times A_2 \times \cdots \times A_k) = \mu(A_0 \cap (\gamma_n^{(1)})^{-1}A_1 \cap \cdots \cap (\gamma_n^{(k)})^{-1}A_k).$$

The fact that k-fold mixing implies our condition now follows from the definitions by the nature of the topology on $\hat{J}^{(k+1)}(\mathbf{X})$ (see Theorem 6.2).

Suppose now that $\mathbf{X}$ satisfies our condition on k-fold graph joinings but is not k-fold mixing. Then there exist sets $A_0, A_1, \ldots, A_k \in \mathcal{X}$ and a sequence $\gamma_n \in \Gamma^{k+1}$, with $\gamma_n^{(1)} \equiv I$ and $(\gamma_n^{(i)})^{-1}\gamma_n^{(i)} \to \infty$ for $i \neq j$, such that

$$\lim_{n\to\infty} \mu\left(\bigcap_{j=0}^{k} (\gamma_n^{(j)})^{-1} A_j\right) = a,$$

where a is a number different from $\prod_{j=0}^{k} \mu(A_j)$.

Now take a weak* convergent subsequence $\lim_{i\to\infty} \mathrm{gr}\,(\mu, \gamma_{n_i}) = \lambda \in \hat{J}^{(k+1)}(\mathbf{X})$. By assumption we have $\lambda = \mu^{k+1}$. On the other hand, by Theorem 6.2 we have

$$\lambda(A_0 \times A_1 \times \cdots \times A_k) = a$$

and this contradiction completes the proof. □

7.4. THEOREM. *Suppose Γ is abelian and let $\pi : \mathbf{X} \to \mathbf{Y}$ be a homomorphism of Γ-dynamical systems such that (i) $\mathbf{Y}$ is k-fold mixing, (ii) $\mathbf{X}$ is weakly mixing and (iii) the extension π is a homogeneous skew product. Then $\mathbf{X}$ is k-fold mixing.*

PROOF. We are given a weakly mixing homogeneous skew-product,

$$\mathbf{X} = \mathbf{Y} \underset{\beta}{\times} K/L = (Y \times K/L, \nu \times \tilde{m}),$$

where K is a compact group, L a closed subgroup, m is Haar measure on the K and $\tilde{m}$ its image on K/L. As in Corollary 3.27 we can assume that $\beta : Y \to K$ is a minimal cocycle with $K = K_\beta$. Then by this corollary there exists an ergodic group skew-product $\hat{\mathbf{X}} = \mathbf{Y} \underset{\beta}{\times} K$ such that the diagram

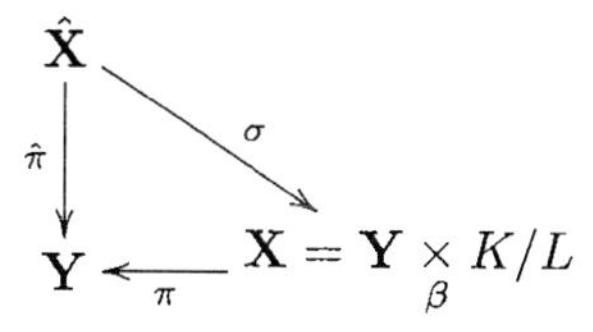

commutes. Here $\hat{\mathbf{X}} = \mathbf{Y} \underset{\beta}{\times} K$ is the group skew-product defined by the cocycle β, i.e. $\hat{\mu} = \nu \times m$ where m is Haar measure on K, and $\hat{\mu}$ is ergodic. We note that $\hat{\mathbf{X}}$ need not be weakly mixing.

Given a sequence $\gamma_n = (\gamma_n^{(0)}, \gamma_n^{(1)}, \ldots, \gamma_n^{(k)}) \in \Gamma^{k+1}$, with $\gamma_n^{(0)} \equiv e$ and $(\gamma_n^{(i)})^{-1}\gamma_n^{(i)} \to \infty$, for $i \neq j$, we have, by Proposition 7.3, $\lim_{n\to\infty} \mathrm{gr}\,(\nu, \gamma_n) = \nu^{k+1}$. Therefore, if $\hat{\lambda} = \lim_{i\to\infty} \mathrm{gr}\,(\hat{\mu}, \gamma_{n_i})$ is some limit point in $\hat{J}^{(k+1)}(\hat{\mathbf{X}})$, we must have $\hat{\pi}^{k+1}(\hat{\lambda}) = \nu^{k+1}$.

Set $\lambda = \sigma^{k+1}(\hat{\lambda})$; by Proposition 7.3 it suffices to show that $\lambda = \mu^{k+1}$. Of course the extension $\hat{\pi}^{k+1} : \hat{\mathbf{X}}^{k+1} \to \mathbf{Y}^{k+1}$ is also a group extension (with group K^{k+1}). We next follow, more or less, the proof of Theorem 3.30.

Set

$$\xi = \int_{K^{k+1}} R_q \hat{\lambda}\, dm^{k+1}(q).$$

The measure ξ projects to ν^{k+1} on Y^{k+1} and is K^{k+1}-invariant, hence must coincide with $\hat{\mu}^{k+1}$. Thus

$$\begin{aligned}\sigma^{k+1}(\xi) &= \sigma^{k+1}(\hat{\mu}^{k+1}) = \mu^{k+1} \\ &= \int \sigma^{k+1}(R_q\hat{\lambda})\, dm^{k+1}(q).\end{aligned} \tag{2.1}$$

Since, by assumption, $\mathbf{X}$ is weakly mixing it follows that the system $\mathbf{X}^{k+1}$ is ergodic, hence the measure μ^{k+1} is ergodic and it follows that in the representation (2.1), $\mu^{k+1} = \sigma^{k+1}(R_q\hat{\lambda})$, for m^{k+1} a.e. q. The action of K^{k+1} on the WAP dynamical system $\hat{J}^{(k+1)}(\hat{\mathbf{X}})$ is continuous and so is the homomorphism $\sigma^{k+1} : \hat{J}^{(k+1)}(\hat{\mathbf{X}}) \to \hat{J}^{(k+1)}(\mathbf{X})$. Thus $\mu^{k+1} = \sigma^{k+1}(R_q\hat{\lambda})$, for every q and in particular $\mu^{k+1} = \sigma^{k+1}(\hat{\lambda}) = \lambda$. □

3. Mixing of all orders of horocycle flows

In this section we prove Mozes' theorem on the mixing of all orders of measure preserving actions of the group $\mathbb{G} = PSL(2,\mathbb{R})$, [**183**]. We have defined k-fold mixing for discrete group actions (Definition 7.2) but the definition is meaningful for an arbitrary locally compact second countable group G. Note also that if $(X, \mathfrak{X}, \mu, G)$ is a k-fold mixing G-system and $H \subset G$ is any infinite subgroup then $(X, \mathfrak{X}, \mu, H)$ is k-fold mixing as well. Applying Mozes' theorem to the action of $\mathbb{G} = PSL(2,\mathbb{R})$ on a homogeneous space $\mathbb{G}/\Sigma$, with $\Sigma \subset \mathbb{G}$ a co-compact discrete subgroup, and then the above observation, we obtain Marcus' theorem about the mixing of all orders of the horocycle flows, [**179**]. Mozes proves his theorem for a general semisimple Lie group with finite center and no compact factors, however the main ideas of the proof are already present in the $PSL(2,\mathbb{R})$ case. We refer to the discussion of the geodesic and horocycle flows in Chapter 4, Section 6, for notations and facts that we shall use in the sequel. For the general facts we use about $PSL(2,\mathbb{R})$, its Lie algebra and the exponential map the reader is referred to any textbook on Lie groups; e.g. [**249**].

Let

$$\mathfrak{g} = \mathfrak{sl}(2,\mathbb{R}) = \left\{\left(\begin{smallmatrix} a & b \\ c & d\end{smallmatrix}\right) : a, b, c, d \in \mathbb{R}, a + d = 0\right\}$$

denote the Lie algebra of the group $\mathbb{G} = PSL(2,\mathbb{R})$ and let $\exp : \mathfrak{g} \to \mathbb{G}$ be the exponential map:

$$\exp(a) = I + a + \frac{1}{2!}a^2 + \frac{1}{3!}a^3 + \cdots$$

(here and in the sequel we identify a matrix in $SL(2,\mathbb{R})$ with its image in $\mathbb{G}$; I is the unit matrix $\left(\begin{smallmatrix} 1 & 0 \\ 0 & 1\end{smallmatrix}\right)$). Let $\|\cdot\|$ be any compatible norm on $\mathfrak{g}$ and recall that there exists an $r > 0$ such that $\exp \restriction B_r(0)$ is an analytic homeomorphism of the open ball $B_r(0)$, of radius r around 0 in $\mathfrak{g}$, onto an open neighborhood V of $I \in \mathbb{G}$.

Given $a \in \mathfrak{g}$ the linear map $\mathrm{ad}(a) : \mathfrak{g} \to \mathfrak{g}$ is defined by $\mathrm{ad}(a)(b) = ab - ba$. Given $g \in \mathbb{G}$ the conjugation map $\phi_g : \mathbb{G} \to \mathbb{G}$ defined by $\phi_g(h) = g^{-1}hg$ is an analytic isomorphism and its derivative at zero $(d\phi_g)_0$ is denoted by $\mathrm{Ad}(g) : \mathfrak{g} \to \mathfrak{g}$. We now have the formula

$$\exp(\mathrm{Ad}(g)b) = \phi_g(\exp b) = g^{-1}(\exp b)g. \tag{3.1}$$

7.5. LEMMA. *The map $Ad : \mathbb{G} \to GL(\mathfrak{g})$ is proper; i.e. the preimage of a compact set in $GL(\mathfrak{g})$ is a compact set in $\mathbb{G}$.*

PROOF. Suppose to the contrary that there exists a sequence $\mathbb{G} \ni g_n \to \infty$ such that

$$\sup_{n\in\mathbb{N}} \|\mathrm{Ad}(g_n)\| < \infty.$$

Then, for some $0 < s < r$

$$\sup_{n\in\mathbb{N}} \sup\{\|\mathrm{Ad}(g_n)b\| : b \in \mathfrak{g},\ \|b\| \le s\} < r.$$

Applying exp to $B_r(0)$ we get by (3.1), $g_n^{-1}Ug_n \subset V = \exp(B_r(0))$, for every $n \in \mathbb{N}$, where $U = \exp(B_s(0))$. Writing the Cartan decomposition of g_n as $g_n = k_n a_n l_n \in KAK$ we see that all the sets $(a_n)^{-1}KUKa_n$ lie in a compact neighborhood KVK of $I \in \mathbb{G}$. Now every neighborhood of I in $\mathbb{G}$ contains matrices of the form either $\left(\begin{smallmatrix}1 & b\\ 0 & 1\end{smallmatrix}\right)$, $b \ne 0$, or $\left(\begin{smallmatrix}1 & 0\\ c & 1\end{smallmatrix}\right)$, $c \ne 0$. With a slight abuse of notation, writing $a_n = \begin{pmatrix} a_n & 0\\ 0 & a_n^{-1}\end{pmatrix}$, we see from the equation

$$\begin{pmatrix} a_n & 0\\ 0 & a_n^{-1}\end{pmatrix}\begin{pmatrix} 1 & b\\ 0 & 1\end{pmatrix}\begin{pmatrix} a_n^{-1} & 0\\ 0 & a_n\end{pmatrix} = \begin{pmatrix} 1 & ba_n^2\\ 0 & 1\end{pmatrix},$$

and the analogous equation for the other case, that our assumption leads to a contradiction. □

7.6. THEOREM. *Any ergodic $\mathbb{G}$-system is mixing of all orders.*

PROOF. Let $\mathbf{X} = (X, \mathcal{X}, \mu, \mathbb{G})$ be an ergodic $\mathbb{G}$-system. It will be convenient here to evoke Theorem 2.18. Thus we now assume that X is a compact metric space and the map $(g, x) \mapsto gx$, $\mathbb{G} \times X \to X$ is continuous.

By Exercise 4.21 the system $\mathbf{X}$ is mixing (i.e. 1-mixing). Let $k > 1$ be given and, proving the theorem by induction, we assume that $\mathbf{X}$ is $k-1$-mixing. Let

$$\lambda = \lim_{n\to\infty} \mathrm{gr}\,(\mu, \boldsymbol{g}_n^{-1}) = ((g_n^{(0)})^{-1}, (g_n^{(1)})^{-1}, \dots, (g_n^{(k)})^{-1})\Delta_\mu^{(k+1)}$$

for a sequence $\boldsymbol{g}_n = (g_n^{(0)}, g_n^{(1)}, \dots, g_n^{(k)}) \in \mathbb{G}^{k+1}$ with $g_n^{(0)} \equiv I$ and $(g_n^{(i)})^{-1}g_n^{(j)} \to \infty$ for $i \ne j$. By Proposition 7.3 it suffices to show that $\lambda = \mu^{k+1}$. We proceed to show this by steps.

1. By Lemma 7.5 there exists a sequence of matrices $b_n \in \mathfrak{g}$ with $\|b_n\| = r$ such that

$$\lim_{n\to\infty} \|\mathrm{Ad}(g_n^{(1)})b_n\| = \infty.$$

We choose a sequence of positive numbers $t_n \to 0$ such that

$$\max_{0\le j\le k} \|\mathrm{Ad}(g_n^{(j)})t_n b_n\| = r.$$

2. Define the sequence $v_n = \exp(t_n b_n) \in \mathbb{G}$ and set

$$\boldsymbol{v}_n = (v_n, (g_n^{(1)})^{-1}v_n g_n^{(1)}, \dots, (g_n^{(k)})^{-1}v_n g_n^{(k)}).$$

We have, for each j,

$$\exp(\mathrm{Ad}(g_n^{(j)})t_n b_n) = (g_n^{(j)})^{-1}v_n g_n^{(j)},$$

and therefore conclude that the sequence $\{\boldsymbol{v}_n\}$ is contained in the compact set

$$M = \{(h_0, h_1, \ldots, h_k) \in \mathbb{G}^{k+1} : \alpha \leq \max_{0 \leq j \leq k} d(h_j, I) \leq \beta\},$$

where d is a compatible metric on $\mathbb{G}$ and

$$\alpha = \min\{d(\exp(b), I) : b \in \mathfrak{g}, \|b\| = r\}, \quad \beta = \max\{d(\exp(b), I) : b \in \mathfrak{g}, \|b\| = r\}.$$

Passing to a subsequence we assume now that $\lim_{n\to\infty} \boldsymbol{v}_n = \boldsymbol{h} = (h_0, h_1, \ldots, h_k) \in M$ exists. Since we have $g_n^{(0)} \equiv I$ and $\lim_{n\to\infty} v_n = I$, it follows that $h_0 = I$. Since $(h_0, h_1, \ldots, h_k) \in M$ there is at least one $j > 1$ with $h_j \neq I$. Rearranging, we can assume that for some $0 \leq l < k$, $h_0 = \cdots = h_0 = I$ while $h_{l+1}, \ldots, h_k$ are all different from I.

3. We claim that each $h_j, j > l$ is a unipotent matrix; i.e. all its eigenvalues are equal to 1. In fact, since $v_n = \exp(t_n b_n) \to I$ and we have $h_j = \lim_{n\to\infty}(g_n^{(k)})^{-1} v_n g_n^{(k)}$, this follows by continuity.

4. Writing $X^{k+1} = X^{l+1} \times X^{k-l} = X_1 \times X_2$ and $\pi_1 : X^{k+1} \to X_1$, $\pi_2 : X^{k+1} \to X_2$, we conclude, by Proposition 7.3 and our induction hypothesis, that $\pi_1(\lambda) = \mu^{l+1}$, $\pi_2(\lambda) = \mu^{k-l}$.

5. Since each $h_j, j > l$ is a unipotent matrix not equal to I it follows that $\{h_j^n : n \in \mathbb{Z}\}$ is unbounded and by Exercise 4.21 the action of each $h_j, j > l$ on $\mathbf{X}$ is mixing. It follows that the action of $(h_{l+1}, \ldots, h_k)$ on X^{k-l} is ergodic.

6. As explained at the beginning of Section 2, the measure λ is an element of $\hat{J}^{k+1}(\mathbf{X})$; however we don't know that it is a $\mathbb{G}$-invariant measure. Now, whenever a locally compact second countable topological group G acts on a compact metric space X as a group of homeomorphisms so that the map $(g, x) \mapsto gx$, $G \times X \to X$ is continuous, then the induced map $(g, \theta) \mapsto g_*(\theta)$, $G \times M(X) \to M(X)$ on the compact space $M(X)$ of probability measures on X is also continuous. (In fact if $f \in C(X)$, $g_n \to g$ in G and $\theta_n \to \theta$ in $M(X)$, then

$$\begin{aligned}|g_n\theta_n(f) - g\theta(f)| &\leq |\theta_n(f \circ g_n^{-1}) - \theta_n(f \circ g^{-1})| + |\theta_n(f \circ g^{-1}) - \theta(f \circ g^{-1})| \\ &\leq \|f \circ g_n^{-1} - f \circ g^{-1}\|_\infty + |\theta_n(f \circ g^{-1}) - \theta(f \circ g^{-1})|.)\end{aligned}$$

Now for every n,

$$\boldsymbol{v}_n \mathrm{gr}\,(\mu, \boldsymbol{g}_n^{-1}) = \mathrm{gr}\,(\mu, \boldsymbol{g}_n^{-1})$$

is a $\boldsymbol{v}_n$-invariant measure on X^k and it follows that

$$\begin{aligned}\lambda &= \lim_{n\to\infty} \mathrm{gr}\,(\mu, \boldsymbol{g}_n^{-1}) \\ &= \lim_{n\to\infty} \boldsymbol{v}_n \mathrm{gr}\,(\mu, \boldsymbol{g}_n^{-1}) \\ &= \boldsymbol{h}\lambda\end{aligned}$$

is an $\boldsymbol{h}$-invariant measure.

7. By Theorem 6.26 the "identity" system $(X_1, \mu^{l+1}, \mathrm{id})$ and the ergodic system $(X_2, \mu^{k-l}, (h_{l+1}, \ldots, h_k))$ are disjoint and we conclude that their joining λ must be the product measure μ^{k+1}. ☐

The discussion at the beginning of this section now yields the following corollary.

7.7. COROLLARY. *Let $\Sigma \subset PSL(2, \mathbb{R})$ be a co-compact lattice. Then the horocycle flow $(PSL(2, \mathbb{R})/\Sigma, \omega, \{h_s\}_{s\in\mathbb{R}})$ is mixing of all orders.*

4. α-weak mixing

7.8. DEFINITION. A $\mathbb{Z}$ dynamical system $\mathbf{X}$ is called *α-weak mixing* (with respect to the sequence n_k) if there are an $0 \leq \alpha \leq 1$ and a sequence $n_k \nearrow \infty$ such that in $J(\mu)$

$$\lim_{k\to\infty} \mathrm{gr}\,(\mu, T^{n_k}) = \alpha(\mu \times \mu) + (1-\alpha)\mathrm{gr}\,(\mu, \mathrm{id}).$$

The property 0-weak mixing is also called rigidity with respect to the sequence n_k (see Exercise 3.38).

Recall the notation, $\mathfrak{C}$ for the one dimensional space of constant functions, and $P_{\mathfrak{C}} : \mathfrak{H} \to \mathfrak{C}$ for the corresponding orthogonal projection. $I : L^2(\mu) \to L^2(\mu)$ is the identity operator.

7.9. THEOREM. *The following conditions on the dynamical system $\mathbf{X}$ are equivalent:*

1. *$\mathbf{X}$ is α-weak mixing with respect to the sequence n_k.*
2. *For σ the spectral type of $\mathbf{X}$ and for every $\phi \in L^1(\sigma)$*

$$\lim_{k\to\infty} \hat{\phi}(n_k) = \lim_{k\to\infty} \int z^{n_k}\phi(z)\, d\sigma = (1-\alpha)\int \phi\, d\sigma.$$

3. *The sequence of unitary operators $U_T^{n_k}$ converges in the weak operator topology to the operator*

$$\alpha \cdot P_{\mathfrak{C}} + (1-\alpha)\cdot I.$$

PROOF. 3 $\Rightarrow$ 2. Let $0 \leq \phi \in L^1(\sigma)$, then there exists $f = f_\phi \in L_0^2(\mu)$ with

$$\langle U^{n_k} f, f\rangle = \int z^{n_k}\phi(z)\, d\sigma(z).$$

Now

$$\lim_{k\to\infty} \langle U^{n_k} f, f\rangle = \alpha\langle \int f, f\rangle + (1-\alpha)\langle f, f\rangle = (1-\alpha)\langle f, f\rangle = (1-\alpha)\int \phi\, d\sigma.$$

2 $\Rightarrow$ 3. For every $f \in L_0^2(\mu)$ the corresponding spectral measure σ_f satisfies $\sigma_f \ll \sigma$, hence $\sigma_f = \phi\sigma$ for some $0 \leq \phi \in L^1(\sigma)$. By assumption:

$$\lim_{k\to\infty} \hat{\phi}(n_k) = \int z^{n_k}\phi(z)\, d\sigma = (1-\alpha)\int \phi\, d\sigma = (1-\alpha)\langle f, f\rangle.$$

For every $f, g \in L_0^2(\mu)$:

$$\begin{aligned}\langle U^{n_k} f, g\rangle =& \frac{1}{4}\{\langle U^{n_k}(f+g), f+g\rangle - \langle U^{n_k}(f-g), f-g\rangle \\ &+ i\langle U^{n_k}(f+ig), f+ig\rangle - i\langle U^{n_k}(f-ig), f-ig\rangle\}\end{aligned}$$

The right hand side tends to

$$\begin{aligned}&\frac{1-\alpha}{4}\{||f+g||^2 - ||f-g||^2 + i||f+ig||^2 - i||f-ig||^2\} \\ &= (1-\alpha)\langle f, g\rangle.\end{aligned}$$

Now for $f, g \in L^2(\mu)$, let $f_0 = f - P_{\mathfrak{C}}(f)$ and $g_0 = f - P_{\mathfrak{C}}(g)$, then:

$$\begin{aligned}\lim_{k\to\infty} \langle U^{n_k} f, g\rangle &= \lim_{k\to\infty} (\langle U^{n_k} f_0, g_0\rangle + \langle P_{\mathfrak{C}}(f), g_0\rangle + \langle U^{n_k} f_0, P_{\mathfrak{C}}(g)\rangle + P_{\mathfrak{C}}(f)P_{\mathfrak{C}}(\bar{g})) \\ &= (1-\alpha)\langle f_0, g_0\rangle + P_{\mathfrak{C}}(f)P_{\mathfrak{C}}(\bar{g}) \\ &= (1-\alpha)\langle f, g\rangle + \alpha P_{\mathfrak{C}}(f)P_{\mathfrak{C}}(\bar{g});\end{aligned}$$

i.e. the sequence U^{n_k} tends to $\alpha \cdot P_{\mathfrak{C}} + (1-\alpha) \cdot I$ in the weak operator topology on $\mathcal{L}(L^2(\mu))$, the space of bounded linear operators on $L^2(\mu)$.
$3 \Rightarrow 1$. We first observe that

$$\langle U^{n_k}\mathbf{1}_A, \mathbf{1}_B\rangle = \mathrm{gr}\,(\mu, T^{n_k})(A \times B).$$

In fact, testing the weak convergence for $f = \mathbf{1}_A$ and $g = \mathbf{1}_B$, we obtain

$$\begin{aligned} \mathrm{gr}\,(\mu, T^{n_k})(A \times B) &= \mu(A \cap T^{-n_k}B) \\ &= \int \mathbf{1}_A(x)\mathbf{1}_B(T^{n_k}x)\,d\mu(x) \\ &= \int \mathbf{1}_A(T^{-n_k}x)\mathbf{1}_B(x)\,d\mu(x) \\ &= \langle U^{n_k}\mathbf{1}_A, \mathbf{1}_B\rangle. \end{aligned}$$

By assumption

$$\lim_{k\to\infty} \langle U^{n_k}\mathbf{1}_A, \mathbf{1}_B\rangle = \langle \alpha\mu(A)\mathbf{1} + (1-\alpha)\mathbf{1}_A, \mathbf{1}_B\rangle.$$

Thus

$$\lim_{k\to\infty} \mathrm{gr}\,(\mu, T^{n_k})(A \times B) = \alpha\mu(A)\mu(B) + (1-\alpha)\mu(A \cap B),$$

as required.
$1 \Rightarrow 3$. Going backwards in the previous argument we get:

$$\lim_{k\to\infty} \langle U^{n_k}f, g\rangle = \alpha\langle f, \mathbf{1}\rangle\langle \mathbf{1}, g\rangle + (1-\alpha)\langle f, g\rangle,$$

for $f = \mathbf{1}_A$ and $g = \mathbf{1}_B$. By linearity and continuity we get it for all $f, g \in L^2(\mu)$. □

7.10. LEMMA. *For every $\alpha > 0$, α-weak mixing implies weak mixing. Conversely, weak mixing implies 1-weak mixing for some sequence n_k.*

PROOF. In view of Theorem 3.5 and Corollary 3.9 it suffices to show that $u = 0 \in \mathbf{E}$. If this is not the case then there exists an $0 \neq f \in L_0^2(\mu)$ with $uf = f$. Now by Proposition 3.1.5, the weak and strong operator topologies coincide on $\mathbf{I}$ the unique minimal ideal in $\mathbf{E}$ and since $U_{T^{n_k}}u \in \mathbf{I}$ we have

$$\|f\| = \|\lim_{k\to\infty} U_T^{n_k}uf\|.$$

On the other hand by α-weak mixing

$$\|\lim_{k\to\infty} U_T^{n_k}uf\| = (1-\alpha)\|f\|.$$

This leads to contradiction when $\alpha > 0$.

Now suppose $\mathbf{X}$ is weakly mixing; i.e. $u = 0 \in \mathbf{E}$. Choose a dense sequence $\{h_i : i \in \mathbb{N}\}$ in $L_0^2(\mu)$ and let $f_i(n) = \langle U_T^n h_i, h_i\rangle$. Since for every i, $uh_i = 0$, there exists for every $\epsilon > 0$ and $j, m_0 \in \mathbb{N}$ an integer $m > m_0$ with $|f_i(m)| < \epsilon$ for $i = 1, 2, \ldots, j$. Using a diagonal process we can find a sequence n_k such that $\lim_{k\to\infty} f_i(n_k) = 0$ for every $i \in \mathbb{N}$. It is now easy to deduce that $\lim_{k\to\infty} U_T^{n_k} = 0$ in the weak operator topology (for operators on $L_0^2(\mu)$); i.e. that $\mathbf{X}$ is 1-weakly mixing. □

For two measures σ and τ on $\mathbb{C}$ we define their *convolution* $\sigma * \tau$ as the image of $\sigma \times \tau$ under the map $(z, w) \mapsto zw$ from $\mathbb{C} \times \mathbb{C}$ to $\mathbb{C}$. We write σ^n for the measure $\sigma * \sigma * \cdots * \sigma$ (n times).

7.11. LEMMA. *Suppose* **X** *is* α*-weak mixing with respect to the sequence* n_k *and let* $\sigma = \sigma_{\mathbf{X}}$, *then for every positive integer* m *and* $\phi \in L^1(\sigma)$

$$\lim_{k\to\infty} \int z^{n_k} \phi \, d\sigma^m = (1-\alpha)^m \int \phi \, d\sigma^m.$$

PROOF. Since σ^m is the image of $\sigma^{\times m} := \sigma \times \sigma \times \cdots \times \sigma$ under the map $\mathbb{T}^m \to \mathbb{T}, (z_1, z_2, \ldots, z_m) \mapsto z_1 z_2 \cdots z_m$, it is enough to show that for every $\Phi \in L^1(\mathbb{T}^m, \sigma^{\times m})$,

$$\lim_{k\to\infty} \int_{\mathbb{T}^m} z_1^{n_k} z_2^{n_k} \cdots z_m^{n_k} \Phi \, d\sigma^{\times m} = (1-\alpha)^m \int \Phi \, d\sigma^{\times m}.$$

Now for Φ of the form $\Phi = \phi_1 \otimes \phi_2 \otimes \cdots \otimes \phi_m$, with $\phi_i \in L^1(\sigma)$, this follows from Theorem 7.9. Since linear combinations of such Φ's are dense in $L^1(\mathbb{T}^m, \sigma^{\times m})$, this completes our proof. □

For an isometric system **X**, the spectrum $\Gamma = \operatorname{spec}(\mathbf{X}) \subset \mathbb{T}$ (on $L^2(\mu)$ rather than $L^2_0(\mu)$) is a countable subgroup of $\mathbb{T}$ and accordingly, the spectral type $\sigma = \sigma_{\mathbf{X}}$ — in fact any measure of the form $\sigma = \sum_{\gamma\in\Gamma} a_\gamma \gamma$, $a_\gamma > 0$ — has the "group property", $\sigma * \sigma \ll \sigma$. It was expected to hold for every ergodic system when A. Stepin produced an example of an α-weakly mixing system and showed that the spectral type of such a system does not satisfy the group property in a very strong sense, [**241**]:

7.12. THEOREM. *If* **X** *is* α*-weak mixing for some* $0 < \alpha < 1$ *then for every* $m \neq n$ *in* $\mathbb{Z}$,

$$\sigma^m \perp \sigma^n,$$

where $\sigma = \sigma_{\mathbf{X}}$ *is the spectral type of* **X**.

PROOF. Suppose to the contrary that **X** is an α-weakly mixing system but σ^m and σ^n are not mutually singular. Then there exits a positive measure ρ with $0 < \rho \ll \sigma^m \wedge \sigma^n$. Denoting

$$\phi_m = \frac{d\rho}{d\sigma^m}, \qquad \phi_n = \frac{d\rho}{d\sigma^n}$$

we have by the previous lemma:

$$\begin{aligned}\lim_{k\to\infty} \int z^{n_k} \, d\rho &= \lim_{k\to\infty} \int z^{n_k} \phi_m \, d\sigma^m \\ &= (1-\alpha)^m \int \phi_m \, d\sigma^m \\ &= (1-\alpha)^m \int d\rho.\end{aligned}$$

Similarly

$$\lim_{k\to\infty} \int z^{n_k} \, d\rho = (1-\alpha)^n \int d\rho.$$

Therefore $(1-\alpha)^m = (1-\alpha)^n$, hence $m = n$. This contradiction completes the proof. □

7.13. COROLLARY. *The spectral type of an* α*-weakly mixing system with* $0 < \alpha < 1$ *is singular with respect to Lebesgue measure.*

5. Rudolph's counterexamples machine

Let $\mathbf{X} = (X, \mathcal{X}, \mu, T)$ be a dynamical system. Let $\mathbf{X}^\infty = (X^{\mathbb{N}}, \mathcal{X}^{\mathbb{N}}, \mu^{\mathbb{N}}, T^{\times\mathbb{N}})$ be the corresponding $\mathbb{N}$-product ($\mathcal{X}^{\mathbb{N}} = \bigotimes_1^\infty \mathcal{X}$). For $n \in \mathbb{N}$, $\mathcal{X}_n$ will denote the sub-σ-algebra $\pi_n^{-1}(\mathcal{X}) \subset \mathcal{X}^{\mathbb{N}}$, where $\pi_n : X^{\mathbb{N}} \to X$ is the projection on the n-th coordinate.

7.14. THEOREM. *Suppose* $\mathbf{X}$ *is* α*-weakly mixing with* $0 < \alpha < 1$. *Suppose that* $\mathcal{F}$ *is a* $T^{\times\mathbb{N}}$*-invariant sub-*σ*-algebra of* $\mathcal{X}^{\mathbb{N}}$ *and* $\mathcal{G}$ *a* T*-invariant sub-*σ*-algebra of* $\mathcal{X}$ *such that the corresponding dynamical systems* $(\mathcal{F}, \mu^{\mathbb{N}}, T^{\times\mathbb{N}})$ *and* $(\mathcal{G}, \mu, T)$ *are isomorphic, then there exists* $n \in \mathbb{N}$ *such that* $\mathcal{F} \subset \mathcal{X}_n$.

PROOF. The Hilbert space $\mathfrak{H} = L_0^2(\mu^{\mathbb{N}})$ has a natural decomposition:

$$\mathfrak{H} = \bigoplus_{\substack{E \subset \mathbb{N} \\ E \text{ finite}}} H_E,$$

where H_E is the closed subspace of $\mathfrak{H}$ spanned by the elements of the form $\bigotimes_{i\in\mathbb{N}} f_i$ with $f_i \in L_0^2(\mathcal{X}_i) = H_i$ for $i \in E$ and $f_i = 1$ otherwise. Now

$$\begin{aligned}\hat{\sigma}_{f\otimes g}(n) &= \langle U^n_{T\times T}(f \otimes g), (f \otimes g)\rangle = \langle U^n_T f, f\rangle\langle U^n_T g, g\rangle \\ &= \hat{\sigma}_f(n)\hat{\sigma}_g(n) = \widehat{\sigma_f * \sigma_g}(n).\end{aligned}$$

It follows that also for the spectral measure $\sigma_{\mathbf{X}} = \sigma_T$ we have $\sigma_{T\times T} = \sigma_T * \sigma_T$, and by induction $\sigma_{T^{\times n}} = \sigma_T^n$. We conclude that the spectral type of $T^{\times\mathbb{N}}$ restricted to H_E is σ_T^n where $n = |E|$.

Therefore, Theorem 7.12 implies that σ_T is singular to $\sigma_T^{|E|}$ — the spectral type of $T^{\times\mathbb{N}}$ restricted to H_E — whenever $|E| > 1$. Since the spectral type of T on $\mathcal{F}$ is absolutely continuous with respect to σ_T, we conclude that

$$L_0^2(\mathcal{F}) \subset \bigoplus_{i\in\mathbb{N}} H_i,$$

Next for $A \in \mathcal{F}$, consider the function $f = \mathbf{1}_A - \mu(A) \in L_0^2(\mathcal{F})$ and its decomposition

$$f = \sum_{i\in\mathbb{N}} f_i.$$

If there are at least two non-zero components, say $f_{i_1} \in H_{i_1}$ and $f_{i_2} \in H_{i_2}$, then we write $f = f_{i_1} + f_{i_2} + F$, and consider f_{i_1}, f_{i_2} and F as three independent random variables, with mean zero. It follows that their distribution measures (on $\mathbb{C}$) satisfy:

$$D(f) = D(f_{i_1}) * D(f_{i_2}) * D(F).$$

However the measure $D(f)$ is supported on two values while the support of the measure on the right hand side must have at least four points. Thus, for each nontrivial $A \in \mathcal{F}$ (say $0 < \mu(A) < 1/2$), there is a unique $i \in \mathbb{N}$ with $A \in \mathcal{X}_i$. If $A_1 \in \mathcal{X}_{i_1}$ and $A_2 \in \mathcal{X}_{i_2}$, with $i_1 \neq i_2$, then $A_1 \cup A_2 \in \mathcal{F}$ but cannot be an element of any $\mathcal{X}_i$. This shows that there is a unique $i \in \mathbb{N}$ such that $\mathcal{F} \subset \mathcal{X}_i$. □

7.15. EXAMPLE (Nondisjoint systems that have no common factors). We can now construct two non-disjoint ergodic systems $\mathbf{Y}$ and $\mathbf{Z}$ with no nontrivial common factor. For a system $\mathbf{X}$ and a positive integer q, let S_q be the group of $q!$ permutations of coordinates on X^q, and let $\mathbf{X}^{\langle q\rangle} = \mathbf{X}^q/S_q = (X^{\langle q\rangle}, \mathcal{X}^{\langle q\rangle}, \mu^{\langle q\rangle}, T^{\times q})$

be the symmetrized dynamical system. Observe that the systems $\mathbf{Y} = \mathbf{X}$ and $\mathbf{Z} = \mathbf{X}^{\langle 2 \rangle}$ are never disjoint since they both are factors of the product system $(X \times X, \mathcal{X} \otimes \mathcal{X}, \mu \times \mu, T \times T)$, but the corresponding σ-algebras $\mathcal{X}_1$ and $\mathcal{X}^{\langle 2 \rangle}$ — the σ-algebra of S_2-invariant sets in $\mathcal{X} \otimes \mathcal{X}$ — are not independent. Now let $\mathbf{X}$ be any α-weakly mixing system with $0 < \alpha < 1$. If $\mathcal{F}$ is a $T \times T$-invariant sub-σ-algebra of $\mathcal{Z} = \mathcal{X}^{\langle 2 \rangle}$ and $\mathcal{G}$ a T-invariant sub-σ-algebra of $\mathcal{Y} = \mathcal{X}_1$, such that the corresponding dynamical systems are isomorphic, then by Theorem 7.14 (where we embed $\mathcal{X}^{\langle 2 \rangle}$ as a factor of $\mathcal{X}^2$, itself identified as the algebra corresponding to the first two coordinates of $\mathcal{X}^{\mathbb{N}}$), there exists an $n \in \mathbb{N}$ such that $\mathcal{F} \subset \mathcal{X}_n$. However, as $\mathcal{X}_n \cap \mathcal{X}^{\langle 2 \rangle}$ is trivial, this implies that $\mathcal{F}$ is trivial.

7.16. Example (Nonisomorphic weakly isomorphic systems). Here we produce two ergodic systems $\mathbf{Y}$ and $\mathbf{Z}$ which are *weakly isomorphic* (i.e. there are factor maps $\mathbf{Y} \to \mathbf{Z}$ and $\mathbf{Z} \to \mathbf{Y}$) but not isomorphic. Again we let $\mathbf{X}$ be any α-weakly mixing system with $0 < \alpha < 1$. Let $\mathbf{Y} = \mathbf{X}^{\mathbb{N}}$ and let $\mathbf{Z} = \mathbf{X}^{\langle 2 \rangle} \times \mathbf{X}^{\{3,4,\dots\}}$, then $\mathcal{Z}$, the σ-algebra of $\mathbf{Z}$, is actually presented as a sub-σ-algebra of $\mathcal{X}^{\mathbb{N}}$. On the other hand, $\mathbf{Y}$ is the image of $\mathbf{Z}$ under the projection $\pi_{\{3,4,\dots\}}$. Thus $\mathbf{Y}$ and $\mathbf{Z}$ are weakly isomorphic. If $\phi : \mathbf{Y} \to \mathbf{Z}$ is an isomorphism, then for every $j \in \mathbb{N}$, $\phi(\mathcal{X}_j)$ is a factor σ-algebra of $\mathcal{Z} = \mathcal{X}^{\langle 2 \rangle} \otimes \mathcal{X}^{\{3,4,\dots\}}$ and Theorem 7.14 implies that for some $n \in \mathbb{N}$, $\phi(\mathcal{X}_j) \subset \mathcal{X}_n$. As in the previous example the values $n = 1$ and $n = 2$ are precluded and we therefore conclude that for every $j \in \mathbb{N}$, $\phi(\mathcal{X}_j) \subset \mathcal{X}^{\{3,4,\dots\}}$. This however contradicts the fact that ϕ is an isomorphism.

6. Notes

Halmos-von Neumann's theorem: This joining proof is from the lecture notes [**165**] by Lemańczyk.

A joining characterization of mixing: Theorem 7.4 is due to Rodulph, [**218**].

α-weak mixing: As mentioned above α-weak mixing was introduced in Stepin [**241**]. Here I mainly follow del Junco and Lemańczyk, [**134**], who extend results of Stepin and Katok, [**241**] and [**145**].

Rudolph's counterexamples machine: Rudolph's original results [**217**], were based on systems with minimal self-joinings. The idea to get most of Rudolph's counter examples using α-weakly mixing instead of minimal self-joining is due to del Junco and Lemańczyk. In fact they show in [**134**] that in a certain sense these "counter examples" are "typical". Using Ratner's results on horocycle flows, [**210**], Glasner and Weiss describe in [**94**] a countable subclass of these systems for which no pair have a non-trivial common factor yet no pair is disjoint.

CHAPTER 8

Quasifactors

For a dynamical system $\mathbf{X} = (X, \mathcal{X}, \mu, \Gamma)$, a factor system $\mathbf{Y} = (Y, \mathcal{Y}, \nu, \Gamma)$ with a factor map $\pi : \mathbf{X} \to \mathbf{Y}$, can be viewed as the Γ-invariant subalgebra $\pi^{-1}(\mathcal{Y}) \subset \mathcal{X}$. One can also retrieve the factor $\mathbf{Y}$ as a measure-preserving transformation on the space $M(X)$ of probability measures on X as follows. Disintegrate the measure μ along the fibers of $\pi^{-1}(\mathcal{Y})$,

$$\mu = \int_Y \mu_y d\nu(y) \tag{0.1}$$

and observe that the Γ-invariance of μ implies that $\gamma\mu_y = \mu_{\gamma y}$, $(\gamma \in \Gamma)$. Denoting by $\phi : Y \to M(X)$ the map $\phi(y) = \mu_y$ and letting $\kappa = \phi_*(\nu)$, we see that $\phi : (Y, \mathcal{Y}, \nu, \Gamma) \to (M(X), \kappa, \Gamma)$ is an isomorphism. The connection with μ is given by (0.1) which says that μ is the barycenter of κ. A general quasifactor of $(X, \mathcal{X}, \mu, \Gamma)$ is any Γ-invariant measure on $M(X)$ whose barycenter is μ. This notion was introduced in Glasner [**86**] and further studied by Glasner and Weiss in [**100**] and [**103**], where it was shown that, like factors, quasifactors inherit some dynamical properties. E.g. zero entropy and distality are preserved under a passage to a quasifactor. We shall prove these results in Chapters 16 and 10 respectively. On the other hand, as was shown by Glasner and Weiss in [**103**], every ergodic $\mathbb{Z}$-system of positive entropy, say $\mathbf{Y}$, admits every ergodic system of positive entropy, say $\mathbf{X}$, as a quasifactor (we shall not prove this result here). Of course a factor of a factor is still a factor, however a factor of a quasifactor need not be a quasifactor. Indeed if we take in the above statement $\mathbf{Y}$ to be a Bernoulli system and $\mathbf{X} = \mathbf{Y} \times \mathbf{Z}$ with $\mathbf{Z}$ any zero entropy system, we can find an isomorphic copy of $\mathbf{X}$ as a quasifactor of $\mathbf{Y}$, and we then see that the factor $\mathbf{Z}$ of the quasifactor $\mathbf{X}$ of the system $\mathbf{Y}$ can not appear as a quasifactor of $\mathbf{Y}$.

Another result that demonstrates this phenomenon is given in Section 9, where we produce an example of a weakly mixing system $\mathbf{X}$ with a non weakly mixing quasifactor $\mathbf{Y}$. The latter has a non trivial factor with discrete spectrum which can not occur as a factor or even a quasifactor of $\mathbf{X}$. Quasifactors are introduced in Section 1 and in Section 2 we use Choquet's theorem to prove the ergodic decomposition theorem for the general Γ-system $\mathbf{X} = (X, \mathcal{X}, \mu, \Gamma)$. The quasifactor theory helps us here to identify this decomposition with the decomposition of the measure μ over the σ-algebra $\mathcal{I}$ of Γ-invariant sets.

A joining λ of two systems $\mathbf{X}$ and $\mathbf{Y}$ similarly gives rise to a quasifactor of $\mathbf{X} = (X, \mathcal{X}, \mu, \Gamma)$ by disintegrating λ over ν:

$$\lambda = \int_Y \mu_y \times \delta_y \, d\nu(y),$$

(and of course, symmetrically, a quasifactor of $\mathbf{Y}$). In fact every quasifactor of $\mathbf{X}$ can be obtained in this way from a suitable joining, but in general the joining λ

carries more information than there is in the corresponding quasifactor. Moreover, even when the quasifactor we start from is ergodic, this need not be the case with the associated joining. Those ergodic quasifactors for which the associated joining is ergodic are called joining quasifactors and are the more interesting ones. They are studied in Section 6.

In Sections 3 to 5 we clarify the close connection between quasifactors and symmetric infinite selfjoinings. This arises from the de Finetti-Hewitt-Savage theorem (Theorem 8.11) which is discussed in Section 4. As a corollary we obtain a general statement on quasifactors (Corollary 8.13) that will allow us later on to deduce the fact that quasifactors of zero entropy systems have zero entropy and that quasifactors of distal systems are distal. In Section 7 we analyze symmetric product quasifactors, and in Section 8 we deal with the question of lifting of quasifactors. Finally in Section 9 we show that a weakly mixing system may have a non weakly mixing quasifactor.

1. Factors and quasifactors

Let $\mathbf{X} = (X, \mathcal{X}, \mu, \Gamma)$ be a dynamical system on a standard Borel space $(X, \mathcal{X})$. Let $M = M(X)$ be the space of probability measures on X with the Borel structure determined by the maps $\mu \mapsto \mu(A)$ for a fixed $A \in \mathcal{X}$. The action of Γ on the standard Borel space X defines a measurable action of Γ on $M(X)$, namely $(\gamma\theta)(A) = \theta(\gamma^{-1}A)$ for $\gamma \in \Gamma, \theta \in M(X)$ and $A \in \mathcal{X}$. A *quasifactor* of the dynamical system $\mathbf{X}$ is any Γ-invariant probability measure κ on $M(X)$ whose *barycenter* is μ; i.e. κ satisfies the barycenter equation:

$$\mu = \int_{M(X)} \theta \, d\kappa(\theta). \tag{1.1}$$

This equation means that — choosing any compact topology on X compatible with its Borel structure — for every continuous function $f \in C(X)$,

$$\int_M \int_X f(x) \, d\theta(x) \, d\kappa(\theta) = \int_X f(x) \, d\mu(x).$$

Let $Q(\mu) = Q(\mathbf{X})$ be the collection of all quasifactors of μ. It is easy to check that $Q(\mu)$ is a convex subset of $M(M(X))$. The set $Q(\mu)$ always contains at least two elements, the *trivial* one point quasifactor $(\{\mu\}, \delta_\mu)$ and the *points quasifactor*

$$(\{\delta_x : x \in X\}, \mu) \cong \mathbf{X}.$$

It is sometimes convenient to dissociate the quasifactor from the original system by writing W for $M(X)$, with the bijection $w \leftrightarrow \theta_w$. In this notation the barycenter equation becomes $\int_W \theta_w \, d\kappa(w) = \mu$ and we write $\mathbf{W} = (W, \kappa, \Gamma)$ for the system $(M(X), \kappa, \Gamma)$. When $\pi : \mathbf{X} \to \mathbf{Y}$ is a factor map and $\kappa \in Q(\mathbf{X})$ is such that its image in $Q(\mathbf{Y})$ is the points quasifactor $\mathbf{Y}$ (i.e. $\mathbf{Y} \cong (\{\delta_y : y \in Y\}, \nu)$), we say that κ is a *quasifactor over* $\mathbf{Y}$ (with respect to π). The collection of all of these is denoted by $Q(\pi)$. The next theorem shows that the quasifactors of a dynamical system do not depend on the choice of a Polish model for $\mathbf{X}$.

8.1. THEOREM. *Let $\mathbf{X}$ and $\mathbf{X}'$ be two Polish models of a dynamical system and let $\phi : X_0 \to X_0'$ be a Borel isomorphism, where X_0 and X_0' are Borel Γ-invariant subsets of full measure of the Polish spaces X and X' respectively, such that for every $\gamma \in \Gamma$, $\gamma \circ \phi = \phi \circ \gamma$ on X_0. Then $\phi_{**} : Q(\mu) \to Q(\mu')$ is a*

Borel affine isomorphism, such that for every $\kappa \in Q(\mu)$, denoting $\kappa' = \phi_{**}(\kappa)$, $\phi_* : (M(X), \kappa, \Gamma) \to (M(X'), \kappa', \Gamma)$ is an isomorphism of dynamical systems.

PROOF. Let M_0 be the subset of $M(X)$ consisting of the measures θ with $\theta(X_0) = 1$, and define M_0' analogously. Now define $\phi_* : M_0 \to M_0'$ by $\theta' = \phi_*(\theta) = \theta \circ \phi^{-1}$. Clearly ϕ_* is a Γ-equivariant Borel isomorphism. For every bounded continuous function $g \in C_b(X')$ and $\theta \in M_0$ we have

$$\int_{X'} g\, d\theta' = \int_X g \circ \phi\, d\theta.$$

Let $\kappa \in M(M(X))$ be a quasifactor of $\mathbf{X}$. Computing the integral of $\mathbf{1}_{X_0}$ we have

$$1 = \int_X \mathbf{1}_{X_0}\, d\mu = \int_{M(X)} \int_X \mathbf{1}_{X_0}\, d\theta\, d\kappa(\theta).$$

Hence $\theta(X_0) = 1$ for κ a.e. θ and $\kappa(M_0) = 1$. Now let $\kappa' = \phi_{**}(\kappa) = \kappa \circ \phi_*^{-1}$, then κ' is an invariant probability measure in $M(M(X'))$ with $\kappa'(M_0') = 1$. For every $g \in C_b(X')$

$$\begin{aligned}\int_{M(X')} \int_{X'} g\, d\theta'\, d\kappa'(\theta') &= \int \int g\, d\theta'\, d(\phi_{**}\kappa)(\theta') = \int \int g\, d(\phi_*(\theta))\, d\kappa(\theta) \\ &= \int \int g \circ \phi\, d\theta\, d\kappa(\theta) = \int_X g \circ \phi\, d\mu = \int_{X'} g\, d\mu'.\end{aligned}$$

Thus κ' is a quasifactor of $\mathbf{X}'$ and

$$\phi_* : (M(X), \kappa, \Gamma) \to (M(X'), \kappa', \Gamma),$$

is an isomorphism. By symmetry we see that $\phi_{**} : Q(\mu) \to Q(\mu')$ is onto. □

Here are some important examples of quasifactors.

8.2. EXAMPLES.

- **Factors:** These are the quasifactors $\kappa \in Q(\mu)$ which are obtained from a factor $\pi : \mathbf{X} \to \mathbf{Y}$ and the corresponding disintegration

$$\mu = \int_Y \mu_y\, d\nu(y);$$

i.e. $\kappa = \phi_*(\nu)$, where ϕ is the map $y \mapsto \mu_y$. Let us warn the reader that it is not usually easy to decide whether a quasifactor is a factor, e.g. the condition that θ and θ' are mutually singular for $\kappa \times \kappa$ a.e. $(\theta, \theta') \in M(X) \times M(X)$ is not a sufficient condition (see Theorem 8.3 below).

- **Joining-quasifactors:** Let $\mathbf{X}$ and $\mathbf{Y}$ be ergodic dynamical systems. Let $\lambda \in J_e(\mathbf{X}, \mathbf{Y})$ be an ergodic joining and let

$$\lambda = \int_Y (\lambda_y \times \delta_y)\, d\nu(y),$$

be its disintegration over ν. Let $\phi : Y \to M = M(X)$ be the measurable map associated with this disintegration, $\phi : y \mapsto \lambda_y$ from Y to the space of probability measures on X and let $\kappa = \phi_*(\nu)$ be the image of ν under ϕ. The Γ invariance of λ and the uniqueness of disintegration imply that for every $\gamma \in \Gamma$, ν almost everywhere $\gamma\lambda_y = \lambda_{\gamma y}$. Thus ϕ is a homomorphism of the system $\mathbf{Y}$ onto the system $\mathbf{W} = (M(X), \kappa, \Gamma)$.

We have the barycenter equation for κ:

$$\int_{M(X)} \theta \, d\kappa(\theta) = \int_Y \lambda_y \, d\nu(y) = \mu.$$

Thus the system $\mathbf{W} = (M(X), \kappa, \Gamma)$ is a factor of $\mathbf{Y}$ and a quasifactor of $\mathbf{X}$. As we shall shortly see, every quasifactor arises from some joining in this manner. We shall call those quasifactors that arise from ergodic joinings "joining quasifactors" (JQF, Definition 8.19 below). For a simple example of a quasifactor which is not a joining quasifactor consider the following example. Let $\mathbf{X} = (X, \mu, R_\alpha)$ be an irrational rotation on the circle $X = \mathbb{R}/\mathbb{Z}$ with Lebesgue measure μ. Fix $\beta \in X$ and for every $x \in X$ let $\theta_x = \frac{1}{2}(\delta_x + \delta_{x+\beta})$. The map $\psi : x \mapsto \theta_x$ is a Borel homomorphism $\psi : (X, R_\alpha) \to (M(X), R_\alpha)$ and it is easy to check that the measure $\kappa = \psi_*(\mu)$ on $M(X)$ satisfies the barycenter equation. Thus $(M(X), \kappa, R_\alpha)$ is a quasifactor of $\mathbf{X}$ and being isomorphic to $\mathbf{X}$ it is ergodic. However, as we shall see, it is not a JQF.

Given a dynamical system $\mathbf{X}$ and a quasifactor $\kappa \in Q(\mathbf{X})$ set

$$\mathcal{F}(\kappa) = \{A \in \mathcal{X} : \text{for } \kappa \text{ a.e. } \theta \in M(X) \text{ either } \theta(A) = 0 \text{ or } \theta(A) = 1\}.$$

The following theorem addresses the question: when is a quasifactor a factor?

8.3. THEOREM. *Let $\mathbf{X}$ be a dynamical system and $\kappa \in Q(\mathbf{X})$ a quasifactor.*

1. *The associated family of sets $\mathcal{F}(\kappa)$ is a Γ-invariant σ-algebra.*
2. *The corresponding factor $\pi : \mathbf{X} \to \mathbf{Y} = (Y, \mathcal{Y}, \nu)$, with $\pi^{-1}(\mathcal{Y}) = \mathcal{F}(\kappa)$, is the largest factor such that $\pi(\kappa)$ is the points quasifactor on $\mathbf{Y}$ (we write $\pi(\kappa) = \nu$). That is, if $\alpha : \mathbf{X} \to \mathbf{Z} = (Z, \mathcal{Z}, \eta)$ is another factor for which $\alpha(\kappa) = \eta$, then $\alpha^{-1}(\mathcal{Z}) \subset \mathcal{F}(\kappa)$.*
3. *The quasifactor κ is a factor iff*

$$(M(X), \kappa) = (\{\mu_y : y \in Y\}, \kappa),$$

where $\mu = \int \mu_y \, d\nu(y)$ is the disintegration of μ over ν; iff the σ-algebra $\mathcal{F}(\kappa)$ separates the points of $(M(X), \kappa)$ — that is, there exists a subset $M_0 \subset M(X)$ with $\kappa(M_0) = 1$ such that $\theta, \theta' \in M_0$, $\theta \neq \theta'$ implies there exists an $A \in \mathcal{F}(\kappa)$ with $\theta(A) = 1$ and $\theta'(A) = 0$.

PROOF. 1. For $A \in \mathcal{X}$ set

$$\Theta(A) = \{\theta \in M(X) : \text{either } \theta(A) = 0 \text{ or } \theta(A) = 1\},$$

thus $A \in \mathcal{F}(\kappa)$ iff $\kappa(\Theta(A)) = 1$. Clearly $\Theta(A) = \Theta(A^c)$ and it is easy to see that for a sequence $A_n \in \mathcal{X}$ both $\Theta(\bigcup_n A_n)$ and $\Theta(\bigcap_n A_n)$ contain the set $\bigcap_n \Theta(A_n)$. It now follows that $\mathcal{F}(\kappa)$ is a σ-algebra. Since $\Theta(\gamma A) = \gamma^{-1}\Theta(A)$, the Γ-invariance of κ implies that $\mathcal{F}(\kappa)$ is Γ-invariant. Write $\pi : \mathbf{X} \to \mathbf{Y} = (Y, \mathcal{Y}, \nu)$ for the corresponding factor.

2. Since $\pi^{-1}(\mathcal{Y}) = \mathcal{F}(\kappa)$ it follows that for every $A \in \mathcal{Y}$ and κ a.e. $\theta \in M(X)$ $\pi(\theta)(A) = \theta(\pi^{-1}A)$ is either zero or one and we conclude that $\pi(\theta)$ is a point mass on Y; that is $\pi(\kappa) = \nu$ — the points quasifactor on $\mathbf{Y}$. If $\alpha : \mathbf{X} \to \mathbf{Z} = (Z, \mathcal{Z}, \eta)$ is another factor for which $\alpha(\kappa) = \eta$, then clearly $\alpha^{-1}(A) \in \mathcal{F}(\kappa)$ for every $A \in \mathcal{Z}$, whence $\alpha^{-1}(\mathcal{Z}) \subset \mathcal{F}$.

3. If $(M(X), \kappa) = (\{\mu_y : y \in Y\}, \kappa)$ then, by definition, κ is a factor and it is evident that $\mathcal{F} = \pi^{-1}(\mathcal{Y})$ separates the points of $(M(X), \kappa)$. In order to prove the converse note that always for κ a.e. $\theta \in M(X)$ the measure $\pi(\theta)$ is a point mass δ_y for some $y \in Y$ or equivalently, that $\theta(\pi^{-1}(y)) = 1$. Now the assumption that $\mathcal{F}(\kappa)$ separates the points of $(M(X), \kappa)$ implies that the map $\theta \mapsto \pi(\theta)$ is an isomorphism of $(M(X), \kappa)$ onto $\mathbf{Y}$. Moreover

$$\mu = \int_Y \mu_y \, d\nu(y) = \int_{M(X)} \theta \, d\kappa(\theta)$$

are two representations of μ where $\mu_y(\pi^{-1}(y)) = 1$ and $\theta_y(\pi^{-1}(y)) = 1$ (for θ_y with $\pi(\theta_y) = \delta_y$). The uniqueness of the disintegration implies that ν a.e. $\theta_y = \mu_y$, hence $(M(X), \kappa) = (\{\mu_y : y \in Y\}, \kappa)$. ☐

8.4. THEOREM. *Let $\mathbf{X}$ and $\mathbf{Y}$ be dynamical systems, then $\mathbf{X}$ is not disjoint from $\mathbf{Y}$ iff $\mathbf{Y}$ has a factor which is isomorphic to a nontrivial quasifactor of $\mathbf{X}$.*

PROOF. By definition $\mathbf{X}$ is not disjoint from $\mathbf{Y}$ iff there exists a joining $\lambda \in J(\mathbf{X}, \mathbf{Y})$ such that $\lambda \neq \mu \times \nu$. Also, the trivial quasifactor of $\mathbf{X}$ is the one point system $(M(X), \delta_\mu)$. If $\psi : (Y, \nu) \to (M(X), \kappa)$ is a homomorphism of $\mathbf{Y}$ onto a nontrivial quasifactor $(M(X), \kappa)$ of $\mathbf{X}$, we can form the measure

$$\lambda = \int_Y (\psi(y) \times \delta_y) \, d\nu(y).$$

Clearly λ is a joining and by the uniqueness of disintegration, $\lambda = \mu \times \nu$ iff $\psi(y) = \mu$ ν-a.e.; i.e. iff $\kappa = \delta_\mu$.

Conversely, if $\lambda \in J(\mathbf{X}, \mathbf{Y})$ is not $\mu \times \nu$ then disintegrating λ over ν we get

$$\lambda = \int_Y (\lambda_y \times \delta_y) \, d\nu,$$

and the function $\psi(y) = \lambda_y$ is not the constant function $\psi(y) = \mu$. Now put $\kappa = \psi_*(\nu)$, a measure on $M(X)$. It is easy to see that $(M(X), \kappa)$ is a nontrivial quasifactor of $\mathbf{X}$. ☐

8.5. EXERCISE. Show that for dynamical systems $\mathbf{X}$ and $\mathbf{Y}$ with common factor $\mathbf{X} \xrightarrow{\pi} \mathbf{Z}$, $\mathbf{Y} \xrightarrow{\sigma} \mathbf{Z}$, $\mathbf{X}$ is not disjoint from $\mathbf{Y}$ over $\mathbf{Z}$ iff $\mathbf{Y}$ has a factor which is isomorphic to a nontrivial quasifactor in $Q(\pi)$.

2. A proof of the ergodic decomposition theorem

In this section we prove the existence and uniqueness of the ergodic decomposition for a general dynamical system with the aid of Theorems A.7, A.13, Theorem 8.3 and the following proposition (see [**206**], Chapter 10, or [**53**], page 163).

8.6. PROPOSITION. *Let (X, Γ) be a compact metric topological system, then the compact convex set $Q = M_\Gamma(X)$ of Γ-invariant Borel probability measures on X is a simplex. This means that the pair $(C^*(X), P)$ is a lattice and that Q is a base for the cone P. Here the topological vector space $C^*(X)$ is the dual of the Banach space $C(X)$ — identified by Riesz' theorem with the Banach space of all finite signed Borel measures on X with the norm of total variation — and P is the cone of positive* *measures in* $C^*(X)$. *For* $\mu, \nu \in P$ *their greatest lower bound* $\mu \wedge \nu$ *is the measure* $(f \wedge g)(\mu + \nu)$, *where* f *and* g *are the Radon-Nikodým derivatives* $f = d\mu/d(\mu+\nu)$ *and* $g = d\nu/d(\mu+\nu)$.

8.7. THEOREM (The ergodic decomposition). *Let* $\mathbf{X}$ *be a dynamical system* $\mathcal{I} \subset \mathcal{X}$ *the* σ*-algebra of* Γ*-invariant sets. Let* $\pi : \mathbf{X} \to \mathbf{Y}$ *be the factor defined by* $\mathcal{I}$, *so that* $\mathcal{I} = \pi^{-1}(\mathcal{Y})$ *(Theorem 2.15). Let* $\mu = \int_Y \mu_y \, d\nu(y)$ *be the disintegration of* μ *over* ν *and for every* $y \in Y$ *denote* $X_y = \pi^{-1}(y)$ *and* $\mathcal{X}_y = \mathcal{X} \cap X_y$, *then:*

1. Γ *acts as the identity on* Y.
2. *For* ν *a.e.* $y \in Y$, *the system* $(X_y, \mathcal{X}_y, \mu_y, \Gamma)$ *is ergodic.*
3. *This decomposition is unique in the following sense. If* $(Z, \mathcal{Z}, \eta)$ *is a standard probability space and* $\psi : z \mapsto \tilde{\mu}_z$ *a measurable map from* Z *into* $M_\Gamma^{\mathrm{erg}}(X)$ *such that* $\mu = \int_Z \tilde{\mu}_z \, d\eta(z)$, *then there exists a measurable map* $\phi : (Z, \mathcal{Z}, \eta) \to (Y, \mathcal{Y}, \nu)$ *such that* η*-a.e.* $\mu_{\phi(z)} = \tilde{\mu}_z$.

PROOF. Part 1 of the theorem is clear, as $\gamma A = A$ for every $\gamma \in \Gamma$ and $A \in \mathcal{I}$. With no loss of generality we assume that X is a compact metric space and that Γ acts on X by homeomorphisms. By Proposition 8.6 $M_\Gamma(X)$ is a simplex and Choquet's theorem A.13 yields a unique representation $\mu = \int_{M_\Gamma^{\mathrm{erg}}(X)} \theta \, d\kappa(\theta)$ with κ a Borel probability measure on the G_δ subset $M_\Gamma^{\mathrm{erg}}(X)$ of $M_\Gamma(X)$. This means that κ is a quasifactor of the system $\mathbf{X}$ (with Γ acting as the identity). If A is an element of $\mathcal{I}$ then we can assume that A is a Borel subset and, replacing A by $\bigcap_{\gamma \in \Gamma} \gamma A$, we can assume that $\gamma A = A$ as sets (rather than as elements of the μ measure algebra). This implies that for every $\theta \in M_\Gamma^{\mathrm{erg}}(X)$ the set A is θ-measurable and that $\theta(A)$ is either zero or one. In other words $A \in \mathcal{F}(\kappa)$, where

$$\mathcal{F}(\kappa) = \{A \in \mathcal{X} : \text{for } \kappa \text{ a.e. } \theta \in M(X) \text{ either } \theta(A) = 0 \text{ or } \theta(A) = 1\}$$

(see Theorem 8.3). We have shown that $\mathcal{I} \subset \mathcal{F}(\kappa)$ and it only remains to show that $\mathcal{F}(\kappa)$ separates points on $(M(X), \kappa)$ since by Theorem 8.3 this is a sufficient condition for the equality $(M(X), \kappa) = (\{\mu_y : y \in Y\}, \kappa)$, which in particular implies that ν a.e. $\mu_y \in M_\Gamma^{\mathrm{erg}}(X)$, so that $(X_y, \mathcal{X}_y, \mu_y, \Gamma)$ is ergodic.

Now if $\theta \neq \theta'$ are elements of $M_\Gamma^{\mathrm{erg}}(X)$ then there exists a Borel subset $B \in \mathcal{X}$ with $\theta(B) = 1$ and $\theta'(B) = 0$ (Theorem 4.2) and if we replace B by $A = \bigcap_{\gamma \in \Gamma} \gamma B$ we see that $A \in \mathcal{I} \subset \mathcal{F}(\kappa)$ and $\theta(A) = 1$, while $\theta'(A) = 0$. In particular we see that $\mathcal{F}(\kappa)$ separates points on $(M(X), \kappa)$ and the proof of part 2 is complete.

To prove the uniqueness, with notation as in part 3 of the theorem, let $\tilde{\kappa} = \psi_*(\eta)$ and observe that $\tilde{\kappa}$ is supported on $M_\Gamma^{\mathrm{erg}}(X)$. Hence the uniqueness of the representation in Choquet's theorem implies that $\tilde{\kappa} = \kappa$ so that η-a.e. $\psi(z) \in \{\mu_y : y \in Y\}$. Now $\phi : z \mapsto \psi(z) = \mu_y \mapsto y$ is the required map $\phi : (Z, \mathcal{Z}, \eta) \to (Y, \mathcal{Y}, \nu)$. □

3. The order of orthogonality of a quasifactor

Given $\kappa \in Q(\mathbf{X})$ set

$$\kappa' = \int_{M(X)} (\theta \times \delta_\theta) \, d\kappa(\theta),$$

a probability measure on $X' := X \times M(X)$ which is a joining of μ and κ. (Note that this shows that, in fact, every quasifactor arises as a quasifactor associated with a joining.)

For a positive integer $k \geq 1$ let

$$\kappa^{(k)} = \int_{M(X)} (\theta \times \theta \times \cdots \times \theta \times \delta_\theta)\, d\kappa(\theta),$$

where $\theta \times \theta \times \cdots \times \theta$ is a k-fold product, and

$$\kappa^{(\infty)} = \int_{M(X)} ((\cdots \theta \times \theta \times \theta \cdots) \times \delta_\theta)\, d\kappa(\theta).$$

Let

$$\bar{\kappa}^{(k)} = \int_{M(X)} \theta \times \theta \times \cdots \times \theta\, d\kappa(\theta),$$

and

$$\bar{\kappa}^{(\infty)} = \int_{M(X)} (\cdots \theta \times \theta \times \theta \cdots)\, d\kappa(\theta),$$

be the corresponding projections onto X^k and X^∞ respectively. The measures $\bar{\kappa}^{(k)}$ and $\bar{\kappa}^{(\infty)}$ are selfjoinings of the system (X, μ). In particular $\bar{\kappa}^{(1)} = \bar{\kappa}' = \mu$ and $\bar{\kappa}^{(2)} = \int_{M(X)} \theta \times \theta\, d\kappa(\theta)$.

In Chapter 6, Section 2 the joinings σ of $\mathbf{X}$ and $\mathbf{Y}$, and τ^* of $\mathbf{Z}$ and $\mathbf{Y}$ are called orthogonal or independent relative to Y if $\tau \circ \sigma$ is the product measure $\mu \times \eta$. The joining σ is orthogonal to σ^* relative to Y iff it is independent (Proposition 6.11). Following this terminology we say that the quasifactor κ of $\mathbf{X}$ is *orthogonal* when $\bar{\kappa}^{(2)} = \mu \times \mu$.

8.8. PROPOSITION. *The only orthogonal quasifactor is the quasifactor $\kappa = \delta_\mu$. I.e. $\bar{\kappa}^{(2)} = \mu \times \mu$ implies $\kappa = \delta_\mu$.*

PROOF. Our assumption here is that

$$\bar{\kappa}^{(2)} = \int_{M(X)} (\theta \times \theta)\, d\kappa(\theta) = \mu \times \mu.$$

Thus for every bounded $f \in L^2(\mu)$, denoting by F the function on $M(X)$ given by

$$F(\theta) = \int_X f(x)\, d\theta(x),$$

we obtain

$$\begin{aligned}
\int_{X \times X} f \otimes f\, d\bar{\kappa}^{(2)} &= \int_{M(X)} \int_{X \times X} f \otimes f\, d(\theta \times \theta)\, d\kappa(\theta) \\
&= \int_{M(X)} \left(\int_X f\, d\theta\right)^2 d\kappa(\theta) = \int_{M(X)} (F(\theta))^2\, d\kappa(\theta) \\
&= \int_{X \times X} f \otimes f\, d(\mu \times \mu) = \left(\int_X f(x)\, d\mu(x)\right)^2 \\
&= \left(\int_{M(X)} \int_X f(x)\, d\theta(x)\, d\kappa(\theta)\right)^2 = \left(\int_{M(X)} F(\theta)\, d\kappa(\theta)\right)^2.
\end{aligned}$$

Equality in Cauchy-Schwartz's inequality implies that $F \equiv \int F\, d\kappa$ is a constant κ a.e., and since the functions of the form F above distinguish between the points of $M(X)$, we conclude that κ is supported on a single element of $M(X)$; i.e. $\kappa = \delta_\mu$. □

If κ is a quasifactor of the ergodic system $\mathbf{X}$ and $\kappa' \in J(\kappa, \mu)$ the associated joining we shall say that a joining λ of r copies $\mathbf{X}_1, \mathbf{X}_2, \ldots, \mathbf{X}_r$ of $\mathbf{X}$ and $(M(X), \kappa)$ is *κ-admissible* if $\kappa' = \mathrm{proj}_{X_j \times M(X)} \lambda$ for every $1 \le j \le k$.

8.9. Proposition. *To every non-trivial quasifactor κ of the ergodic system $\mathbf{X}$ (i.e. $\kappa \ne \delta_\mu$) corresponds a greatest integer $r \ge 1$ such that there exists a κ-admissible $\lambda \in J(\mathbf{X}_1, \mathbf{X}_2, \ldots, \mathbf{X}_k, (M(X), \kappa))$ with $\mathrm{proj}_{\mathbf{X}_1 \times \mathbf{X}_2 \times \cdots \times \mathbf{X}_k} \lambda = \mu^r$.*

Proof. This follows from the fact that every ergodic system is disjoint from any identity system. In fact, negating the conclusion of the proposition and using the compactness of $M(X^{\mathbb{Z}} \times M(X))$, we see that there exists a joining $\lambda \in J(\kappa, \mu^{\mathbb{Z}})$ with the property that all the projections $\mathrm{proj}_{X_j \times M(X)} \lambda$, coincide with the joining κ'. However we can also consider λ as a joining of the system $(X^{\mathbb{Z}}, S, \mu^{\mathbb{Z}})$, where S is the shift on $X^{\mathbb{Z}}$, and the identity transformation on $(M(X), \kappa)$. Since the former is certainly ergodic (as a Bernoulli shift) and since ergodic transformations are disjoint from the identity transformation on any space (Theorem 6.26) we conclude that $\lambda = \mu^{\mathbb{Z}} \times \kappa$ and in particular

$$\kappa' = \mu \times \kappa.$$

Thus

$$\kappa' = \int_{M(X)} (\theta \times \delta_\theta)\, d\kappa(\theta) = \int_{M(X)} (\mu \times \delta_\theta)\, d\kappa(\theta),$$

and the uniqueness of disintegration implies that $\kappa = \delta_\mu$. □

8.10. Definition. We call the number r given by Proposition 8.9 *the order of orthogonality* of the quasifactor κ.

4. The de Finetti-Hewitt-Savage theorem

Let X be a compact metric space. As usual $X^{\mathbb{Z}}$ is the infinite product space and we let $\mathfrak{S}$ be the group of permutations of coordinates in $X^{\mathbb{Z}}$ which fix all but finitely many of them. Let $M(X)$ be the compact metric simplex of all probability Borel measures on X and let $\mathfrak{Q} = M(M(X))$ be the space of probability Borel measures on $M(X)$. Since $\mathrm{ext}\,((M(X))) = \{\delta_x : x \in X\}$ is a closed subset of $M(X)$ we see that both $M(X)$ and $\mathfrak{Q}$ are Bauer simplices (see Theorem A.13).

Let $\mathfrak{J} \subset M(X^{\mathbb{Z}})$ be the convex closed subset of $M(X^{\mathbb{Z}})$ consisting of all probability measures on $X^{\mathbb{Z}}$ which are invariant under $\mathfrak{S}$, these measures are called *symmetric*. We set

$$\phi : \mathfrak{Q} \to \mathfrak{J},$$

$$\kappa \mapsto \phi(\kappa) = \lambda = \bar{\kappa}^{(\infty)} = \int_{M(X)} (\cdots \theta \times \theta \times \theta \cdots)\, d\kappa(\theta). \tag{4.1}$$

Clearly ϕ is a continuous affine map of $\mathfrak{Q}$ into $\mathfrak{J}$. The de Finetti-Hewitt-Savage theorem, [**121**], in the version proved by Weiss and Dubins (see [**52**]), states that this map is an affine homeomorphism of $\mathcal{Q}$ onto $\mathfrak{J}$. In particular, as in the Bauer simplex $\mathcal{Q}$ the set of extreme points is the closed set

$$\{\delta_\theta : \theta \in M(X)\},$$

it follows that in $\mathfrak{J}$, the set of extreme points is the closed set

$$\{\phi(\delta_\theta) : \theta \in M(X)\} = \{\hat{\theta} = \cdots \theta \times \theta \times \theta \cdots : \theta \in M(X)\},$$

and that (4.1) is the unique Choquet representation of an element λ of the simplex $\mathfrak{J}$ as an integral over the set of extreme points; or in other words its $\mathfrak{S}$ ergodic decomposition (Theorem 3.22).

8.11. THEOREM (de Finetti-Hewitt-Savage). *The map $\phi : \mathcal{Q} \to \mathfrak{J}$ is an affine homeomorphism.*

PROOF. To begin with we observe that $\mathfrak{J}$ is a simplex, the simplex of $\mathfrak{S}$-invariant Borel probability measures on the compact space $X^{\mathbb{Z}}$. Hence a measure $\gamma \in \mathfrak{J}$ is an extreme point iff it is a $\mathfrak{S}$-ergodic measure. It is therefore enough to show that the closed set

$$\{\phi(\delta_\theta) : \theta \in M(X)\} = \{\hat{\theta} = \cdots \theta \times \theta \times \theta \cdots : \theta \in M(X)\},$$

coincides with the set of $\mathfrak{S}$-ergodic measures in $\mathfrak{J}$. In fact two Bauer simplices are affinelly homeomorphic iff their closed sets of extreme points are homeomorphic compact spaces.

First we show that every $\hat{\theta}$ is Γ-ergodic. We let $\mathcal{F}_n$ be the σ-algebra of sets which are measurable with respect to the coordinates $-n \le k \le n$. Let E be a $\mathfrak{S}$-invariant measurable set. Then there is a sequence $E_n \in \mathcal{F}_n$ with $\hat{\theta}(E \triangle E_n) \to 0$. Let σ_n be any permutation that exchanges the intervals $[-n, n]$ and $[n+1, 3n+2]$ and fixes all other integers, then

$$\hat{\theta}(E \triangle E_n) \to 0 \quad \text{and} \quad \hat{\theta}(\sigma_n E \triangle \sigma_n E_n) \to 0,$$

imply

$$\hat{\theta}(\sigma_n E \triangle E_n) \to 0 \quad \text{hence} \quad \hat{\theta}(\sigma_n E \cap E_n) \to \hat{\theta}(E).$$

However, since clearly the events E_n and $\sigma_n E_n$ are independent we also have

$$\hat{\theta}(\sigma_n E \cap E_n) = \hat{\theta}(\sigma_n E)\hat{\theta}(E_n) \to \hat{\theta}(E)^2.$$

Thus $\hat{\theta}(E)$ is either 0 or 1; i.e. $\hat{\theta}$ is indeed $\mathfrak{S}$-ergodic.

It remains to show that conversely, every $\mathfrak{S}$-ergodic measure in $\mathfrak{J}$ is of the form $\hat{\theta}$ for some $\theta \in M(X)$. We fix a measurable function $f : X \to \mathbb{R}$ and observe that with respect to any measure $\zeta \in \mathfrak{J}$ the sequence of random variables $X_n = f \circ \pi_n$, where $\pi_n : X^{\mathbb{Z}} \to X$ are the coordinate projection maps, is identically distributed. In particular the measure $\theta = (\pi_n)_*(\zeta)$ is well defined and does not depend on n. Our proof will be complete when we show that for ζ ergodic these random variables are also independent.

Let α and β be two non-empty finite disjoint subsets of $\mathbb{Z}$, and let A and B be two subsets of $X^{\mathbb{Z}}$ measurable with respect to $\{X_n : n \in \alpha\}$ and $\{X_n : n \in \beta\}$ respectively. For every $N \in \mathbb{N}$ let S_N be the subgroup of $\mathfrak{S}$ consisting of the permutations on $[-N, N]$. The group $\mathfrak{S}$ is the union of the finite groups S_N and is therefore an amenable group with $\{S_N\}$ as a Følner sequence. Let

$$H_N = \{\sigma \in S_N : \sigma\beta \cap \alpha \ne \emptyset\} = \bigcup_{i \in \beta, j \in \alpha} \{\sigma \in S_N : \sigma(i) = j\},$$

then clearly

$$|H_N| \leq |\alpha||\beta|(2N)!.$$

Now for any $\sigma \in S_N$ such that $\sigma \notin H_N$, there exists a $\sigma' \in S_N$ which fixes each element of α and such that $\sigma' \circ \sigma(i) = i$ for every $i \in \beta$. For this σ' we have

$$\zeta(A \cap \sigma B) = \zeta(\sigma'(A \cap \sigma B)) = \zeta(A \cap B).$$

Applying the mean ergodic theorem, Theorem 3.33, with the Følner sequence S_N we get

$$\begin{aligned}
\zeta(A)\zeta(B) &= \lim_{N\to\infty} \frac{1}{(2N+1)!} \sum_{\sigma \in S_N} \zeta(A \cap \sigma B) \\
&= \lim_{N\to\infty} \frac{1}{(2N+1)!} \left(\sum_{\sigma \notin H} \zeta(A \cap \sigma B) + \sum_{\sigma \in H} \zeta(A \cap \sigma B) \right) \\
&= \lim_{N\to\infty} \left(\frac{(2N+1)! - |H_N|}{(2N+1)!} \zeta(A \cap B) + O(\frac{|\alpha||\beta|}{2N+1}) \right) \\
&= \zeta(A \cap B).
\end{aligned}$$

Thus the sequence $\{X_n\}_{n\in\mathbb{Z}}$ is indeed an independent sequence of random variables and we conclude that

$$\zeta = \hat{\theta} = \cdots \theta \times \theta \times \theta \cdots ,$$

as required. □

5. Quasifactors and infinite order symmetric selfjoinings

Let $\mathbf{X} = (X, \mathcal{X}, \mu, \Gamma)$ be an ergodic system with X a compact metric space and Γ acting on X as a group of homeomorphisms. Let Γ act on $X^{\mathbb{Z}}$ by the diagonal action. Let $\mathfrak{S}$ be the group of permutations of coordinates in $X^{\mathbb{Z}}$ which fix all but finitely many of them. Denote by $J^{\infty}_{\mathrm{sym}}(\mathbf{X})$ the set of all symmetric (i.e. $\mathfrak{S}$-invariant), Γ-invariant probability measures on $X^{\mathbb{Z}}$ with marginal μ on the 0 (hence also every $n \in \mathbb{Z}$) coordinate. Clearly for every $\kappa \in Q(\mathbf{X})$ the corresponding infinite joining $\bar{\kappa}^{(\infty)}$ is in $J^{\infty}_{\mathrm{sym}}(\mathbf{X})$.

8.12. THEOREM. 1. *Every quasifactor κ of the system $\mathbf{X}$ is canonically isomorphic to a factor of the infinite order symmetric selfjoining $(X^{\mathbb{Z}}, \bar{\kappa}^{(\infty)}) \in J^{\infty}_{\mathrm{sym}}(\mathbf{X})$ of $\mathbf{X}$. The factor map $\alpha : (X^{\mathbb{Z}}, \bar{\kappa}^{(\infty)}) \to (W, \kappa) = (M(X), \kappa)$ is the factor map which corresponds to the σ-algebra of $\mathfrak{S}$-invariant sets; i.e. to the $\mathfrak{S}$-ergodic decomposition. Thus, denoting $\zeta_w = \cdots \theta_w \times \theta_w \times \theta_w \cdots$, the $\mathfrak{S}$-ergodic decomposition*

$$\bar{\kappa}^{(\infty)} = \int_W \zeta_w \, d\kappa(w),$$

coincides with the disintegration of $\bar{\kappa}^{(\infty)}$ over κ with respect to α, and for κ a.e. w, $\zeta_w(\alpha^{-1}(w)) = 1$.

2. *The map $\phi : \kappa \mapsto \bar{\kappa}^{(\infty)}$, $\phi : Q(\mathbf{X}) \to J^{\infty}_{\mathrm{sym}}(\mathbf{X})$ is an affine homeomorphism onto.*

PROOF. 1. Apply the de Finetti-Hewitt-Savage theorem (Theorem 8.11) as follows. Fix a compact metric topology on X and observe that $Q(\mathbf{X})$ is a closed convex subset of $\mathfrak{Q}$ and that $J^{\infty}_{\mathrm{sym}}(\mathbf{X})$ is a closed convex subset of $\mathfrak{J}$. Moreover, clearly $\phi : \mathfrak{Q} \to \mathfrak{J}$ maps $Q(\mathbf{X})$ into $J^{\infty}_{\mathrm{sym}}(\mathbf{X})$. If $\lambda \in J^{\infty}_{\mathrm{sym}}(\mathbf{X})$, then λ as an element of $\mathfrak{J}$ has a unique representation (4.1) (see Section 4). Applying $\gamma \in \Gamma$ to (4.1) we get:

$$\begin{aligned}\gamma\lambda &= \int_{M(X)} (\cdots\gamma\theta \times \gamma\theta \times \gamma\theta\cdots)\, d\kappa(\theta)\\ &= \int_{M(X)} (\cdots\theta \times \theta \times \theta\cdots)\, d\gamma\kappa(\theta).\end{aligned}$$

On the other hand

$$\gamma\lambda = \lambda = \int_{M(X)} (\cdots\theta \times \theta \times \theta\cdots)\, d\kappa(\theta),$$

and the uniqueness of the representation implies that $\gamma\kappa = \kappa$. By projecting the representation (4.1) of λ on, say the zero coordinate, we see that κ satisfies the barycenter equation (1.1) and conclude that $\kappa \in Q(\mathbf{X})$ and that $\phi(\kappa) = \lambda$. Thus $\phi : \mathfrak{Q} \to \mathfrak{J}$ maps $Q(\mathbf{X})$ onto $J^{\infty}_{\mathrm{sym}}(\mathbf{X})$.

It is now clear that the map $\alpha : (X^Z, \bar{\kappa}^{(\infty)}, \Gamma) \to (M(X), \kappa, \Gamma)$ which corresponds to the $\mathfrak{S}$ ergodic decomposition

$$\bar{\kappa}^{(\infty)} = \int_W \zeta_w \, d\kappa(w),$$

is indeed a Γ homomorphism. In particular κ a.e. the product measures, ζ_w are mutually singular and $\zeta_w(\alpha^{-1}(w)) = 1$.

2. This follows from the fact that $\phi : \mathfrak{Q} \to \mathfrak{J}$ is an affine homeomorphism. □

8.13. COROLLARY. *Let P be a property of ergodic systems which is preserved by infinite ergodic selfjoinings as well as factors. Then every ergodic quasifactor of a system with property P also has property P. By Theorem 8.4 it follows that a system $\mathbf{Y}$ with no non-trivial property P factor is disjoint from all property P systems.*

As we shall see later, distality (which is disjoint from weak mixing) and zero-entropy (which is disjoint from K-systems) are two such properties. Here we consider the properties rigidity (Definition 3.2) and mild mixing.

8.14. EXERCISE. Show that rigidity is a property that satisfies the conditions of Corollary 8.13.

8.15. DEFINITION. An ergodic system is *mildly mixing* if it has no non-trivial rigid factor.

8.16. COROLLARY. *An ergodic dynamical system is mildly mixing iff it is disjoint from every rigid system.*

8.17. EXERCISE. Suppose Γ is commutative. Show that an ergodic system $\mathbf{X}$ is mildly mixing iff

$$\limsup_{\gamma\to\infty} \phi(\gamma) < 1,$$

for every matrix coefficient $\phi = \phi_f$, where $\phi_f(\gamma) := \langle U_\gamma f, f\rangle$, $f \in L^2(X,\mu), \|f\| = 1$. [[If $\mathbf{X} \to \mathbf{Y}$ is a rigid factor, then there exists a sequence $\Gamma \ni \gamma_n \to \infty$ such that $U_{\gamma_n} \to \mathrm{id}$ strongly on $L^2(Y,\nu)$. For any function $f \in L^2_0(Y,\nu)$ with $\|f\| = 1$, we have $\lim_{n\to\infty} \phi_f(\gamma_n) = 1$. Conversely, if $\lim_{n\to\infty} \phi_f(\gamma_n) = 1$ for some $\gamma_n \to \infty$ and $f \in L^2_0(X,\mu), \|f\| = 1$, then $\lim_{n\to\infty} U_{\gamma_n} f = f$. Show that f can be replaced by a bounded function and let A be the sub-algebra of $L^\infty(X,\mu)$ generated by $\{U_\gamma f : \gamma \in \Gamma\}$. By Theorem A.10, A defines a non-trivial factor $\mathbf{X} \to \mathbf{Y}$. Use the fact that Γ is commutative to deduce that $U_{\gamma_n} \to \mathrm{id}$ strongly on $L^2(Y,\nu)$.]]

6. Joining quasifactors

Let $\mathbf{X}$ and $\mathbf{Y}$ be ergodic dynamical systems. Let $\lambda \in J_e(\mu,\nu)$ be an ergodic joining and let

$$\lambda = \int_Y (\lambda_y \times \delta_y)\, d\nu(y),$$

be its disintegration over ν. Let $\phi : Y \to M = M(X)$ be the measurable map associated with this disintegration, $\phi : y \mapsto \lambda_y$ from Y to the space of probability measures on X and let $\kappa = \phi_*(\nu)$ be the image of ν under ϕ.

Thus the system $\mathbf{W} = (M(X), \kappa, \Gamma)$ is a factor of $\mathbf{Y}$ and a quasifactor of $\mathbf{X} = (X, \mu, \Gamma)$. With the factor map $\phi : Y \to M = M(X)$, we associate the disintegration of ν over κ:

$$\nu = \int_M \nu_\theta d\kappa(\theta) = \int_M \nu_{\lambda_y}\, d\kappa(\lambda_y) = \int_M \nu_w\, d\kappa(w),$$

where we write w for λ_y.

Two natural joinings now arise.

$$\kappa' = \int_M (\theta \times \delta_\theta)\, d\kappa(\theta) = \int_Y (\lambda_y \times \delta_{\lambda_y})\, d\nu(y),$$

on $X' := X \times M$, and

$$\lambda' = \int_Y (\lambda_y \times \delta_{\lambda_y} \times \delta_y)\, d\nu(y),$$

on $X' \times Y$.

Clearly the map $\psi : (x, \lambda_y, y) \mapsto (x, y)$ from $X' \times Y$ to $X \times Y$ is 1-1 and equivariant. Applying ψ to the definition of λ' we get

$$\begin{aligned}\psi(\lambda') &= \psi\Big(\int_Y (\lambda_y \times \delta_{\lambda_y} \times \delta_y)\, d\nu(y)\Big)\\ &= \int_Y (\lambda_y \times \delta_y)\, d\nu(y) = \lambda,\end{aligned}$$

so that ψ is a canonical isomorphism of $(X' \times Y, \lambda')$ onto $(X \times Y, \lambda)$.

But we also see that

$$\begin{aligned}
\lambda' &= \int_Y (\lambda_y \times \delta_w \times \delta_y)\, d\nu(y) \\
&= \int_M \int_Y (\lambda_y \times \delta_w \times \delta_y)\, d\nu_w(y)\, d\kappa(w) \\
&= \int_M (\lambda_w \times \delta_w \times \nu_w)\, d\kappa(w) \\
&= \kappa' \underset{\kappa}{\times} \nu.
\end{aligned}$$

So that finally:

$$(X' \times Y, \lambda') \cong (X' \underset{M}{\times} Y, \kappa' \underset{\kappa}{\times} \nu) \cong (X \times Y, \lambda).$$

We have shown:

8.18. PROPOSITION. *Let* $\mathbf{X}$ *and* $\mathbf{Y}$ *be ergodic systems, and* λ *an ergodic joining. Form the dynamical systems* $\mathbf{W}$ *and* $\mathbf{X}'$ *as above. Then* $\mathbf{W}$ *is a quasifactor of* $\mathbf{X}$*, a factor of* $\mathbf{Y}$*, and in the commutative diagram:*

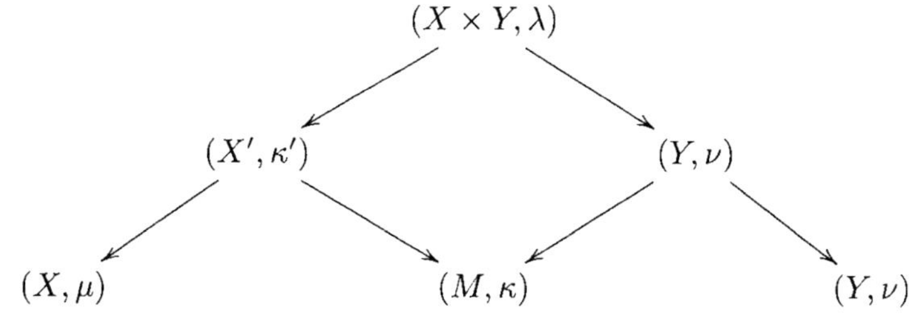

we have $\lambda = \kappa' \underset{\kappa}{\times} \nu$.

8.19. DEFINITION. Let $\mathbf{X}$ be an ergodic system, κ an ergodic quasifactor of $\mathbf{X}$. We say that κ is a *joining quasifactor*, JQF for short, if the joining

$$\kappa' = \int_M (\theta \times \delta_\theta)\, d\kappa(\theta),$$

of the systems (X, μ) and (M, κ), is ergodic. We denote the collections of joining quasifactors and ergodic JQF of $\mathbf{X}$ by $Q_J(X, \mu) = Q_J(\mathbf{X})$ and $Q_J^e(\mathbf{X})$ respectively. When $\pi : \mathbf{X} \to \mathbf{Y}$ is a factor map the collection of all JQF of $\mathbf{X}$ over $\mathbf{Y}$ is denoted by $Q_J(\pi)$.

Thus the ergodic systems $\mathbf{Y}$ that appear as JQFs of an ergodic system $\mathbf{X}$ are those that admit an ergodic joining $\lambda \in J_e(\mu, \nu)$ for which the map $y \mapsto \lambda_y$ is an embedding of $\mathbf{Y}$ as an ergodic quasifactor of $\mathbf{X}$. Proposition 8.18 shows that in order to understand the general ergodic joining $\lambda \in J_e(\mu, \nu)$ it is enough to study the corresponding JQF $\kappa \in Q_J(\mu)$ and the joining $\kappa' \in J_e(\mu, \kappa)$. We make this precise in the following proposition.

8.20. PROPOSITION. *Let* $\mathbf{X}$ *be an ergodic system*

1. *The ergodic quasifactor* κ *of the system* $\mathbf{X}$ *is a JQF iff there exists an ergodic system* $\mathbf{Y}$ *and an ergodic joining* λ *of* (X, μ) *and* (Y, ν) *such that* $\kappa = \phi_*(\nu)$*; where*
$$\lambda = \int_Y (\lambda_y \times \delta_y)\, d\nu(y),$$
is the disintegration of λ *over* ν *and* $\phi : Y \to M = M(X)$, $\phi : y \mapsto \lambda_y$ *is the associated map. When this is the case then* $\lambda = \kappa' \underset{\kappa}{\times} \nu$, *with* $\kappa' = \int_M (\theta \times \delta_\theta)\, d\kappa(\theta)$.
2. *The ergodic system* $\mathbf{Y}$ *appears as a JQF of* $\mathbf{X}$ *iff there is an ergodic joining* $\lambda \in J_e(\mu, \nu)$ *such that in the disintegration*
$$\lambda = \int_Y (\lambda_y \times \delta_y)\, d\nu(y),$$
the map $y \mapsto \lambda_y$ *is 1-1.*

PROOF. 1. If κ is a JQF, we take $(Y, \nu) = (M(X), \kappa)$ and
$$\lambda = \kappa' = \int_M (\theta \times \delta_\theta)\, d\kappa(\theta),$$
which by assumption is ergodic. Conversely, if an ergodic system $(Y, \mathcal{Y}, \nu, \Gamma)$ and an ergodic joining λ of (X, μ) and (Y, ν) are given, Proposition 8.18 shows that $\kappa = \phi(\nu)$ is an ergodic quasifactor of (X, μ) and that κ' is ergodic. Part 2 is clear. □

8.21. DEFINITION. Let (X, μ) be an ergodic system and (W, κ) a JQF. We say that κ is:

1. *ergodically embedded* if the extension $(X \times W, \kappa') \to (W, \kappa)$ is an ergodic extension.
2. *weak mixingly embedded* if this extension is a weakly mixing extension.
3. *mixingly embedded* if this extension is a mixing extension.

8.22. REMARK. As one can easily see, for a dynamical system $\mathbf{X}$ with factor $\pi : \mathbf{X} \to \mathbf{Y}$, the system $\mathbf{X}$ is ergodic iff both the system $\mathbf{Y}$ and the extension π are ergodic. It thus follows that an ergodic quasifactor is ergodically embedded iff it is a JQF.

7. The symmetric product quasifactors

8.23. DEFINITION. Call a JQF κ, *continuously embedded* if κ a.e. θ is a continuous measure (has no atoms) and *finitely embedded* or of *finite type* if there exists a positive integer r such that κ a.e. θ is an equidistributed measure on a set of r points. In the latter case we say that κ is of *type* r.

8.24. PROPOSITION. *Let* κ *be a JQF of the ergodic system* $\mathbf{X}$, *then either* κ *is continuously embedded or* κ *is of type* r *for some positive integer* r*. In the latter case the system* $\mathbf{W} = (M(X), \kappa)$ *is isomorphic to a symmetric self-joining* $(X^{\langle r \rangle}, \kappa^*, \Gamma)$ *and the corresponding ergodic joining*
$$\kappa' = \int_M (\theta \times \delta_\theta)\, d\kappa(\theta),$$

is a joining of type r over $\mathbf{W}$.

PROOF. The function $f(x,\theta) = \theta(\{x\})$, $(x,\theta) \in X \times M(X)$, is clearly Γ-invariant and therefore the set $\{(x,\theta) : f(x,\theta) > 0\}$ has either κ' measure 0 or 1. In the first case κ is continuously embedded. In the second, the ergodicity of κ' implies that there exists a positive integer r such that κ a.e. θ has the form $\theta = \frac{1}{r}(\delta_{x_1} + \delta_{x_2} + \cdots + \delta_{x_r})$ for $x_1, x_2, \ldots, x_r$ distinct points in X. The map $\theta \mapsto \langle x_1, x_2, \ldots, x_r\rangle$, $M(X) \to X^{\langle r\rangle}$ defines the isomorphism of $\mathbf{W} = (M(X), \kappa, \Gamma)$ with $(X^{\langle r\rangle}, \kappa^*, \Gamma)$. □

A special case of a finitely embedded quasifactor is the following natural embedding of the r-symmetric product system (see Chapter 2, Section 3).

8.25. DEFINITION (Symmetric product quasifactors). The symmetric product $\mathbf{X}^{\langle r\rangle} = (X^{\langle r\rangle}, \mu^{\langle r\rangle}, \Gamma)$ can be realized as a quasifactor as follows. Define $\phi : X^{\langle r\rangle} \to M(X)$ by $\phi(\langle x_1, x_2, \ldots, x_r\rangle) = \frac{1}{r}(\delta_{x_1} + \delta_{x_2} + \cdots + \delta_{x_r})$ and $\kappa = \phi_*(\mu^{\langle r\rangle})$. Clearly then $(X^{\langle r\rangle}, \mu^{\langle r\rangle}, \Gamma)$ and $(M(X), \kappa, \Gamma)$ are isomorphic. More generally, if $\mathbf{X} \to \mathbf{Y}$ is a factor map with $\mu = \int \mu_y \, d\nu(y)$ the corresponding disintegration, then the map $\phi : Y^{\langle r\rangle} \to M(X)$ defined by $\phi(\langle y_1, y_2, \ldots, y_r\rangle) = \frac{1}{r}(\mu_{y_1} + \mu_{y_2} + \cdots + \mu_{y_r})$, defines a quasifactor $(M(X), \kappa, \Gamma)$ of $\mathbf{X}$ (where $\kappa = \phi(\mu^{\langle r\rangle})$) isomorphic to the symmetric product $\mathbf{Y}^{\langle r\rangle}$. We call these quasifactors the *r-symmetric product quasifactor* and the *r-symmetric product quasifactor associated with the factor* $\mathbf{Y}$ respectively.

8.26. EXAMPLE. (See Lemma 6.20) Let $\mathbf{X} = (X, \mu, \Gamma)$ be an ergodic system and κ the r-symmetric product quasifactor on $\mathbf{X}$, then for $\theta = \frac{1}{r}(\delta_{x_1} + \delta_{x_2} + \cdots + \delta_{x_r})$ and $\kappa' = \int (\theta \times \delta_\theta) \, d\kappa(\theta)$, we have

$$\begin{aligned}
\kappa'^* \circ \kappa' &= \int (\theta \times \theta) \, d\kappa(\theta) = \frac{1}{r^2} \sum_{i=1}^{r} \sum_{j=1}^{r} \int (\delta_{x_i} \times \delta_{x_j}) \, d\mu(x_1) \cdots d\mu(x_r) \\
&= \frac{1}{r^2} \left(\sum_{i=1}^{r} \int (\delta_{x_i} \times \delta_{x_i}) \, d\mu(x_i) + \sum_{i \neq j} \int (\delta_{x_i} \times \delta_{x_j}) \, d\rho(x_i) \, d\rho(x_j) \right) \\
&= \frac{1}{r} \Delta_\mu + (1 - \frac{1}{r})(\mu \times \mu).
\end{aligned}$$

Hence, $\kappa'^* \circ \kappa'(\Delta) = \frac{1}{r}$.

In the next proposition we examine the possibility of embedding self-joinings as quasifactors.

8.27. PROPOSITION. *Let $\mathbf{Y} = (Y, \mathcal{Y}, \nu, \Gamma)$ be an ergodic system and let $(Y \times Y \cdots \times Y, \lambda)$ be a q-fold selfjoining of $\mathbf{Y}$, q a positive integer or ∞.*

1. *Let $\{\alpha_1, \alpha_2, \ldots, \alpha_q\}$ be a set of q distinct positive real numbers with sum 1 and define a map*

$$\phi : Y \times Y \cdots \times Y \to M(Y)$$

$$\phi : (y_1, y_2, \ldots, y_q) \mapsto \sum_{j=1}^{q} \alpha_j \delta_{y_j}.$$

Clearly ϕ *is a Borel equivariant isomorphism and we set* $\kappa = \phi_*(\lambda)$*; then* $(M(Y), \kappa)$ *is a quasifactor of* $\mathbf{Y}$ *which is isomorphic to* (Y^q, λ)*. Thus every self-joining of* $\mathbf{Y}$*, is isomorphic to a quasifactor of* $\mathbf{Y}$*.*

2. *For* $q < \infty$*, let* $\sigma : Y^q \to Y^{\langle q \rangle} = Y^q/S_q$ *be the quotient map, and set* $\hat{\lambda} =: \sigma_*(\lambda)$*. Define a map*

$$\psi : Y^{\langle q \rangle} \to M(Y)$$

$$\psi : \langle y_1, y_2, \ldots, y_q \rangle \mapsto \gamma_{\langle y_1, y_2, \ldots, y_q \rangle} := \frac{1}{q} \sum_{j=1}^{q} \delta_{y_j}.$$

Clearly ψ *is a Borel isomorphism and we set* $\kappa = \psi_*(\hat{\lambda})$*. Then* $(M(Y), \kappa)$ *is a quasifactor of* $\mathbf{Y}$ *which is isomorphic to* $(Y^{\langle q \rangle}, \hat{\lambda})$*. It is a JQF iff the measure* λ *is* symmetric *(i.e.* λ *is invariant under the symmetric group* S_q *of permutations of coordinates in* Y^q*). Thus every symmetric q-fold self-joining of* $\mathbf{Y}$*, has a* $q!$ *to 1 factor (an* S_q*-quotient) which is isomorphic to a JQF of* (Y, ν)*.*

PROOF. 1. The barycenter of κ is

$$b(\kappa) = \int \phi(y_1, y_2, \ldots, y_q)\, d\lambda(y_1, y_2, \ldots, y_q)$$
$$= \sum_{j=1}^{q} \alpha_j \int \delta_{y_j}\, d\nu(y_j) = (\sum_{j=1}^{q} \alpha_j)\nu = \nu.$$

Thus $(M(Y), \kappa)$ is a quasifactor of $\mathbf{Y}$ which is isomorphic to λ.

2. If λ is symmetric then the disintegration of λ over $\hat{\lambda}$ is given by:

$$\lambda = \int_{Y^{\langle q \rangle}} \frac{1}{q!} \sum \{\delta_{(y_{\pi(1)}, y_{\pi(2)}, \ldots, y_{\pi(q)})} : \pi \in S_q\}\, d\hat{\lambda}(\langle y_1, y_2, \ldots, y_q \rangle).$$

To check the JQF property we form the measures:

$$\kappa' = \int \gamma_{\langle y_1, y_2, \ldots, y_q \rangle} \times \delta_{\langle y_1, y_2, \ldots, y_q \rangle}\, d\hat{\lambda},$$

and

$$\hat{\kappa} = \int \delta_{y_1} \times \delta_{(y_1, y_2, \ldots, y_q)}\, d\lambda.$$

Now the map

$$\mathrm{id} \times \sigma : Y \times Y^q \to Y \times Y^{\langle q \rangle},$$
$$(y, (y_1, y_2, \ldots, y_q)) \mapsto (y, \langle y_1, y_2, \ldots, y_q \rangle)$$

is an equivariant map and one can easily check that $(\mathrm{id} \times \sigma)_*(\hat{\kappa}) = \kappa'$. It follows that also κ' is ergodic so that κ is indeed a JQF.

On the other hand if λ is not symmetric then there exists $\pi \in S_q$ with $\pi_*(\lambda)$ singular to λ and it is easy to check that in this case

$$\kappa' = \int \gamma_{\langle y_1, y_2, \ldots, y_q \rangle} \times \delta_{\langle y_1, y_2, \ldots, y_q \rangle}\, d\hat{\lambda},$$

is not ergodic. □

8. A weakly mixing system with a non-weakly mixing quasifactor

Let $\mathbf{X} = (X, \mathcal{X}, \mu, T)$ be a weakly mixing $\mathbb{Z}$-system and K the circle group with normalized Lebesgue measure m. With a measurable function $\phi : X \to K$ we associate the *skew-product* system $\mathbf{X}_\phi = (X \times K, \mu \times m, T_\phi)$, where $T_\phi(x, k) = (Tx, \phi(x)k)$, $(x, k) \in X \times K$. Let $k_0 \in K$ be a non-root of unity, we define new systems $\mathbf{X}_S = (X \times K \times K, \mu \times m \times m, S)$ and $\mathbf{X}_R = (X \times K \times K \times K, \mu \times m \times m \times m, R)$ by the formulas

$$S(x, k, k') = (Tx, \phi(x)k, kk'), \qquad R(x, k, k', k'') = (Tx, \phi(x)k, kk', k_0kk'').$$

8.28. LEMMA. *If $\mathbf{X}_\phi$ is weakly mixing so is $\mathbf{X}_S$.*

PROOF. Suppose $f \in L^2(\mu \times m \times m)$ is an eigenfunction of T_ϕ with eigenvalue λ. We write $f = \sum_{n=-\infty}^{\infty} c_n(x, k)k'^n$, the Fourier decomposition of f over $X \times K$, then

$$\lambda \sum_{n=-\infty}^{\infty} c_n(x, k)k'^n = \sum_{n=-\infty}^{\infty} c_n(Tx, \phi(x)k)k^n k'^n.$$

Hence, for every n, $\lambda c_n(x, k) = k^n c_n(Tx, \phi(x)k)$. Now fix n and in turn expand $g(x, k) = c_n(x, k)$ as a Fourier series $g(x, k) = \sum_{j=-\infty}^{\infty} b_j(x)k^j$. We then get

$$\lambda \sum_{j=-\infty}^{\infty} b_j(x)k^j = \sum_{j=-\infty}^{\infty} b_j(Tx)\phi(x)^j k^{j+n}$$

and therefore, for every j, $\lambda b_{j+n}(x) = b_j(Tx)\phi(x)^j$. Since $|\lambda| = |\phi(x)| = 1$, we have $\int |b_{j+n}(x)|^2 \, d\mu(x) = \int |b_j(x)|^2 \, d\mu(x)$. But $\int |g|^2 \, d\mu \times dm = \sum \int |b_j(x)|^2 \, d\mu$. Hence, for non-zero g this can hold only for $n = 0$. Thus $f(x, k, k') = c_0(x, k)$ and $\lambda c_0(x, k) = c_0(T_\phi(x, k))$. By weak mixing of T_ϕ, f is a constant and our claim is proved. □

We leave the easy proofs of the following lemma and proposition as an exercise.

8.29. LEMMA. *The function $f(x, k, k', k'') = (k')^{-1}k''$ is an eigenfunction of the system $\mathbf{X}_R$ with eigenvalue k_0. In fact $\mathbf{X}_R$ is isomorphic to the product system $\mathbf{X}_S \times (K, m, R_{k_0})$, where (K, m, R_{k_0}) is the k_0 rotation on K, via the map $(x, k, k', k'') \mapsto (x, k, k', f(x, k, k', k''))$.*

8.30. PROPOSITION. *Let $\mathbf{X}$ be a weakly mixing system, $K, \phi, \mathbf{X}_\phi, k_0, \mathbf{X}_S$ and $\mathbf{X}_R$ as above and assume that $\mathbf{X}_S$ is weakly mixing. The map*

$$F : (X \times K \times K \times K, R_\phi) \to (M(\mathbf{X}_S), S), \quad F(x, k, k', k'') = \frac{1}{2}(\delta_{(x,k,k')} + \delta_{(x,k_0k,k'')}),$$

is a Borel isomorphism. The measure $\kappa = F_*(\mu \times m \times m)$ is a JQF of the system $\mathbf{X}_S$ and the map $F : \mathbf{X}_S \to (M(\mathbf{X}_S), \kappa, S)$ is an isomorphism. Thus $(M(\mathbf{X}_S), \kappa, S)$ is an ergodic non-weakly mixing JQF of the weakly mixing system $\mathbf{X}_S$.

8.31. REMARK. Given any weakly mixing system $\mathbf{X}$ there always exists $\phi : X \to K$ such that the associated skew-product system $\mathbf{X}_\phi = (X \times K, \mu \times m, T_\phi)$ is also weakly mixing; see for example Glasner and Weiss [**95**].

9. Notes

The sources for this chapter are mainly the works Glasner [**86**] and Glasner and Weiss [**100**] and [**103**]. See also [**90**] for a review of results on quasifactors in ergodic theory and topological dynamics. The notion of mild mixing for $\mathbb{Z}$ actions was introduced by Furstenberg and Weiss [**82**]; see also the works of Schmidt and Walters, [**232**], Schmidt [**229**], Rudolph, [**219**] and Lemańczyk and Lesigne [**166**]. Exercise 8.17 is from Proposition 2 in [**166**].

CHAPTER 9

Isometric and Weakly Mixing Extensions

Two common examples of ergodic $\mathbb{Z}$ dynamical systems, an irrational rotation of the circle with Lebesgue measure, and the Bernoulli shift on $\{0,1\}^{\mathbb{Z}}$ with the $\{\frac{1}{2},\frac{1}{2}\}$ product measure are typical members of the classes of isometric and weakly mixing systems respectively. We consider these examples, and to a less extent also these classes of dynamical systems, as well understood. Can one hope to be able to describe the general dynamical system in terms of such building blocks? The surprising answer to this question is that if one allows for the relative notions of isometry and weak mixing to play the part of building blocks then this is indeed the case. This vague statement is made precise in the Furstenberg-Zimmer structure theorem (Theorem 10.15) that we shall prove in the next chapter. The present chapter is devoted to the definitions and the detailed study of the notions of isometric and weakly mixing extensions.

We start by defining an isometric extension. The characterization of isometric systems as those having discrete spectrum is the one we use in order to define relative discrete spectrum or isometric extensions. This is done, given an extension $\mathbf{X} \to \mathbf{Y}$ of dynamical systems, by viewing the Hilbert space $L^2(\mathbf{X})$ as a direct integral of a Hilbert bundle over $L^\infty(\mathbf{Y})$. First we define, in this Hilbert bundle, generalized or $\mathbf{Y}$-eigenfunctions and then call the extension isometric (or with relative discrete spectrum) if the whole bundle is spanned by $\mathbf{Y}$-eigenfunctions. The extension is called weakly mixing if no non-trivial $\mathbf{Y}$-eigenfunction exists. This work is done in the first two sections. The third section is even more technical; it deals with the auxiliary Hilbert bundle of Hilbert-Schmidt operators and prepares the ground for the key theorem (Theorem 9.21), proved in Section 4, which describes the $\mathbf{Y}$-eigenfunction space of a relative product $\mathbf{X}_1 \underset{\mathbf{Y}}{\times} \mathbf{X}_2$ in terms of the $\mathbf{Y}$-eigenfunction spaces of $\mathbf{X}_1$ and $\mathbf{X}_2$. The last section (Section 5) deals with weakly mixing extensions.

1. $L^2(\mathbf{X})$ as a direct integral of the Hilbert bundle $\dot{\mathfrak{H}}(\mathbf{Y})$

Let $\pi : \mathbf{X} \to \mathbf{Y}$ be a factor map of dynamical systems and assume that $\mathbf{Y}$ is ergodic. Let $\mu = \int \mu_y \, d\nu(y)$ be the corresponding disintegration. We write $\mathfrak{H} = L^2(X,\mu)$ and $\mathfrak{H}_y = L^2(X,\mu_y)$ and view the collection of Hilbert spaces $\mathfrak{H}_y$ as a "Hilbert bundle" $\dot{\mathfrak{H}} = \dot{\mathfrak{H}}(\mathbf{Y})$ over $\mathbf{Y}$ and the Hilbert space $\mathfrak{H}$ as a "direct integral" of the bundle over $\mathbf{Y}$. Each $f \in \mathfrak{H}$ defines a cross-section $Y \to \dot{\mathfrak{H}}$ of the bundle, $y \mapsto f_y$, $y \to \mathfrak{H}_y$, where for almost all $y \in Y$, f_y is just f regarded as an element of $\mathfrak{H}_y$. The relations $\gamma\mu_y = \mu_{\gamma y}$, $\gamma \in \Gamma$ which follow a.e. by the invariance of μ and the uniqueness of the disintegration, imply that for almost all y, the maps $U(\gamma, y) : \mathfrak{H}_y \to \mathfrak{H}_{\gamma y}$ are isometries onto. Explicitly, for $\gamma \in \Gamma$ and $y \in Y$, the operator $U(\gamma, y) : \mathfrak{H}_y \to \mathfrak{H}_{\gamma y}$ is defined by

$$\big(U(\gamma, y)h\big)(x) = (h \circ \gamma^{-1})(x) \qquad (h \in \mathfrak{H}_y,\ x \in \pi^{-1}(\gamma y)).$$

It maps $\mathfrak{H}_y$ isometrically onto $\mathfrak{H}_{\gamma y}$. In fact

$$\begin{aligned}\langle U(\gamma, y)h, U(\gamma, y)k\rangle_{\gamma y} &= \int h(\gamma^{-1}x)\overline{k(\gamma^{-1}x)}\, d\mu_{\gamma y}(x) \\ &= \int h(x)\overline{k(x)}\, d\gamma^{-1}\mu_{\gamma y}(x) = \int h(x)\overline{k(x)}\, \mu_y(x) \\ &= \langle h, k\rangle_y, \qquad (h, k \in \mathfrak{H}_y).\end{aligned}$$

Clearly $U(\gamma^{-1}, \gamma y)$ is an inverse for $U(\gamma, y)$, so that $U(\gamma, y)$ is onto.

A convenient way of constructing the Hilbert bundle $\dot{\mathfrak{H}}$ is as follows. Fix a compact metric model for $\mathbf{Y}$. By Rohlin's skew-product Theorem, 3.18, we have $\mathbf{X} \cong \mathbf{Y} \underset{\alpha}{\times} (U, \rho) = (Y \times U, \mathcal{Y} \otimes \mathcal{U}, \mu \times \rho, \Gamma)$, with $(U, \mathcal{U}, \rho)$ a compact metric space with a full probability measure and $\alpha : \Gamma \times Y \to \mathrm{Aut}\,(U, \rho)$ a measurable cocycle, so that $\gamma(y, u) = (\gamma y, \alpha(\gamma, y)u)$. The Hilbert space $\mathfrak{H} = L^2(X, \mu) = L^2(Y, \nu) \otimes L^2(U, \rho)$ and identifying $\pi^{-1}(y)$ with $\{y\} \times U$ we have $\mu_y = \delta_y \times \rho$, hence a canonical isomorphism $\mathfrak{H}_y = L^2(X, \mu_y) \cong L^2(U, \rho)$. Let $\mathcal{C}_0 = \{f_n : n \in \mathbb{N}\}$ be a fixed countable dense subset of the algebra $C(X)$ of continuous complex valued functions on X. For convenience we assume that $0 \in \mathcal{C}_0$. The *Hilbert bundle* $\dot{\mathfrak{H}}(\mathbf{Y})$ *over* Y is the disjoint union of the Hilbert spaces $\mathfrak{H}_y$ endowed with the metric

$$d(h_1, h_2) = d(\pi(h_1), \pi(h_2)) + \sum_{n=1}^{\infty} 2^{-n}\big|\|h_1 - f_n\|_{y_1} - \|h_2 - f_n\|_{y_2}\big|,$$

where $\pi : \dot{\mathfrak{H}} \to Y$ is the map that sends an element in $\mathfrak{H}_y$ to the point $y \in Y$.

In the following proposition we enumerate the main properties of the Hilbert bundle $\dot{\mathfrak{H}}$. The details are not hard to check and are left as an exercise (see also [**64**], [**76**] and [**78**]).

9.1. PROPOSITION. 1. *The map* $(g, h) \mapsto \langle g, h\rangle = \langle g, h\rangle_y$, *where* $y = \pi(g) = \pi(h)$, *from* $\dot{\mathfrak{H}} \underset{Y}{\times} \dot{\mathfrak{H}} = \bigcup_{y \in Y} \mathfrak{H}_y \times \mathfrak{H}_y$ *to* $\mathbb{C}$ *is continuous, and so are the maps* $(g, h) \mapsto g + h$ *and* $(h, \lambda) \mapsto \lambda h$ *from* $\dot{\mathfrak{H}} \underset{Y}{\times} \dot{\mathfrak{H}}$ *to* $\mathbb{C}$ *and* $\dot{\mathfrak{H}} \times \mathbb{C}$ *to* $\mathbb{C}$ *respectively.*

2. *If* $y_0 \in Y$ *and* h_i *is a sequence in* $\dot{\mathfrak{H}}$ *such that* $y_i = \pi(h_i) \to y_0$ *and* $\|h_i\| \to 0$, *then* $h_i \to 0_{y_0}$, *where the latter is the zero element of* $\mathfrak{H}_y$. *More generally* $h_i \to h$ *in* $\dot{\mathfrak{H}}$ *iff* $\pi(h_i) \to \pi(h)$ *and* $\|h_i - f\| \to \|h - f\|$ *for every* $f \in \mathcal{C}_0$.
3. d *is a complete metric on* $\dot{\mathfrak{H}}$ *in which* $\dot{\mathfrak{H}}$ *is a separable, hence a Polish space.*
4. *The metric* d *induces the* L^2*-norm metric on each fiber* $\mathfrak{H}_y$.
5. $\pi : \dot{\mathfrak{H}} \to Y$ *is continuous and open.*
6. *Up to a* ν*-null set of* Y *the corresponding Borel structure on* $\dot{\mathfrak{H}}$ *does not depend on the various choices (the compact model for* $\mathbf{Y}$, *the Rohlin skew-product model for* X, *the collection* $\mathcal{C}_0$, *etc.) made in the construction of* $\dot{\mathfrak{H}}$.
7. *The* weak topology *on* $\dot{\mathfrak{H}}$ *is defined as follows: A sequence* $h_n \in \dot{\mathfrak{H}}$ *tends weakly to* $h \in \dot{\mathfrak{H}}$ *(we write* $h_n \xrightarrow{w} h$*) iff the sequence* $\|h_n\|$ *is bounded,*

$y_n = \pi(h_n) \to y_0 = \pi(h)$ and for every $f \in \mathfrak{C}_0$, $\langle h_n, f\rangle_{y_n} \to \langle h, f\rangle_{y_0}$. We then have $h_n \to h$ iff $h_n \xrightarrow{w} h$ and $\|h_n\| \to \|h\|$.

9.2. DEFINITION. 1. A closed subspace $M \subset \mathfrak{H}$ is called a $\mathbf{Y}$-*module* if for every $f \in M$ and any measurable function h on Y such that $hf \in \mathfrak{H}$ we have $hf \in M$. (This condition is equivalent to the requirement that M is an $L^\infty(\mathbf{Y})$-module; that is: $hf \in M$ whenever $f \in M$ and $h \in L^\infty(\mathbf{Y})$.) If M is a $\mathbf{Y}$-module denote by $M_y \subset \mathfrak{H}_y$ the image of M in $\mathfrak{H}_y$. We say that M is of *finite rank* if a.e. $\dim M_y \le r < \infty$ for some positive integer r. It is of *rank* r if a.e. $\dim M_y = r < \infty$.

2. Let M be a $\mathbf{Y}$-module, we say that a subset $L \subset M$ *spans* M if every $f \in M$ can be represented as a convergent sum $\sum c_n(\pi(x))f_n(x)$ with $f_n \in L$ and $c_n(y)$ measurable functions on Y. We say that L is an *orthonormal* set if $\langle f, g\rangle = \delta_{f,g}$ for every $f, g \in L$. If L is both spanning and orthogonal we call it an *orthonormal* $\mathbf{Y}$-*basis* for M.
3. A function $f \in \mathfrak{H}$ is a *generalized eigenfunction* or a $\mathbf{Y}$-*eigenfunction* if the module spanned by $\{f \circ \gamma : \gamma \in \Gamma\}$ is of finite rank.

Our next goal is to show that a $\mathbf{Y}$-module M with $\dim M_y$ a constant, admits an orthonormal $\mathbf{Y}$-basis.

9.3. LEMMA. *Let M be a closed $\mathbf{Y}$-module such that $\dim M_y \ge 1$ a.e., then there exists $f \in M$ with $\int |f|^2\, d\mu_y = 1$ for almost all y.*

PROOF. For $f \in \mathfrak{H}$ let $S(f) = \{y \in Y : \int |f|^2\, d\mu_y > 0\}$. Given $f \in M$ let $g(y) = \int |f|^2\, d\mu_y$ and define the function h on Y as $h(y) = g(y)^{-1}$ on $S(f)$ and 0 elsewhere, then $\hat{f} = hf \in M$ and $\int |\hat{f}|^2\, d\mu_y = \mathbf{1}_{S(f)}$. If $\nu(A) > 0$ then our assumption on M implies that $\nu(A \cap S(f)) > 0$ for some $f \in M$. Moreover if $f_i \in M$ with $A_i = S(f_i)$ then for $\hat{f}_i$ as above we have $\mathbf{1}_{A_1 \cup A_2} = \int |\hat{f}_1 + \mathbf{1}_{A_2 \setminus A_1} \hat{f}_2|^2\, d\mu_y$. A routine exhaustion argument shows that $\mathbf{1}_Y = \int |f|^2\, d\mu_y$ for a function f that can be represented as a convergent series of functions in M (see for example Lemma 3.17). Since M is closed, $f \in M$. □

9.4. LEMMA. *Let M be a closed $\mathbf{Y}$-module of rank $r < \infty$, then there exists an orthonormal $\mathbf{Y}$-basis $\{f_1, f_2, \cdots, f_r\}$ for M.*

PROOF. Let f_1 be the function constructed in Lemma 9.3 and set $M' = \{g \in M : \langle g, f_1\rangle_y = 0 \text{ a.e.}\}$. M' is again a $\mathbf{Y}$-module and clearly $\dim M_y = r - 1$ a.e. By induction we proceed to obtain $\{f_1, f_2, \cdots, f_r\} \subset M$ such that for a.e. y $\{f_{1,y}, f_{2,y}, \cdots, f_{r,y}\}$ form a basis for M_y and thus, for $g \in M$

$$g = \sum_{j=1}^{r} \langle g, f_j\rangle_y \cdot f_j = \sum_{j=1}^{r} c_j(y) f_j.$$

This completes the proof of the lemma. □

9.5. LEMMA. *Let M be a closed $\mathbf{Y}$-module such that $\dim M_y \ge r$ a.e., and let $g_1, g_2, \ldots, g_r$ be r functions in M, then there exists a Y-submodule $N \subset M$ with $\dim N_y = r$ a.e. and $\{g_1, g_2, \ldots, g_r\} \subset N$.*

PROOF. If $r = 1$ we let f be the function constructed in Lemma 9.3 and set $f_1 = \hat{g}_1 + \mathbf{1}_{Y\setminus S(g_1)} f$. We have $\int |f_1|^2 \, d\mu_y = \mathbf{1}_Y$ and if we let N be the $\mathbf{Y}$-module generated by f_1 then $\dim N_y = 1$ a.e. and $g_1 \in N$. Now let M be a $\mathbf{Y}$-module satisfying the assumptions of the lemma and, by induction, we assume that there is a Y-submodule $N' \subset M$ with $\dim N'_y = r-1$ a.e. and $\{g_1, g_2, \ldots, g_{r-1}\} \subset N'$. Let $M' = \{g \in M : \langle g, h\rangle_y = 0 \text{ a.e. for all } h \in N'\}$. By Lemma 9.4 there is an orthonormal $\mathbf{Y}$-basis $\{f_1, f_2, \ldots, f_{r-1}\}$ for N'.

Now for every $g \in M$

$$g = \sum_{j=1}^{r-1} c_j f_j + \left(g - \sum_{j=1}^{r-1} c_j f_i\right) \in N' + M', \qquad c_j(y) = \int g \bar{f}_j \, d\mu_y.$$

It follows that $\dim M'_y \geq 1$ and, by the case $r = 1$, we can find a Y-submodule $M'' \subset M'$ containing $g_r - \sum_{j=1}^{r-1} c_j f_i$ $(c_j(y) = \int g_r \bar{f}_j \, d\mu_y)$ with $\dim M''_y = 1$. The Y-submodule $N = N' + M'' \subset M$ satisfies $\dim N_y = r$ a.e. and $\{g_1, g_2, \ldots, g_r\} \subset N$. □

9.6. THEOREM. *Every closed* $\mathbf{Y}$*-module* M *in* $L^2(\mathbf{X})$ *with* $\dim M_y = r \leq \infty$ *a.e. has an orthonormal* $\mathbf{Y}$*-basis. In particular* $L^2(\mathbf{X})$ *admits an orthonormal* $\mathbf{Y}$*-basis.*

PROOF. Lemma 9.4 deals with the case $r < \infty$ and we now assume $r = \infty$. Let $\{g_n : n \in \mathbb{N}\}$ be a dense sequence in $\mathfrak{H} = L^2(\mathbf{X})$ and use Lemma 9.5 to construct inductively Y-submodules N_n, $N_n \subset N_{n+1}$, with $\dim N_{n,y} = n$ a.e. and $g_1, g_2, \ldots, g_n \in N_n$. From the way this lemma was proved we see that we now have a sequence $\{f_1, f_2, \ldots\} \subset \mathfrak{H}$ such that for every n, $\{f_1, f_2, \ldots, f_n\}$ is an orthonormal $\mathbf{Y}$-basis for N_n. Let N be the $\mathbf{Y}$-module generated by the sequence $\{f_1, f_2, \ldots\}$, then $N \subset M$ and $N \supset \{g_n : n \in \mathbb{N}\}$, hence $N = M$ and $\{f_1, f_2, \ldots\}$ is an orthonormal $\mathbf{Y}$-basis for M. □

9.7. DEFINITION. A function $s : Y \to \dot{\mathfrak{H}}$ is called a *section* if $\pi \circ s = \mathrm{id}_Y$. A measurable section s is *square integrable* if $\int_Y \|s(y)\|^2 \, d\nu(y) < \infty$.

The property of $L^2(\mathbf{X})$ given in the next proposition is usually described by saying that $L^2(\mathbf{X})$ is the *direct integral* of the Hilbert bundle $\dot{\mathfrak{H}}$.

9.8. PROPOSITION. *There is a 1-1 surjective correspondence between the space* $\mathfrak{H} = L^2(X, \mu)$ *and the space of square integrable sections* $s : Y \to \dot{\mathfrak{H}}$*. This correspondence is given by* $f \mapsto s_f$ *where* $s_f(y) = f_y$*.*

PROOF. Since $\|f\|^2 = \int_Y \int_X |f|^2 \, d\mu_y(x) \, d\nu(y) = \int_Y \|f\|_y^2 \, d\nu(y)$, it follows that every s_f is indeed a square integrable section. It is clear that the map $f \mapsto s_f$ is 1-1. Let $s : Y \to \dot{\mathfrak{H}}$ be a square integrable section. Let $\{e_n : n \in \mathbb{N}\}$ be an orthonormal $\mathbf{Y}$-basis for $L^2(\mathbf{X})$, then the functions $c_n(y) = \langle s(y), e_{n,y}\rangle$ are elements of $L^2(\mathbf{Y})$

and we can form the sum $f = \sum_{n=1}^{\infty} c_n(y)e_n$, which satisfies

$$\begin{aligned}\|f\|^2 &= \int_Y \|\sum_{n=1}^{\infty} c_n(y)e_n\|_y^2\, d\nu(y)\\ &= \int_Y \sum_{n=1}^{\infty} \|c_n(y)\|_y^2\, d\nu(y)\\ &= \int_Y \|s(y)\|_y^2\, d\nu(y) < \infty.\end{aligned}$$

Thus $f \in L^2(\mathbf{X})$ and clearly $s = s_f$ a.e. □

Let $y \mapsto A(y) \in \mathrm{End}(\mathfrak{H}_y)$ be an operator valued cross-section. We shall say that it is *measurable* if the cross-section $y \mapsto A(y)s(y) \in \mathfrak{H}_y$ is measurable whenever s is square integrable cross-section. When $\|A(y)\|$ is uniformly bounded (i.e. ess-sup $\|A(y)\| < \infty$), in order to show that $A(y)$ is measurable it suffices to check that $y \mapsto A(y)e_n(y)$ is measurable, for an orthonormal Y basis $\{e_n : n \in \mathbb{N}\}$.

Suppose that $V(y)$ is an r-dimensional subspace of $\mathfrak{H}_y$. We say that V is a *measurable r-plane cross-section* when the corresponding operator valued cross-section $y \mapsto P_{V(y)}$ is measurable (here $P_{V(y)}$ is the orthogonal projection on $V(y)$). The following is an analogue of Proposition 9.8 for r-plane cross-section.

9.9. PROPOSITION. *For any fixed $r < \infty$, there is a 1-1 surjective correspondence between the collection of closed $\mathbf{Y}$-modules in $\mathfrak{H} = L^2(X,\mu)$ and the space of measurable r-plane cross-sections. This correspondence is given by $M \mapsto V_M$, where $V_M(y) = M_y$.*

PROOF. Let M be a rank r $\mathbf{Y}$-module. Choose an orthonormal $\mathbf{Y}$-basis $\{f_1, f_2, \dots, f_r\}$ for M and observe that the projection valued cross-section $y \mapsto P_{M_y}$ is measurable, since for $f \in L^2(\mathbf{X})$, $P_{M_y} f_y = \sum_{j=1}^{r} (\int f \bar{f}_j \, d\mu_y) f_{j,y}$. Conversely, let $y \mapsto V(y)$ be a r-plane cross-section. Then, for a fixed orthonormal $\mathbf{Y}$-basis of $\mathfrak{H}$, set $s_n(y) = P_{V(y)}e_n$, a square integrable cross-section, and let M be the $\mathbf{Y}$-module generated by the functions $f_n \in \mathfrak{H}$ which correspond to s_n. Clearly $M_y = V(y)$ a.e. □

2. Generalized eigenfunctions and isometric extensions

We keep the setup of the previous section and suppose that M is a Γ-invariant $\mathbf{Y}$-module. Since clearly $U(\gamma, y)(M_y) = M_{\gamma y}$ a.e., the function $y \mapsto \dim M_y$ is Γ-invariant and measurable and the ergodicity of $\mathbf{Y}$ implies that it is a constant $r \le \infty$. In particular any finite rank Γ-invariant module is of rank $r < \infty$ for some r.

9.10. DEFINITION. Recall that a function $f \in \mathfrak{H}$ is a generalized eigenfunction (or a $\mathbf{Y}$-eigenfunction) when the $\mathbf{Y}$-module generated by $\{f \circ \gamma : \gamma \in \Gamma\}$ is of finite rank, that is, if f belongs to some Γ-invariant finite rank $\mathbf{Y}$-module M. Choosing an orthonormal $\mathbf{Y}$-basis $\{h_1, h_2, \dots, h_r\}$ for M we see that $\{U_\gamma h_1, U_\gamma h_2, \dots, U_\gamma h_r\}$ is also an orthonormal $\mathbf{Y}$-basis for M and we have, for every $\gamma \in \Gamma$

$$U_\gamma h_i(x) = \sum \lambda_{ij}(\gamma, y)h_j(x) \quad \text{or} \quad U_\gamma H = \Lambda(\gamma, y)H,$$

where $H(x) = (h_1, h_2, \ldots, h_r)$ is a vector "eigenfunction" and $\Lambda(\gamma, y) = (\lambda_{ij}(\gamma, y))$ the *unitary matrix* $\Lambda(\gamma, y)$ as a matrix eigenvalue. Clearly the $\mathbf{Y}$-eigenfunctions form a subspace and we let $\mathcal{E}(\mathbf{X}/\mathbf{Y})$ be its closure in $L^2(\mathbf{X})$. $\mathcal{E}(\mathbf{X}/\mathbf{Y})$ is a $\mathbf{Y}$-module and it contains $L^2(\mathbf{Y})$ as well as $\mathfrak{I} = \mathrm{Inv}\,(L^2(\mathbf{X}))$, the closed subspace of Γ-invariant functions in $L^2(\mathbf{X})$. We say that the extension $\pi : \mathbf{X} \to \mathbf{Y}$ is an *isometric extension* (or that $\mathbf{X}$ has a *relatively discrete spectrum over* $\mathbf{Y}$) when $L^2(\mathbf{X}) = \mathcal{E}(\mathbf{X}/\mathbf{Y})$.

9.11. EXAMPLE. We recall that the topological system (X, T) described in Examples 1.6.1, and 4.15, namely the transformation $T(\xi, \eta) = (\xi + \alpha, \eta + \xi)$ on the torus $X = \mathbb{T}^2$, is minimal, topologically distal and strictly ergodic. Let $\mathbf{X} = (X, \mathcal{X}, \mu, T)$, where $\mu = \nu \times \nu$ and ν is Lebesgue measure on $\mathbb{T}$, be the corresponding measure-preserving system. Let $\mathbf{Y} = (\mathbb{T}, \nu, R_\alpha)$ and $\pi : \mathbf{X} \to \mathbf{Y}$ be the factor system and factor map corresponding to the projection on the first coordinate. For $0 \neq m \in \mathbb{Z}$ consider the function $f_m(\xi, \eta) = \exp(2\pi i m\eta)$, $0 \neq m \in \mathbb{Z}$. We have for $h \in L^\infty(\nu)$,

$$U_T(h(\xi) f_m(\xi, \eta)) = \exp(2\pi i m\xi) h(\xi + \alpha) f_m(\xi, \eta),$$

and it follows that the rank one $L^\infty(\nu)$ module $M_m \subset L^2(\mu)$ spanned by f_m is U_T invariant. Clearly each $(M_m)_\xi$ is one-dimensional and since the collection $\{f_m : m \in \mathbb{Z}\}$ together with $L^\infty(\nu)$ span $L^2(\mu)$ it follows that $\pi : \mathbf{X} \to \mathbf{Y}$ is an isometric extension. (In fact $\mathbf{X}$ is an example of a system with quasi-discrete spectrum, see the Notes at the end of Chapter 10.)

9.12. LEMMA. *If $\pi : \mathbf{X} \to \mathbf{Y}$ is an isometric extension and $\mathbf{X} \overset{\sigma}{\to} \mathbf{Z} \overset{\rho}{\to} \mathbf{Y}$, with $\pi = \rho \circ \sigma$, is an intermediate factor, then $\sigma : \mathbf{X} \to \mathbf{Z}$ and $\rho : \mathbf{Z} \to \mathbf{Y}$ are also isometric extensions.*

PROOF. The fact that $\sigma : \mathbf{X} \to \mathbf{Z}$ is an isometric extension follows directly from the definition. Let $P = \mathbb{E}^{\mathbf{Z}} : L^2(\mathbf{X}) \to L^2(\mathbf{Z})$, then P commutes with the operators U_γ, $\gamma \in \Gamma$ and with the operators R_f defined by multiplication with $f \in L^\infty(\mathbf{Y})$. It therefore follows that the image of a Γ-invariant, finite rank, $\mathbf{Y}$-module $M \subset L^2(\mathbf{X})$ under P is a Γ-invariant, finite rank, $\mathbf{Y}$-module in $L^2(\mathbf{Z})$. Since the union of such $\mathbf{Y}$-modules M is dense in $L^2(\mathbf{X})$, it follows that the union of their images, $P(M)$ is dense in $L^2(\mathbf{Z})$. ☐

9.13. THEOREM. *The closed subspace $\mathcal{E}(\mathbf{X}/\mathbf{Y})$ contains a dense subalgebra of bounded $\mathbf{Y}$-eigenfunctions. Thus $\mathcal{E}(\mathbf{X}/\mathbf{Y})$ corresponds to an isometric extension $\rho : \mathbf{Z} \to \mathbf{Y}$. I.e. there exists an intermediate factor $\mathbf{X} \overset{\sigma}{\to} \mathbf{Z} \overset{\rho}{\to} \mathbf{Y}$, with $\pi = \rho \circ \sigma$ and $\rho : \mathbf{Z} \to \mathbf{Y}$ an isometric extension, such that $\mathcal{E}(\mathbf{X}/\mathbf{Y}) = L^2(\mathbf{Z}) \circ \sigma$.*

PROOF. Let M be a Γ-invariant, rank r, $\mathbf{Y}$-module and let $\{h_1, h_2, \ldots, h_r\}$ be an orthonormal $\mathbf{Y}$-basis for M. Define the function $p(x) = \sum_{j=1}^r |h_j(x)|^2$, then for every $\gamma \in \Gamma$ $p(\gamma x) = p(x)$ a.e. and it follows that the sets $A_N = \{x \in X : p(x) \leq N\}$ are Γ-invariant. Moreover it is easy to check that the functions $\mathbf{1}_{A_N} h_j$ are bounded $\mathbf{Y}$-eigenfunctions with the same eigenvalue matrix and of course $h_j = \lim_{N\to\infty} \mathbf{1}_{A_N} h_j$. The spaces $M_N = \mathrm{span}\{\mathbf{1}_{A_N} h_j : 1 \leq j \leq r\}$ are therefore $\mathbf{Y}$-modules with a bounded orthogonal $\mathbf{Y}$-basis. (Note that when $\mathbf{X}$ is ergodic, the function p is the constant r, so that in this case the functions h_j are necessarily bounded.) Now if M_i is a $\mathbf{Y}$-module with with bounded orthonormal $\mathbf{Y}$-basis $\{h_{i,1}, h_{i,2}, \ldots, h_{i,r_i}\}$, $(i = 1, 2)$, then the set $\{h_{1,j} \cdot h_{2,k} : 1 \le j \le r_1,\ 1 \le k \le r_2\}$ forms a spanning set for a finite rank $\mathbf{Y}$-module N which contains the product of any two bounded eigenfunctions $f \cdot g$ $(f \in M_1, g \in M_2)$. This proves the first part of the theorem. The second follows from Theorem A.9. □

9.14. THEOREM. *An extension $\pi : \mathbf{X} \to \mathbf{Y}$, with $\mathbf{X}$ ergodic, is an isometric extension iff it is a homogeneous skew-product extension. That is, iff there exists a compact group K, a closed subgroup $L \subset K$, and a cocycle $\alpha : \Gamma \times Y \to K$ such that $\mathbf{X} \cong (Y \times K/L, \nu \times \tilde{m}, \Gamma)$, where m is Haar measure on K, $\tilde{m}$ its image on K/L, and for $\gamma \in \Gamma$ and $(y, kL) \in Y \times K/L$, $\gamma(y, kL) = (\gamma y, \alpha(\gamma, y)kL)$. In particular, an ergodic system $\mathbf{X}$ is isometric iff it is a homogeneous system.*

PROOF. Suppose first that $\pi : \mathbf{X} \to \mathbf{Y}$ is an isometric extension. Let $M \subset L^2(\mathbf{X})$ be a Γ-invariant $\mathbf{Y}$-module of finite rank r. Choosing an orthonormal $\mathbf{Y}$-basis $\{h_1, h_2, \ldots, h_r\}$ for M we have, for every $\gamma \in \Gamma$

$$U_\gamma h_i(x) = \sum \lambda_{ij}(\gamma, y) h_j(x) \quad \text{or} \quad U_\gamma H = \Lambda(\gamma, y) H,$$

where $H(x) = (\sqrt{r})^{-1}(h_1, h_2, \ldots, h_r)$ and $\Lambda(\gamma, y) = (\lambda_{ij}(\gamma, y))$. As we have seen in the proof of Theorem 9.13, the function $p(x) = \sum_{j=1}^r |h_j(x)|^2$ is the constant function r, so that the map $x \mapsto H(x)$ can be viewed as a map $H : X \to S^{2r-1} \cong \mathbb{U}(r)/\mathbb{U}(r-1)$, where $\mathbb{U}(r)$ is the unitary group of $r \times r$ matrices. We now observe that the function $\Lambda : \Gamma \times Y \to \mathbb{U}(r)$ is a cocycle and therefore defines a homogeneous skew-product extension $\mathbf{X}_M \to \mathbf{Y}$, where $\mathbf{X}_M = Y \underset{\Lambda}{\times} S^{2r-1}$ and $\gamma(y, v) = (\gamma y, \Lambda(\gamma, y)v)$, $\gamma \in \Gamma, v \in S^{2r-1}$. Moreover, the map $\sigma_M : \mathbf{X} \to \mathbf{X}_M$, defined by $\sigma_M(x) = (\pi(x), H(x))$, is a factor map such that $M \subset L^2(\mathbf{X}_M) \circ \sigma_M = \{f \circ \sigma_M : f \in L^2(\mathbf{X}_M)\}$.

Let M_n be a sequence of Γ-invariant finite rank $\mathbf{Y}$-modules which together span $L^2(\mathbf{X})$. With each n we associate the corresponding homogeneous skew-product factor $\sigma_n : \mathbf{X} \to \mathbf{X}_{M_n}$. Now form the group $G = \prod_{n \in \mathbb{N}} \mathbb{U}(r_n)$, the relative product system $\prod_{\mathbf{Y}} \mathbf{X}_{M_n} = \mathbf{Y} \times \prod_n S^{2r_n - 1}$, with $\gamma(y, (v_n)_{n \in \mathbb{N}}) = (\gamma y, (\Lambda_n(\gamma, y)v_n)_{n \in \mathbb{N}})$, and define a homomorphism $\sigma : \mathbf{X} \to \prod_{\mathbf{Y}} \mathbf{X}_{M_n}$, $\sigma(x) = (\pi(x), (H_n(x))_{n \in \mathbb{N}})$. By Theorem 3.26 the image system $\mathbf{Z} = \phi(\mathbf{X}) = \mathbf{Y} \underset{\alpha}{\times} K/L$ is a homogeneous skew-product over $\mathbf{Y}$, corresponding to some closed subgroups $L \subset K \subset G$ and a cocycle $\alpha : \Gamma \times Y \to K$. Since $M_n \subset L^2(\mathbf{X}_{M_n}) \circ \sigma_n$, we conclude that $\mathbf{Z} \cong \mathbf{X}$.

Conversely, suppose $\mathbf{X} = \mathbf{Y} \underset{\alpha}{\times} K/L$ is a skew product extension with cocycle α. We want to show that $L^2(\mathbf{X})$ is spanned by $\mathbf{Y}$-eigenfunctions. By Lemma 9.12 and Corollary 3.27 it suffices to show this for an ergodic group extension $\mathbf{X} = \mathbf{Y} \underset{\alpha}{\times} K$. By Peter-Weyl's theorem A.15, a countable family of irreducible finite dimensional unitary representations $\Lambda_n : K \to \mathbb{U}(r_n) = \mathrm{Aut}\,(\mathbb{C}^{r_n})$ separate points on K. For each n, the map $\phi_n : \mathbf{X} \to \mathbf{Y} \underset{\Lambda_n \circ \alpha}{\times} \mathbb{U}(r_n)$ defined by $\phi_n(y, k) = (y, \Lambda_n(k))$, is a factor map and we now observe that the functions $h_{i,j}(y, V) = v_{ij} \in L^2(\mathbf{Z}_n)$, where v_{ij} is the (i, j)-th entry of the unitary matrix $V \in \mathbb{U}(r_n)$, are generalized eigenfunctions over $\mathbf{Y}$. The rest of the proof is now clear. □

3. The Hilbert-Schmidt bundle

For a better understanding of the structure of isometric extensions we need a theorem that describes the **Y**-eigenfunctions of a relative product over **Y** (Theorem 9.21 below). For the proof of this theorem we need to build a machinery for handling Hilbert-Schmidt operators. We do this in the present section whose main purpose is to show that certain sections are measurable. The discussion is a bit technical and can be omitted on a first reading if the reader is ready to believe the plausible (in view of the ergodicity of **Y**) claim of Proposition 9.20 below. We begin by recalling some facts about Hilbert-Schmidt operators on a Hilbert space $\mathfrak{H}$. We consider these compact operators as the elements of the Hilbert space $\mathfrak{G} = \mathfrak{H} \otimes \mathfrak{H}$. When $\mathfrak{H} = L^2(X, \mu)$ then $\mathfrak{G}$ can be identified with the Hilbert space $L^2(X \times X, \mu \times \mu)$ via the correspondence $\psi \mapsto A_\psi$, where

$$A_\psi(h)(x) = \int \psi(x, x')h(x')\, d\mu(x'), \qquad \psi \in L^2(X \times X, \mu \times \mu), h \in L^2(X, \mu).$$

We then have $\|A_\psi\| = \|\psi\| = \left(\int |\psi|^2\, d\mu\right)^{\frac{1}{2}}$. Clearly the *uniform norm*

$$|A| = \sup\{\|Ah\| : h \in \mathfrak{H},\ \|h\| = 1\}$$

satisfies $|A| \leq \|A\|$.

If A is a self-adjoint element of $\mathfrak{G}$, there exists a unique decomposition

$$A = \sum_{n=1}^{\infty} \lambda_n P_n, \tag{3.1}$$

where the λ_n are distinct real numbers and the sequence $|\lambda_n|$ decreases to zero. The P_n are projections ($P_n^* = P_n$, and $P_n^2 = P_n$) on finite dimensional subspaces of $\mathfrak{H}$ and $P_m P_n = 0$ for $m \neq n$. The Hilbert-Schmidt norm of A satisfies $\|A\|^2 = \sum \lambda_n^2 \dim P_n$. We refer to this decomposition as the *spectral decomposition* of A.

We enlarge $\mathfrak{G}$ by considering the set $\mathfrak{G}^+$ of all operators on $\mathfrak{H}$ of the form $\lambda I + A$ where $\lambda \in \mathbb{C}$ and $A \in \mathfrak{G}$. Define $\|\lambda I + A\| = |\lambda| + \|A\|$; then $\mathfrak{G}^+$ becomes a Banach algebra with a unit. An element $\lambda I + A$ of this Banach algebra has an inverse iff $\lambda I + A$, as an operator on $\mathfrak{H}$, has an inverse. (This follows from the fact that $\mathfrak{G}$ is an ideal in the algebra of all bounded operators on $\mathfrak{H}$.) Thus the spectrum of A is the same whether A is considered as an element of the Banach algebra $\mathfrak{G}^+$ or as an operator on $\mathfrak{H}$. If f is a single valued analytic function on a neighborhood of the spectrum of $A \in \mathfrak{G}$, then

$$\frac{1}{2\pi i} \int_\rho f(\lambda)(\lambda I - A)^{-1}\, d\lambda,$$

exists in $\mathfrak{G}^+$ in the Hilbert-Schmidt norm, where ρ is a positively oriented rectifiable Jordan curve, contained in the domain of f, which does not intersect the spectrum of A. If moreover f vanishes at 0 and U is a bounded open set in $\mathbb{C}$ containing the spectrum of A whose boundary ρ is a finite union of closed positively oriented rectifiable Jordan curves, then this integral equals $f(A)$ and it is an element of $\mathfrak{G}$. In particular if σ is a *spectral set* (that is σ is a clopen subset of the spectrum of A), U an open set whose intersection with the spectrum of A is σ, then for $f = \mathbf{1}_\sigma$ we get

$$\frac{1}{2\pi i} \int_\rho (\lambda I - A)^{-1}\, d\lambda = P,$$

where P is the spectral projection corresponding to the spectral set σ (see e.g. [**53**], VII, 3 and [**54**], XI, 6.7).

9.15. LEMMA. *Let $A \in \mathfrak{G}$ be self-adjoint and let $\sigma(A)$ be the spectrum of A. Let λ_0 be a real number and ρ a circle around λ_0 such that there exists a neighborhood U of ρ and its interior, with $\sigma(A) \cap U = \{\lambda_0\}$ or $\emptyset$. Let $\delta > 0$ be given, then there are points z_+ and z_- in the complex plane such that*

$$(\lambda I - A)^{-1} = \frac{1}{\lambda - z_+} \sum_{k=0}^{\infty} \left(\frac{z_+ I - A}{z_+ - \lambda} \right)^k$$

uniformly for $\lambda \in \rho_+ = \{\lambda \in \rho : \Im\lambda \geq \delta\}$, the convergence being in the uniform norm topology. An analogous equation, with z_- replacing z_+, holds for $\lambda \in \rho_- = \{\lambda \in \rho : \Im\lambda \leq -\delta\}$.

PROOF. For $z \in \mathbb{C}$, we have

$$\begin{aligned} |zI - A| &= \sup\{\|(zI - A)h\| : h \in \mathfrak{H},\ \|h\| = 1\} \\ &= \sup\{\|\sum_i (z - \lambda_i) a_i e_i\| : \sum_i |a_i|^2 = 1\} \\ &= \sup \left\{ \left(\sum_i |z - \lambda_i|^2 |a_i|^2 \right)^{\frac{1}{2}} : \sum_i |a_i|^2 = 1 \right\} \\ &= \sup_i |z - \lambda_i|, \end{aligned}$$

where $\{e_i\}$ is an orthonormal basis for $\mathfrak{H}$ and $Ae_i = \lambda_i e_i$. It is now clear that a point z_+ ($\Im z_+ < 0$) can be found such that for every $\lambda \in \rho_+$,

$$|z_+ I - A| = \sup_i |\lambda_i - z_+| < |\lambda - z_+|.$$

Similarly a point z_- ($\Im z_- > 0$) exists such that for every $\lambda \in \rho_-$,

$$|z_- I - A| = \sup_i |\lambda_i - z_-| < |\lambda - z_-|.$$

We therefore have

$$(\lambda I - A)^{-1} = \frac{1}{\lambda - z_+} \sum_{k=0}^{\infty} \left(\frac{z_+ I - A}{z_+ - \lambda} \right)^k, \qquad \text{uniformly for } \lambda \in \rho_+,$$

and

$$(\lambda I - A)^{-1} = \frac{1}{\lambda - z_-} \sum_{k=0}^{\infty} \left(\frac{z_- I - A}{z_- - \lambda} \right)^k, \qquad \text{uniformly for } \lambda \in \rho_-,$$

where convergence is in the uniform norm. □

Now in our usual setup, where $\pi : \mathbf{X} \to \mathbf{Y}$ is an extension of dynamical systems with $\mathbf{Y}$ ergodic, we form, for each $y \in Y$, the Hilbert space $\mathfrak{G}_y = \mathfrak{H}_y \otimes \mathfrak{H}_y = L^2(X, \mu_y) \underset{\mathbf{Y}}{\otimes} L^2(X, \mu_y)$ and define the *Hilbert-Schmidt bundle* $\dot{\mathfrak{G}} = \bigcup_{y \in Y} \mathfrak{G}_y$, whose direct integral is the Hilbert space

$$\mathfrak{H} \underset{\mathbf{Y}}{\otimes} \mathfrak{H} = L^2(X, \mu) \underset{\mathbf{Y}}{\otimes} L^2(X, \mu) = L^2(X \underset{Y}{\times} X, \mu \underset{\nu}{\times} \mu).$$

We choose a countable dense family $\mathcal{C}_1$ in $C(X \times X)$; e.g. one can take $\mathcal{C}_1$ to be the (countable) vector space generated by $\{f \otimes g : f, g \in \mathcal{C}_0\}$ over the rational numbers

$\mathbb{Q}$. Using $\mathfrak{C}_1$ we define a metric on $\dot{\mathfrak{G}}$ in the same way it was defined on $\dot{\mathfrak{H}}$, and then prove the analogue for $\dot{\mathfrak{G}}$ of the list in Proposition 9.1. In particular we have $A_i \to A$ in $\dot{\mathfrak{G}}$ iff $\pi(A_i) \to \pi(A)$ and $\|A_i - F\| \to \|A - F\|$ for every $F \in \mathfrak{C}_1$.

We shall use two other topologies on $\dot{\mathfrak{G}}$, the *weak* and the *uniform norm* topologies. A sequence $A_n \in \dot{\mathfrak{G}}$ converges *weakly* to $A \in \dot{\mathfrak{G}}$ iff the sequence $\|A_n\|$ is bounded, $y_n = \pi(A_n) \to y_0 = \pi(A)$ and for every $F \in \mathfrak{C}_1$,

$$\langle A_n, F\rangle_{y_n} \to \langle A, F\rangle_{y_0}.$$

We denote this convergence by $A_n \xrightarrow{w} A$. It is easy to check that this topology coincides with the *weak operator topology*; i.e. replacing the condition $\langle A_n, F\rangle_{y_n} \to \langle A, F\rangle_{y_0}$ by the condition $\langle A_n f, g\rangle_{y_n} \to \langle Af, g\rangle_{y_0}$, for every $f, g \in \mathfrak{C}_0$. We say that a sequence $A_n \in \dot{\mathfrak{G}}$ converges to $A \in \dot{\mathfrak{G}}$ in *uniform norm topology*, and denote $A_n \xrightarrow{u} A$, if $y_n = \pi(A_n) \to y_0 = \pi(A)$ and for every $F \in \mathfrak{C}_1$,

$$|A_n - F|_{y_n} \to |A - F|_{y_0},$$

where $|A| = \sup\{\|Ah\| : \|h\| = 1\}$ denotes the norm of the operator A.

We leave the proof of the following proposition as an exercise.

9.16. PROPOSITION. 1. *If $A_n \to A$ in $\dot{\mathfrak{G}}$ and $f \in \mathfrak{C}_0$ then $A_n f \to Af$.*

2. *If $h_n \to h$ in $\dot{\mathfrak{H}}$ and $A_n \to A$ in $\dot{\mathfrak{G}}$ (where $\pi(A_n) = \pi(h_n) = y_n$), then $A_n h_n \to Ah$ in $\dot{\mathfrak{H}}$.*
3. *$A_n \to A$ in $\dot{\mathfrak{G}}$ iff $A_n \xrightarrow{w} A$ and $\|A_n\| \to \|A\|$.*
4. *If $A_n \xrightarrow{u} A$ in $\dot{\mathfrak{G}}$ then $A_n \xrightarrow{w} A$.*
5. *If $A_n \to A$ and $B_n \to B$ in $\dot{\mathfrak{G}}$ (with $\pi(A_n) = \pi(B_n) = y_n$), then $A_n + B_n \to A + B$ and $A_n B_n \to AB$. The same is true with uniform norm convergence.*

9.17. LEMMA. *Let $A_n \to A$ be a convergent sequence of self-adjoint elements of $\dot{\mathfrak{G}}$. Let λ_0 be a real number and ρ a positively oriented circle around λ_0 such that there exists a neighborhood U of the closed disk defined by ρ, with $\sigma(A) \cap U = \{\lambda_0\}$ or $\emptyset$. Moreover assume that the distance between ρ and $\cup_n \sigma(A_n)$ is positive, then*

$$\frac{1}{2\pi i}\int_\rho \lambda(\lambda I - A_n)^{-1}\, d\lambda \xrightarrow{u} \frac{1}{2\pi i}\int_\rho \lambda(\lambda I - A)^{-1}\, d\lambda.$$

PROOF. Since the distance between ρ and $\cup_n \sigma(A_n)$ is positive, there exists a constant M such that $|(\lambda I - A_n)^{-1}| < M$ for every $\lambda \in \rho$ and every n. Given $\eta > 0$, let $\delta > 0$ be such that for every n

$$\left|\frac{1}{2\pi i}\int_{\rho_\delta} \lambda(\lambda I - A_n)^{-1}\, d\lambda\right| < \eta,$$

where $\rho_\delta = \{\lambda \in \rho : |\Im\lambda| \le \delta\}$. As in Lemma 9.15, we choose z_+ such that

$$(\lambda I - A_n)^{-1} = \frac{1}{\lambda - z_+}\sum_{k=0}^{\infty}\left(\frac{z_+ I - A_n}{z_+ - \lambda}\right)^k, \qquad \text{uniformly for } \lambda \in \rho_+,$$

and

$$(\lambda I - A)^{-1} = \frac{1}{\lambda - z_+}\sum_{k=0}^{\infty}\left(\frac{z_+ I - A}{z_+ - \lambda}\right)^k, \qquad \text{uniformly for } \lambda \in \rho_+,$$

where $\rho_+ = \{\lambda \in \rho : \Im\lambda \geq \delta\}$. Multiplying by λ and integrating over ρ_+, we have

$$\frac{1}{2\pi i}\int_{\rho_+} \lambda(\lambda I - A_n)^{-1}\,d\lambda = \frac{1}{2\pi i}\sum_{k=0}^{\infty}(z_+I - A_n)^k \int_{\rho_+} c_k\,d\lambda \quad \text{and,}$$

$$\frac{1}{2\pi i}\int_{\rho_+} \lambda(\lambda I - A)^{-1}\,d\lambda = \frac{1}{2\pi i}\sum_{k=0}^{\infty}(z_+I - A)^k \int_{\rho_+} c_k\,d\lambda,$$

where $c_k = \left(\frac{\lambda}{z_+ - \lambda}\right)^k$. With respect to the uniform norm this series converges uniformly in n. Therefore, given $\epsilon > 0$, we can find an l such that for every n

$$\left|\frac{1}{2\pi i}\int_{\rho_+} \lambda(\lambda I - A_n)^{-1}\,d\lambda - \frac{1}{2\pi i}\sum_{k=0}^{l}(z_+I - A_n)^k \int_{\rho_+} c_k\,d\lambda\right| < \epsilon,$$

and also

$$\left|\frac{1}{2\pi i}\int_{\rho_+} \lambda(\lambda I - A)^{-1}\,d\lambda - \frac{1}{2\pi i}\sum_{k=0}^{l}(z_+I - A)^k \int_{\rho_+} c_k\,d\lambda\right| < \epsilon.$$

By Lemma 9.16,

$$\sum_{k=0}^{l}(z_+I - A_n)^k \to \sum_{k=0}^{l}(z_+I - A)^k.$$

This implies convergence in the uniform norm topology. Thus if $F \in \mathcal{C}_1$, then eventually

$$\begin{aligned}
&\left|\left|\frac{1}{2\pi i}\int_{\rho_+} \lambda(\lambda I - A_n)^{-1}\,d\lambda - F\right| - \left|\frac{1}{2\pi i}\int_{\rho_+} \lambda(\lambda I - A)^{-1}\,d\lambda - F\right|\right| \\
&\leq \left|\left|\frac{1}{2\pi i}\int_{\rho_+} \lambda(\lambda I - A_n)^{-1}\,d\lambda - F\right| - \left|\frac{1}{2\pi i}\sum_{k=0}^{l}(z_+I - A_n)^k \int_{\rho_+} c_k\,d\lambda - F\right|\right| \\
&+ \left|\left|\frac{1}{2\pi i}\sum_{k=0}^{l}(z_+I - A_n)^k \int_{\rho_+} c_k\,d\lambda - F\right| - \left|\frac{1}{2\pi i}\sum_{k=0}^{l}(z_+I - A)^k \int_{\rho_+} c_k\,d\lambda - F\right|\right| \\
&+ \left|\left|\frac{1}{2\pi i}\sum_{k=0}^{l}(z_+I - A)^k \int_{\rho_+} c_k\,d\lambda - F\right| - \left|\frac{1}{2\pi i}\int_{\rho_+} \lambda(z_+I - A)^{-1}\,d\lambda - F\right|\right| < 2\epsilon.
\end{aligned}$$

Thus

$$\frac{1}{2\pi i}\int_{\rho_+} \lambda(\lambda I - A_n)^{-1}\,d\lambda \xrightarrow{u} \frac{1}{2\pi i}\int_{\rho_+} \lambda(\lambda I - A)^{-1}\,d\lambda.$$

Similarly, using z_-, we can show that the same holds for ρ_-, and by Lemma 9.16,

$$\frac{1}{2\pi i}\int_{\rho_+\cup\rho_-} \lambda(\lambda I - A_n)^{-1}\,d\lambda \xrightarrow{u} \frac{1}{2\pi i}\int_{\rho_+\cup\rho_-} \lambda(\lambda I - A)^{-1}\,d\lambda.$$

Since η was arbitrary, we finally get

$$\frac{1}{2\pi i}\int_{\rho} \lambda(\lambda I - A_n)^{-1}\,d\lambda \xrightarrow{u} \frac{1}{2\pi i}\int_{\rho} \lambda(\lambda I - A)^{-1}\,d\lambda.$$

$\square$

9.18. LEMMA. *Let $A_n \to A$ be a convergent sequence of self-adjoint elements of $\dot{\mathfrak{G}}$.*

1. *If $\lambda^{(n)}$ is an eigenvalue of A_n, and $\lambda^{(n)} \to \lambda_0 \neq 0$, then λ_0 is an eigenvalue of A.*
2. *Let $A = \sum_k \lambda_k P_k$ and $A_n = \sum_k \lambda_k^{(n)} P_k^{(n)}$ be the spectral decompositions of A and A_n respectively; then for any fixed k, the sequence $\dim P_k^{(n)}$ is bounded.*

PROOF. We have

$$\|A\|^2 = \sum_k |\lambda_k|^2 \dim P_k, \qquad \|A_n\|^2 = \sum_k |\lambda_k^{(n)}|^2 \dim P_k^{(n)}, \tag{3.2}$$
$$\text{and} \qquad \|A_n\| \to \|A\|.$$

It therefore follows that for every $\epsilon > 0$ there exists a positive integer N such that for every n the number of $\lambda_k^{(n)}$ with $|\lambda_k^{(n)}| \geq \epsilon$ is less than N. Suppose $\lambda_0 \notin \sigma(A)$; we choose $\epsilon > 0$ such that $|\lambda_0| > \epsilon$ and let N correspond to ϵ as above. It is now possible to choose a subsequence of A_n (which we continue to denote by A_n) for which the sequence of sets $\Lambda_n = \{\lambda_k^{(n)} : |\lambda_k^{(n)}| > \epsilon\}$, each containing at most N elements, converges to a finite subset of real numbers. Necessarily λ_0 is an element of this set.

Now we can find a circle ρ around λ_0 and a neighborhood U of ρ and the points inside it, such that $U \cap \sigma(A) = \emptyset$ and such that the distance of ρ from $\cup\sigma(A_n)$ is positive. By Lemma 9.17

$$\frac{1}{2\pi i}\int_\rho \lambda(\lambda I - A_n)^{-1}\,d\lambda = \sum_{\lambda_k^{(n)} \text{ inside } \rho} \lambda_k^{(n)} P_k^{(n)} \xrightarrow{u} \frac{1}{2\pi i}\int_\rho \lambda(\lambda I - A)^{-1}\,d\lambda = 0.$$

In particular $\lambda^{(n)}P^{(n)} \xrightarrow{u} 0$, where $P^{(n)}$ is the projection which corresponds to $\lambda^{(n)}$ in the spectral decomposition of A_n. However, $\lambda^{(n)} \to \lambda_0 \neq 0$ and by taking $F = 0 \in \mathfrak{C}_1$ we get $|\lambda^{(n)}P^{(n)} - F| = |\lambda^{(n)}P^{(n)}| = |\lambda^{(n)}| \to |F| = 0$, a contradiction. We therefore conclude that $\lambda_0 \in \sigma(A)$. This proves part 1 of the lemma and part 2 now follows from (3.2). □

9.19. LEMMA. *Let P_n be a sequence of projections in $\dot{\mathfrak{G}}$ which converges in the uniform norm topology to a projection $P \in \dot{\mathfrak{G}}$. Suppose further that $\dim P \leq r$ and for every n, $\dim P_n \leq r$; then $P_n \to P$.*

PROOF. By Lemma 9.16 $P_n \xrightarrow{w} P$; hence by the same lemma it is enough to show that $\|P_n\| \to \|P\|$. Given $\epsilon > 0$, choose $F \in \mathfrak{C}_1$ such that

$$\big|\|F\|_{y_0} - \|P\|\big| \leq \|F - P\|_{y_0} < \epsilon,$$

and such that in a neighborhood of $y_0 = \pi(P)$, the section $s_F(y)$ vanishes on the orthogonal complement of an r-dimensional subspace of $\mathfrak{H}_y$. (For example, choose an orthonormal basis $\{e_1, e_2, \ldots, e_r\}$ for $P\mathfrak{H}_{y_0}$ and choose $f_1, f_2, \ldots, f_r$ in $\mathfrak{C}_0$ close in norm on $\mathfrak{H}_{y_0}$ to $e_1, e_2, \ldots, e_r$ respectively, then put $F = \sum_{i=1}^r f_i \otimes f_i$.)

For large n we then have $|F - P_n| < \epsilon$ and

$$\big|\|F\|_{y_0} - \|P_n\|\big|^2 \leq \|P_n - F\|^2 = \sum_{i=1}^\infty \|(P_n - F)e_i^{(n)}\|^2,$$

where $\{e_i^{(n)}\}_{i=1}^{\infty}$ is an orthonormal basis for $\mathfrak{H}_{y_n}$.

We fix such n and, by an appropriate choice of basis, can assume that on $(\text{span}\{e_1^{(n)}, e_2^{(n)}, \ldots, e_{2r}^{(n)}\})^{\perp}$, both P_n and F vanish. Therefore

$$\left|\|F\|_{y_0} - \|P_n\|\right|^2 \leq \sum_{i=1}^{2r} \|(P_n - F)e_i^{(n)}\|^2 \leq 2r \cdot |P_n - F| < 2r \cdot \epsilon.$$

Since $\|F\|_{y_n} \to \|F\|_{y_0}$ and since ϵ is arbitrary, our lemma follows. $\square$

Let $\psi(x, x')$ be a self-adjoint Γ-invariant element of $L^2(X \underset{Y}{\times} X, \mu \underset{\nu}{\times} \mu)$ and $A = A_\psi$ the corresponding Hilbert-Schmidt operator, then the section $y \mapsto A_y$, with $A_y : \mathfrak{H}_y \to \mathfrak{H}_y$, is "intertwined" by the Koopman representation. For $x \in \pi^{-1}(\gamma y)$ and $h \in L^2(X, \mu_y)$ we have

$$\begin{aligned} U(\gamma, y)A_y h(x) &= \int \psi(\gamma^{-1}x, x')h(x')\, d\mu_y(x') = \int \psi(x, \gamma x')h(x')\, d\mu_y(x') \\ &= \int \psi(x, x')h(\gamma^{-1}x')\, d\mu_{\gamma y}(x') = A_{\gamma y}U(\gamma, y)h(x), \end{aligned}$$

so that $U(\gamma, y)A_y = A_{\gamma y}U(\gamma, y)$. We now have the following proposition.

9.20. PROPOSITION. *For ψ and A as above, in the spectral decomposition*

$$A_y = \sum_{n=1}^{\infty} \lambda_n P_n(y),$$

up to a null set, the spectrum $\{\lambda_n : n \in \mathbb{N}\}$ does not depend on y. For each n the section $y \mapsto P_n(y)$ is measurable and $\dim P_n(y)$ *is a constant a.e.. Moreover, the section $y \mapsto V_n(y)$, where $V_n(y) = P_n(y)\mathfrak{H}_y$ is the eigenspace corresponding to the eigenvalue $\lambda_n \neq 0$, defines a Γ-invariant finite rank $\mathbf{Y}$-module M_n in $L^2(\mathbf{X})$ with $M_{n,y} = V_n(y)$.*

PROOF. By Lusin's theorem there exists, for every $\epsilon > 0$, a closed subset $Y_\epsilon \subset Y$, where the section $y \mapsto A_y$, as a map from Y_ϵ to $\dot{\mathfrak{G}}$, is continuous. Denoting $M = \|\psi\|$, we have $\sigma(A_y) \subset [-M, M]$ a.e. We consider the map $\text{spec}_A : y \mapsto \sigma(A_y)$ as a function from Y into the Polish space $2^{[-M,M]}$ of closed subsets of the interval $[-M, M]$ equipped with the Hausdorff metric. From Lemma 9.18 it now follows that for each closed interval $[a, b] \subset [-M, M]$, with $0 \notin [a, b]$, the set

$$\{y \in Y : \sigma(A_y) \cap [a, b] \neq \emptyset\} \cap Y_\epsilon,$$

is a closed subset of Y. It therefore follows that the function $\text{spec}_A : Y \to 2^{[-M,M]}$ is measurable. Since for every $\gamma \in \Gamma$, $\sigma(A_y) = \sigma(A_{\gamma y})$, the ergodicity of $\mathbf{Y}$ implies that $\sigma(A_y) = \text{constant} = \{\lambda_n\}_{n=1}^{\infty}$, ν-a.e.

Fix k and let ρ be a circle in $\mathbb{C}$ around λ_k such that no other member of the sequence λ_n meets the closed disk defined by ρ. We then have a.e.

$$P_k(y) = \frac{1}{2\pi i} \int_\rho (zI - A_y)^{-1}\, dz.$$

Now consider the map $\text{proj}_k : y \mapsto P_k(y)$ as a map from Y to the Polish space $\dot{\mathfrak{G}}$. From Lemmas 9.17, 9.18 and 9.19 it follows that for each closed subset C of $\dot{\mathfrak{G}}$, the set

$$\{y \in Y : P_k(y) \in C\} \cap Y_\epsilon,$$

is a closed subset of Y. It therefore follows that the function $\mathrm{proj}_k : Y \to \mathfrak{G}$ is measurable. Since for every $\gamma \in \Gamma$, $P_k(\gamma y) = U(\gamma, y) P_k(y) U(\gamma, y)^{-1}$, we have $\dim P_k(\gamma y) = \dim P_k(y)$ and the ergodicity of $\mathbf{Y}$ implies that $P_k(y) =$ constant a.e. In view of Proposition 9.9 it follows that M_k is a Γ-invariant finite rank $\mathbf{Y}$-module in $L^2(\mathbf{X})$ with $M_{k,y} = V_k(y) = P_k(y)\mathfrak{H}_y$, a.e. □

4. The Y-eigenfunctions of a relative product

Let $\pi_i : \mathbf{X}_i \to \mathbf{Y}$, $i = 1, 2$ be two dynamical systems with a common ergodic factor $\mathbf{Y}$. As usual we let $\mathbf{X}_1 \underset{\mathbf{Y}}{\times} \mathbf{X}_2$ be the relatively independent product over $\mathbf{Y}$ and set $\mathfrak{H} = L^2(\mathbf{X}_1 \underset{\mathbf{Y}}{\times} \mathbf{X}_2)$. If $M_i \subset \mathfrak{H}_i = L^2(\mathbf{X}_i)$ are $\mathbf{Y}$-modules, we let $M_1 \underset{\mathbf{Y}}{\otimes} M_2$ be the closed $\mathbf{Y}$-module of $\mathfrak{H}$ spanned by products $f_1 \cdot f_2$ of bounded functions in M_1 and M_2 respectively (the boundedness is required to ensure that the product is in fact an element of $\mathfrak{H}$).

Our goal in this section is to prove the following important theorem.

9.21. THEOREM. *Let $\rho_i : \mathbf{Z}_i \to \mathbf{Y}$ be the maximal intermediate isometric extensions corresponding to the factors $\pi_i : \mathbf{X}_i \to \mathbf{Y}$, $i = 1, 2$, and $\rho : \mathbf{Z} \to \mathbf{Y}$ the one corresponding to the factor $\pi : \mathbf{X}_1 \underset{\mathbf{Y}}{\times} \mathbf{X}_2 \to \mathbf{Y}$, then $\mathbf{Z} \cong \mathbf{Z}_1 \underset{\mathbf{Y}}{\times} \mathbf{Z}_2$; that is*

$$\mathcal{E}(\mathbf{X}_1 \underset{\mathbf{Y}}{\times} \mathbf{X}_2/\mathbf{Y}) = \mathcal{E}(\mathbf{X}_1/\mathbf{Y}) \underset{\mathbf{Y}}{\otimes} \mathcal{E}(\mathbf{X}_2/\mathbf{Y}).$$

PROOF. In view of Theorem 9.13 $\mathcal{E}(\mathbf{X}_1/\mathbf{Y}) \underset{\mathbf{Y}}{\otimes} \mathcal{E}(\mathbf{X}_2/\mathbf{Y}) \subset \mathcal{E}(\mathbf{X}_1 \underset{\mathbf{Y}}{\times} \mathbf{X}_2/\mathbf{Y})$. Conversely, denote $\mathbf{W} = \mathbf{Z}_1 \underset{\mathbf{Y}}{\times} \mathbf{Z}_2$ and note that if $g \in \mathcal{E}(\mathbf{X}_1 \underset{\mathbf{Y}}{\times} \mathbf{X}_2/\mathbf{Y})$ then — since for every $\gamma \in \Gamma$, $U_\gamma \mathbb{E}^{\mathbf{W}}(g) = \mathbb{E}^{\mathbf{W}}(U_\gamma g)$ — the function $\mathbb{E}^{\mathbf{W}}(g)$ is also a $\mathbf{Y}$-eigenfunction and $f = g - \mathbb{E}^{\mathbf{W}}(g)$ is a $\mathbf{Y}$-eigenfunction in $\mathcal{E}(\mathbf{X}_1 \underset{\mathbf{Y}}{\times} \mathbf{X}_2/\mathbf{Y})$ which is orthogonal to every function in $\mathcal{E}(\mathbf{X}_1/\mathbf{Y}) \underset{\mathbf{Y}}{\otimes} \mathcal{E}(\mathbf{X}_2/\mathbf{Y})$. Thus it is enough to show that such an f is necessarily 0.

Let M be the finite rank Γ-invariant $\mathbf{Y}$-module generated by f. For $\gamma \in \Gamma$, bounded $h \in L^2(\mathbf{Y})$ and $g \in \mathcal{E}(\mathbf{X}_1/\mathbf{Y}) \underset{\mathbf{Y}}{\otimes} \mathcal{E}(\mathbf{X}_2/\mathbf{Y})$ we have

$$\int h({}_\gamma f)\bar{g}\, d\mu = \mathbb{E}\big(\mathbb{E}^{\mathbf{Y}}(h({}_\gamma f)\bar{g})\big) = \mathbb{E}\big(h\mathbb{E}^{\mathbf{Y}}({}_\gamma f\bar{g})\big) = 0,$$

hence every function in M is orthogonal to $\mathcal{E}(\mathbf{X}_1/\mathbf{Y}) \underset{\mathbf{Y}}{\otimes} \mathcal{E}(\mathbf{X}_2/\mathbf{Y})$. We choose an orthonormal $\mathbf{Y}$-basis $\{f_1, f_2, \ldots, f_r\}$ for M and write for each $\gamma \in \Gamma$,

$$U_\gamma f_i(x) = \sum \lambda_{ij}(\gamma, y) f_j(x),$$

where $\big(\lambda_{ij}(\gamma, y)\big)$ is a unitary matrix. As in the proof of Theorem 9.13 we can assume that the functions f_i are bounded. Define a function ψ on $\mathbf{X}_1 \underset{\mathbf{Y}}{\times} \mathbf{X}_1$ by

$$\psi(x_1, x_1') = \int \sum_{j=1}^{r} f_j(x_1, x_2)\overline{f_j(x_1', x_2)}\, d\mu_{2,y}(x_2).$$

We clearly have $\psi(x_1, x_1') = \overline{\psi(x_1', x_1)}$ and

$$|\psi(x_1, x_1')| \leq \sum_{j=1}^{r} \left(\int |f_j(x_1, x_2)|^2 \, d\mu_{2,y}(x_2) \int |f_j(x_1', x_2)|^2 \, d\mu_{2,y}(x_2) \right)^{\frac{1}{2}},$$

hence

$$\int |\psi(x_1, x_1')|^2 \, d\mu_{1,y}(x_1) \, d\mu_{2,y}(x_2) \leq$$
$$r \sum_{j=1}^{r} \left(\int \int |f_j(x_1, x_2)|^2 \, d\mu_{1,y}(x_1) \, d\mu_{1,y}(x_1') \right)^2 = r^2.$$

It follows that ψ defines an element $A \in \mathfrak{G} = L^2(\mu_1) \underset{\mathbf{Y}}{\otimes} L^2(\mu_1)$ as well as a section $A_y \in \dot{\mathfrak{G}}$, where $\mathfrak{H}_y = L^2(\mu_{1,y})$, $\mathfrak{G}_y = \mathfrak{H}_y \otimes \mathfrak{H}_y$ and $\dot{\mathfrak{G}}$ is the corresponding Hilbert-Schmidt bundle, with Hilbert-Schmidt norm bounded by r. Moreover we have for each $\gamma \in \Gamma$,

$$\begin{aligned}
\psi(\gamma x_1, \gamma x_1') &= \int \sum_{j=1}^{r} f_j(\gamma x_1, x_2) \overline{f_j(\gamma x_1', x_2)} \, d\mu_{2,\gamma y}(x_2) \\
&= \int \sum_{j=1}^{r} f_j(\gamma x_1, \gamma x_2) \overline{f_j(\gamma x_1', \gamma x_2)} \, d\mu_{2,y}(x_2) \\
&= \int \sum_{j,k,l=1}^{r} \lambda_{jl}(\gamma, y) \overline{\lambda_{jk}(\gamma, y)} f_l(x_1, x_2) \overline{f_k(x_1', x_2)} \, d\mu_{2,y}(x_2) \\
&= \int \sum_{j} f_j(x_1, x_2) \overline{f_j(x_1', x_2)} \, d\mu_{2,y}(x_2) = \psi(x_1, x_1'),
\end{aligned}$$

so that $U_\gamma A = A$ and $U_\gamma A_y = A_{\gamma y}$ a.e.

By Proposition 9.20 we have

$$A = \sum_{n=1}^{\infty} \lambda_n P_n(y),$$

where the spectrum $\{\lambda_n : n \in \mathbb{N}\}$ does not depend on y. As in Proposition 9.20 we let, for each n, $y \mapsto V_n(y)$ be the eigenspace section with $V_n(y) = P_n(y)\mathfrak{H}_y$ the eigenspace corresponding to the eigenvalue $\lambda_n \neq 0$, and let M_n be the corresponding "direct integral", a finite rank **Y**-module in $L^2(\mathbf{X})$.

We now fix n, choose an orthonormal **Y**-basis $\{\phi_i : 1 \leq i \leq s\}$ for M, and write $U_\gamma \phi_i = \sum_{j=1}^{s} \zeta_{ij}(\gamma, y) \phi_j$, with $\zeta_{ij}(\gamma, y)$ a unitary matrix. Defining

$$g_{ij}(x_2) = \int f_i(x_1, x_2) \overline{\phi_j(x_1)} \, d\mu_{1,y}(x_1),$$

we have for ν a.e. y

(4.1)

$$\begin{aligned}
\lambda_n &= \int A_y\phi_1(x_1)\overline{\phi_1(x_1)}\,d\mu_{1,y}(x_1)\\
&= \int\int \psi(x_1,x_1')\phi_1(x_1')\overline{\phi_1(x_1)}\,d\mu_{1,y}(x_1)\,d\mu_{1,y}(x_1')\\
&= \int\int\int \sum_{j=1}^{r} f_j(x_1,x_2)\overline{f_j(x_1',x_2)}\phi_1(x_1')\overline{\phi_1(x_1)}\,d\mu_{1,y}(x_1)\,d\mu_{1,y}(x_1')\,d\mu_{2,y}(x_2)\\
&= \int\int \sum_{j=1}^{r} f_j(x_1',x_2)\overline{g_{j1}(x_2)\phi_1(x_1)}\,d\mu_{1,y}(x_1)\,d\mu_{2,y}(x_2).
\end{aligned}$$

Now for g_{ij} and $\gamma \in \Gamma$ we have

$$\begin{aligned}
g_{ij}(\gamma x_2) &= \int f_i(x_1,\gamma x_2)\overline{\phi_j(x_1)}\,d\mu_{1,\gamma y}(x_1)\\
&= \int f_i(\gamma x_1,\gamma x_2)\overline{\phi_j(\gamma x_1)}\,d\mu_{1,y}(x_1)\\
&= \sum \lambda_{il}(\gamma,y)\overline{\zeta_{jk}(\gamma,y)}\int f_l(x_1,x_2)\overline{\phi_k(x_1)}\,d\mu_{1,y}(x_1)\\
&= \sum \lambda_{il}(\gamma,y)\overline{\zeta_{jk}(\gamma,y)}g_{lk}(x_2).
\end{aligned}$$

This shows that $g_{j1} \in \mathcal{E}(\mathbf{X}_2/\mathbf{Y})$, whence $g_{j1}(x_2)\phi_1(x_1) \in \mathcal{E}(\mathbf{X}_1/\mathbf{Y}) \underset{Y}{\otimes} \mathcal{E}(\mathbf{X}_2/\mathbf{Y})$ and since f_j is orthogonal to every element in the latter subspace, we get $\lambda_n = 0$ from equation (4.1). This holds for every n and we conclude that $A_y = 0$ ν-a.e.; i.e. $\psi(x_1,x_1') = 0$.

Defining $F(x_1) = (f_1(x_1,x_2), f_2(x_1,x_2), \dots, f_r(x_1,x_2))$ we have, for almost all $y = \pi_1(x_1) \in Y$, a map $F : (X_1, \mu_{1,y}) \to \bigoplus_{j=1}^{r} L^2(\mu_{2,y})$, and we can express the equation $\psi(x_1,x_1') = 0$ as $\langle F(x_1), F(x_1')\rangle = 0$ for $\mu_1 \underset{\nu}{\times} \mu_1$-a.e $(x_1,x_1') \in X_1 \underset{Y}{\times} X_1$. If we let $\theta = F_*(\mu_1 \underset{\nu}{\times} \mu_1)$ — a probability measure on $\bigoplus_{j=1}^{r} L^2(\mu_{2,y})$ — then for every point u in the support of θ, every neighborhood of u contains a pair of orthogonal vectors. This implies $\theta = \delta_0$, so that $F \equiv 0$, that is $M = 0$, whence $f = 0$. □

5. Weakly mixing extensions

9.22. DEFINITION. Let $\pi : \mathbf{X} \to \mathbf{Y}$ be a factor of the ergodic system $\mathbf{X}$. We say that this factor map (or extension) is *relatively weakly mixing* if the system $\mathbf{X} \underset{\mathbf{Y}}{\times} \mathbf{X}$ is ergodic.

9.23. THEOREM. 1. *The ergodic system* $\mathbf{X}$ *is a relatively weakly mixing extension of* $\mathbf{Y}$ *iff* $\mathcal{E}(\mathbf{X}/\mathbf{Y}) = L^2(\mathbf{Y})$.

2. *If* $\mathbf{X}$ *is a relatively weakly mixing extension of* $\mathbf{Y}$ *and* $\mathbf{Z}$ *any ergodic system with* $\mathbf{Z} \to \mathbf{Y}$, *then* $\mathbf{X} \underset{\mathbf{Y}}{\times} \mathbf{Z}$ *is ergodic.*

PROOF. 1. Assuming $\mathcal{E}(\mathbf{X}/\mathbf{Y}) \supsetneqq L^2(\mathbf{Y})$ we can find a Γ-invariant $\mathbf{Y}$-module M of finite rank $r \geq 1$. Let $\{h_1, h_2, \dots, h_r\}$ be an orthonormal $\mathbf{Y}$-basis for M, and form the function $f = \sum_{i=1}^{r} h_i \otimes \overline{h_i} \in M \otimes \overline{M} \subset L^2(\mathbf{X} \underset{\mathbf{Y}}{\times} \mathbf{X})$. We have for every $\gamma \in \Gamma$ and $\mu \underset{\nu}{\times} \mu$ a.e.

$$\begin{aligned} f(\gamma x, \gamma x') &= \sum_{i=1}^{r} h_i(\gamma x)\overline{h_i(\gamma x')} \\ &= \sum_{i=1}^{r}\Big(\sum_{j=1}^{r} \lambda_{ij} h_j(x)\Big)\Big(\sum_{k=1}^{r} \overline{\lambda_{ik} h_k(x')}\Big) \\ &= \sum_{j=1}^{r} h_j(x)\overline{h_j(x')} = f(x, x'). \end{aligned}$$

Thus the system $\mathbf{X} \underset{\mathbf{Y}}{\times} \mathbf{X}$ is not ergodic and the extension is not relatively weakly mixing.

Conversely, if $\mathcal{E}(\mathbf{X}/\mathbf{Y}) = L^2(\mathbf{Y})$, then by Theorem 9.21

$$\operatorname{Inv}(L^2(\mathbf{X} \underset{\mathbf{Y}}{\times} \mathbf{X})) \subset \mathcal{E}(\mathbf{X} \underset{\mathbf{Y}}{\times} \mathbf{X}/\mathbf{Y}) = \mathcal{E}(\mathbf{X}/\mathbf{Y}) \underset{\mathbf{Y}}{\otimes} \mathcal{E}(\mathbf{X}/\mathbf{Y}) = L^2(\mathbf{Y}).$$

Since $\mathbf{Y}$ is ergodic this implies that $\mathbf{X} \underset{\mathbf{Y}}{\times} \mathbf{X}$ is ergodic, that is, the extension $\mathbf{X} \to \mathbf{Y}$ is relatively weakly mixing.

2. By part 1, $\mathcal{E}(\mathbf{X}/\mathbf{Y}) = L^2(\mathbf{Y})$ and as above

$$\begin{aligned} \operatorname{Inv}(L^2(\mathbf{X} \underset{\mathbf{Y}}{\times} \mathbf{Z})) &\subset \mathcal{E}(\mathbf{X} \underset{\mathbf{Y}}{\times} \mathbf{Z}/\mathbf{Y}) \\ &= \mathcal{E}(\mathbf{X}/\mathbf{Y}) \underset{\mathbf{Y}}{\otimes} \mathcal{E}(\mathbf{Z}/\mathbf{Y}) \\ &= L^2(\mathbf{Y}) \underset{\mathbf{Y}}{\otimes} \mathcal{E}(\mathbf{Z}/\mathbf{Y}). \end{aligned}$$

Thus every invariant function in $L^2(\mathbf{X} \underset{\mathbf{Y}}{\times} \mathbf{Z})$ is in $L^2(\mathbf{Z})$ and the ergodicity of $\mathbf{Z}$ implies it is a constant. □

6. Notes

The main sources for this chapter are R. Zimmer's papers [**267**] and [**268**] as well as H. Furstenberg's original paper on Szemerédi's theorem [**76**] and his book [**77**]. Section 3 is based on Furstenberg and Glasner [**78**].

CHAPTER 10

The Furstenberg-Zimmer Structure Theorem

In 1977 H. Furstenberg came up with a new proof of the recent celebrated theorem of Szemerédi ([**242**], 1975), which was based on ergodic theory ([**76**],[**77**]). Szemerédi's theorem, which generalizes van der Waerden's (Theorem 1.57), asserts that in any subset A of the natural numbers $\mathbb{N}$ (or the integers $\mathbb{Z}$) with positive upper density, one can find arithmetic progressions of any length. In fact his new proof went much further and yielded the same result for the groups $\mathbb{Z}^d$, $d \geq 1$. In a series of outstanding papers culminating in Furstenberg and Katznelson's [**80**], and Bergelson and Leibman's [**20**] these authors obtained spectacular new results in combinatorial number theory. All of these proofs, including the original proof of Furstenberg, rely on some version of the structure theorem for ergodic systems that we shall prove in the present chapter.

The idea of this ergodic proof, in a nutshell, is as follows. Using a *correspondence principle*, with the set $A \subset \mathbb{Z}$ one can associate an ergodic $\mathbb{Z}$ dynamical system $\mathbf{X} = \mathbf{X}(A) = (\{0,1\}^{\mathbb{Z}}, \mu, S)$, where S is the shift and μ an invariant probability measure with the property that $\{n \in \mathbb{Z} : \mu(S^n[0] \cap [0]) > 0\} \subset A$, where $[0] = \{x \in \{0,1\}^{\mathbb{Z}} : x_0 = 0\}$. Now as in the topological dynamical proof of van der Waerden's theorem (Theorem 1.57), Szemerédi's theorem will follow from the multiple recurrence theorem:

10.1. THEOREM (Multiple recurrence theorem). *For every positive integer n there exists a positive integer k for which:*

$$\mu([0] \cap S^k[0] \cap S^{2k}[0] \cap \cdots \cap S^{(n-1)k}[0]) > 0.$$

Furstenberg's book [**77**] is the best, and most enjoyable source for learning this fascinating subject.

The reader will recall that a topologically distal dynamical system is a topological system (X, Γ) with the property that the only proximal pairs in X are the diagonal pairs; i.e. if $\lim \gamma_i x = \lim \gamma_i x'$ for a pair $x, x' \in X$ and a sequence $\gamma_i \in \Gamma$ then $x = x'$. Furstenberg's distal structure theorem asserts that every minimal distal system (X, Γ) has, uniquely, a structure of an inverse limit of a family of factors $\{(X_\alpha, \Gamma) : \alpha < \eta\}$ directed by a countable ordinal η such that for every $\alpha < \eta$ the extension $X_{\alpha+1} \to X_\alpha$ is a maximal topologically isometric extension.

Now with the concepts of isometric system and isometric extension in the context of ergodic theory, as developed in the previous chapter, we can define a measure distal system to be one that can be built from an isometric system by a (countable) transfinite succession of isometric extensions and inverse limits (we call such systems I-systems; Zimmer's nomenclature for this property is "generalized discrete spectrum"). In fact this is the course that was taken by Furstenberg in his proof of the multiple recurrence theorem — the ergodic theoretical analogue of Szemerédi's theorem.

Furstenberg noticed that the multiple recurrence property holds for weakly mixing systems as well as for isometric ones. He then based his proof of the multiple recurrence theorem on a general structure theorem for ergodic systems according to which every ergodic system $\mathbf{X}$ is a weakly mixing extension $\mathbf{X} \to \mathbf{Y}$ of a distal system $\mathbf{Y}$. The multiple recurrence property is then proved for the distal factor $\mathbf{Y}$ — essentially by induction on its "height" or rank — and then, using the fact that the extension $\mathbf{X} \to \mathbf{Y}$ is a weakly mixing extension, this property is lifted to $\mathbf{X}$.

For the purpose of the multiple recurrence theorem this definition of measure distality is perfectly adequate. However, from the point of view of ergodic theory and for the deeper study of the class of distal systems it is not the "right" definition and one would like to have an intrinsic definition of this property, similar to the one we have in topological dynamics.

Now W. Parry in his 1967 paper **[200]** suggested an intrinsic definition of measure distality. He defines in this paper a property of measure dynamical systems, called "admitting a separating sieve", which imitates the intrinsic definition of topological distality. He showed that every measure dynamical system admitting a separating sieve has zero entropy and that any Γ-invariant measure on a minimal topologically distal system gives rise to a measure dynamical system admitting a separating sieve. The variational principle then implies that every topologically distal system has zero topological entropy.

In 1976 in two outstanding papers **[267]**, **[268]** R. Zimmer developed the theory of distal systems and distal extensions for a general locally compact acting group. He showed that, as in the topologically distal case, systems admitting Parry's separating sieve are exactly those with generalized discrete spectrum, that is those systems which are exhausted by their Furstenberg towers of isometric extensions. With these two characterizations of measure distality (and measure distal extensions) he was able to achieve an elegant and complete theory of such systems. In this chapter we use a mixture of Furstenberg's and Zimmer's approaches to develop an important part of this theory.

1. I-extensions

10.2. DEFINITION. Let $\pi : \mathbf{X} \to \mathbf{Y}$ be a factor map of an ergodic system $\mathbf{X}$. We call the extension π an *I-extension* (or say that $\mathbf{X}$ has a *generalized discrete spectrum over* $\mathbf{Y}$) if there exists a countable ordinal η and a directed family of factors $\mathbf{X}_\theta$, $\theta \le \eta$ (see Chapter 6, Section 1) such that

1. $\mathbf{X}_0 = \mathbf{Y}$ and $\mathbf{X}_\eta = \mathbf{X}$.
2. For $\theta < \eta$ the extension $\mathbf{X}_{\theta+1} \to \mathbf{X}_\theta$ is isometric and non-trivial (i.e. not an isomorphism).
3. For a limit ordinal $\lambda \le \eta$, $\mathbf{X}_\lambda = \text{inv lim}_{\theta<\lambda} \mathbf{X}_\theta$; (i.e. $\mathcal{X}_\lambda = \bigvee_{\theta<\lambda} \mathcal{X}_\theta$).

When $\pi : \mathbf{X} \to \mathbf{Y}$ is an I-extension as above we write $(\mathbf{X}; \mathbf{X}_\theta, \theta \le \eta(\mathbf{X}))$ for the associated directed family of factors and factor maps. If $\mathbf{X}$ is an I-extension of the trivial one point system, then we say that $\mathbf{X}$ is an *I-system*.

10.3. PROPOSITION. *Let $\pi : \mathbf{X} \to \mathbf{Y}$ be a factor map of an ergodic system $\mathbf{X}$, then there exists an intermediate factor $\mathbf{X} \xrightarrow{\phi} \mathbf{Z} \xrightarrow{\rho} \mathbf{Y}$ (with $\pi = \phi \circ \rho$) such that ρ is a (maximal) I-extension and ϕ is a weakly mixing extension.*

PROOF. Let $\Sigma = \{(\mathbf{X}^\sigma; \mathbf{X}^\sigma_\theta, \theta \le \eta(\mathbf{X}^\sigma))\}$ be the collection of all the intermediate factors $\mathbf{X} \overset{\phi^\sigma}{\to} \mathbf{X}^\sigma \overset{\rho^\sigma}{\to} \mathbf{Y}$ with ρ^σ an I-extension, together with the associated directed family of factors. Since in the definition of an I-extension we require the isometric extensions to be non-trivial, the collection Σ is clearly a set (in the sense of set theory). We define a partial order on Σ: $\sigma_2 \ge \sigma_1$ iff $\eta_2 = \eta(\mathbf{X}^{\sigma_2}) \ge \eta_1 = \eta(\mathbf{X}^{\sigma_1})$ and $(\mathbf{X}^{\sigma_1}; \mathbf{X}^{\sigma_1}_\theta, \theta \le \eta_1) = (\mathbf{X}^{\sigma_2}_{\eta_1}; \mathbf{X}^{\sigma_2}_\theta, \theta \le \eta_1)$.

If $\Xi = \{(\mathbf{X}^\xi; \mathbf{X}^\xi_\theta, \theta \le \eta(\mathbf{X}^\xi))\} \subset \Sigma$ is a chain in Σ, we let $\eta = \sup_{\xi\in\Xi}\eta(\mathbf{X}^\xi)$. Since for $\alpha < \beta < \eta$ we have $L^2(\mathbf{X}^\beta_\alpha) \subsetneq L^2(\mathbf{X}^\beta)$ we deduce that η is a countable ordinal. If η is not a limit ordinal then clearly $(\mathbf{X}^\eta; \mathbf{X}^\eta_\theta, \theta \le \eta)$ is a maximal element of Ξ. Otherwise set

$$\mathbf{X}^\eta = \operatorname{inv}\lim_{\xi\in\Xi} \mathbf{X}^\xi.$$

It is easy to see that $(\mathbf{X}^\eta; \mathbf{X}^\xi_\theta, \theta \le \eta(\xi), \xi \in \Xi)$ defines an upper bound for elements of Ξ.

Thus Zorn's lemma applies and we obtain a maximal element $(\mathbf{Z}; \mathbf{Z}_\theta, \theta \le \eta(\mathbf{Z}))$ in Σ. Denoting by $\mathbf{X} \overset{\phi}{\to} \mathbf{Z} \overset{\rho}{\to} \mathbf{Y}$ the corresponding factorization of $\pi = \rho \circ \phi$, we apply Theorem 9.23.1 and the maximality of $\mathbf{Z}$ to deduce that ϕ is a weakly mixing extension. As ρ is an I-extension this completes the proof. □

2. Separating sieves, distal and I-extensions

10.4. DEFINITION. Let $\pi : \mathbf{X} \to \mathbf{Y}$ be a factor map of the ergodic system $\mathbf{X}$. A sequence $A_1 \supset A_2 \supset \cdots$ of sets in $\mathcal{X}$ with $\mu(A_n) > 0$ and $\mu(A_n) \to 0$, is called a *separating sieve over* Y if there exists a subset $X_0 \subset X$ with $\mu(X_0) = 1$ such that for every $x, x' \in X_0$ with $\pi(x) = \pi(x')$, the condition "for every $n \in \mathbb{N}$ there exists $\gamma \in \Gamma$ with $\gamma x, \gamma x' \in A_n$" implies $x = x'$, or in symbols:

$$\bigcap_{n=1}^{\infty} \left(\bigcup_{\gamma\in\Gamma} \gamma(A_n \underset{Y}{\times} A_n) \right) \cap (X_0 \underset{Y}{\times} X_0) \subset \Delta,$$

where, say, $X_0 \underset{Y}{\times} X_0 = \{(x, x') \in X_0 \times X_0 : \pi(x) = \pi(x')\}$. We also say that the extension π has a *separating sieve*. We say that the extension π is a *(measure) distal extension* if either $\mathbf{X}$ is finite or there exists a separating sieve over Y. Taking $\mathbf{Y}$ to be the trivial one point system in these definitions yields the definitions of a *separating sieve* for $\mathbf{X}$ and of $\mathbf{X}$ being *(measure) distal* .

10.5. EXAMPLES. If (X, Γ) is a minimal topological system, which is distal in the topological sense (see Chapter 1, Section 3), then any decreasing sequence of open sets U_n with $\bigcap_n U_n = \{x_0\}$ will form a separating sieve for any invariant probability measure on X. Thus for any such μ the corresponding measure system $\mathbf{X} = (X, \mathcal{X}, \mu, \Gamma)$ is measure distal. In particular the topologically distal systems presented in Exercises 1.6, and the various nil-systems of Theorem 4.11.2, are also examples of measure distal systems. There are however minimal topological systems (X, T) and invariant measures $\mu \in M_T(X)$ such that the corresponding measure-preserving systems $\mathbf{X} = (X, \mathcal{X}, \mu, T)$ are measure but not topologically distal (see e.g. [**268**], Example 9.4, or use Lehrer's theorem, [**164**], which provides a topologically mixing uniquely ergodic model to every ergodic measure-preserving $\mathbb{Z}$-system).

10.6. PROPOSITION. *An ergodic, isometric system is distal.*

PROOF. By Theorem 9.14 $\mathbf{X}$ is a homogeneous system, say $\mathbf{X} = (K/L, \tilde{m}, \Gamma)$, where K is a compact group, $L \subset K$ a closed subgroup, $\tilde{m}$ the projection of Haar measure m_K on K/L and $\alpha : \Gamma \to K$ a homomorphism of Γ into K with a dense image, with $\gamma kL = \alpha(\gamma) \cdot kL$. If K/L is not finite, choose a right invariant metric on K/L and let U_n be a decreasing sequence of neighborhoods of the identity $x_0 = \{L\} \in K/L$ such that $\bigcap U_n = \{x_0\}$, hence diam $(U_n) \to 0$ and $\tilde{m}(U_n) \to 0$. Clearly the sequence U_n is a separating sieve for $\mathbf{X}$. $\square$

The next lemma follows directly from the definition of a separating sieve.

10.7. LEMMA. *If* $\mathbf{X} \xrightarrow{\phi} \mathbf{Y} \xrightarrow{\theta} \mathbf{Z}$ *(with* $\pi = \theta \circ \phi$*) are factors of the ergodic system* $\mathbf{X}$ *and* A_n *is a separating sieve for* $\mathbf{X}$ *over* $\mathbf{Z}$*, then it is also a separating sieve for* $\mathbf{X}$ *over* $\mathbf{Y}$*. Hence if* $\pi : \mathbf{X} \to \mathbf{Z}$ *is a distal extension so is* $\mathbf{X} \to \mathbf{Y}$*.*

10.8. THEOREM (Zimmer). *An extension* $\pi : \mathbf{X} \to \mathbf{Y}$*, with* $\mathbf{X}$ *ergodic, is an* I*-extension iff it is a distal extension.*

For the proof we shall need the following lemmas.

10.9. LEMMA. *Let* $\pi : \mathbf{X} \to \mathbf{Y}$ *be an isometric extension with* $\mathbf{X}$ *ergodic,* $\mathbf{Y}$ *finite and* $\mathbf{X}$ *non-atomic, then* $\mathbf{X}$ *has a separating sieve.*

PROOF. As $\mathbf{Y}$ is ergodic and finite it has the form Γ/Γ_0 where $\Gamma_0 \subset \Gamma$ is a subgroup of finite index. By Theorem 9.14 $\mathbf{X}$ is a skew-product of the form $\mathbf{Y} \underset{\alpha}{\times} K/L$. Choosing a decreasing sequence of sets $U_n \subset K/L$ as in Proposition 10.6 and denoting $y_0 = \Gamma_0 \in Y$ it is now easy to see that $\{y_0 \times U_n\}_{n \in \mathbb{N}}$ is a separating sieve for $\mathbf{X}$. $\square$

10.10. LEMMA. *If* $\mathbf{X} \xrightarrow{\phi} \mathbf{Z} \xrightarrow{\theta} \mathbf{Y}$ *(with* $\pi = \theta \circ \phi$*) are factors of the ergodic system* $\mathbf{X}$*, such that* ϕ *is an isometric extension and the extension* θ *has a separating sieve, then also* π *has a separating sieve.*

PROOF. Again Theorem 9.14 implies that $\mathbf{X}$ is a skew-product of the form $\mathbf{Z} \underset{\alpha}{\times} K/L$, and we choose a decreasing sequence of sets $U_n \subset K/L$ as in Proposition 10.6. Let $A_n \subset Z$ be a separating sieve for θ with respect to the conull set $Z_0 \subset Z$. We show that the sequence $B_n = A_n \times U_n$ forms a separating sieve for π with respect to $X_0 := \phi^{-1}(Z_0)$. In fact, if for $(z, kL), (z', k'L) \in X_0$ and for every n there is a $\gamma_n \in \Gamma$ such that $\gamma_n(z, kL), \gamma_n(z', k'L) \in B_n$, then $z, z' \in Z_0$, $\gamma_n z, \gamma_n z' \in A_n$ and therefore $z = z'$. Now we have $\alpha(\gamma_n, z)kL, \alpha(\gamma_n, z)k'L \in U_n$ for every n and we conclude that also $kL = k'L$. $\square$

10.11. LEMMA. *Let $\pi : \mathbf{X} \to \mathbf{Y}$ be an extension with $\mathbf{X}$ ergodic. Let $\mathbf{X} \xrightarrow{\phi_n} \mathbf{X}_n \xrightarrow{\theta_n} \mathbf{Y}$, $n = 1, 2, \ldots$ be a family of intermediate factors such that for every m, n the diagram*

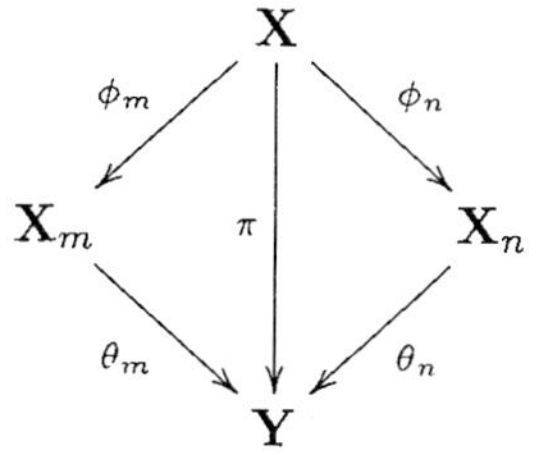

commutes. Suppose further that for every n the extension $\mathbf{X}_n \xrightarrow{\theta_n} \mathbf{Y}$ has a separating sieve and that there exists a conull set $Z \subset X$ such that $x, x' \in X_0$, $x \neq x'$ implies there exists an n with $\phi_n(x) \neq \phi_n(x)$, then the extension π has a separating sieve.

PROOF. Let $\{A_n^i\}_{n=1}^{\infty}$ be a separating sieve for $\mathbf{X}_i$, with respect to the conull sets $X_{i0} \subset X_i$. By ergodicity of $\mathbf{X}$ we can define inductively a triangular array $\gamma_{in} \in \Gamma, 1 \leq i \leq n$ such that for every n, $\mu(B_n) > 0$, $n = 1, 2, \ldots$, where $B_n = \bigcap_{i=1}^n \phi_i^{-1}(\gamma_{in} A_n^i)$. Set $A_1 = B_1$ and define inductively a sequence of sets $A_n \in \mathcal{X}$ as follows. Given A_{n-1}, choose $\delta_n \in \Gamma$ with $\mu(A_{n-1} \cap \delta_n B_n) > 0$ and set $A_n = A_{n-1} \cap \delta_n B_n$. We next check that $\{A_n\}_{n=1}^{\infty}$ is a separating sieve for $\mathbf{X}$ with respect to the conull set

$$X_0 = Z \cap \bigcap_{i=1}^{\infty} \phi_i^{-1}(X_{i0}) \subset X.$$

We have

$$A_n \subset \delta_n B_n \subset \phi_1^{-1}(\delta_n \gamma_{1n} A_n^1),$$

and $\mu_1(A_n^1) \to 0$, whence $\mu(A_n) \to 0$.

Now let $x, x' \in X_0$ with $\pi(x) = \pi(x')$ and assume that for every n there exists a $\zeta_n \in \Gamma$ with $\zeta_n x, \zeta_n x' \in A_n$. We define, for each i, a sequence $\tau_{in} \in \Gamma$, $n \geq i$ by $\tau_{in} = \gamma_{in}^{-1} \delta_n^{-1} \zeta_n$. Then for $n \geq i$ we have

$$\tau_{in} x = \gamma_{in}^{-1} \delta_n^{-1} \zeta_n x \in \gamma_{in}^{-1} \delta_n^{-1} A_n \subset \gamma_{in}^{-1} B_n \subset \gamma_{in}^{-1} \phi_i^{-1}(\gamma_{in} A_n^i).$$

It therefore follows that $\tau_{in} \phi_i(x) \in A_n^i$. Similarly, for $n \geq i$ we get $\tau_{in} \phi_i(x') \in A_n^i$. Since $\phi_i(x), \phi_i(x') \in X_{i0}$, it follows that $\phi_i(x) = \phi_i(x')$ and as $x, x' \in Z$ we finally get $x = x'$. □

10.12. COROLLARY. *Let $\pi_n : \mathbf{X}_n \to \mathbf{Y}$, $n \in \mathbb{N}$ be a collection of distal extensions with $\mathbf{X}_n$ ergodic and let $\lambda \in J_e^{\mathbf{Y}}(\mathbf{X}_n : n \in \mathbb{N})$ be an ergodic joining over $\mathbf{Y}$. Then the extension $\pi : (\prod_{n \in \mathbb{N}} X_n, \lambda) \to \mathbf{Y}$ is a distal extension. In particular if $\{\mathbf{X}_\alpha : \alpha \in D\}$ is a (countable) directed family of ergodic systems sharing a common factor $\pi_\alpha : \mathbf{X}_\alpha \to \mathbf{Y}$ then the induced extension $\pi : \mathbf{X} = \mathrm{inv}\ \lim_D \mathbf{X}_\alpha \to \mathbf{Y}$ is also a distal extension.*

PROOF OF THE FIRST PART OF THEOREM 10.8. We assume that the extension $\pi : \mathbf{X} \to \mathbf{Y}$ is an I-extension and that $\mathbf{X}$ is not finite. We shall show that π has a separating sieve. There are two cases to consider.

Case 1: **$\mathbf{Y}$ is infinite.** Using the notation of Definition 10.2, we let

$$\Lambda = \{\theta \leq \eta : \mathbf{X}_\theta \text{ has a separating sieve over } \mathbf{Y}\}.$$

Since $\mathbf{Y}$ is non-atomic, $0 \in \Lambda$. If $\theta \in \Lambda$ and $\theta < \eta$, then $\theta+1 \in \Lambda$ by Lemma 10.10. If $\lambda \leq \eta$ is a limit ordinal and $\theta \in \Lambda$ for $\theta < \lambda$, then (as η is countable) $\lambda \in \Lambda$ by Corollary 10.12. Now transfinite induction implies that $\eta \in \Lambda$, that is, $\mathbf{X}$ has a separating sieve over $\mathbf{Y}$.

Case 2: **Y is finite**. Let $\Sigma = \{\theta \leq \eta : \mathbf{X}_\theta \text{ is finite}\}$ and let $\eta_0 = \sup\Sigma$. Again we consider two cases:

(i) $\eta_0 \in \Sigma$: We then have $\eta_0 < \eta$ and by Lemma 10.9, $\mathbf{X}_{\eta_0+1}$ has a separating sieve. We can now repeat the argument of Case 1 to deduce that $\mathbf{X}$ has a separating sieve over $\mathbf{Y}$.

(ii) $\eta_0 \notin \Sigma$: It follows that η_0 is a limit ordinal, so that $\mathbf{X}_{\eta_0} = \text{inv}\lim_{\theta<\eta_0} \mathbf{X}_\theta$. By Proposition 6.4 $\mathbf{X}_{\eta_0}$ is an infinite isometric system and by Proposition 10.6 it has a separating sieve. Again we repeat the argument of Case 1 to deduce that $\mathbf{X}$ has a separating sieve over $\mathbf{Y}$. □

For the proof of the other direction of Theorem 10.8 we need the following lemma.

10.13. LEMMA. *Let $\pi : \mathbf{X} \to \mathbf{Y}$ be an extension with $\mathbf{X}$ ergodic. If π has a separating sieve and is a weakly mixing extension, then it is an isomorphism.*

PROOF. Let $\{A_n\}_{n=1}^\infty$ be a separating sieve for π with respect to the conull set $X_0 \subset X$. This means that the set

$$\bigcap_{n=1}^{\infty}\left(\bigcup_{\gamma\in\Gamma}\gamma(A_n \underset{Y}{\times} A_n)\right) \cap (X_0 \underset{Y}{\times} X_0)$$

is contained in the diagonal $\Delta = \{(x,x) : x \in X\}$. Now π is an isomorphism iff $\mu \underset{Y}{\times} \mu(\Delta) = 1$. Assuming π is not an isomorphism we have either $0 < \mu \underset{Y}{\times} \mu(\Delta) < 1$ or $\mu \underset{Y}{\times} \mu(\Delta) = 0$. In the first case Δ is a non-trivial invariant set contradicting the weak mixing of the extension π. In the latter, we have

$$\lim_{n\to\infty} \mu \underset{Y}{\times} \mu\left(\bigcup_{\gamma\in\Gamma}\gamma(A_n \underset{Y}{\times} A_n)\right) = 0.$$

Since for every n, $\mu(A_n) > 0$, we have also $\mu \underset{Y}{\times} \mu(A_n \underset{Y}{\times} A_n) > 0$, hence for some n,

$$0 < \mu \underset{Y}{\times} \mu\left(\bigcup_{\gamma\in\Gamma}\gamma(A_n \underset{Y}{\times} A_n)\right) < 1.$$

Again this is a non-trivial Γ-invariant set, contradicting the weak mixing of π and we conclude that π is an isomorphism. □

PROOF OF THE SECOND PART OF THEOREM 10.8. We assume that the extension $\pi : \mathbf{X} \to \mathbf{Y}$ is a distal extension. We shall show that π is an I-extension. If $\mathbf{X}$ is finite, then π is a finite, hence isometric, hence an I-extension. So we now assume that $\mathbf{X}$ is infinite and therefore that π has a separating sieve. Let $\mathbf{X} \xrightarrow{\sigma} \mathbf{Z} \xrightarrow{\rho} \mathbf{Y}$ be the factorization of Proposition 10.3. By Lemma 10.7 also σ has a separating sieve; however, σ is also a weakly mixing extension and we apply Lemma 10.13 to deduce that σ is an isomorphism. Thus $\pi = \sigma \circ \rho = \rho$ is an I-extension. $\square$

3. The structure theorem

Given a factor map $\pi : \mathbf{X} \to \mathbf{Y}$ of an ergodic system $\mathbf{X}$ we now construct, via transfinite induction, a canonical sequence of intermediate factors as follows. Let $\mathbf{X}_0 = \mathbf{Y}$. The first term of the sequence, $\mathbf{X}_1$, is the factor of $\mathbf{X}$ determined by the subspace $\mathcal{E}(\mathbf{X}/\mathbf{Y}) \subset L^2(\mathbf{X})$. Thus (by Theorem 9.13) $\mathbf{X}_1 \to \mathbf{X}_0$ is the largest intermediate isometric extension of $\mathbf{Y}$ in $\mathbf{X}$. If for any (countable) ordinal θ the intermediate factor $\mathbf{X}_\theta$ is defined, we let $\mathbf{X}_{\theta+1}$ be the factor of $\mathbf{X}$ determined by the subspace $\mathcal{E}(\mathbf{X}/\mathbf{X}_\theta)$, the largest intermediate isometric extension of $\mathbf{X}_\theta$ in $\mathbf{X}$. For a limit ordinal λ with $\mathbf{X}_\theta$ defined for all $\theta < \lambda$, we set $\mathbf{X}_\lambda = \text{inv}\lim_{\theta<\lambda} \mathbf{X}_\theta$. As we have already seen, this procedure stabilizes at a certain countable ordinal $\eta = \eta(\mathbf{X}, \mathbf{Y})$; i.e. η is the first ordinal with $\mathbf{X}_\eta = \mathbf{X}_{\eta+1}$. We call the directed family

$$(\mathbf{X}; \mathbf{X}_\theta, \theta \leq \eta),$$

the canonical distal tower of the extension $\pi : \mathbf{X} \to \mathbf{Y}$.

In the sequel we shall need the following proposition.

10.14. PROPOSITION. *Let* $\mathbf{X} \xrightarrow{\sigma} \mathbf{Z} \xrightarrow{\rho} \mathbf{Y}$, *with* $\pi = \rho \circ \sigma$ *a distal extension. If* $\rho : \mathbf{Z} \to \mathbf{Y}$ *is not an isomorphism then it is not a weakly mixing extension. In particular, an extension which is both distal and weakly mixing is an isomorphism.*

PROOF. Let $(\mathbf{X}; \mathbf{X}_\theta, \theta \leq \eta(\mathbf{X}))$ be the directed family of factors and factor maps associated with the distal extension $\mathbf{X} \xrightarrow{\pi} \mathbf{Y}$. Let

$$\Lambda = \{\theta \leq \eta : (L^2(\mathbf{X}_\theta) \ominus L^2(\mathbf{Y})) \perp (L^2(\mathbf{Z}) \ominus L^2(\mathbf{Y}))\}.$$

As $\rho : \mathbf{Z} \to \mathbf{Y}$ is not an isomorphism, $\eta \notin \Lambda$ and we define $\lambda = \sup\Lambda$. Clearly $\lambda \in \Lambda$, hence $\lambda < \eta$. We denote $\mathbf{W} = \mathbf{X}_\lambda$ and $\mathbf{V} = \mathbf{X}_{\lambda+1}$. As $\lambda \in \Lambda$, the σ-algebras $\mathcal{W} = \mathcal{X}_\lambda$ and $\mathcal{Z}$ are relatively independent over $\mathcal{Y}$ and hence the factor determined by $\mathcal{W} \vee \mathcal{Z}$ is isomorphic to the relative product system $\mathbf{W} \underset{Y}{\times} \mathbf{Z}$ (see Exercise 6.7). Denoting $\mathbf{U} = \mathbf{W} \underset{Y}{\times} \mathbf{Z}$ we have the commutative diagram:

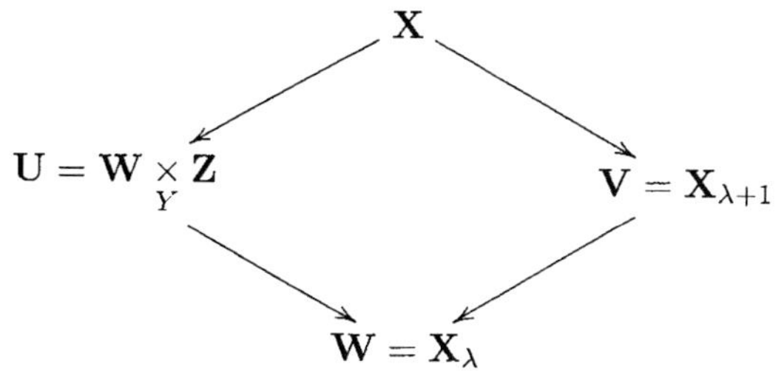

Let $P : L^2(\mathbf{V}) \to L^2(\mathbf{U})$ denote the restriction of $\mathbb{E}^{\mathbf{U}} : L^2(\mathbf{X}) \to L^2(\mathbf{U})$ to $L^2(\mathbf{V})$. Then P commutes with the operators U_γ, $\gamma \in \Gamma$ and with the operators R_f defined by multiplication with $f \in L^\infty(\mathbf{W})$. It therefore follows that the image of a Γ-invariant, finite rank, $\mathbf{W}$-module $M \subset L^2(\mathbf{V})$ under P is a Γ-invariant, finite rank, $\mathbf{W}$-module in $L^2(\mathbf{U})$. Since the extension $\mathbf{V} = \mathbf{X}_{\lambda+1} \to \mathbf{W} = \mathbf{X}_\lambda$ is isometric, the union of such $\mathbf{W}$-modules M is dense in $L^2(\mathbf{V})$. Therefore, if for every such M

$P(M) \subset L^2(\mathbf{W})$, then $P(L^2(\mathbf{V})) \subset L^2(\mathbf{W})$, contradicting the fact that $\lambda + 1 \notin \Lambda$. Thus there exists a Γ-invariant, finite rank, $\mathbf{W}$-module, $M \subset L^2(\mathbf{V})$ with $L^2(\mathbf{W}) \subsetneq P(M) \subset L^2(\mathbf{U})$. Theorem 9.23.1 implies that the extension $\mathbf{U} \to \mathbf{W}$ is not weakly mixing; i.e.

$$\mathbf{U} \underset{\mathbf{W}}{\times} \mathbf{U} = (\mathbf{W} \underset{Y}{\times} \mathbf{Z}) \underset{\mathbf{W}}{\times} (\mathbf{W} \underset{Y}{\times} \mathbf{Z})$$

is not ergodic. Now

$$(\mathbf{W} \underset{Y}{\times} \mathbf{Z}) \underset{\mathbf{W}}{\times} (\mathbf{W} \underset{Y}{\times} \mathbf{Z}) \cong (\mathbf{W} \underset{Y}{\times} \mathbf{Z}) \underset{Y}{\times} \mathbf{Z}.$$

If $\rho : \mathbf{Z} \to \mathbf{Y}$ is a weakly mixing extension, then Theorem 9.23.2 implies that $(\mathbf{W} \underset{Y}{\times} \mathbf{Z})$ is ergodic and a second application of the same theorem implies that also $(\mathbf{W} \underset{Y}{\times} \mathbf{Z}) \underset{Y}{\times} \mathbf{Z} \cong \mathbf{U} \underset{\mathbf{W}}{\times} \mathbf{U}$ is ergodic. This contradiction completes the proof. □

We are now ready to state and prove the structure theorem.

10.15. THEOREM (The Furstenberg-Zimmer structure theorem). *Let $\pi : \mathbf{X} \to \mathbf{Y}$ be a factor map of an ergodic system $\mathbf{X}$.*

1. *There exists an intermediate factor $\mathbf{X} \xrightarrow{\phi} \mathbf{Z} \xrightarrow{\rho} \mathbf{Y}$ (with $\pi = \rho \circ \phi$) such that ρ is a (maximal) distal extension and ϕ is a weakly mixing extension.*
2. *This factorization is unique.*
3. *The factor $\mathbf{Z}$ coincides with the factor $\mathbf{X}_\eta$ from the canonical distal tower of the extension $\pi : \mathbf{X} \to \mathbf{Y}$.*
4. *The extension π is a weakly mixing extension iff ρ is trivial.*
5. *The extension π is a distal extension iff ϕ is trivial.*

PROOF. 1. This follows from Proposition 10.3 and the characterization of distal extensions as I-extensions (Theorem 10.8).

2. If $\mathbf{X} \xrightarrow{\phi_i} \mathbf{Z}_i \xrightarrow{\rho_i} \mathbf{Y}$, $i = 1, 2$, are two such factorizations, we let $\mathbf{Z} = \mathbf{Z}_1 \vee \mathbf{Z}_2$ (that is, the factor of $\mathbf{X}$ defined by the σ-algebra $\mathcal{Z} = \mathcal{Z}_1 \vee \mathcal{Z}_2$). By Theorem 10.8 and Corollary 10.12, the extension $\rho : \mathbf{Z} \to \mathbf{Y}$ is a distal extension and unless $\mathbf{Z}_1 = \mathbf{Z}_2$ we get a contradiction to the maximality of, say, the extension $\rho_1 : \mathbf{Z}_1 \to \mathbf{Y}$ as a distal extension.

3. If the distal extension $\mathbf{X}_\eta \to \mathbf{Y}$ is not a maximal distal intermediate extension then there exists a non-trivial intermediate distal extension α, $\mathbf{X} \to \mathbf{Z}_1 \xrightarrow{\alpha} \mathbf{X}_\eta$. However, this extension α is both distal and weakly mixing and therefore is an isomorphism by Lemma 10.13 (or by Proposition 10.14). Thus the extension $\mathbf{X}_\eta \to \mathbf{Y}$ is a maximal distal intermediate extension and by the uniqueness proved in part 2 we get $\mathbf{X}_\eta = \mathbf{Z}$.

The proofs of parts 4 and 5 are now straightforward. □

For a distal extension $\pi : \mathbf{X} \to \mathbf{Y}$ we have $\mathbf{X} = \mathbf{X}_\eta$ and the ordinal $\eta = \eta(\pi)$ is called the *distal order* of the extension π. In particular when $\mathbf{Y}$ is the trivial one point system and $\mathbf{X} = \mathbf{X}_\eta$, η is the *distal order* of $\mathbf{X}$.

10.16. EXERCISE. Let $\pi : \mathbf{X} \to \mathbf{Y}$ be a non-trivial weakly mixing extension with $\mathbf{X}$ ergodic. Let $\mathbf{X} \cong \mathbf{Y} \underset{\alpha}{\times} (U, \rho) = (Y \times U, \mathcal{Y} \otimes \mathcal{U}, \nu \times \rho, \Gamma)$ be the corresponding

Rohlin decomposition (Theorem 3.18). Then in the standard probability space $(U, \mathfrak{U}, \rho)$ the measure ρ has no atoms. Equivalently, the relative product measure

$$\lambda = \int_Z (\mu_z \times \mu_z)\, d\nu(y) \cong \nu \times \rho \times \rho,$$

considered as a measure on $X \times X$, satisfies $\lambda(\Delta_X) = 0$. [[Show that the existence of atoms for ρ implies that the extension $\psi : \mathbf{X} \to \mathbf{Y}$ is a non-trivial r to one extension for some finite number r, so that π is both distal and weakly mixing hence a trivial extension.]]

10.17. EXERCISE. The measure-preserving system $\mathbf{X} = (X, \mathfrak{X}, \mu, T)$ described in Example 9.11, namely the transformation $T(\xi, \eta) = (\xi + \alpha, \eta + \xi)$ on the torus $X = \mathbb{T}^2$, is measure distal of order 2. Show that its Koopman operator U_T has countable Lebesgue spectrum on the orthogonal complement of $L^2(\mathbf{Y})$, where $\mathbf{Y}$ is the largest isometric factor: $T\xi = \xi + \alpha$ on $\mathbb{T}$ with factor map $\pi(\xi, \eta) = \xi$. [[For the countable Lebesgue spectrum part consider the functions $f_m(\xi, \eta) = \exp(2\pi i m \eta)$, $0 \neq m \in \mathbb{Z}$.]]

4. Factors and quasifactors of distal extensions

Our first goal in this section is to show that distality is preserved by factors. Surprisingly this requires both the I-extension and the separating sieves characterizations of distality.

10.18. THEOREM. *Let $\pi : \mathbf{X} \to \mathbf{Y}$ be a distal extension with $\mathbf{X}$ ergodic. Then for any intermediate factor $\mathbf{X} \xrightarrow{\phi} \mathbf{Z} \xrightarrow{\psi} \mathbf{Y}$ (with $\pi = \psi \circ \phi$) the extension $\psi : \mathbf{Z} \to \mathbf{Y}$ is also a distal extension. In particular, any factor of a distal system is distal.*

PROOF. Let $\mathbf{Z} \xrightarrow{\sigma} \mathbf{Z}_1 \xrightarrow{\rho} \mathbf{Y}$ be the factorization of the structure theorem applied to the extension $\mathbf{Z} \xrightarrow{\psi} \mathbf{Y}$. Now consider the commutative diagram:

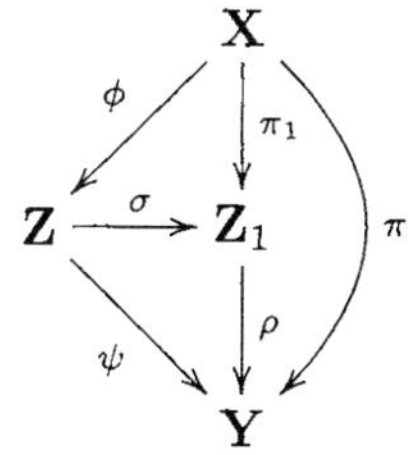

In this diagram π_1 is a distal extension (by Lemma 10.7) and σ is a weakly mixing extension. Thus by Proposition 10.14, σ is an isomorphism, so that $\psi = \rho$ is indeed a distal extension. □

10.19. THEOREM. 1. *Let $\pi : \mathbf{X} \to \mathbf{Y}$ be a distal extension with $\mathbf{X}$ ergodic and $\kappa \in Q^e_J(\pi)$, an ergodic JQF over $\mathbf{Y}$. Then the ergodic system $\mathbf{W} = (M(x), \kappa)$ is a distal extension of $\mathbf{Y}$. In particular a JQF of a distal system is distal.*

2. *Let $\pi : \mathbf{X} \to \mathbf{Y}$ be a distal extension and $\sigma : \mathbf{Z} \to \mathbf{Y}$ a weakly mixing extension, with $\mathbf{X}$ and $\mathbf{Y}$ ergodic systems, then $\mathbf{X}$ and $\mathbf{Z}$ are disjoint over $\mathbf{Y}$. In particular every ergodic distal system is disjoint from every weakly mixing system.*

PROOF. 1. By corollary 10.12 and Theorem 10.18 the family of ergodic systems which are distal extensions of $\mathbf{Y}$ is preserved by infinite ergodic selfjoinings and factors (over $\mathbf{Y}$). Corollary 8.13 applies and proves the first part of the theorem.

2. Since a homomorphic image of a weakly mixing extension (over $\mathbf{Y}$) is a weakly mixing extension, the second part of the theorem follows from the first part, Exercise 8.5, and the fact that a quasifactor in $Q(\pi)$ which is both distal and weakly mixing extension of $\mathbf{Y}$ is trivial (Proposition 10.14). (Alternatively one can use Theorem 6.25.) □

5. Notes

As in the previous chapter my main sources here are R. Zimmer's papers [**267**] and [**268**] and H. Furstenberg's works [**76**] and [**77**]. Between the classes of isometric and distal systems there exists a natural intermediate class, namely the systems with quasi-discrete spectrum. For a $\mathbb{Z}$-system $\mathbf{X}$ let $D_0 \subset L^2(\mu)$ denote the set of constant complex valued functions of modulus 1, and for $n \geq 1$ set $D_n = \{f \in L^2(\mu) : |f| = 1 \text{ a.e. and } U_T f/f \in D_{n-1}\}$. The ergodic system $\mathbf{X}$ has *quasi-discrete spectrum* if the closed linear subspace spanned by $D = \cup_{n\geq 0} D_n$ is all of $L^2(\mu)$. This class was defined and studied by Abramov [**2**]. In [**268**], Zimmer shows that a totally ergodic $\mathbb{Z}$-system $\mathbf{X}$ has quasi-discrete spectrum iff it is distal and isomorphic to a totally ergodic *affine transformation* on a compact connected abelian group X (i.e. $T : X \to X$ has the form $Tx = x_0 A(x)$, where $A : X \to X$ is an automorphism of the group X and $x_0 \in X$). As was observed in Example 9.11 the strictly ergodic skew-product $T(x, y) = (x + \alpha, y + x)$ on $\mathbb{T}^2$ has quasi-discrete spectrum. On the other hand one can show that the Heisenberg nil-system (see Theorem 4.17) is a measure distal system which does not have quasi-discrete spectrum.

In [**171**] E. Lindenstrauss proves the following:

10.20. THEOREM. *Every ergodic measure distal $\mathbb{Z}$-system $\mathbf{X} = (X, \mathfrak{X}, \mu, T)$ can be represented as a minimal topologically distal system (X, T, μ) with $\mu \in M_T^{\mathrm{erg}}(X)$.*

Note however that this topological model need not, in general, be uniquely ergodic. In other words there are measure distal systems for which no uniquely ergodic topologically distal model exists.

CHAPTER 11

Host's Theorem

Our goal in this chapter is to present the most significant progress in recent years on the open question whether, for $\mathbb{Z}$-systems, mixing implies mixing of all orders. This is B. Host's theorem that gives an affirmative answer to that question when the spectral type of the system is singular with respect to Lebesgue measure. (See also S. Kalikow's work [**141**], where he shows that for rank one transformations mixing implies mixing of all orders, and V. Ryzhikov's extension of this result to finite rank systems [**222**], using joining techniques.)

1. Pairwise independent joinings

11.1. THEOREM. *Let $\mathbf{X}_i = (X_i, \mathfrak{X}_i, \mu_i, T)$, $i = 1, 2, \ldots, n$ be weakly mixing dynamical systems with spectral types σ_i singular with respect to Lebesgue measure on $\mathbb{T}$. If λ is an n-fold joining of these systems which is pairwise independent then λ is the independent joining $\lambda = \prod_{i=1}^n \mu_i$.*

Host's proof of this theorem is via a reduction to a statement about "correlation" measures on tori. In the Section 3 we shall prove the following "purity" theorem:

11.2. THEOREM (Purity Theorem). *Let θ be a probability measure on $\mathbb{T}$ and ρ a finite signed measure on $\mathbb{T}^3$. Let*

$$H = \{(s,t,u) \in \mathbb{T}^3 : s + t + u = 0\}.$$

Suppose

1. *ρ is supported on H*
2. *ρ is invariant under permutations of coordinates in $\mathbb{T}^3$*
3. *ρ is invariant under the map $J : x \mapsto -x$ in $\mathbb{T}^3$*
4. *$(\pi_1)_*\rho = \theta$ and $\eta := (\pi_{1,2})_*\rho \ll \theta \times \theta$,*

where $\pi_1, \pi_{1,2}$ are the projections of $\mathbb{T}^3$ onto the corresponding one and two dimensional tori, then θ is a convex combination of a discrete measure and an absolutely continuous one.

In this section we shall see how Theorem 11.1 can be deduced from Theorem 11.2

PROOF OF THEOREM 11.1. Using the weak mixing assumption it suffices to prove the theorem for the case $n = 3$. For functions $f_i \in L_0^2(\mu_i)$ define the *correlation function*

$$c(n_1, n_2, n_3) = \int f_1(T^{n_1}x_1) f_2(T^{n_2}x_2) f_3(T^{n_3}x_3)\, d\lambda(x_1, x_2, x_3).$$

We show that $c(n_1, n_2, n_3) = \hat{\rho}(n_1, n_2, n_3)$ is the Fourier transform of a probability measure ρ supported on the subgroup H of $\mathbb{T}^3$. In fact, put $G(x_1, x_2) = \mathbb{E}_{\lambda}^{\mathfrak{X}_1 \otimes \mathfrak{X}_2 \times X_3}(f_3)$; i.e.

$$\int\int F(x_1,x_2)G(x_1,x_2)\,d\mu_1(x_1)\,d\mu_2(x_2) = \int F(x_1,x_2)f_3(x_3)\,d\lambda(x_1,x_2,x_3),$$

for every $F(x_1, x_2) \in L^2(\mu_1 \times \mu_2)$.

Set

$$\rho_0 = \sigma_{f_1 \otimes f_2, \bar{G}};$$

i.e.

$$\hat{\rho}_0(k,l) = \int\int f_1(T^k x_1) f_2(T^l x_2) G(x_1,x_2)\,d\mu_1(x_1)\,d\mu_2(x_2),$$

and let $\rho_1 = \alpha_*(\rho_0)$, where $\alpha : \mathbb{T}^2 \to \mathbb{T}^3$ is the map $\alpha : (s,t) \mapsto (s,t,-(s+t))$.

Now

$$\begin{aligned}
\hat{\rho}_1(n_1,n_2,n_3) &= \int e^{2\pi i(n_1 s + n_2 t + n_3 u)}\,d\alpha_*\rho_0(s,t,u) \\
&= \int e^{2\pi i(n_1 s + n_2 t - n_3(s+t))}\,d\rho_0(s,t) = \hat{\rho}_0(n_1 - n_3, n_2 - n_3) \\
&= \int\int f_1(T^{n_1 - n_3}x_1) f_2(T^{n_2 - n_3}x_2) G(x_1,x_2)\,d\mu_1(x_1)\,d\mu_2(x_2) \\
&= \int f_1(T^{n_1 - n_3}x_1) f_2(T^{n_2 - n_)}x_2) f_3(x_3)\,d\lambda(x_1,x_2,x_3) \\
&= \int f_1(T^{n_1}x_1) f_2(T^{n_2}x_2) f_3(T^{n_3}x_3)\,d\lambda(x_1,x_2,x_3) \\
&= c(n_1,n_2,n_3),
\end{aligned}$$

as required.

Of course the measure ρ_1 is supported on H and $(\pi_{1,2})_*\rho_1 = \rho_0$. It therefore follows that

$$(\pi_{1,2})_*\rho_1 = \rho_0 = \sigma_{f_1 \otimes f_2, \bar{G}} \ll \sigma_{f_1 \otimes f_2} = \sigma_{f_1} \times \sigma_{f_2} \ll \sigma_1 \times \sigma_2$$

(see Exercise 5.5; we worn the reader however that these spectral measures correspond to $\mathbb{Z}^2$ rather than $\mathbb{Z}$ actions). The argument above can be repeated with f_1 or f_2 instead of f_3 and we therefore conclude that also

$$(\pi_{2,3})_*\rho_1 \ll \sigma_2 \times \sigma_3 \quad \text{and} \quad (\pi_{1,3})_*\rho_1 \ll \sigma_1 \times \sigma_3.$$

Next set $\sigma_0 = \frac{1}{3}(\sigma_1 + \sigma_2 + \sigma_3)$ and $\sigma = \frac{1}{2}(\sigma_0 + \check{\sigma}_0)$ where $\check{\sigma}_0$ is the image of σ_0 under the map $J : t \mapsto -t$ on $\mathbb{T}$. Also let ρ_2 be the symmetrization of ρ_1 under the group of permutations of the coordinates s, t, u in $\mathbb{T}^3$. Finally let $\rho = \frac{1}{2}(\rho_2 + \check{\rho}_2)$, where $\check{\rho}_2$ is the image of ρ_2 under J (on $\mathbb{T}^3$), and let $\theta = (\pi_1)_*\rho = (\pi_2)_*\rho = (\pi_3)_*\rho$.

It is now clear that the measures ρ and θ satisfy the assumptions of Theorem 11.2 except for the requirement $(\pi_{1,2})_*\rho \ll \theta \times \theta$. To see that this requirement is satisfied denote $\eta = (\pi_{1,2})_*\rho$ and observe first that

$$\eta \ll \sigma \times \sigma, \quad \text{hence} \quad \theta = (\pi_1)_*\eta \ll \sigma.$$

Let $g(t) = \frac{d\theta}{d\sigma}$ and $A = \{t : g(t) > 0\}$, then

$$\eta(\mathbb{T} \times A) = \theta(A) = \eta(A \times \mathbb{T}) \Rightarrow \eta(A \times A) = 1,$$

and on A, $d\sigma \restriction A = g^{-1}d\theta$. Let $K \subset A \times A$ with $\theta \times \theta(K) = 0$, then denoting

$$F(s,t) = \frac{d\eta}{d(\sigma \times \sigma)},$$

we now have

$$\begin{aligned}\eta(K) &= \int_{A\times A} \mathbf{1}_K \, d\eta = \int_A \int_A \mathbf{1}_K(s,t)F(s,t)\, d\sigma(s)\, d\sigma(t) \\ &= \int_A \int_A \mathbf{1}_K(s,t)F(s,t)g^{-1}(s)g^{-1}(t)\, d\theta(s)\, d\theta(t) \\ &= \int \int_K F(s,t)g^{-1}(s)g^{-1}(t)\, d(\theta \times \theta)(s,t) = 0.\end{aligned}$$

Thus Theorem 11.2 applies and we conclude that θ is a convex combination of a discrete measure and an absolutely continuous one. However, since $\theta \ll \sigma$ and by assumption σ is continuous and singular, we conclude that $\theta = 0$. Therefore $\rho = 0$ and finally also $\rho_0 = 0$. Thus $\int f_1 f_2 f_3 \, d\lambda = 0$ for every $f_i \in L_0^2(\mu_i)$, $i = 1,2,3$, and for any $h_i \in L^2(\mu_i)$ with $\int h_i = c_i$, we have for $\tilde{h}_i = h_i - c_i$:

$$\begin{aligned}\int h_1h_2h_3\, d\lambda = &\int \tilde{h}_1\tilde{h}_2\tilde{h}_3\, d\lambda + c_1 \int \tilde{h}_2\tilde{h}_3\, d\lambda + c_2 \int \tilde{h}_1\tilde{h}_3\, d\lambda + c_3 \int \tilde{h}_1\tilde{h}_2\, d\lambda \\ &+ c_1c_2c_3.\end{aligned}$$

By pairwise independence the integrals of the products of two functions vanish and by our proof $\int \tilde{h}_1\tilde{h}_2\tilde{h}_3\, d\lambda = \hat{\rho}_0(0,0,0) = 0$. Thus

$$\int h_1h_2h_3\, d\lambda = \int h_1\, d\mu_1 \int h_2\, d\mu_2 \int h_3\, d\mu_3;$$

i.e. $\lambda = \mu_1 \times \mu_2 \times \mu_3$. □

2. Mandrekar-Nadkarni's theorem

In this section m denotes the Lebesgue measure on the circle $\mathbb{T}$, and for each $x \in \mathbb{T}$, $R_x : \mathbb{T} \to \mathbb{T}$ denotes the map $t \mapsto t+x$. A measure λ on a Borel space $(X, \mathcal{X})$ equipped with a Borel action of a group Γ is called Γ-*quasi-invariant* if $\gamma_*(\mu) \sim \mu$ for every $\gamma \in \Gamma$. Such a measure is *ergodic* if in addition $\mu(A)$ is either zero or one for every Γ-invariant $A \in \mathcal{X}$.

The main tool used in the proof of the theorem of Mandrekar and Nadkarni is Mackey-Weil's theorem (for the proof see [**174**]):

11.3. THEOREM (Mackey-Weil). *Let H be a Polish topological group, λ a probability measure which is H-quasi-invariant and ergodic with respect to the action of H on itself by left multiplication, then H is locally compact. If moreover λ is left invariant then λ is a left Haar measure for H.*

11.4. THEOREM (Mandrekar-Nadkarni, [**177**]). *Let $D \subset \mathbb{T}$ be a countable group, λ a probability measure on $\mathbb{T}$ which is D-quasi-invariant and D-ergodic. Let*

$$H(\lambda) = \{x \in \mathbb{T} : R_x\lambda \sim \lambda\},$$

and suppose that λ is supported on the group $H(\lambda)$, then λ is either discrete or absolutely continuous with respect to Lebesgue measure on $\mathbb{T}$

PROOF. For $x \in H(\lambda)$ let

$$h_x = \frac{dR_x\lambda}{d\lambda} \in L^1(\lambda),$$

and define

$$V_x : L^2(\lambda) \to L^2(\lambda),$$
$$V_x f(t) = \sqrt{h_x} f(t+x).$$

We then have

$$\begin{aligned} \|V_x f\|^2 &= \int h_x(t) f^2(t+x)\, d\lambda(t) = \int f^2(t+x) dR_x\lambda(t) \\ &= \int f^2(t)\, d\lambda(t) = \|f\|^2. \end{aligned}$$

Thus V_x is an isometry and since $V_{-x} = V_x^{-1}$, we conclude that V_x is a unitary operator. Let $\mathcal{U}$ be the group of all unitary operators on $L^2(\lambda)$ equipped with the following topology: the sequence U_n converges to U iff

$$\|U_n f - Uf\| \to 0,$$

for every $f \in L^2(\lambda)$. For a dense sequence $f_k \in L^2(\lambda)$, the metric $d(U,V) = \sum_{k=1}^{\infty} 2^{-k} \left(\|Uf_k - Vf_k\| + \|U^{-1}f_k - V^{-1}f_k\|\right)$ is a compatible complete metric. Whence with this topology $\mathcal{U}$ is a Polish topological group.

11.5. LEMMA. *The group $\mathcal{V}(\lambda) = \{V_x : x \in H(\lambda)\}$ is a closed subgroup of $\mathcal{U}$.*

PROOF. Let $\mathcal{V}(\lambda) \ni V_{x_n} \to V \in \mathcal{U}$, then V is a unitary operator which is also positive. By Theorem A.11 V has the form: $Vf(t) = \sqrt{\frac{dT\lambda}{d\lambda}}(t) f(Tt)$, where $T : \mathbb{T} \to \mathbb{T}$ is an invertible measurable map. Note that in particular $V_{x_n}\mathbf{1} \to V\mathbf{1}$, implies that $\sqrt{\frac{dR_{x_n}\lambda}{d\lambda}} \to \sqrt{\frac{dT\lambda}{d\lambda}}$ in $L^2(\lambda)$, and hence also in measure.

For every $x \in H(\lambda)$ we have

$$VV_x = \lim V_{x_n} V_x = \lim V_x V_{x_n} = V_x V.$$

Therefore V commutes with every element of $\{V_x : x \in H(\lambda)\}$ and thus, by first applying this commutation relation to the function $\mathbf{1}$, we see that $f(T(t+x)) = f(Tt + x)$, for every $f \in L^2(\lambda)$. This, in turn, implies $T(t+x) = Tt + x$, and denoting $g(t) = Tt - t$, $g : \mathbb{T} \to \mathbb{T}$, we get

$$g(t+x) = T(t+x) - (t+x) = Tt - t = g(t).$$

It follows that g is an $H(\lambda)$-invariant function and the D-ergodicity of λ implies that $g(t)$ is a constant λ almost everywhere, say $g(t) \equiv x_0$. This leads to $Tt = t + x_0$, hence to $V = V_{x_0}$; in particular $x_0 \in H(\lambda)$. □

11.6. LEMMA. *The map $x \mapsto V_x$ from $H(\lambda) \to \mathcal{V}(\lambda)$ is 1-1 and its inverse $V_x \mapsto x$ is continuous.*

PROOF. The injectivity is clear. Suppose that $V_{x_n} \to V_x$, for $x_n, x \in H(\lambda)$. As we have already seen, this implies that $h_{x_n} \to h_x$ in measure. By compactness of $\mathbb{T}$ we can assume that $x_n \to y \in \mathbb{T}$. Now for every continuous f we have on one hand, by assumption, the $L^2(\lambda)$ convergence

$$V_{x_n} f = \sqrt{h_{x_n}} f(t + x_n) \to \sqrt{h_x} f(t+x),$$

and on the other hand we get the convergence in measure

$$V_{x_n} f = \sqrt{h_{x_n}} f(t + x_n) \to \sqrt{h_x} f(t + y).$$

It is easy to see that this implies $x = y$, so that $V_x \mapsto x$ is continuous. □

An application of the Mackey-Weil theorem (Theorem 11.3) now shows that $\mathcal{V}(\lambda)$ is a separable locally compact abelian group and we can apply the structure theory of these groups to $\mathcal{V}(\lambda)$. According to this theory $\mathcal{V}(\lambda)$ contains an open and closed subgroup $\mathcal{V}_0$ such that $\mathcal{V}/\mathcal{V}_0$ is a countable discrete group and moreover the group $\mathcal{V}_0$ has the form

$$\mathcal{V}_0 = \mathbb{R}^n \oplus K,$$

where K is a compact abelian group (see [**216**], Theorem 2.4.1).

Next, by Lemma 11.6, $V_x \mapsto x$ is a continuous map of $\mathcal{V}_0$ into $\mathbb{T}$ and it follows that the image of the summand $\mathbb{R}^n$ is a locally compact subgroup of $\mathbb{T}$ which is isomorphic, as an abstract group, to $\mathbb{R}^n$. This can only happen when $n = 0$; i.e. when $\mathcal{V}_0 = K$ is a compact group. Thus we can now think of $\mathcal{V}_0 = K$ as a compact subgroup of $\mathbb{T}$.

We have two cases to consider, either $K = \mathbb{T}$ or K is finite. In the first case it follows that $H(\lambda) = \mathbb{T}$ and that λ is equivalent to Lebesgue measure m. In the second case we conclude that $H(\lambda)$ is a countable group and therefore that λ is a discrete measure. This completes the proof of the Theorem. □

3. Proof of the purity theorem

PROOF OF THE PURITY THEOREM (11.2). We split the proof into a number of lemmas. The first lemma associates with θ an auxiliary countable subgroup of $\mathbb{T}$, as follows:

11.7. LEMMA. *Let θ be a probability measure on $\mathbb{T}$, then there exists a countable subgroup $\Gamma \subset \mathbb{T}$ with the following property: For every $\lambda \ll \theta$ and every $t \in \mathbb{T}$ with $R_t\lambda \ll \theta$,*

$$R_t\lambda(A) = \lambda(A) \text{ and } \lambda(A \setminus (A - t)) = 0, \text{ forevery } A \in \mathcal{D},$$

where $\mathcal{D}$ denotes the σ-algebra of Γ-invariant Borel sets.

PROOF. As usual we identify the space of finite complex measures on $\mathbb{T}$ which are absolutely continuous with respect to θ with the Banach space $L^1(\theta)$. We then define — for each $t \in \mathbb{T}$ and $\lambda \ll \theta$ — $L_t\lambda := \lambda_{t,a}$, where $R_t\lambda = \lambda_{t,a} + \lambda_{t,s}$ is the decomposition of the translated measure $R_t\lambda$ into an absolutely continuous and a singular part with respect to θ. Clearly $L_t : L^1(\theta) \to L^1(\theta)$ is a linear operator of norm ≤ 1. Using the separability of $L^1(\theta)$ it is an easy exercise to show that there is a countable subset $\Gamma_0 \subset \mathbb{T}$ such that, with respect to the strong operator topology on the space of bounded linear operators on $L^1(\theta)$, $\{L_t : t \in \Gamma_0\}$ is dense in $\{L_t : t \in \mathbb{T}\}$. Let Γ be the countable subgroup of $\mathbb{T}$ generated by Γ_0. Now suppose $\lambda \ll \theta$ and $t \in \mathbb{T}$ with $R_t\lambda \ll \theta$. Choose $\gamma_n \in \Gamma$ with $L_t = \lim L_{\gamma_n}$ and let $A \in \mathcal{D}$, then

$$R_t\lambda(A) = L_t\lambda(A) = \lim L_{\gamma_n}\lambda(A) \leq \lim R_{\gamma_n}\lambda(A) = \lambda(A).$$

Similarly $R_t\lambda(\mathbb{T}\setminus A) \le \lambda(\mathbb{T}\setminus A)$, hence $R_t\lambda(A) \ge \lambda(A)$ and $R_t\lambda(A) = \lambda(A)$. Now $\mathbf{1}_A \cdot \lambda \ll \lambda$ implies $R_t(\mathbf{1}_A \cdot \lambda) \ll R_t\lambda \ll \theta$, hence also $R_t(\mathbf{1}_A \cdot \lambda)(A) = (\mathbf{1}_A \cdot \lambda)(A)$; i.e. $\lambda(A \cap (A-t)) = \lambda(A)$, whence $\lambda(A \setminus (A-t)) = 0$. ☐

Next observe that the map $\pi_{1,2} : \mathbb{T}^3 \to \mathbb{T}^2$, when restricted to the subgroup H, is in fact an isomorphism. We let

$$\alpha = \pi_{1,2} \restriction H : H \to \mathbb{T}^2, \quad \alpha(s,t,-s-t) = (s,t),$$

and $\eta := \alpha_*\rho$. The chain of maps

$$V : (s,t) \xrightarrow{\alpha^{-1}} (s,t,-s-t) \xrightarrow{(2,3)} (s,-s-t,t) \xrightarrow{\alpha} (s,-s-t),$$

sends η to itself:

$$V_* : \eta \xrightarrow{\alpha_*^{-1}} \rho \xrightarrow{(2,3)_*} \rho \xrightarrow{\alpha_*} \eta.$$

We also have $(\pi_1)_*\eta = \theta$.

11.8. LEMMA. *Let*

$$\eta = \int_{\mathbb{T}} \delta_s \times \eta_s \, d\theta(s),$$

be the disintegration of η over θ, then for θ a.e. s,

1. $0 \neq \eta_s \ll \theta$.
2. $(R_s)_*\eta_s = J_*\eta_s$,
3. $\eta_s(A \setminus (A-s)) = 0$ *for every* $A \in \mathcal{D}$.

PROOF. 1. If $\theta(E) > 0$ and $\eta_s = 0$ for every $s \in E$, then $\theta(E) = \eta(E \times \mathbb{T}) = \int \eta_s(\mathbb{T})\, d\theta(s) = 0$, a contradiction. By assumption (Theorem 11.2(4)), $\eta \ll \theta \times \theta$ and if we let

$$F(s,t) = \frac{d\eta}{d(\theta\times\theta)},$$

then $d\eta_s = F(s,\cdot)d\theta$, so that θ a.e., $\eta_s \ll \theta$.

2. Now for any continuous function ϕ on $\mathbb{T}^2$:

$$\begin{aligned}\int \phi(s,t)\, d\eta &= \int \phi(s,t)\, dV_*\eta = \int \phi(s,-s-t)\, d\eta \\ &= \int\int \phi(s,-s-t)\, d\eta_s(t) d\theta(s) = \int\int \phi(s, J(s+t))\, d\eta_s(t) d\theta(s) \\ &= \int\int \phi(s,t)\, dJ_*(R_s)_*\eta_s(t) d\theta(s).\end{aligned}$$

Hence, by the uniqueness of disintegration, for θ a.e. s, $J_*(R_s)_*\eta_s = \eta_s$, or $(R_s)_*\eta_s = J_*\eta_s$.

3. Since $J_*\theta = \theta$, we get $(R_s)_*\eta_s = J_*\eta_s \ll J_*\theta = \theta$, and by Lemma 11.7, $\eta_s(A \setminus (A-s)) = 0$ for every $A \in \mathcal{D}$. ☐

11.9. LEMMA. *For $A, B \in \mathcal{D}$ with $A \cap B = \emptyset$ we have $\eta(A \times B) = 0$.*

PROOF. Let $Q : \mathbb{T}^2 \to \mathbb{T}$ be the sum map $Q(s,t) = s+t$, then for $A, B \in \mathcal{D}$, we have

$$A \times B \setminus Q^{-1}B = \bigcup_{s\in A} \{(s,t) : t \in B \setminus (B-s)\},$$

hence, by Lemma 11.8.3,

$$\eta\left(A \times B \setminus Q^{-1}B\right) = \int_A \eta_s(B \setminus (B-s))\, d\theta(s) = 0.$$

Symmetrically we also get $\eta\left(A \times B \setminus Q^{-1}A\right) = 0$. However $Q^{-1}A \cap Q^{-1}B = \emptyset$, hence $\eta(A \times B) = 0$. □

11.10. LEMMA. *The restriction of θ to the σ-algebra $\mathcal{D}$ is purely atomic.*

PROOF. Suppose to the contrary that $A \in \mathcal{D}$, $\theta(A) > 0$ and $\theta \restriction \mathcal{D} \cap A$ has no atoms, then for every $n \geq 1$ there is a finite partition of A, $\{A_1, A_2, \ldots, A_{k_n}\}$ with $A_j \in \mathcal{D}$, $A = \bigcup_{j=1}^{k_n} A_j$ and $\theta(A_j) < 1/n$ for all j. Let $K_n = \bigcup_{j=1}^{k_n} A_j \times A_j$ and $K = \cap_{n\geq 1} K_n$, then

$$\theta \times \theta(K_n) = \sum_{j=1}^{k_n} \theta(A_j)^2 < (1/n) \sum_{j=1}^{k_n} \theta(A_j) \leq 1/n,$$

hence $\theta \times \theta(K) = 0$. Since $\eta \ll \theta \times \theta$ we have also $\eta(K) = 0$. On the other hand by Lemma 11.9,

$$\eta(K_n) = \sum_{j=1}^{k_n} \eta(A_j \times A_j) = \eta(A \times A),$$

and it follows that $\eta(A \times A) = \eta(K) = 0$. We get

$$0 < \theta(A) = \eta(A \times \mathbb{T}) = \eta(A \times A) + \eta(A \times (\mathbb{T} \setminus A)) = \eta(A \times A) = 0,$$

and this contradiction finishes the proof. □

11.11. LEMMA. *Let E be an atom of $\theta \restriction \mathcal{D}$ and define the probability measure*

$$\nu = \frac{1}{\theta(E)} \mathbf{1}_E \theta,$$

then ν is supported on the set

$$A(\nu) = \{s \in \mathbb{T} : R_s\nu \not\perp \nu\}.$$

PROOF. From $\eta(E \times (\mathbb{T} \setminus E)) = 0$, we deduce that for ν a.e. s, η_s is supported on E and this implies $\eta_s \ll \nu$. Also, by Lemma 11.8, $\eta_s(E \setminus (E-s)) = 0$. Thus η_s is supported on $E - s$, hence $R_s\eta_s$ is supported on E. It now follows that also $R_s\eta_s \ll \nu$, hence $\eta_s \ll R_{-s}\nu$. Finally, since θ a.e. $\eta_s \neq 0$, we conclude that for ν a.e. S, $\nu \not\perp R_s\nu$. □

11.12. LEMMA. *Let E be an atom of $\theta \restriction \mathcal{D}$, then the measure*

$$\nu = \frac{1}{\theta(E)} \mathbf{1}_E \theta,$$

is either discrete or absolutely continuous with respect to Lebesgue measure.

PROOF. We define a probability measure λ as follows:

$$\lambda = \sum_{\gamma \in \Gamma} a_\gamma \gamma \nu,$$

where for each γ, $a_\gamma > 0$ and $\sum_{\gamma \in \Gamma} a_\gamma = 1$. Clearly λ is a Γ-quasi-invariant measure. If $A \in \mathcal{D}$ then

$$\lambda(A) = \sum_{\gamma \in \Gamma} a_\gamma \gamma \nu(A) = \sum_{\gamma \in \Gamma} a_\gamma \nu(A) = \nu(A),$$

so the fact that E is an atom of $\theta \upharpoonright \mathcal{D}$ implies that $\lambda(A)$ is either 0 or 1. Thus λ is also Γ-ergodic.

From these properties of λ we deduce that $\lambda \not\perp R_s\lambda$ implies $\lambda \sim R_s\lambda$; i.e. $A(\lambda) = H(\lambda)$. In fact if $\lambda(A) > 0$ and $R_s\lambda(A) = 0$, then $\lambda(\Gamma A) = 1$ and $R_s\lambda(\Gamma A) = 0$, a contradiction to $\lambda \not\perp R_s\lambda$.

We now find that all the assumptions of Theorem 11.4 are satisfied by λ and conclude that λ, hence also ν, is either discrete or absolutely continuous with respect to Lebesgue measure as required. □

Since θ is purely atomic on $\mathcal{D}$ and has there at most a countable number of atoms, we have now completed the proof of Theorem 11.2. □

4. Mixing systems of singular type are mixing of all orders

11.13. THEOREM. *A mixing system of singular type is mixing of all orders.*

PROOF. Suppose that $\mathbf{X}$ is a mixing system with singular spectral type which is not mixing of order $k \geq 2$; then there exist sets $A_0, A_1, A_2, \ldots, A_k \in \mathcal{X}$ and a sequence $n_i = (n_{i,1}, n_{i,2}, \ldots, n_{i,k}) \in \mathbb{N}^k$, with $n_{i,0} = 0$ and

$$\text{for all} \quad 0 \leq j \leq k-1 \quad \lim_{i \to \infty} (n_{i,j+1} - n_{i,j}) = \infty$$

such that

$$\lim_{i \to \infty} \mu \left(\bigcap_{j=0}^{k} T^{n_{i,j}} A_j \right) = a,$$

where a is a number different from $\prod_{j=0}^{k} \mu(A_j)$.

Denoting the $(k+1)$-fold self-joining of $\mathbf{X}$

$$\mathrm{gr}\,(\mu, I, T^{-n_{i,1}}, \ldots, T^{-n_{i,k}}) = (I, T^{-n_{i,1}}, \ldots, T^{-n_{i,k}})_*(\mu) \in J(\mu, \mu, \ldots, \mu)$$

by λ_i, we can take a weak* convergent subsequence $\lambda_{i_p} \to \lambda \in J(\mu, \mu, \ldots, \mu)$, with

$$\lambda(A_0 \times A_1 \times \cdots \times A_k) = a$$

(see Theorem 6.2). Since $\mathbf{X}$ is mixing (i.e. 1-mixing) it follows that the joining λ is pairwise independent, and Theorem 11.1 implies that $\lambda = \mu^{k+1}$, contradicting our assumption that

$$a \neq \prod_{j=0}^{k} \mu(A_j).$$

□

5. Notes

More or less I followed B. Host's original paper [**123**] (as well as a preliminary version of that paper).

Mandrekar-Nadkarni's theorem: This proof of the Mandrekar-Nadkarni theorem was devised by Aaronson and Glasner. For the Mackey-Weil theorem see [**258**] page 146, [**112**] Chapter 12 and [**174**].

CHAPTER 12

Simple Systems and Their Self-Joinings

Given three ergodic systems and an ergodic pairwise independent joining λ of the three, it is a basic problem in ergodic theory to find conditions under which λ is independent. We treated the case of singular spectrum in the previous chapter, here we deal with another special case; that of simple systems.

Let $k \geq 1$ be an integer, a k-fold self-joining λ of an ergodic system $\mathbf{X} = (X, \mathcal{X}, \mu, \Gamma)$ (i.e. λ is a measure on a product of k copies of X denoted $X_1, X_2, \dots, X_k$ invariant under the product transformation and projecting onto μ in each coordinate) is called an *off-diagonal* if

$$\lambda = \mathrm{gr}\,(\mu, \phi_1, \dots, \phi_k);$$

i.e. λ is the image of μ under the map $x \mapsto \big(\phi_1(x), \phi_2(x), \dots, \phi_k(x)\big)$ of X into $\prod_{i=1}^{k} X_i$, where each ϕ_i is an element of the group $\mathrm{Aut}\,(\mathbf{X})$ of automorphisms of $\mathbf{X} = (X, \mathcal{X}, \mu, \Gamma)$. The joining λ is a *product of off-diagonals* (POOD) if there exists a partition $(J_1, \dots, J_m)$ of $\{1, \dots, k\}$ such that (i) For each l, the projection of λ on $\prod_{i \in J_l} X_i$ is an off-diagonal, (ii) The systems $(\prod_{i \in J_l} X_i, 1 \leq l \leq m)$ are independent.

An ergodic system $\mathbf{X}$ is *simple of order* k (or k*-simple*) if every k-fold ergodic self-joining of $\mathbf{X}$ is a product of off-diagonals. We say that it is *simple* when it is simple of all orders (see [**138**]). When $\mathbf{X}$ is a k-simple $\mathbb{Z}$-system with $\mathrm{Aut}\,(\mathbf{X}) = \{T^n : n \in \mathbb{Z}\}$ we say that it has *minimal self-joinings of order* k.

In this chapter we show that, for $\mathbb{Z}$-systems, weak mixing and simple of order 3 imply simple of all orders (Theorem 12.16). When $\mathbf{X}$ is rigid and simple of order 2 then it is simple of all orders (Theorem 12.18).

The main lemma (Lemma 12.11) used in the proofs of Theorems 12.16 and 12.18 deals with the following situation. We are given a 3-fold ergodic, pairwise independent joining λ of a simple system $\mathbf{X} = (X, \mathcal{X}, \mu, T)$, an ergodic system $\mathbf{Y} = (Y, \mathcal{Y}, \nu, T)$ and a weakly mixing system $\mathbf{Z} = (Z, \mathcal{Z}, \eta, T)$. The lemma shows that if for some $n \neq 0$ the joinings λ and $\lambda_n = (\mathrm{id} \times \mathrm{id} \times T^n)\lambda$ are not orthogonal over $Y \times Z$ (see Definition 6.9), then λ is the independent joining; i.e., $\lambda = \mu \times \nu \times \eta$. It is then shown (Theorem 12.13) that under these circumstances, when λ is not independent, $\mathbf{X}$ admits a nontrivial factor with absolutely continuous (with respect to Lebesgue measure on $\mathbb{T}$) spectral type.

In Section 1 we note that the non weakly mixing simple systems are precisely the group systems. In Sections 2 to 5 the factors, the joinings and the quasifactors of a simple system are studied. The pairwise independent joinings are treated in Section 6, and Theorem 12.16 is proved in Section 7. The last section relates the problem whether 2-simple implies simple of higher orders to Rohlin's question whether mixing implies mixing of all orders.

1. Group systems

12.1. THEOREM. *An ergodic system* $\mathbf{X}$ *is a group system iff it is simple and not weakly mixing.*

PROOF. Assuming $\mathbf{X} = (K, m_K)$ is a group system, the facts that it is not weakly mixing and that it is 2-simple follow from Theorem 10.15 and Proposition 6.15, with $\mathbf{Y}$ being the trivial one point system. Since any product system $\mathbf{X}^r = (K^r, m^r)$ is a group extension of the trivial system, we deduce r-simplicity, for every $r \in \mathbb{N}$, in the same fashion. For the converse claim, use the fact that for a non-weakly mixing system $\mathbf{X}$ the product measure $\lambda = \mu \times \mu$ is not ergodic, and then Theorem 6.18, again with $\mathbf{Y}$ trivial, implies that $\mathbf{X}$ is a group system. $\square$

2. Factors of simple systems

12.2. PROPOSITION. *For a 2-simple system* $\mathbf{X}$ *we have* $\mathrm{End}\,(\mathbf{X}) = \mathrm{Aut}\,(\mathbf{X})$.

PROOF. Let ϕ be an element of $\mathrm{End}\,(\mathbf{X})$ and consider the corresponding self-joining $\mathrm{gr}\,(\mu, \phi)$. Clearly $\mathrm{gr}\,(\mu, \phi) \neq \mu \times \mu$ and therefore $\mathrm{gr}\,(\mu, \phi) = \mathrm{gr}\,(\mu, \psi)$ for some $\phi \in \mathrm{Aut}\,(\mathbf{X})$. We then have for any $A \in \mathcal{X}$, $\mathrm{gr}\,(\mu, \psi)(\phi^{-1}A \times A) = \mu(\phi^{-1}A \cap \psi^{-1}A) = \mathrm{gr}\,(\mu, \phi)(\phi^{-1}A \times A) = \mu(\phi^{-1}A \cap \phi^{-1}A) = \mu(A)$, and we get $\phi^{-1}A = \psi^{-1}A$ for every $A \in \mathcal{X}$. It follows that $\phi^{-1} = \psi^{-1}$ and therefore $\phi = \psi \in \mathrm{Aut}\,(\mathbf{X})$. $\square$

12.3. THEOREM (Veech). *Let* $\pi : \mathbf{X} \to \mathbf{Y}$ *be a factor of the 2-simple system* $\mathbf{X}$ *with* $\mathbf{Y}$ *nontrivial. Then there is a compact subgroup* $K \subset \mathrm{Aut}\,(\mathbf{X})$ *such that* $\mathbf{Y} \cong \mathbf{X}/K$.

PROOF. Let

$$\mu = \int_Y \mu_y \, d\nu(y) \qquad \text{and} \qquad \lambda = \mu \underset{\nu}{\times} \mu = \int_Y \mu_y \times \mu_y \, d\nu(y)$$

be the corresponding disintegration and relatively independent joining. Let

$$\lambda = \int_\Omega \omega \, dm(\omega), \tag{2.1}$$

be the ergodic decomposition of λ. Since m a.e. ω is an ergodic self-joining it follows by 2-simplicity that for some $0 \le a \le 1$

$$\lambda = a(\mu \times \mu) + \int_{\Omega_0} \mathrm{gr}\,(\mu, \phi_\omega) \, dm(\omega), \tag{2.2}$$

where Ω is the union of Ω_0 and an atom of mass a. However, for $A \in \mathcal{Y}$ we have on one hand

$$\lambda(A \times A^c) = \int \mathbb{E}^Y(\mathbf{1}_A)\mathbb{E}^Y(\mathbf{1}_{A^c}) \, d\nu(y) = \int \mathbf{1}_A \cdot \mathbf{1}_{A^c} \, d\nu(y) = 0$$

and on the other

$$\lambda(A \times A^c) = a\mu(A)\mu(A^c) + \int_{\Omega_0} \mu(A \cap \phi_\omega^{-1} A^c) \, dm(\omega).$$

As $\mathbf{Y}$ is non-trivial, there exists an $A \in \mathcal{Y}$ with $0 < \mu(A) < 1$ and we must have $a = 0$.

Thus

$$\lambda = \int_\Omega \mathrm{gr}\,(\mu, \phi_\omega)\, dm(\omega),$$

and we can apply Veech's theorem (Theorem 6.18) to conclude the proof. □

12.4. THEOREM. *Let* $\mathbf{X}$ *be a 2-simple system and* $\pi : \mathbf{X} \to \mathbf{Y}$ *a non-trivial factor map, then the system* $\mathbf{Y}$ *is 2-simple iff the compact subgroup* $K \subset \mathrm{Aut}\,(\mathbf{X})$ *for which, according to Theorem 12.3,* $\mathbf{Y} \cong \mathbf{X}/K$*, is a normal subgroup of* $\mathrm{Aut}\,(\mathbf{X})$.

PROOF. Suppose K is a normal subgroup and let $\sigma \in J_e(\mathbf{Y})$ be an ergodic self-joining such that $\sigma \neq \nu \times \nu$. Choose a joining $\lambda \in J_e(\mathbf{X})$ which projects onto σ. (To see that such λ exists, take any $\rho_0 \in M(X \times X)$ with $(\pi \times \pi)(\rho_0) = \sigma$, set $\rho = \int_{K\times K}(k \times k')\rho_0$ and observe that ρ, the "Haar-lift" of σ, is Γ-invariant; finally take λ to be a "typical" ergodic component of ρ.) Clearly $\lambda \neq \mu \times \mu$ and as $\mathbf{X}$ is 2-simple there exists an automorphism $\psi \in \mathrm{Aut}\,(\mathbf{X})$ such that $\lambda = \mathrm{gr}\,(\mu, \psi)$. It is now easy to check that, since K is a normal subgroup of $\mathrm{Aut}\,(\mathbf{X})$, the automorphism $\psi \in \mathrm{Aut}\,(\mathbf{X})$ defines an automorphism $\hat{\psi} \in \mathrm{Aut}\,(\mathbf{Y})$ with $\sigma = \mathrm{gr}\,(\nu, \hat{\psi})$, and it follows that $\mathbf{Y}$ is 2-simple. (The same argument will show that $\mathbf{Y}$ is simple when $\mathbf{X}$ is simple.)

Conversely, assume now that $\mathbf{Y}$ is 2-simple and let $\psi \in \mathrm{Aut}\,(\mathbf{X})$. Set $\lambda = \mathrm{gr}\,(\mu, \psi)$ and $\sigma = (\pi \times \pi)(\lambda) \in J_e(\mathbf{Y})$. We claim that $\sigma \neq \nu \times \nu$. If $\mathbf{Y}$ is not weakly mixing, this is clear since σ is ergodic. Suppose $\mathbf{Y}$ is weakly mixing, and assume that $\sigma = \nu \times \nu$. Consider the commutative diagram

$$\begin{array}{ccc}
(X, \mu) \cong (X \times X, \lambda) & & \\
\quad\downarrow \pi & \searrow^{\pi\times\pi} & \\
(Y, \nu) & \xleftarrow[\pi_1]{} & (Y \times Y, \sigma = \nu \times \nu)
\end{array}$$

Here π is a distal (in fact a group) extension, while π_1 is clearly a weakly mixing extension. This contradicts Proposition 10.14 and we conclude that in any case $\sigma \neq \nu \times \nu$.

Now use the assumption that $\mathbf{Y}$ is 2-simple to deduce that $\sigma = (\pi \times \pi)(\lambda) = \mathrm{gr}\,(\nu, \hat{\psi})$ for some $\hat{\psi} \in \mathrm{Aut}\,(\mathbf{Y})$. Considering the σ-algebra

$$\pi^{-1}(\mathcal{Y}) = \mathcal{A}(K) = \{A \in \mathcal{X} : \phi(A) = A,\ \forall \phi \in K\}$$

this means that $\psi(A) = \hat{\psi}(A) \in \mathcal{A}(K)$ for every $A \in \mathcal{A}(K)$ and therefore $\psi^{-1}\phi\psi(A) = A$. Thus $\psi^{-1}\phi\psi \in H(\mathcal{A}) = K$, where $H(\mathcal{A}) = \{\chi \in \mathrm{Aut}\,(\mathbf{X}) : \chi(A) = A,\ \forall A \in \mathcal{A}\}$ (Proposition 6.16), and we conclude that K is normal in $\mathrm{Aut}\,(\mathbf{X})$. □

We have the following corollary for systems with minimal self-joinings (originally due to Rudolph [**217**]).

12.5. COROLLARY. *Let* $\mathbf{X}$ *be a non-atomic* $\mathbb{Z}$ *dynamical system with minimal self-joinings of order* 2*, then*

1. $\mathrm{Aut}\,(\mathbf{X}) = \{T^n : n \in \mathbb{Z}\}$,
2. $\mathbf{X}$ *is a prime system,*
3. $\mathbf{X}$ *is weakly mixing.*

PROOF. 1. Since for each $\phi \in \mathrm{Aut}(\mathbf{X})$, $\mathrm{gr}(\mu, \phi)$ is an ergodic self-joining, it follows that $\mathrm{gr}(\mu, \phi) = \mathrm{gr}(\mu, T^n)$ for some $n \in \mathbb{Z}$, whence $\phi = T^n$.

2. The only compact subgroup of $\mathrm{Aut}(\mathbf{X}) \cong \mathbb{Z}$ is the trivial group $\{\mathrm{id}\}$ and it follows from Theorem 12.3 that $\mathbf{X}$ admits no non-trivial factors.

3. Apply Theorem 3.15, part 2, and the assumption that $\mathbf{X}$ is non-atomic. □

3. Joinings of simple systems I

12.6. THEOREM. *Let λ and λ' be two ergodic joinings of $\mathbf{X} = (X, \mathcal{X}, \mu, \Gamma)$ and $\mathbf{Y} = (Y, \mathcal{Y}, \nu, \Gamma)$, where $\mathbf{X}$ is 2-simple and $\mathbf{Y}$ ergodic, then either λ and λ' are orthogonal over Y or there exists $\phi \in \mathrm{Aut}(\mathbf{X})$ such that $\lambda' = (\phi \times \mathrm{id})\lambda$.*

PROOF. Let $\hat{\lambda} = \lambda \underset{Y}{\times} \lambda'^*$ be considered as a measure on $X \times Y \times X'$, where X' is a copy of X. Let $\hat{\lambda} = \int_\Omega \omega \, dP(\omega)$ be the ergodic decomposition of $\hat{\lambda}$. The elements of Ω are ergodic joinings of X, Y and X'. Let $\Omega_0 \subset \Omega$ be the subset of those $\omega \in \Omega$ for which the projection of ω on $X \times X'$ is not the product measure $\mu \times \mu'$. Clearly $P(\Omega_0) = 0$ implies $\lambda'^* \circ \lambda = \mu \times \mu$; i.e. λ and λ' are orthogonal over Y.

Assume now $P(\Omega_0) > 0$ and let $\pi_{1,2} : X \times Y \times X' \to X \times Y$, $\pi_{3,2} : X \times Y \times X' \to X' \times Y$ and $\pi_{1,3} : X \times Y \times X' \to X \times X'$ be the natural projections; then for P-a.e. $\omega \in \Omega_0$ we have

(i) $\pi_{1,3}(\omega)$, the projection of ω on $X \times X'$, is an ergodic joining $\neq \mu \times \mu'$, hence of the form $\Delta_\mu^\phi = \mathrm{id} \times \phi(\Delta_\mu)$, where $\phi = \phi_\omega \in \mathrm{Aut}(\mathbf{X})$ is considered as an isomorphism of X onto X',

(ii) the projections of ω on $X \times Y$ and $Y \times X'$ are λ and λ'^* respectively.

It follows that

$$\pi_{3,2}(\omega) = (\phi_\omega \times \mathrm{id}_Y)\pi_{1,2}(\omega)$$

and since for P almost every $\omega \in \Omega_0$ we have $\pi_{1,2}(\omega) = \lambda$, $\pi_{3,2}(\omega) = \lambda'$, we conclude that

$$P\text{-a.e.} \qquad \lambda' = (\phi_\omega \times \mathrm{id})\lambda, \tag{3.1}$$

□

12.7. THEOREM. *Let λ be a non-independent ergodic joining of the 2-simple system $\mathbf{X}$ and the ergodic system $\mathbf{Y}$. Then there exist a positive integer r and a compact subgroup K of $C(X)$ such that for the corresponding group factor $\pi : (X, \mu, \Gamma) \to (U, \rho, \Gamma) \cong (X/K, \mu, \Gamma)$ we have*
(i) λ is the relatively independent product of (X, μ, Γ) and $(U \times Y, \tau, \Gamma)$ over (U, ρ, Γ) where τ is the image of λ under $\pi \times \mathrm{id}$,
(ii) τ is a joining of finite type r and for every $f, g \in L_0^2(U, \rho)$

$$\int P_\tau f(y) P_\tau g(y) d\nu(y) = \frac{1}{r} \int f(u) g(u) \, d\rho(u) \, .$$

PROOF. We use the notation introduced in the proof of Theorem 12.6, and let $\lambda = \lambda'$. Since λ is not orthogonal to itself (Proposition 6.11) we conclude that $P(\Omega_0) > 0$. Let K be the set of those $\phi \in C(X)$ for which $(\phi \times \mathrm{id})\sigma = \lambda$. Clearly K is a closed subgroup of $C(X)$.

Via the correspondence $\omega \mapsto \phi_\omega$, where $\pi_{1,3}\omega = (\phi_\omega \times \mathrm{id})\Delta_\mu = \Delta_\mu^{\phi_\omega}$, $P\,|_{\Omega_0}$ induces a measure ξ on $C(X)$. By (3.1) we conclude that P-a.e. $\lambda = (\phi_\omega \times \mathrm{id})\lambda$, whence that ξ is supported on K. Moreover, for $\psi \in K$ we clearly have $(\mathrm{id} \times \mathrm{id} \times \psi)\hat{\lambda} = \hat{\lambda}$ and it follows from the uniqueness of the ergodic decomposition that P is $\mathrm{id} \times \mathrm{id} \times \psi$-invariant. As

$$\pi_{1,3}(\mathrm{id} \times \mathrm{id} \times \psi)\omega = (\psi \circ \phi_\omega)\Delta_\mu = \Delta_\mu^{\psi\circ\phi_\omega}$$

we conclude that ξ is a K-invariant measure.

The theorem of A. Weil (Theorem 11.3) implies that K is compact and that $\xi = c \cdot m$ with m the normalized Haar measure on K and $c = P(\Omega_0)$. Denote by $\pi : \mathbf{X} \to \mathbf{U}$ the quotient map and quotient system obtained modulo K, and let τ be the image of λ under $\pi \times \mathrm{id}$. If $\tau = \int \delta_u \times \tau_u \, d\rho(u)$ is the disintegration of τ over (U, ρ) then for $f \in L^2(X, \mu)$, $g \in L^2(Y, \nu)$ and $\phi \in K$

$$\int f \otimes g \, d\lambda = \int f(x)g(y)\, d\lambda(x,y) = \int (f \circ \phi)(x) g(y)\, d\lambda(x,y) \;;$$

hence

$$\begin{aligned}
\int f \otimes g \, d\lambda &= \iint (f \circ \phi)(x) g(y)\, d\lambda(x,y) dm(\phi) \\
&= \int P_{\mathbf{U}} f(u) g(y)\, d\tau(u,y) \\
&= \int \left(\int f(\phi x) dm(\phi) \right) \left(\int g(y) d\tau_u(y) \right) d\rho(u) \\
&= \int f \otimes g \, d\mu \underset{U}{\times} \tau,
\end{aligned}$$

where $P_{\mathbf{U}} : L^2(X,\mu) \to L^2(U,\rho)$ is the conditional expectation operator. Thus $\lambda = \mu \underset{U}{\times} \tau$ and (i) is proved.

Since, with obvious notations, for P-a.e. $\omega \in \Omega \backslash \Omega_0$, $\mathrm{proj}_{U \times U'} \omega = \rho \times \rho'$ and $\mathrm{proj}_{U \times U'} \omega = \Delta_\rho$ for $\omega \in \Omega_0$, we get:

$$\begin{aligned}
\tau'^* \circ \tau = \mathrm{proj}_{U \times U'} \hat{\lambda} &= \int \mathrm{proj}_{U \times U'} \omega \, dP(\omega) \\
&= P(\Omega_0)\Delta_\rho + (1 - P(\Omega_0))\rho \times \rho'.
\end{aligned}$$

Since $P(\Omega_0) > 0$, Lemma 6.20 yields the fact that τ is of finite type r, where $P(\Omega_0) = \frac{1}{r}$. Moreover for $f, g \in L_0^2(U, \rho)$

$$\begin{aligned}
\int P_\tau f(y) \cdot P_\tau g(y) &= \int f(u) g(u')\, d\tau'^* \circ \tau(u,u') \\
&= \frac{1}{r} \int f(u) g(u)\, d\rho(u).
\end{aligned}$$

□

4. JQFs of simple systems

Let $\mathbf{X}$ be a simple system, $\mathbf{W} = (M(X), \kappa)$ a non-trivial JQF (i.e. $\kappa \neq \delta_\mu$) and $\kappa' = \int \theta \times \delta_\theta \, d\kappa(\theta)$ the associated *ergodic* joining in $J_e(\mathbf{X}, \mathbf{W})$. In particular then $\mathbf{X}$ is 2-simple and Theorem 12.7 applies. We let

$$K = \{\phi \in \mathrm{Aut}\,(\mathbf{X}) : (\phi \times \mathrm{id})\kappa' = \kappa'\}$$

be the compact subgroup of $C(\mathbf{X})$ that is associated with the, non-independent, ergodic joining κ' as in Theorem 12.7. Let $\pi : \mathbf{X} \to \mathbf{U} = (X/K, \rho) = \mathbf{X}/K$ be the corresponding group factor, $\tau = \pi(\kappa) \in Q_e(\mathbf{U})$ and $\tau' = (\pi \times \mathrm{id})(\kappa') = \int_{M(U)} \zeta \times \delta_\zeta \, d\tau(\zeta)$.

We know that τ' is a joining of finite type r and that κ' is the relatively independent product of (X, μ, Γ) and $(U \times M(X), \tau', \Gamma)$ over (U, ρ, Γ). Also let $\mu = \int \mu_u \, d\rho(u)$ be the disintegration of μ over ρ.

12.8. THEOREM. *For a simple system $\mathbf{X}$ and a joining quasifactor κ as above we have*

1. *Let r' be the order of orthogonality of κ (see Proposition 8.9), then $r' = r$ and the quasifactor κ is the r-symmetric product quasifactor associated with the factor map $\pi : \mathbf{X} \to \mathbf{U}$. Thus $\mathbf{W} \cong \mathbf{U}^{\langle r \rangle}$ and the measure κ is supported on the set $\{\frac{1}{r}(\delta_{\mu_{u_1}} + \cdots + \delta_{\mu_{u_r}}) : u_1, \ldots, u_r$ are distinct elements of $U\}$ (see Definition 8.25).*
2. *The quasifactor τ is the r-symmetric product quasifactor on $\mathbf{U}$; hence also $(M(U), \tau) \cong \mathbf{U}^{\langle r \rangle}$.*
3. *κ' is the relatively independent product of (X, μ) and $(U \times M(X), \tau')$ over their common factor (U, ρ).*
4. *τ' is a joining of finite type r.*
5. *When $\mathbf{X}$ is not weakly mixing (i.e. when $\mathbf{X}$ is a group system) $r = 1$.*

PROOF. 1. By Proposition 8.9 there exists a κ-admissible joining $\lambda \in J_e(\mathbf{X}_1, \mathbf{X}_2, \ldots, \mathbf{X}_{r'}, \mathbf{W})$, where each $\mathbf{X}_j$, $j = 1, 2, \ldots, r'$, is a copy of $\mathbf{X}$ and $\mathbf{W} = (M(X), \kappa)$, such that

$$\mathrm{proj}_{\mathbf{X}_1 \times \mathbf{X}_2 \times \cdots \times \mathbf{X}_{r'}}(\lambda) = \mu^{r'}. \tag{4.1}$$

Moreover r' is the maximal positive integer for which such λ exists. Now the joining $\lambda \underset{\kappa}{\times} \lambda^*$ is clearly a κ-admissible joining of $\mathbf{W}$ with $2r'$ copies of $\mathbf{X}$. Let

$$\lambda \underset{\kappa}{\times} \lambda^* = \int_\Omega \omega \, dP(\omega).$$

be its ergodic decomposition. As in the proof of Theorem 12.6 we consider the projection

$$\lambda^* \underset{\kappa}{\circ} \lambda = \mathrm{proj}_{X^{r'} \times X^{r'}}(\lambda \underset{\kappa}{\times} \lambda^*) = \int_\Omega \mathrm{proj}_{X^{r'} \times X^{r'}}(\omega) \, dP(\omega).$$

Since $\mathbf{X}$ is simple the ergodic $2r'$-selfjoinings $\mathrm{proj}_{X^{r'} \times X^{r'}}(\omega)$ are POODs and the maximality of r' implies that almost every ergodic component ω identifies each $\mathbf{X}$ in the second $X^{r'}$ factor with some $\mathbf{X}$ in the first $X^{r'}$ factor, via a map ϕ in $\mathrm{Aut}\,(\mathbf{X})$. Since on every $X_i \times M(X) \times X_j$ the joining $\lambda \underset{\kappa}{\times} \lambda^*$ projects onto an isomorphic copy of $\kappa' \underset{\kappa}{\times} \kappa'^*$ we see that $\phi \in K$. In other words, we have

$$\lambda \underset{\kappa}{\times} \lambda^* = \int_{\mathrm{Aut}\,(X^{r'} \times M(X),\lambda)} \mathrm{gr}\,(\lambda, \psi)\, dm'(\psi),$$

where m' is a probability measure supported on the closed subgroup H' of $\mathrm{Aut}\,(X^{r'} \times M(X), \lambda)$, consisting of maps ψ of the form $\psi = ((\phi_1 \times \phi_2 \times \cdots \times \phi_{r'}) \circ \alpha) \times \mathrm{id}$ with $\phi_j \in K$, $j = 1, 2, \ldots, r'$ and α a coordinate permutation on $X^{r'}$.

Now Veech's theorem, Theorem 6.18, applies and we conclude that the map $(X^{r'} \times M(X), \lambda) \to (M(X), \kappa)$ is a group factor with compact group

$$H = H(\mathcal{W}) = \{\psi \in \mathrm{Aut}\,(X^{r'} \times M(X), \lambda) : \phi(A) = A,\ \forall A \in \mathcal{W}\},$$

where $\mathcal{W}$ is the σ-algebra of the system $\mathbf{W}$ considered as a factor of $(X^{r'} \times M(X), \lambda)$, and that $P = m$ is Haar measure on H. Since $H' \subset H$ and $m(H') = 1$, it follows that $H' = H$.

Set

$$\hat{K} = \{(\phi_1 \times \phi_2 \times \cdots \times \phi_{r'}) \circ \alpha : \text{ with } \phi_j \in K,$$
$$j = 1, 2, \ldots, r'; \text{ and } \alpha \text{ a coordinate permutation on } X^{r'}\}.$$

$\hat{K}$ is a closed subgroup of $\mathrm{Aut}\,(\mathbf{X}^{r'})$ (recall that λ restricts to $\mu^{r'}$ on the subalgebra corresponding to $\mathbf{X}^{r'}$). Clearly the groups $\hat{K}$ and $H' \subset \mathrm{Aut}\,(X^{r'} \times M(X), \lambda)$ are isomorphic via the map $\psi \in \hat{K} \mapsto \psi \times \mathrm{id}$.

Now Theorem 6.18 implies that with respect to λ we have $\mathcal{W} = \mathcal{A}(H)$. This yields a Boolean isomorphism of measure σ-algebras

$$F : (\mathcal{B}(X^{r'}/\hat{K}), \mu^{\langle r' \rangle}) \to (\mathcal{W}, \kappa),$$

as follows: each $A \in \mathcal{B}(X^{r'}/\hat{K})$, the Borel σ-algebra of the quotient system $\mathbf{X}^{r'}/\hat{K}$, regarded as a $\hat{K}$-invariant set in $(X^{r'} \times M(X), \lambda)$, agrees λ a.e. with a unique (mod κ) set $A' = F(A) \in \mathcal{W}$. By Theorem 2.15 this Boolean isomorphism defines a system isomorphism $\mathbf{W} \to (X^{r'}, \mu^{r'})/\hat{K}$. From the form of elements in $\hat{K}$ and by equation (4.1) it is clear that $(X^{r'}, \mu^{r'})/\hat{K}$ is canonically isomorphic to $\mathbf{U}^{\langle r' \rangle}$ and we proved the isomorphism $\mathbf{W} \cong \mathbf{U}^{\langle r' \rangle}$. In particular $r = r'$. This completes the proof of part 1, and part 2 clearly follows.

Parts 3 and 4 now follow from Theorem 12.7 parts (i) and (ii) respectively.

5. By Theorem 12.1 $\mathbf{X}$ is an ergodic group system and the claim follows since for $\mathbf{X}$ an ergodic group system, the order of orthogonality of every JQF κ is clearly 1. □

5. Joinings of simple systems II

12.9. THEOREM. *Let* $\mathbf{X}$ *be a simple system,* $\mathbf{Y}$ *an ergodic system and* $\lambda = \int \lambda_y \times \delta_y \, d\nu(y)$ *an ergodic joining in* $J_e(\mathbf{X}, \mathbf{Y})$*. There exists then a compact subgroup* $K \subset \mathrm{Aut}\,(\mathbf{X})$ *and a positive integer* $r \geq 1$ *with the following properties. Let* $\pi : \mathbf{X} \to \mathbf{U} = (X/K, \rho) = \mathbf{X}/K$ *be the corresponding group factor. Let* $\xi_y = \pi(\lambda_y)$*, so that* $\xi = (\pi \times \mathrm{id})(\lambda) = \int \xi_y \times \delta_y \, d\nu(y)$*. Finally let* $\mathbf{U}^{\langle r \rangle}$ *be the symmetric product system (see Examples 2.21) and let* $\zeta \in J_e(\mathbf{U}, \mathbf{U}^{\langle r \rangle})$ *be the corresponding natural* *joining; that is, the image in* $U \times U^{\langle r \rangle}$*, under the map* $\mathrm{id} \times \mathrm{sym}$*, of the measure* $\int \delta_{u_1} \times \delta_{(u_1, u_2, \ldots, u_r)}\, d\rho(u_1)\, d\rho(u_2) \ldots d\rho(u_r)$*.*

1. *There exists a factor map* $\alpha : \mathbf{Y} \to \mathbf{U}^{\langle r \rangle}$ *and the joining* ξ *has the form*
$$\xi = \zeta \underset{U^{\langle r \rangle}}{\times} \nu.$$
2. λ *is the relatively independent product of* (X, μ) *and* $(U \times Y, \xi)$ *over* (U, ρ)*, so that*
$$\lambda = \mu \underset{U}{\times} \xi = \mu \underset{U}{\times} (\zeta \underset{U^{\langle r \rangle}}{\times} \nu).$$
3. *Another description of* λ *is as the projection of the relatively independent joining* $\mu^r \underset{U^{\langle r \rangle}}{\times} \nu$ *onto the first component* $X \times Y$*,*

(5.1)
$$\begin{array}{ccccc}
 & & (X^r \underset{U^{\langle r \rangle}}{\times} Y, \mu^r \underset{U^{\langle r \rangle}}{\times} \nu) & & \\
 & \swarrow \quad \swarrow & & \searrow & \\
(X \times Y, \lambda) & (X^r, \mu^r) & & & (Y, \nu) \\
 & & \searrow & \swarrow & \\
 & & (U^{\langle r \rangle}, \rho^{\langle r \rangle}) & &
\end{array}$$

4. *For* $\mathbf{X}$ *a group system and any ergodic* $\mathbf{Y}$ *and* $\lambda \in J_e(\mathbf{X}, \mathbf{Y})$*, we have* $r = 1$*.*

PROOF. Set $\kappa = \phi_*(\nu) \in Q_J(\mathbf{X})$, where $\phi : y \mapsto \lambda_y$ and $\lambda = \int \lambda_y \times \delta_y\, d\nu(y)$ is the disintegration of λ over ν. We then have have $\kappa' = \lambda$. By Theorem 12.8 $(M(X), \kappa) = \mathbf{W} \cong \mathbf{U}^{\langle r \rangle} = (U^{\langle r \rangle}, \rho^{\langle r \rangle})$ and therefore, denoting $\tau = \pi_{**}(\kappa) \in Q_J(\mathbf{U})$, we have $\zeta = \tau'$. Theorem 8.18 tells us that $\xi = \tau' \underset{U^{\langle r \rangle}}{\times} \nu$ and Theorem 12.8.3 implies
$$\lambda = \mu \underset{U}{\times} \xi = \mu \underset{U}{\times} (\zeta \underset{U^{\langle r \rangle}}{\times} \nu).$$
This proves parts 1 and 2 of the theorem. Part 3 now follows by inspection of the diagram (5.1) and part 4 follows from Theorem 12.8.5. □

12.10. THEOREM. *If* $\mathbf{X}$ *and* $\mathbf{Y}$ *are simple systems, then any ergodic joining* $\lambda \in J_e(\mu, \nu)$ *has the form* $\lambda = \mu \underset{U}{\times} \nu$*, where* $\mathbf{U}$ *is a common group factor of both* $\mathbf{X}$ *and* $\mathbf{Y}$*:*
$$\begin{array}{ccccc}
 & & (X \underset{U}{\times} Y, \mu \underset{U}{\times} \nu) & & \\
 & \swarrow & & \searrow & \\
(X, \mu) & & & & (Y, \nu) \\
 & \searrow & & \swarrow & \\
 & & (U, \rho) & &
\end{array}$$
i.e. $r = 1$ *in Theorem 12.9. In particular, two simple systems are non-disjoint iff they have a non-trivial common (group) factor.*

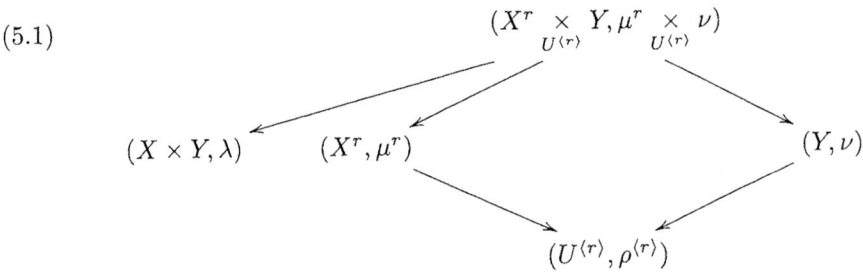

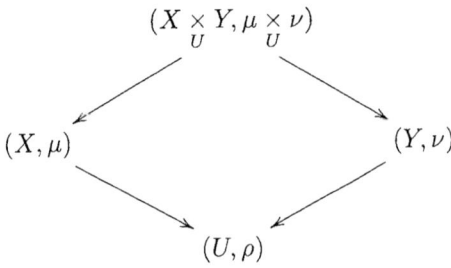

PROOF. When $\mathbf{X}$ is not weakly mixing, it is a group system (Theorem 12.1) and our assertion follows from Theorem 12.9. So we can assume that $\mathbf{X}$ is weakly mixing and then, by Theorems 6.27 and 12.1, if $\lambda \neq \mu \times \nu$, so is $\mathbf{Y}$. In view of Theorem 12.9, it suffices to show that a factor map $\sigma : \mathbf{Y} \to \mathbf{U}^{\langle r \rangle}$ with $\mathbf{U} = (X/K, \rho) = \mathbf{X}/K$ non-trivial, is possible only with $r = 1$.

Consider the self-joining $\xi = \rho^n \underset{U_1}{\times} \rho^n$, the relatively independent product of the system (U^n, ρ^n) with itself over the first factor U_1. Note that ξ is ergodic, being the image of the POOD, $\mu^n \underset{X_1}{\times} \mu^n$ of the simple weakly mixing system $\mathbf{X}$.

The map, sym $\times$ sym : $U^r \underset{U_1}{\times} U^r \to U^{\langle r \rangle} \times U^{\langle r \rangle}$, induces a homomorphism $\beta : (U^r \underset{U_1}{\times} U^r, \xi) \to (U^{\langle r \rangle} \times U^{\langle r \rangle}, \zeta)$, with $\zeta := \beta(\xi) = (\mathrm{sym} \times \mathrm{sym})(\xi)$.

Recall that by assumption there exists a homomorphism $\sigma : \mathbf{Y} \to \mathbf{U}^{\langle r \rangle}$ and we let $\iota : Y \times Y \to U^{\langle r \rangle} \times U^{\langle r \rangle}$ be the product map $\iota = \sigma \times \sigma$. There exists then an ergodic selfjoining $\theta \in J_e(\mathbf{Y})$, with $\iota(\theta) = \zeta$ (see diagram below). Since $\mathbf{Y}$ is 2-simple, this selfjoining θ has the form $\theta = \mathrm{gr}\,(\nu, \phi)$, for some $\phi \in \mathrm{Aut}\,(\mathbf{Y})$. In particular the system $(Y \times Y, \mathrm{gr}\,(\nu, \phi))$ is a simple system (isomorphic to $\mathbf{Y}$) and Veech's theorem, Theorem 12.3, implies that the extension $\iota \circ \delta$

$$(Y \times Y, \mathrm{gr}\,(\nu, \phi)) \xrightarrow{\iota} (U^{\langle 3 \rangle} \times U^{\langle 3 \rangle}, \zeta) \xrightarrow{\delta} (U^{\langle 3 \rangle}, \rho^{\langle 3 \rangle})$$

is a group extension. We conclude that also

$$\delta : (U^{\langle 3 \rangle} \times U^{\langle 3 \rangle}, \zeta) \to (U^{\langle 3 \rangle}, \rho^{\langle 3 \rangle}) \tag{5.2}$$

is an isometric extension (Lemma 9.12).

Next consider the following commutative diagram

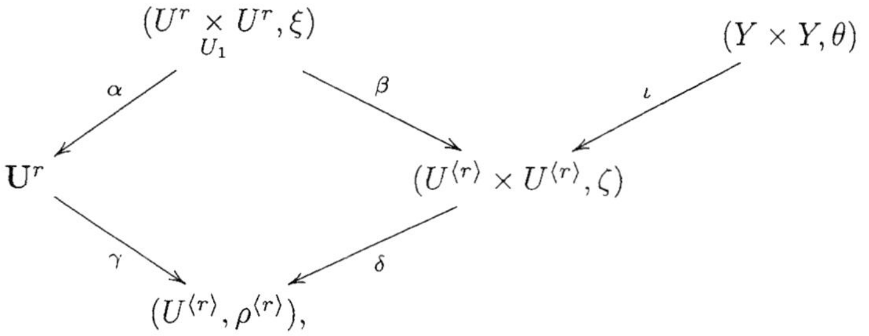

where α and δ are the projections on the first component of the relevant product space and the maps β and γ are the maps sym $\times$ sym , and sym , respectively. Clearly β is an isometric extension and, as we have seen above, so is δ. Thus the extension $\delta \circ \beta = \gamma \circ \alpha$ is a distal extension. By Theorem 10.18, α is also a distal extension. However, as the system $\mathbf{U}^r$ is weakly mixing, the homomorphism α is clearly a weakly mixing extension (provided that $r > 1$). This implies that α is an isomorphism and we conclude that indeed $r = 1$. $\square$

6. Pairwise independent joinings of simple $\mathbb{Z}$-systems

The situation considered in this section is that of an ergodic joining λ of three $\mathbb{Z}$-systems $\mathbf{X} = (X, \mathcal{X}, \mu, T)$, $\mathbf{Y} = (Y, \mathcal{Y}, \nu, T)$ and $\mathbf{Z} = (Z, \mathcal{Z}, \eta, T)$ of which the first is (at least) 2-simple, the second ergodic and the third weakly mixing.

12.11. LEMMA. *Let λ be an ergodic pairwise independent joining of X, Y and Z. If for some $n \neq 0$, λ and $\lambda_n = (\mathrm{id} \times \mathrm{id} \times T^n)\lambda$ are not orthogonal relative to $\mathbf{Y} \times \mathbf{Z}$ (i.e. $\lambda_n^* \circ \lambda \neq \mu \times \mu$), then $\mathbf{X} \times \mathbf{Y}$ and $\mathbf{Z}$ are also independent.*

PROOF. Suppose that for $n \neq 0$, λ and λ_n are not orthogonal over $(Y \times Z, \nu \times \eta, T \times T)$, then according to Theorem 12.6 there exists $\phi \in \mathrm{Aut}\,(\mathbf{X})$ such that $\lambda_n = (\phi \times \mathrm{id} \times \mathrm{id})\lambda$ and it follows that $F\lambda = \lambda$ for $F = \phi \times \mathrm{id} \times T^{-n}$.

Let f be a bounded measurable function on Y; for every bounded measurable function h on $X \times Z$:

$$\int P_\lambda f(x,z)h(x,z)\,d\mu(x)\,d\eta(z) = \int f(y)h(x,z)\,d\lambda(x,y,z)$$

$$\int f(y)h(x,z)dF\lambda(x,y,z) = \int f(y)h(\phi x, T^{-n}z)\,d\lambda(x,y,z)$$

$$= \int P_\lambda f(x,z)h(\phi x, T^{-n}z)\,d\mu(x)\,d\eta(z)$$

$$= \int P_\lambda f(\phi^{-1}x, T^n z)h(x,z)\,d\mu(x)d\eta(z)$$

and $P_\lambda f$ is a $\phi^{-1} \times T^n$-invariant function.

Since $\mathbf{Z}$ is weakly mixing $P_\lambda f$ depends only on x. This means that $P_\lambda f$ is the conditional expectation of f on X, which by independence of $\mathbf{X}$ and $\mathbf{Y}$, is a constant. We thus have shown that $\mathbf{X} \times \mathbf{Z}$ and $\mathbf{Y}$ are independent and the proof is complete. □

Note that in this lemma the assumption that $\mathbf{Y}$ and $\mathbf{Z}$ are λ-independent was used tacitly when it was assumed that λ_n is a joining of $\mathbf{X}$ and $\mathbf{Y} \times \mathbf{Z}$. For this assumption means that the $\mathbf{Y} \times \mathbf{Z}$ projection of λ is $\mathrm{id} \times T^n$-invariant. This is the case iff $\mathbf{Y}$ and $\mathbf{Z}$ are independent.

12.12. COROLLARY. *Let λ be an ergodic pairwise independent joining of $\mathbf{X}, \mathbf{Y}$ and $\mathbf{Z}$ where $\mathbf{X}$ is 2-simple weakly mixing, $\mathbf{Y}$ ergodic and $\mathbf{Z}$ weakly mixing. If in addition we assume that $\mathbf{X}$ is rigid, then λ is the independent joining $\lambda = \mu \times \nu \times \eta$.*

PROOF. Recall that a system $\mathbf{X}$ is rigid if there exists a sequence $n_k \nearrow \infty$ such that $\lim \mu\,(T^{n_k} A \cap A) = \mu(A)$ for every measurable subset A of X. If for some k we have $\lambda_{n_k}^* \circ \lambda \neq \mu \times \mu$, then Lemma 12.11 completes the proof. Otherwise, with $\lambda = \int \lambda_{(y,z)} \times \delta_{(y,z)}\,d\nu \times \eta(y,z)$ the disintegration of λ over $Y \times Z$, we get

$$\mu \times \mu = \lambda_{n_k}^* \circ \lambda = \int T^{n_k}\lambda_{(y,z)} \times \lambda_{(y,z)}\,d\nu(y)\,d\eta(z).$$

Letting $k \to \infty$ we also get, using Exercise 8.14,

$$\mu \times \mu = \lambda^* \circ \lambda = \int \lambda_{(y,z)} \times \lambda_{(y,z)}\,d\nu(y)\,d\eta(z).$$

By Proposition 6.11 this implies $\lambda = \mu \times \nu \times \eta$ as claimed. □

In view of Theorem 5.11 the following theorem is a much stronger result.

12.13. THEOREM. *Let $\mathbf{X}$ be 2-simple, $\mathbf{Y}$ ergodic and $\mathbf{Z}$ weakly mixing. If there exists an ergodic pairwise independent joining λ of the three systems which is not* *independent then* $\mathbf{X}$ *admits a nontrivial factor whose spectral type is absolutely continuous with respect to Lebesgue measure* m *on* $\mathbb{T}$.

PROOF. We consider λ as a joining of $\mathbf{X}$ and $\mathbf{Y} \times \mathbf{Z}$. Let $\mathbf{U} = (U, \rho, T)$ be the factor of $\mathbf{X}$ whose existence is proved in Theorem 12.7 and let τ be the projection of λ on $U \times Y \times Z$. Since $\mathbf{U}$ is non-trivial and since $\mathbf{X}$ and $\mathbf{Y} \times \mathbf{Z}$ are relatively independent over $\mathbf{U}$, it follows that τ is not independent. On the other hand τ is clearly pairwise independent. We are going to show that the spectral type of (U, ρ, T) is absolutely continuous. For $n \in \mathbb{Z}$ let $\tau_n = (\mathrm{id} \times \mathrm{id} \times T^n)\tau$; then Lemma 12.11 implies that for $n \neq 0$, $\tau_n^* \circ \tau = \rho \times \rho$. Let $f \in L_0^2(U, \rho)$, we need to show that the correlation measure α_f corresponding to f (i.e. the measure on $\mathbb{T}$ whose Fourier coefficients are given by $\widehat{\alpha}_f(k) = \int f(T^k u)\overline{f}(u)\, d\rho(u) \quad (k \in \mathbb{Z})$), is absolutely continuous with respect to m.

Let $F = P_\tau f \in L^2(Y \times Z, \nu \times \eta)$ and let ω be its correlation measure for the $\mathbb{Z}^2$ action on $Y \times Z$; i.e. ω is the positive measure on $\mathbb{T}^2$ whose Fourier coefficients are given by

$$\widehat{\omega}(p,q) = \int F\left(T^p y, T^q z\right) \overline{F}(y,z)\, d\nu(y)\, d\eta(z) \quad (p, q \in \mathbb{Z}).$$

Let $\tau = \int \tau_{(y,z)}\, d\nu(y)\, d\eta(z)$ be the disintegration of τ over $Y \times Z$, then for $(p,q) \in \mathbb{Z}^2$ we have

$$\begin{aligned}
F\left(T^p y, T^q z\right) &= P_\tau f\left(T^p y, T^q z\right) = \int f(x) d\tau_{(T^p y, T^q z)}(x) \\
&= \int f(x)\, d\tau_{(T\times T)^p (y, T^{q-p} z)}(x) = \int f(x) d(T \times T)^p \tau_{(y, T^{q-p}z)}(x) \\
&= \int f(T^p x)\, d\tau_{(y, T^{q-p}z)}(x) = P_{\tau_{q-p}}\left(T^p f\right)(y,z).
\end{aligned}$$

Let θ be the image of ω under the map $(s,t) \mapsto s+t$ of $\mathbb{T}^2$ onto $\mathbb{T}$, then for $p \in \mathbb{Z}$ we have $\widehat{\theta}(p) = \widehat{\omega}(p,p)$ and by Theorem 12.7

$$\begin{aligned}
\widehat{\theta}(p) = \widehat{\omega}(p,p) &= \int P_\tau\left(T^p f\right)(y,z) P_\tau \overline{f}(y,z)\, d\nu(y)\, d\eta(z) \\
&= \frac{1}{r} \int f\left(T^p u\right) \overline{f}(u)\, d\rho(u) = \frac{1}{r} \widehat{\alpha}_f(p),
\end{aligned}$$

so that $\alpha_f = r\theta$. It therefore suffices to show that $\theta \ll m$.

Consider now $\widehat{\omega}(p,q)$ for $p \neq q$. In this case, $\tau_{q-p}^* \circ \tau = \rho \times \rho$ and we get

$$\begin{aligned}
\widehat{\omega}(p,q) &= \int P_{\tau_{q-p}}\left(T^p f\right)(y,z) P_\tau \overline{f}(y,z)\, d\nu(y)\, d\eta(z) \\
&= \int f\left(T^p u\right) \overline{f}(v)\, d\tau_{q-p}^* \circ \tau(u,v) \\
&= \int f\left(T^p u\right)\, d\rho(u) \cdot \int \overline{f}(v)\, d\rho(v) = 0.
\end{aligned}$$

It follows that ω is invariant under the maps $(s,t) \mapsto (s+u, t-u)$ of $\mathbb{T}^2$ into itself, for every $u \in \mathbb{T}$, and therefore its two natural projections onto $\mathbb{T}$ are invariant under all translations of $\mathbb{T}$. This means that these projections are constant multiples of m. Since ω is absolutely continuous with respect to a product of two measures on $\mathbb{T}$ (e.g. the product of the spectral types of (Y, ν, T) and (Z, η, T)), we deduce that

ω is absolutely continuous with respect to the product of its two natural projections (an exercise). Combining these results we get $\omega \ll m \times m$, and therefore, finally $\theta \ll m$. $\square$

12.14. THEOREM. *If* $\mathbf{X}$ *is* 2*-simple weakly mixing and does not admit a non-trivial factor with absolutely continuous spectral type, then* $\mathbf{X}$ *is simple of all orders.*

PROOF. By induction using Theorem 12.13 (see the proof of Theorem 12.16 below). $\square$

12.15. REMARK. Theorem 11.1 states that every pairwise independent joining of $r \geq 3$ weakly mixing systems with purely singular spectrum is independent. Of course this also implies that a weakly mixing, 2-simple system with purely singular spectral type is simple of all orders.

7. Simplicity of higher orders

12.16. THEOREM. *A weakly mixing system which is simple of order* 3 *is simple of all orders.*

12.17. LEMMA. *Let* $\mathbf{X}$ *be w.m. and simple of order* 3, $\mathbf{X}' = (X', \mu', T)$ *a copy of* $\mathbf{X}$ *and* $\mathbf{Y}$ *an ergodic system, then a pairwise independent ergodic joining of these three systems is necessarily independent.*

PROOF. Let λ be a pairwise independent ergodic joining of $\mathbf{X}', \mathbf{X}$ and $\mathbf{Y}$ such that $\lambda \neq \mu' \times \mu \times \nu$. Let $n \neq 0$ and let $\lambda_n = (\mathrm{id} \times T^n \times \mathrm{id})\, \lambda$. We consider the joining $\omega = \lambda \underset{X \times Y}{\times} \lambda_n^*$; ω is a measure on $X' \times X \times Y \times X''$ where X'' is another copy of X. The projection of ω on $X' \times X''$ is $\lambda_n^* \circ \lambda$ which by Lemma 12.11 is equal to $\mu' \times \mu''$. It follows that the projection of ω onto $X' \times X \times X''$ is pairwise independent and therefore, by assumption, is independent. Thus for bounded functions f, g, h on X, X' and X'' respectively (and using the fact that ω is the relatively independent joining of λ and λ_n over $(X \times Y, \mu \times \nu)$), we have

$$\int f(x)\, d\mu(x) \int g(x')\, d\mu'(x') \int h(x'')\, d\mu''(x'')$$
$$= \int f(x) g(x') h(x'')\, d\omega(x', x, y, x'')$$
$$= \int P_\lambda g(x, y) P_\lambda h\,(T^n x, y)\, f(x)\, d\mu(x)\, d\nu(y).$$

In particular for $n \neq 0$ and every bounded function h in $L^2_0\,(X'', \mu'')$ and bounded function u on X we get, by taking $f(x) = u\,(T^n x)\, \overline{u}(x)$ and $g = \overline{h}$,

$$0 = \int P_\lambda h\,(T^n x, y)\;\; u\,(T^n x)\, P_\lambda \overline{h}(x, y) \overline{u}(x)\, d\mu(x)\, d\nu(y)\;.$$

Summing from 1 to N

$$0 = \int \frac{1}{N} \sum P_\lambda h\,(T^n x, y)\, u\,(T^n x)\, P_\lambda \overline{h}(x, y) \overline{u}(x)\, d\mu(x)\, d\nu(y)\;,$$

and by the ergodic theorem

$$0 = \int \left| \int P_\lambda h(x,y)u(x)\, d\mu(x) \right|^2 d\nu(y) \ .$$

Thus $\int P_\lambda h(x,y)u(x)\, d\mu(x) = 0$ ν-a.e. and for every bounded v on Y we have

$$\int P_\lambda h(x,y)u(x)v(x)\, d\mu(x)\, d\nu(y) = \int h(x')u(x)v(y)\, d\lambda(x',x,y) = 0.$$

This means that λ is independent contradicting our assumption. □

PROOF OF THEOREM 12.16. Suppose $\mathbf{X}$ is simple of order $k \geq 3$; we shall show that it is simple of order $k+1$. Let λ be an ergodic joining of $k+1$ copies of $\mathbf{X}$ denoted $\mathbf{X}_1, \dots, \mathbf{X}_{k+1}$. If for some $i \neq j$ the projection of λ on $X_i \times X_j$ is an off-diagonal, then by an induction hypotheses, λ is a product of off-diagonals and we are done. Thus we may assume that λ is pairwise independent. By the induction hypothesis the projections of λ on $X_1 \times \dots \times X_k$ and $X_2 \times \dots \times X_{k+1}$ are the independent joinings. Hence λ can be viewed as a pairwise independent joining of $\mathbf{X}_1$, $\mathbf{X}_2 \times \dots \times \mathbf{X}_k$ and $\mathbf{X}_{k+1}$. Lemma 12.17 applies and we conclude that λ is independent. This completes the proof. □

12.18. THEOREM. *An ergodic 2-simple rigid system is simple of all orders.*

PROOF. Same as the proof of Theorem 12.16; only, at the end, use Corollary 12.12 instead of Lemma 12.17. □

Next we follow Ryzhikov [**224**] and show that when a weakly mixing 2-simple transformation is embedable in a flow then it is simple of all orders. Let us call two joinings $\lambda, \tau \in J(\mathbf{X},\mathbf{Y})$ *equivalent* if there exists an automorphism $R \in \mathrm{Aut}\,(\mathbf{X})$ such that $\tau = (R \times \mathrm{id})\lambda$.

12.19. LEMMA. *If $\mathbf{X}$ is 2-simple and $\mathbf{Y}$ is ergodic, then there are at most countably many equivalence classes in $J_e(\mathbf{X},\mathbf{Y})$.*

PROOF. This is a direct corollary of Proposition 6.14 and Theorem 12.6. In fact by Theorem 12.6 if $\lambda, \lambda' \in J_e(\mathbf{X},\mathbf{Y})$ are not equivalent then they are orthogonal and by Proposition 6.14, $P_\lambda L_0^2(\mu)$ and $P_{\lambda'} L_0^2(\mu)$ are orthogonal subspaces of $L^2(\nu)$. Taking a representative in each equivalence class in $J_e(\mathbf{X},\mathbf{Y})$ and denoting the collection thus formed Λ, we see that the family $\{P_\lambda L_0^2(\mu) : \lambda \in \Lambda\}$ of subspaces of $L^2(\nu)$ is pairwise orthogonal, hence countable. □

12.20. THEOREM. *Let $\mathbf{X} = (X, \mathcal{X}, \mu, T)$ be a 2-simple weakly mixing system which is embedded in a flow; i.e. there exists a one-parameter group $\{S_t : t \in \mathbb{R}\} \subset$ $\mathrm{Aut}\,(\mathbf{X})$ such that $T = S_1$. Then $\mathbf{X}$ is 3-simple (hence simple of all orders).*

PROOF. Let $\lambda \in J_e(\mathbf{X},\mathbf{X},\mathbf{X})$ be a pairwise independent ergodic joining. We consider λ as an element of $J_e(\mathbf{X}, \mathbf{X}\times\mathbf{X})$ and observe that Lemma 12.19 implies that there exists some $\tau \in J_e(\mathbf{X},\mathbf{X},\mathbf{X})$ such that for uncountably many $t > 0$ there exists $R_t \in \mathrm{Aut}\,(\mathbf{X})$ with $(\mathrm{Id} \times (\mathrm{Id} \times S_t))\lambda = (R_t \times (\mathrm{Id} \times \mathrm{Id}))\tau$, or $(R_t^{-1} \times (\mathrm{Id} \times S_t))\lambda = \tau$.

It follows that there are $t, s > 0$, $t \neq s$ such that $(R_s R_t^{-1} \times (\mathrm{Id} \times S_s^{-1} S_t))\lambda = \lambda$. Denoting $R = R_s R_t^{-1}$ and $S = S_s^{-1} S_t$ we finally have

$$(R \times (\mathrm{Id} \times S))\lambda = \lambda.$$

Since T is weakly mixing it can be shown that for some t, s as above the transformation S is weakly mixing as well (use an argument similar to that indicated in Exercise 4.24). The weak mixing of S implies the existence of a sequence $n_i \nearrow \infty$ such that S^{n_i} converges weakly in $\mathbf{E}(\mathbf{X})$ to $P_{\mathfrak{C}}$. We can further assume that $R^{n_i} \to \hat{R}$ for some $\hat{R} \in \mathbf{E}(\mathbf{X})$. For any measurable sets $A, B, C \in \mathfrak{X}$ we now have

$$\begin{aligned}
\int \mathbf{1}_A \otimes \mathbf{1}_B \otimes \mathbf{1}_C \, d\lambda &= \lim_{i\to\infty} \int R^{n_i}(\mathbf{1}_A) \otimes \mathbf{1}_B \otimes S^{n_i}\mathbf{1}_C \, d\lambda \\
&= \int \hat{R}(\mathbf{1}_A) \otimes \mathbf{1}_B \otimes P_{\mathfrak{C}}(\mathbf{1}_C) \, d\lambda \\
&= \mu(C) \cdot \int \hat{R}(\mathbf{1}_A) \otimes \mathbf{1}_B \, d\mu \times \mu \\
&= \left(\int \hat{R}(\mathbf{1}_A) \, d\mu \right) \cdot \mu(B)\mu(C) = \mu(A)\mu(B)\mu(C).
\end{aligned}$$

Thus $\lambda = \mu \times \mu \times \mu$ and the proof is complete. □

8. About 2-simple but not 3-simple systems

Using Theorem 12.13 and a result of Ryzhikov (that we shall not prove here) we show in this section that an example of a system with 2 but not 3 minimal self-joinings (if such exists) is also an example of a mixing system which is not mixing of all orders.

12.21. THEOREM. *If the system* $\mathbf{X}$ *is* 2*-simple but not* 3*-simple, then* $\mathbf{X}$ *is mixing.*

PROOF. By Theorem 12.13 it follows that $\mathbf{X}$ admits a factor $\pi : \mathbf{X} \to \mathbf{Y}$ whose spectral type is absolutely continuous. By Exercise 5.10.5 the system $\mathbf{Y}$ is mixing. By Theorem 12.3 the map π is a group extension. Finally, since $\mathbf{X}$ is necessarily weakly mixing, Theorem 7.4 implies that $\mathbf{X}$ is mixing. □

A theorem of Ryzhikov [**224**] asserts that an ergodic system $\mathbf{X}$ which has minimal self-joinings of order 2 (hence is 2-simple) and is in addition 2-mixing has already minimal self-joinings of all orders (hence, is simple of all orders). Together with Theorem 12.21 these results lead to the following surprising corollary.

12.22. COROLLARY. *If there exists an ergodic system* $\mathbf{X}$ *which has minimal self-joinings of order* 2 *but not of order* 3*, then* $\mathbf{X}$ *is mixing (of order* 1*) but not mixing of order* 2*.*

9. Notes

Most of the material in this chapter is from del-Junco and Rudolph's important paper [**138**], and from Glasner, Host and Rudolph's paper [**91**]. See also Thouvenot's survey [**246**] and Ryzhikov's paper [**224**].

Factors of simple systems: Theorem 12.3 may suggest that every simple system has a (unique) structure as a group extension of a *prime* factor (i.e. a system with no proper factors). That this is not the case is shown, first, in Glasner and Weiss [**99**], where a simple system with non-unique prime factor is constructed, and then in del Junco [**132**], where a simple system with no non-trivial prime factors is constructed.

Joinings of simple systems II: Theorem 12.9 is due to del Junco and Rudolph [**138**]. In Glasner [**86**] there is a weak version of this theorem.

Simplicity of higher orders: A simple rigid system is constructed by del Junco and Rudolph in [**139**].

CHAPTER 13

Kazhdan's Property and the Geometry of $M_\Gamma(\mathbf{X})$

Unlike most other chapters in this book the present one requires some more than standard undergraduate knowledge in probability theory (more specifically in the theory of Gaussian processes) and in group theory (Kazhdan groups). Since, however, the results established here are not used in any of the subsequent chapters the readers can skip it if they are inclined to do so.

A convex closed metrizable subset K of a locally convex linear space E is a *simplex* if each point in K is the barycenter of a unique probability measure supported on the G_δ subset $\partial_e K$ of extreme points of K. For a topological system (X, Γ), the uniqueness of the ergodic decomposition for every element μ of $M_\Gamma(X)$, shows that $M_\Gamma(X)$is always a simplex.

A metrizable simplex K is called a *Bauer* simplex if $\partial_e K$ is closed in K. Of course any compact metric space X is homeomorphic to the set $\partial_e K$ of some Bauer simplex K; simply take $K = M(X)$. In a sense the opposite kind of simplices are those for which $\partial_e K$ is dense in K. Such a simplex is called a *Poulsen* simplex and the surprising fact that up to affine isomorphism only one Poulsen simplex exists, is proved in Lindenstrauss, Olsen and Sternfeld [**173**]. An old and well known result is that for the topological action $(\Omega, \mathbb{Z})$, where $\Omega = \{0, 1\}^{\mathbb{Z}}$ and $\mathbb{Z}$ acts by translations on Ω, the simplex $M_{\mathbb{Z}}(\Omega)$ is the Poulsen simplex.

We now briefly remind the reader of the definition of Kazhdan's property T, and some basic properties of Kazhdan's groups, [**148**]. Let π be a unitary representation of a locally compact second countable group Γ on a Hilbert space H. Given $\epsilon > 0$ and a compact subset $K \subset \Gamma$ a unit vector $v \in H$ is (ϵ, K)*-invariant* if

$$\sup\{\|\pi(\gamma)v - v\| : \gamma \in K\} < \epsilon.$$

We say that π *almost has invariant vectors* if for every pair (ϵ, K) there is an (ϵ, K)-invariant unit vector $v \in H$. For a countable Γ this is the same as the existence of a sequence v_n of unit vectors in H such that, for every $\gamma \in \Gamma$, $\lim_{n\to\infty} \|\pi(\gamma)v_n - v_n\| = 0$. We call such a sequence an *almost invariant sequence*.

The group Γ is a *Kazhdan group*, or has *property* T, if every unitary representation π of Γ which almost has invariant vectors actually has a non-zero invariant vector. An equivalent seemingly weaker condition is the following (see e.g. [**116**], page 16). Whenever π is a unitary representation of Γ on a Hilbert space H with an almost invariant sequence as above and with the additional property that the representations $\pi_n = \pi \restriction H_n$ are irreducible, then π has a non-zero invariant vector in H. Here H_n denotes the cyclic subspace of H generated by v_n (that is H_n is the closed subspace spanned by the set $\{\gamma v_n : \gamma \in \Gamma\}$).

The class of Kazhdan's groups includes, among other examples, all connected semisimple Lie groups with finite center each of whose factors has $\mathbb{R}$-rank at least 2, as well as any lattice subgroup of such a group. A group Γ which is both Kazhdan and amenable is necessarily compact. If Γ is discrete and Kazhdan then it is finitely generated. For information concerning Kazhdan's property T, refer to the books [**269**] and [**116**]. Other dynamical characterizations of groups having property T appear in Connes and Weiss [**42**] and Schmidt [**228**].

As usual, we shall deal here only with countable discrete groups Γ. For such a group, the natural action of Γ on the space of functions $\Omega = \{0,1\}^\Gamma$ — the topological Bernoulli system — will be denoted by (Ω, Γ). Section 1 presents a dynamical characterization of property T. Our goal in Section 2 is to show that a seemingly weaker condition on the unitary representations of Γ is actually equivalent to property T, a theorem due to Bekka and Valette [**16**]. In order to be self contained the section contains a mini-course on affine actions (following [**116**]); a reader who is ready to accept Theorem 13.11 without a proof can skip the rest of the section.

In Sections 3 and 4 we prove the following results.

(i) If Γ has property T then for every topological action (X, Γ), $M_\Gamma(X)$ is a Bauer simplex.

(ii) Γ has property T iff the simplex $M_\Gamma(\Omega)$ is a Bauer simplex.

(iii) Γ does not have property T iff the simplex $M_\Gamma(\Omega)$ is the Poulsen simplex.

In the last section (Section 5) we discuss the analogous results for the Haagerup property, or rather its negation, property T_w.

1. Strong ergodicity and property T

13.1. DEFINITION. Let $\mathbf{X} = (X, \mathcal{X}; \mu, \Gamma)$ be a dynamical system. A sequence of subset $B_n \in \mathcal{X}$ is called *asymptotically invariant* if for all $\gamma \in \Gamma$, $\lim_{n\to\infty} \mu(B_n \triangle \gamma B_n) = 0$, and such a sequence is *trivial* when $\lim_{n\to\infty} \mu(B_n)(1 - \mu(B_n)) = 0$. We say that the dynamical system $\mathbf{X}$ is *strongly ergodic* if any asymptotically invariant sequence in $\mathbf{X}$ is trivial.

13.2. LEMMA. *If* $\mathbf{X}$ *is not strongly ergodic then there exists an asymptotically invariant sequence* $B_n \in \mathcal{X}$ *with* $\mu(B_n) = \frac{1}{2}$.

PROOF. Let $C_n \in \mathcal{X}$ be an asymptotically invariant sequence, then if $f \in L^\infty(\mu)$ is a weak* limit point of the set of functions $\{\mathbf{1}_{C_n} : n \in \mathbb{N}\}$ in the unit ball of $L^\infty(\mu)$, say $f = \lim_{i\to\infty} \mathbf{1}_{C_{n_i}}$, we have

$$\int h(\gamma f - f)\, d\mu = \lim_{i\to\infty} \int h(\mathbf{1}_{\gamma^{-1}C_{n_i}} - \mathbf{1}_{C_{n_i}})\, d\mu = 0, \qquad (h \in L^\infty(\mu), \gamma \in \Gamma)$$

and the ergodicity implies that $f = a$ is a constant. From this fact we deduce that for every $A \in \mathcal{X}$

$$\lim_{i\to\infty} (\mu(C_{n_i} \cap A) - \mu(C_{n_i})\mu(A)) = \lim_{n\to\infty} \int \mathbf{1}_{C_{n_i}} (\mathbf{1}_A - \mu(A))\, d\mu$$
$$= a \int (\mathbf{1}_A - \mu(A))\, d\mu = 0,$$

hence

$$\lim_{n\to\infty} (\mu(C_n \cap A) - \mu(C_n)\mu(A)) = 0, \qquad (A \in \mathcal{X}). \tag{1.1}$$

With no loss of generality we assume that the limit $c = \lim_{n\to\infty} \mu(C_n)$ exists and that $0 < c \le \frac{1}{2}$. If $c = \frac{1}{2}$ we are done; otherwise we can construct an asymptotically invariant sequence $D_n \in \mathcal{X}$ with $c < \lim_{n\to\infty} \mu(D_n) = 2(c - c^2) < \frac{1}{2}$, as follows. Use (1.1) to choose, for any $n \geq 1$, an integer $m_n > n$ such that $|\mu(C_{m_n} \cap C_n) - \mu(C_{m_n})\mu(C_n)| < \frac{1}{n}$ and set $D_n = C_{m_n} \triangle C_n$. A simple exhaustion argument completes the proof of the lemma. □

13.3. THEOREM. *The group Γ has the Kazhdan property iff every ergodic Γ-system is strongly ergodic.*

PROOF. 1. Suppose $\mathbf{X} = (X, \mathcal{X}, \mu, \Gamma)$ is an ergodic system and $B_n \in \mathcal{X}$ an asymptotically invariant sequence satisfying $\lim_{n\to\infty} \mu(B_n)(1 - \mu(B_n)) = \delta > 0$. Set

$$f_n = (\mu(B_n) - \mu(B_n)^2)^{-1} (\mu(B_n)\mathbf{1} - \mathbf{1}_{B_n}).$$

Then $f_n \in L_0^2(\mu)$, $\|f_n\| = 1$ and the sequence f_n satisfies $\lim_{n\to\infty} \|f_n - U_\gamma f_n\| = 0$ for every $\gamma \in \Gamma$. It follows that the Koopman representation on $L_0^2(\mu)$ has almost invariant vectors. Since by ergodicity no unit vector in $L_0^2(\mu)$ is invariant it follows that Γ is not a Kazhdan group.

2. Suppose now that Γ does not have property T; then there exists a unitary representation π of Γ on a Hilbert space H which admits an almost invariant sequence of vectors v_n with the property that π_n is irreducible (where $\pi_n = \pi \restriction H_n$ and H_n denotes the cyclic subspace of H generated by v_n), but π does not have a non-zero invariant vector. Using these almost invariant unit vectors we construct positive definite functions

$$\phi_n(\gamma) = \langle \pi(\gamma)v_n, v_n \rangle$$

such that $\phi_n(\gamma) \to 1$ for all $\gamma \in \Gamma$. Consider now a system of centered Gaussian random variables X_n^γ with covariance given by

$$\mathbb{E}(X_n^\gamma X_n^\delta) = \phi_n(\gamma\delta^{-1}), \quad \forall \gamma, \delta \in \Gamma$$

and let (Ω_n, μ_n) be a probability space on which the X_n^γ are defined and generate the σ-algebra. Let $\mathcal{H}_n$ be the first chaos of $L_0^2(\Omega_n, \mu_n)$. Then $\pi_n = \pi \restriction H_n$ and the Koopman representation of Γ restricted to $\mathcal{H}_n$ are unitarily equivalent. Set

$$(\Omega, \mu, \Gamma) = \prod_{n=1}^{\infty} (\Omega_n, \mu_n, \Gamma),$$

the product Γ-system. We next examine two possibilities.

Case 1: For infinitely many n the representation π_n is infinite dimensional. With no loss in generality we then assume that this is the case for all n. The unitary representation of Γ on $L_0^2(\Omega_n, \mu_n)$ is unitarily equivalent to the representation $\bigoplus_{k=1}^{\infty} \pi_n^{\odot k}$. It follows that the Koopman representation of Γ on $L_0^2(\Omega, \mu)$ is equivalent to the (countable) direct sum of representations of the form

$$\pi_{n_1}^{\odot k_1} \otimes \pi_{n_2}^{\odot k_2} \otimes \cdots \otimes \pi_{n_p}^{\odot k_p}. \tag{1.2}$$

Since each π_n is irreducible and infinite dimensional, Theorem 3.5 implies that each direct summand in (1.2) has no non-trivial finite dimensional subrepresentation and we conclude that also the Koopman representation of Γ on $L_0^2(\Omega, \mu)$ has no non-trivial finite dimensional subrepresentations. By Theorem 3.11 the system (Ω, μ, Γ) is ergodic (in fact weakly mixing).

We next show that in Ω there is a non trivial sequence of asymptotically invariant sets. For this define:

$$B_n = \{\omega \in \Omega : X_n^e(\omega) \geq 0\}, \qquad (e \text{ is the unit of } \Gamma).$$

The centering assumption implies that $\mu(B_n) = \frac{1}{2}$. Fix $\gamma_0 \in \Gamma$ and let α_n be defined by

$$\mathbb{E}(X_n^e X_n^{\gamma_0}) = \cos\alpha_n.$$

To calculate $\mu(B_n \triangle \gamma_0 B_n)$ recall that X_n^e and $X_n^{\gamma_0}$ can be represented by two independent standard Gaussian variables X and Y in the form

$$X_n^e = X, \quad X_n^{\gamma_0} = \cos\alpha_n X + \sin\alpha_n Y.$$

It follows then from the circular symmetry of the joint distribution of X and Y that

$$\mu(B_n \triangle \gamma_0 B_n) = \mu(\{X_n^e \geq 0 \text{ and } X_n^{\gamma_0} < 0\} \cup \{X_n^e < 0 \text{ and } X_n^{\gamma_0} \geq 0\}) = \frac{\alpha_n}{\pi}.$$

The almost invariance of the vectors v_n implies that $\alpha_n \to 0$ and it follows that $\{B_n\}$ is a nontrivial asymptotically invariant sequence.

Case 2: We now assume that all the representations π_n are finite dimensional. We claim that for each n there is an element $\gamma_n \in \Gamma$ with $\phi_n(\gamma_n) < 0$. To see this let $K = \mathrm{cls}\{\pi_n(\gamma) : \gamma \in \Gamma\} \subset \mathfrak{U}(H_n)$ be the corresponding compact group of unitary operators and let λ be Haar measure on K. We then see that $\int kv_n \, d\lambda(k)$ is a Γ-invariant vector in H_n, hence $\int_K kv_n \, d\lambda(k) = \int_K kX_n^e \, d\lambda(k) = 0$ and it follows that $\phi_n(\gamma) = \mathbb{E}(X_n^e X_n^\gamma) > 0$ can not hold for all $\gamma \in \Gamma$.

Again we form the product dynamical system $(\Omega, \mu, \Gamma) = \prod_{n=1}^\infty (\Omega_n, \mu_n, \Gamma)$, consider the sets $B_n = \{\omega \in \Omega : X_n^e(\omega) \geq 0\}$ and observe that $\mu(B_n \triangle \gamma_0 B_n) \to 0$, as above. However, we no longer have in this case the ergodicity of the system (Ω, μ, Γ). Let $\mu = \int \mu_t \, dm(t)$ be the ergodic decomposition of μ.

As in Case 1 we write

$$X_n^e = X, \quad X_n^{\gamma_n} = \cos\alpha_n X + \sin\alpha_n Y$$

with X, Y two independent standard Gaussian variables. Now the circular symmetry of the joint distribution of X and Y and the fact that $\phi_n(\gamma_n) = \mathbb{E}(X_n^e X_n^{\gamma_n}) = \cos\alpha_n \leq 0$ imply that

$$\begin{aligned}\mu(B_n \triangle \gamma_n B_n) &= \mu(\{X_n^e \geq 0 \text{ and } X_n^{\gamma_n} < 0\} \cup \{X_n^e < 0 \text{ and } X_n^{\gamma_n} \geq 0\}) \\ &= \frac{\alpha_n}{\pi} \geq \frac{1}{2}.\end{aligned}$$

It then follows that for a positive set of ergodic components μ_t we have $\limsup_{n\to\infty} \mu_t(B_n \triangle \gamma_n B_n) > 0$. On the other hand, for almost every ergodic component μ_t and for any fixed γ we have, $\lim_{n\to\infty} \mu_t(B_n \triangle \gamma B_n) = 0$. It follows that there exists some t for which the ergodic system (Ω, μ_t, Γ) is not strongly ergodic. $\square$

2. A theorem of Bekka and Valette

In representation theory the condition on a unitary representation π of a topological group G that it "has almost invariant vectors" is sometimes expressed as the condition that "the identity representation $\mathbf{1}_G$ is *weakly contained* in π", and one denotes this by $\mathbf{1}_G \prec \pi$. Similarly, the fact that the representation π is "not ergodic" is expressed by saying that π "*contains the trivial representation*" (i.e. there is a non-zero $\pi(G)$-invariant vector); this property is denoted by $\mathbf{1}_G \subset \pi$. In these terms the group G is a

- Kazhdan group iff, $\mathbf{1}_G \prec \pi \;\Rightarrow\; \mathbf{1}_G \subset \pi \quad (\pi \text{ is not ergodic})$,

for every continuous unitary representation π.

Let us say, provisionally, that a topological group G has the

- BV property if, $\mathbf{1}_G \prec \pi \otimes \overline{\pi} \;\Rightarrow\; \mathbf{1}_G \subset \pi \otimes \overline{\pi} \quad (\pi \text{ is not weakly mixing})$,

and the

- BV^- property if, $\mathbf{1}_G \prec \pi \;\Rightarrow\; \mathbf{1}_G \subset \pi \otimes \overline{\pi} \quad (\pi \text{ is not weakly mixing})$,

for every continuous unitary representation π.

In [**16**] Bekka and Valette prove the equivalence of these three properties for locally compact, second countable, topological groups. In the present section we reproduce their proof. *En route* we introduce some rudimentary cohomology theory. As usual we assume, for simplicity, that $G = \Gamma$ is a countable discrete group. Also we shall work throughout this section with real Hilbert spaces (and orthogonal representations); the complex Hilbert space results are then obtained by a standard complexification method (see [**116**]).

13.4. DEFINITION. Let $\mathfrak{H}$ be a Hilbert space and let $\pi : \Gamma \to \mathfrak{U}(\mathfrak{H})$ be a unitary representation. A function $b : \Gamma \to \mathfrak{H}$ satisfying

$$b(\gamma\delta) = \pi(\gamma)b(\delta) + b(\gamma), \qquad \forall \gamma, \delta \in \Gamma, \tag{2.1}$$

is called a *cocycle* (associated to π). A map $\alpha : \Gamma \to \mathrm{Iso}(\mathfrak{H})$ of Γ into the group of isometries of $\mathfrak{H}$ is an *affine action* (associated to π) if there exists a cocycle b such that for every $x \in \mathfrak{H}$ and $\gamma \in \Gamma$, $\alpha(\gamma)x = \pi(\gamma)x + b(\gamma)$. It is easy to check that the cocycle equation (2.1) ensures that α is indeed an action. Given $x \in \mathfrak{H}$, we set $b_x(\gamma) = \pi(\gamma)x - x$, clearly a cocycle. We call such a cocycle a *coboundary*.

13.5. LEMMA (Min-max point). *Let A be a bounded subset of a Hilbert space $\mathfrak{H}$. Define a function $\rho : \mathfrak{H} \to \mathbb{R}$ by $\rho(x) = \sup_{a \in A} \|x - a\|$. Then there exists a unique point $y \in \mathfrak{H}$ such that*

$$\rho(y) = \inf\{\rho(x) : x \in \mathfrak{H}\}.$$

PROOF. For every $t > 0$ set $Q(t) = \{y \in \mathfrak{H} : \rho(y) \leq t\}$. It is easy to check that $Q(t)$, when non-empty, is a convex, norm closed, and bounded subset of $\mathfrak{H}$, hence weakly compact. Thus the intersection $Q = \cap\{Q(t) : Q(t) \text{ is non-empty}\}$ is non-empty. Each $y \in Q$ satisfies $\rho(y) = \inf\{\rho(z) : z \in \mathfrak{H}\}$ and by uniform convexity we see that Q contains a single point $Q = \{y\}$. □

Let us call this point y the *min-max point* of A. (It is sometimes called the *center of mass* of A but this term is not so fortunate since already for a homogeneous triangle the true center of mass may differ from the point defined above.)

13.6. LEMMA. *Let $\alpha : \Gamma \to \mathrm{Iso}(\mathfrak{H}), \alpha = \pi + b$ be an affine action. The following conditions are equivalent.*

1. *The action α has a fixed point.*
2. *There exists a bounded α orbit in $\mathfrak{H}$.*
3. *Every α orbit is bounded.*
4. *The cocycle b is a coboundary; i.e. there exists $x \in \mathfrak{H}$ such that, for every $\gamma \in \Gamma$, $b(\gamma) = \pi(\gamma)x - x$.*
5. *The function $b : \Gamma \to \mathfrak{H}$ is bounded.*

PROOF. The implications $3 \Rightarrow 2$ and $1 \Rightarrow 2$ are obvious and $4 \Rightarrow 3$ is easy. If α has a fixed point, say $\alpha(\gamma)x = x, \forall \gamma \in \Gamma$, then $\alpha(\gamma)x = \pi(\gamma)x + b(\gamma) = x$, implies $b(\gamma) = \pi(\gamma)(-x) - (-x)$, so that b is a coboundary. We next show that $2 \Rightarrow 1$. Assume that for some $x \in \mathfrak{H}$ the set $A = \{\alpha(\gamma)x : \gamma \in \Gamma\}$ is bounded. Since A is an invariant set for the affine Γ action α it follows that y, the set A's min-max point (Lemma 13.5), is an $\alpha(\Gamma)$-fixed point. Finally, considering the α-orbit of 0, we obtain the implications $3 \Rightarrow 5$ and $5 \Rightarrow 2$. □

13.7. PROPOSITION. *If $\sigma : \Gamma \to \mathfrak{U}(\mathfrak{H})$ is a unitary representation such that $\mathbf{1}_\Gamma \prec \sigma$ but $\mathbf{1}_\Gamma \not\subset \sigma$, then for the associated infinite sum representation:*

$$\pi = \bigoplus_{n=1}^{\infty} \sigma : \Gamma \to \mathfrak{H}_\infty = \bigoplus_{n=1}^{\infty} \mathfrak{H},$$

there exists a π-cocycle $b : \Gamma \to \mathfrak{H}_\infty$ which is not a coboundary.

PROOF. Let $\Gamma = \cup_n \Gamma_n$, where $\Gamma_n \nearrow \Gamma$ is an increasing sequence of finite sets. By assumption there is a sequence $\{x_n\}_n \in \mathbb{N}$ of unit vectors in $\mathfrak{H}$ such that for each n, $\|\sigma(\gamma)x_n - x_n\| < 2^{-n}, \forall \gamma \in \Gamma_n$. Define a function $b : \Gamma \to \mathfrak{H}_\infty$ by $b(\gamma) = \bigoplus_n n(\sigma(\gamma)x_n - x_n)$. We have $\|b(\gamma)\|^2 \leq \sum_n n^2 2^{-2n} < \infty$, whence $b(\gamma) \in \mathfrak{H}_\infty$. Clearly $b(\gamma)$ is a cocycle and we claim that it is not bounded; the proof of this claim will complete the proof of our lemma in view of Lemma 13.6. Now if $\sup_{\gamma \in \Gamma} \|b(\gamma)\| \leq R$, then for every $n \in \mathbb{N}$ we have $\sup_{\gamma \in \Gamma} n\|\sigma(\gamma)x_n - x_n\| \leq R$. For sufficiently large n we shall have $R/n \leq 1/2$, hence $\sup_{\gamma \in \Gamma} \|\sigma(\gamma)x_n - x_n\| \leq 1/2$. Fix such n. Setting $A = \{\sigma(\gamma)x_n : \gamma \in \Gamma\}$ and $\rho(x) = \sup_{\gamma \in \Gamma} \|\sigma(\gamma)x_n - x\|$, then applying Lemma 13.5 to the bounded set A we have a unique vector $y \in \mathfrak{H}$ with $\rho(y) = \inf\{\rho(x) : x \in \mathfrak{H}\}$. In particular,

$$\rho(y) = \sup_{\gamma \in \Gamma} \|\sigma(\gamma)x_n - y\| \leq \sup_{\gamma \in \Gamma} \|\sigma(\gamma)x_n - x_n\| \leq 1/2. \tag{2.2}$$

The fact that A is Γ-invariant and the uniqueness of y imply that y is a fixed point for the group $\{\sigma(\gamma) : \gamma \in \Gamma\}$. Taking $\gamma = e$ in (2.2) we have $\|x_n - y\| \leq 1/2$ and, x_n being a unit vector, we conclude that $y \neq 0$. In other words, we have shown that $\mathbf{1}_\Gamma \subset \sigma$ and this conflict with our assumption concludes the proof. □

The following family of canonically defined Hilbert spaces and unitary representations associated with an affine action can be obtained via a standard Gelfand-Naimark-Segal construction ([**116**], page 56-60) or by a direct exponentiation process ([**116**], page 50); we shall use a Gaussian construction instead. Note that the construction we described in Theorem 3.56, of a Γ-stationary Gaussian process with a given continuous positive definite function $c : \Gamma \longrightarrow \mathbb{R}$ as a covariance function, is valid for an arbitrary topological group Γ (see [**187**]). Here we apply it to the topological group that underlies the Hilbert space $\mathfrak{H}$ and the family of positive definite functions $\phi_t(x) = \exp(-t\|x\|^2), t > 0, x \in \mathfrak{H}$.

13.8. PROPOSITION. 1. *With every Hilbert space $\mathfrak{H}$ and real number $t > 0$, there are associated a Hilbert space $\mathfrak{H}_t$ and a continuous map Φ_t from $\mathfrak{H}$ to the unit sphere of $\mathfrak{H}_t$, such that the set $\{\Phi_t(x) : x \in \mathfrak{H}\}$ generates $\mathfrak{H}_t$ (i.e. the smallest closed subspace containing this set of vectors is $\mathfrak{H}_t$), and such that*

$$\langle \Phi_t(x), \Phi_t(y)\rangle_{\mathfrak{H}_t} = \exp(-t\|x - y\|^2_{\mathfrak{H}}), \qquad \forall\ x, y \in \mathfrak{H}.$$

2. *Moreover, up to a unitary equivalence this construction is unique.*
3. *With every affine action $\alpha : \Gamma \longrightarrow \mathrm{Iso}(\mathfrak{H})$ and $t > 0$ there is associated a unique unitary representation $\pi_t^\alpha : \Gamma \to \mathfrak{U}(\mathfrak{H}_t)$ such that*

$$\Phi_t(\alpha(\gamma)x) = \pi_t^\alpha(\gamma)(\Phi_t(x)), \qquad \forall\ x \in \mathfrak{H}, \gamma \in \Gamma.$$

PROOF. Consider the positive definite function $\phi : H \longrightarrow \mathbb{R}$ defined on the additive group H by $\phi(x) = \exp(-t\|x\|^2)$. Let $\{\Phi_t(x) : x \in H\}$ be the Gaussian process defined by the covariance function

$$\mathbb{E}(\Phi_t(x)\Phi_t(y)) = \phi(x - y) = \exp(-t\|x - y\|^2)$$

and let $\mathfrak{H}_t$ be the first chaos for this process; i.e. $\mathfrak{H}_t$ is the L^2 subspace generated by $\{\Phi_t(x) : x \in H\}$. This proves part 1 of the proposition and part 2, the uniqueness, is clear. Finally for part 3 observe that, $\alpha(\gamma)$ being an isometry, we have for all $x, y \in \mathfrak{H}$

$$\begin{aligned}\langle \Phi_t(\alpha(\gamma)x), \Phi_t(\alpha(\gamma)y)\rangle_{\mathfrak{H}_t} &= \exp(-t\|\alpha(\gamma)x - \alpha(\gamma)y\|^2_{\mathfrak{H}})\\ &= \exp(-t\|x - y\|^2_{\mathfrak{H}})\\ &= \langle \Phi_t(x), \Phi_t(y)\rangle_{\mathfrak{H}_t}.\end{aligned}$$

Thus the map $\Phi_t(x) \mapsto \Phi_t(\alpha(\gamma)x)$ defines an isometry on the set $\{\Phi_t(x) : x \in H\}$. Since by construction this set is linearly independent, the map $\Phi_t(x) \mapsto \Phi_t(\alpha(\gamma)x)$ can be uniquely extended from $\{\Phi_t(x) : x \in H\}$ to a linear isometry $\pi_t^\alpha(\gamma) : \mathfrak{H}_t \to \mathfrak{H}_t$, which clearly satisfies the required identity. □

13.9. LEMMA. *Let $\alpha : \Gamma \longrightarrow \mathrm{Iso}(\mathfrak{H})$ be an affine action, $\alpha = \pi + b$.*

1. *If t_n is a sequence of positive real numbers tending to zero, then $\mathbf{1}_\Gamma \prec \bigoplus_n \pi_{t_n}^\alpha$.*
2. *For any fixed $t > 0$, if $\{x_k\}$ is a sequence of vectors in $\mathfrak{H}$ tending to infinity, then for every $u \in \mathfrak{H}_t$ we have $\lim_{k\to\infty}\langle \Phi_t(x_k), u\rangle = 0$.*
3. *For any $t > 0$, $\pi_t^{\alpha\oplus\alpha} = \pi_t^\alpha \otimes \pi_t^\alpha$.*
4. *For any $t > 0$ we have: the α action of Γ on $\mathfrak{H}$ has a fixed point iff $\mathbf{1}_\Gamma \subset \pi_t^\alpha$.*

PROOF. 1. Fix $x \in \mathfrak{H}$ a finite subset $K \subset \Gamma$ and $\epsilon > 0$. Then, when $n \to \infty$ for all $\gamma \in K$

$$\begin{aligned}\langle \pi^\alpha_{t_n}(\gamma)\Phi_{t_n}(x), \Phi_{t_n}(x)\rangle &= \langle \Phi_{t_n}(\alpha(\gamma)x), \Phi_{t_n}(x)\rangle \\ &= \exp(-t_n\|\alpha(\gamma)x - x\|^2) \to 1,\end{aligned}$$

and therefore $\|\pi^\alpha_{t_m}(\gamma)\Phi_{t_m}(x) - \Phi_{t_m}(x)\| < \epsilon$ for sufficiently large m. Thus the unit vector $\Phi_{t_m}(x)$ considered as an element of $\bigoplus_n \mathfrak{H}_{t_n}$ is (ϵ, K)-invariant and we have shown that $\mathbf{1}_\Gamma \prec \bigoplus_n \pi^\alpha_{t_n}$.

2. Fix $t > 0$ and let $\{x_k\}$ be a sequence in $\mathfrak{H}$ such that $\lim_{k\to\infty}\|x_k\| = \infty$. Let $\epsilon > 0$ and $u \in \mathfrak{H}_t$ be given, then — as $\mathfrak{H}_t$ is spanned by $\{\Phi_t(x) : x \in \mathfrak{H}\}$ — there exist a positive integer m and $y_j \in \mathfrak{H}, a_j \in \mathbb{R}, j = 1, \dots, m$ such that $\|u - \sum_{j=1}^m a_j y_j\| < \epsilon/2$. We now have

$$\begin{aligned}|\langle \Phi_t(x_k), u\rangle| &\leq |\langle \Phi_t(x_k), u - \sum_{j=1}^m a_j y_j\rangle| + \sum_{j=1}^m |a_j||\langle \Phi_t(x_k), \Phi_t(y_j)\rangle| \\ &\leq \epsilon/2 + \sum_{j=1}^m |a_j|\exp(-t\|x_k - y_j\|^2).\end{aligned}$$

Since the sequence $\{\|x_k\|\}$ tends to infinity, it follows that $|\langle \Phi_t(x_k), u\rangle| \leq \epsilon$ for sufficiently large k, as required.

3. We have for $x, x', y, y' \in \mathfrak{H}$,

$$\begin{aligned}\langle \Phi_t^{\mathfrak{H}\oplus\mathfrak{H}}(x,y), \Phi_t^{\mathfrak{H}\oplus\mathfrak{H}}(x',y')\rangle &= \exp(-t\|(x,y) - (x',y')\|^2) \\ &= \exp(-t(\|x - x'\|^2 + \|y - y'\|^2)) \\ &= \exp(-t\|x - x'\|^2)\exp(-t\|y - y'\|^2) \\ &= \langle \Phi_t^{\mathfrak{H}}(x), \Phi_t^{\mathfrak{H}}(x')\rangle\langle \Phi_t^{\mathfrak{H}}(y), \Phi_t^{\mathfrak{H}}(y')\rangle \\ &= \langle \Phi_t^{\mathfrak{H}}(x) \otimes \Phi_t^{\mathfrak{H}}(y), \Phi_t^{\mathfrak{H}}(x') \otimes \Phi_t^{\mathfrak{H}}(y')\rangle.\end{aligned}$$

Therefore $\Phi_t^{\mathfrak{H}\oplus\mathfrak{H}} = \Phi_t^{\mathfrak{H}} \otimes \Phi_t^{\mathfrak{H}}$ and a straightforward calculation shows that for every $\gamma \in \Gamma$,

$$\pi_t^{\alpha\oplus\alpha}(\gamma)(\Phi_t^{\mathfrak{H}\oplus\mathfrak{H}}(x,y)) = (\pi_t^\alpha \otimes \pi_t^\alpha)(\gamma)(\Phi_t^{\mathfrak{H}\oplus\mathfrak{H}}(x,y)),$$

hence $\pi_t^{\alpha\oplus\alpha} = \pi_t^\alpha \otimes \pi_t^\alpha$.

4. If $\alpha(\gamma)x = x$ for every $\gamma \in \Gamma$ then

$$\Phi_t(\alpha(\gamma)x) = \pi_t^\alpha(\gamma)(\Phi_t(x)) = \Phi_t(x),$$

and the unit vector $\Phi_t(x)$ is π_t^α-invariant.

Conversely, suppose there is a unit vector $u \in \mathfrak{H}_t$ such that $\pi_t^\alpha(\gamma)u = u$ for all $\gamma \in \Gamma$. By Lemma 13.6 it suffices to show that there exists some $x \in \mathfrak{H}$ such that $\alpha(\Gamma)x$ is bounded. Suppose then that for every $x \in \mathfrak{H}$ the set $\alpha(\Gamma)x$ is unbounded. Choose a sequences $\{\gamma_k\}$ in Γ such that $\alpha(\gamma_k)x$ diverges to infinity. Then

$$\begin{aligned}0 &= \lim_{k\to\infty}\langle(\Phi_t(\alpha(\gamma_k)x), u\rangle && \text{(by part 2)} \\ &= \lim_{k\to\infty}\langle(\pi_t^\alpha(\gamma_k)\Phi_t(x), u\rangle && \text{(by definition of } \pi_t) \\ &= \lim_{k\to\infty}\langle(\Phi_t(x), \pi_t^\alpha(\gamma_k^{-1})u\rangle \\ &= \langle(\Phi_t(x), u\rangle && \text{(by } \Gamma\text{-invariance of } u).\end{aligned}$$

It follows that the unit vector u is orthogonal to every $\Phi_t(x), x \in \mathfrak{H}$. Since these vectors generate $\mathfrak{H}_t$ this is a contradiction. □

13.10. THEOREM. *The group Γ has property T iff every affine Γ action admits a fixed point.*

PROOF. 1. Suppose $\sigma : \Gamma \to \mathfrak{U}(\mathfrak{H})$ is a unitary representation such that $\mathbf{1}_\Gamma \prec \sigma$ but $\mathbf{1}_\Gamma \not\subset \sigma$, then by, Lemma 13.7, for the associated infinite sum representation

$$\pi = \bigoplus_{n=1}^{\infty} \sigma : \Gamma \to \mathfrak{H}_\infty = \bigoplus_{n=1}^{\infty} \mathfrak{H},$$

there exists a π-cocycle $b : \Gamma \to \mathfrak{H}_\infty$ which is not a coboundary. By Lemma 13.6 the associated affine action $\alpha = \pi + b$ does not admit a fixed point.

2. Conversely suppose Γ has property T and let $\alpha = \pi + b$ be an affine Γ action on the Hilbert space $\mathfrak{H}$. Consider the associated family $\pi_t^\alpha : \Gamma \to \mathfrak{U}(\mathfrak{H}_t)$ of unitary representations provided by Proposition 13.8. Apply Lemma 13.9.1 with, say $t_n = 1/n$, to get $\mathbf{1}_\Gamma \prec \bigoplus_n \pi_{t_n}^\alpha$. Property T now implies $\mathbf{1}_\Gamma \subset \bigoplus_n \pi_{t_n}^\alpha$ and therefore also $\mathbf{1}_\Gamma \subset \pi_{t_{n_0}}^\alpha$ for at least one n_0. An application of Lemma 13.9.4 finishes the proof. □

We are now ready to prove the main result of this section.

13.11. THEOREM. *For a discrete countable group Γ the properties Kazhdan, BV and BV^-, are equivalent.*

PROOF. Kazhdan $\Rightarrow$ BV: Assuming $\mathbf{1}_G \prec \pi \otimes \pi$, the Kazhdan property implies that $\pi \otimes \pi$ contains the trivial representation $\mathbf{1}_G$, which, by Theorem 3.5, is equivalent to π being not weakly mixing.

BV $\Leftrightarrow$ BV$^-$: Since clearly $\mathbf{1}_G \prec \pi \Rightarrow \mathbf{1}_G \prec \pi \otimes \pi$, the implication BV $\Rightarrow$ BV$^-$ follows. On the other hand, assuming $\mathbf{1}_G \prec \pi \otimes \pi$, the BV$^-$ property implies that $(\pi \otimes \pi) \otimes (\pi \otimes \pi)$ contains the trivial representation $\mathbf{1}_G$, and again, by Theorem 3.5, π is not weakly mixing.

BV$^-$ $\Rightarrow$ Kazhdan: We assume property BV$^-$ and shall use the criterion provided in Theorem 13.10 to prove property T. Let then $\alpha = \pi + b$ be an arbitrary affine Γ action on a Hilbert space $\mathfrak{H}$; we have to show that α admits a fixed point.

Choosing, say $t_n = 1/n$ and denoting $\pi_{1/n}^\alpha = \pi_n$, we define $\tilde{\pi} = \bigoplus_n \pi_n$. Then by Lemma 13.9.1 we have, $\mathbf{1}_\Gamma \prec \tilde{\pi}$.

Next use the property BV$^-$ to deduce that $\mathbf{1}_\Gamma \subset \tilde{\pi} \otimes \tilde{\pi}$. Now

$$\tilde{\pi} \otimes \tilde{\pi} = \bigoplus_{n_1, n_2} \pi_{n_1} \otimes \pi_{n_2},$$

and it follows that for some pair n_1, n_2 we have $\mathbf{1}_\Gamma \subset \pi_{n_1} \otimes \pi_{n_2}$. Applying Theorem 3.5 we deduce that the representation π_{n_1} is not weakly mixing. In turn, this implies that $\mathbf{1}_\Gamma \subset \pi_{n_1} \otimes \pi_{n_1}$. From Lemma 13.9.3 we have

$$\pi_{n_1} \otimes \pi_{n_1} = \pi_{1/n_1}^\alpha \otimes \pi_{1/n_1}^\alpha = \pi_{1/n_1}^{\alpha \oplus \alpha},$$

and by Lemma 13.9.4 we deduce that the affine Γ action $\alpha \oplus \alpha$ on $\mathfrak{H} \oplus \mathfrak{H}$ has a fixed point. Finally this clearly means that also the affine Γ action α admits a fixed point, as required. □

3. A topological characterization of property T

13.12. THEOREM. *If Γ has property T then for any topological system (X,Γ), with $M_\Gamma(X) \neq \emptyset$, $M_\Gamma(X)$ is a Bauer simplex.*

PROOF. Suppose (X,Γ) is a system for which a non-ergodic measure μ exists which is in the closure of the ergodic measures in $M_\Gamma(X)$. Thus there exists a sequence of ergodic measures $\mu_n \in M_\Gamma^{\rm erg}(X)$ and $\mu_n \to \mu$. Let

$$Y = \prod_1^\infty X \quad \text{and} \quad \nu = \prod_1^\infty \mu_n.$$

Let λ be an ergodic measure in the ergodic decomposition of ν such that for every n, $\pi_n(\lambda) = \mu_n$ (here π_n is the projection of Y onto its n-th component); i.e. λ is an ergodic joining of the systems (X,μ_n). Since μ is non-ergodic there exists a Γ-invariant measurable function f such that $\int_X f d\mu = 0$ and say $\int_X |f|^2 d\mu = 1$. For a fixed $\epsilon > 0$ there exists then a continuous function ϕ on X such that

$$\int_X \phi d\mu = 0, \qquad \int_X |\phi|^2 d\mu = 1$$

and for all $\gamma \in \Gamma$

$$\int_X |\phi - \gamma\phi|^2 d\mu < \epsilon,$$

(recall that the operator induced by each $\gamma \in \Gamma$ on $L^2(\mu)$ is unitary). Define a sequence of continuous functions ϕ_n on Y by: $\phi_n(y) = \phi(\pi_n y)$, then

$$\int_Y |\phi_n - \gamma\phi_n|^2 d\lambda = \int_X |\phi - \gamma\phi|^2 d\mu_n \to \int_X |\phi - \gamma\phi|^2 d\mu < \epsilon.$$

Thus we have shown that for every $\epsilon > 0$ and a finite subset F of Γ, there exists a function $\psi \in L_0^2(Y,\lambda)$ with

$$\|\psi - \gamma\psi\|_{L^2(\lambda)} < \epsilon.$$

In other words, the natural representation of Γ on the subspace $L_0^2(\lambda)$, of zero integral functions of $L^2(\lambda)$, has almost invariant vectors. However by ergodicity of λ, the subspace $L_0^2(\lambda)$ contains no Γ-invariant vectors and we conclude that Γ does not have property T. □

Next consider the Bernoulli action of Γ on the compact space $\Omega = \{0,1\}^\Gamma$ by left translations : for $\gamma, \gamma' \in \Gamma$ and $\omega \in \Omega, \gamma\omega$ is defined by

$$(\gamma\omega)_{\gamma'} = \omega_{\gamma^{-1}\gamma'}.$$

The next result is a special case of Theorem 13.15 which will be stated and proved in the next section. We include it here since its proof is simple and helps clarify the proof of Theorem 13.15.

13.13. PROPOSITION. *If $M_\Gamma(\Omega)$ is a Bauer simplex then Γ has property T. Thus Γ has property T iff $M_\Gamma(\Omega)$ is a Bauer simplex.*

PROOF. If Γ does not have property T then by Theorem 13.3 there exists an ergodic but not strongly ergodic dynamical system $\mathbf{X}$; i.e. an ergodic measure-preserving action of Γ on a probability space $(X, \mathcal{X}, \mu)$ with asymptotically invariant sets, say (Lemma 13.2)

$$(3.1)\qquad B_n \subset X, \qquad \mu(B_n) = \frac{1}{2}, \qquad \forall \gamma \in \Gamma,\ \mu(\gamma B_n \vartriangle B_n) \to 0.$$

For each n, define a mapping

$$\theta_n : X \to \Omega, \qquad (\theta_n(x))_\gamma = \mathbf{1}_{B_n}(\gamma^{-1}x), \qquad (x \in X, \gamma \in \Gamma),$$

and let $\mu_n = \theta_n(\mu)$, then μ_n is an ergodic measure in $M_\Gamma(\Omega)$. Our proof will be complete when we show that the sequence $\{\mu_n\}$ of ergodic measures converges to the non-ergodic measure $\eta = \frac{1}{2}(\delta_\mathbf{0} + \delta_\mathbf{1})$, where $(\mathbf{0})_\gamma = 0$ and $(\mathbf{1})_\gamma = 1$ for all $\gamma \in \Gamma$ and δ_ω is the point mass at ω. For a finite $F \subset \Gamma$ put

$$[0]_F = \{\omega \in \Omega : \omega_\gamma = 0,\ \forall \gamma \in F\}$$
$$[1]_F = \{\omega \in \Omega : \omega_\gamma = 1,\ \forall \gamma \in F\}.$$

Then η is characterized by the property that $\eta([0]_F) = \eta([1]_F) = \frac{1}{2}$ for all finite F.

Now by the properties of the sets B_n and the definition of μ_n,

$$\mu_n([1]_F) = \mu(\bigcap_{\gamma \in F} \gamma B_n) \to \frac{1}{2},$$

and similarly for $[0]_F$. □

4. The Bauer Poulsen dichotomy

Recall that a measure-preserving Γ action on a measure space $(X, \mathfrak{X}, \mu)$ is called weakly mixing if the product diagonal action on $(X \times X, \mathfrak{X} \otimes \mathfrak{X}, \mu \times \mu, \Gamma)$, defined by $\gamma(x, x') = (\gamma x, \gamma x')$, $x, x' \in X, \gamma \in \Gamma$, is ergodic and that this property is equivalent to the condition that the corresponding representation on $L_0^2(X, \mu)$ has no finite dimensional sub-representations (Theorem 3.5).

13.14. PROPOSITION. *Suppose Γ admits a measure-preserving action on a measure space $(X, \mathfrak{X}, \mu)$ with the following properties.*

1. *There exists in X a sequence of asymptotically invariant sets satisfying (3.1).*
2. *The system $\mathbf{X} = (X, \mathfrak{X}, \mu, \Gamma)$ is weakly mixing.*

Then the simplex $M_\Gamma(\Omega)$ is the Poulsen simplex.

PROOF. In order to show that $M_\Gamma(\Omega)$is the Poulsen simplex it is clearly enough to show that the closure of the ergodic measures in $M_\Gamma(\Omega)$is convex. In turn this will follow when we show that for ergodic measures ν_1 and ν_2 in $M_\Gamma(\Omega)$the measure $\frac{1}{2}(\nu_1 + \nu_2)$ is a limit of a sequence of ergodic measures η_n in $M_\Gamma(\Omega)$. So let λ be an ergodic joining of the measures ν_1 and ν_2; i.e. λ is a Γ-invariant ergodic probability measure on $\Omega \times \Omega$ whose projections on the first and second components are ν_1 and ν_2 respectively. The fact that $\mathbf{X}$ is weakly mixing implies now that the Γ action on $X \times \Omega \times \Omega$, is ergodic with respect to the product measure $\sigma = \mu \times \lambda$ (Theorem 3.11).

Now put $[1]_e = \{\omega \in \Omega : \omega_e = 1\}$, where e is the identity element of Γ and

$$D_n = (B_n \times [1]_e \times \Omega) \cup (B_n^c \times \Omega \times [1]_e),$$

and define the maps $\theta_n : X \times \Omega \times \Omega \to \Omega$,

$$(\theta_n(x, \omega_1, \omega_2))_\gamma = \mathbf{1}_{D_n}(\gamma^{-1}x, \gamma^{-1}\omega_1, \gamma^{-1}\omega_2).$$

We claim that the sequence of ergodic measures $\eta_n = \theta_n(\sigma)$ tends weak* to $\frac{1}{2}(\nu_1 + \nu_2)$.

For this it suffices to check the convergence of the sequence $\eta_n([\alpha]_F)$ where F is a finite subset of Γ, $\alpha \in \{0,1\}^F$ and $[\alpha]_F = \{\omega \in \Omega : \omega|_F = \alpha\}$ is the corresponding cylinder set.

Now

$$\eta_n([\alpha]_F) = \sigma(\{(x,\omega_1,\omega_2) : \mathbf{1}_{D_n}(\gamma^{-1}(x,\omega_1,\omega_2))|_F = \alpha\}).$$

Let $F = \{\gamma_j\}_{j=1}^N$, $\alpha(\gamma_j) = \epsilon_j$, and note that $D_n^c = (B_n^c \times \Omega \times [0]_e) \cup (B_n \times [0]_e \times \Omega)$. Then

$$\begin{aligned} E_n =& \{(x,\omega_1,\omega_2) : \mathbf{1}_{D_n}(\gamma^{-1}(x,\omega_1,\omega_2))|_F = \alpha\} \\ =& \bigcap_{\epsilon_j=1} \gamma_j D_n \cap \bigcap_{\epsilon_j=0} \gamma_j D_n^c \\ =& \bigcap_{\epsilon_j=1} \{(\gamma_j B_n \times \gamma_j[1]_e \times \Omega) \cup (\gamma_j B_n^c \times \Omega \times \gamma_j[1]_e)\} \\ & \cap \bigcap_{\epsilon_j=0} \{(\gamma_j B_n^c \times \Omega \times \gamma_j[0]_e) \cup (\gamma_j B_n \times \gamma_j[0]_e) \times \Omega\}. \end{aligned}$$

Since the sets B_n are almost invariant we have for large n,

$$\begin{aligned} \sigma(E_n) \approx& \sigma(\{B_n \times (\bigcap_{\epsilon_j=1} \gamma_j[1]_e \cap \bigcap_{\epsilon_j=0} \gamma_j[0]_e) \times \Omega\} \\ & \cup \{B_n^c \times \Omega \times (\bigcap_{\epsilon_j=1} \gamma_j[1]_e \cap \bigcap_{\epsilon_j=0} \gamma_j[0]_e)\}) \\ =& \sigma(\{(x,\omega_1,\omega_2) : x \in B_n,\ \mathbf{1}_{[1]_e}(\gamma^{-1}\omega_1)|_F = \alpha\}) \\ & + \sigma(\{(x,\omega_1,\omega_2) : x \in B_n^c,\ \mathbf{1}_{[1]_e}(\gamma^{-1}\omega_2)|_F = \alpha\}) \\ =& \frac{1}{2}(\nu_1 + \nu_2)([\alpha]_F). \end{aligned}$$

This completes the proof of the proposition. □

13.15. THEOREM. *If Γ does not have property T then the simplex $M_\Gamma(\Omega)$ is the Poulsen simplex. Thus for every countable group Γ the simplex $M_\Gamma(\Omega)$ is either Bauer or Poulsen depending on whether Γ has property T or not.*

PROOF. By Theorem 13.11 the assumption that the group Γ does not have property T implies that there exists a unitary representation π of Γ which admits almost invariant unit vectors $\{v_n\}$ but does not have a finite dimensional subrepresentation. As in the proof of Theorem 13.3 we use these almost invariant unit vectors to define positive definite functions

$$\phi_n(\gamma) = \langle \pi(\gamma)v_n, v_n \rangle$$

so that $\phi_n(\gamma) \to 1$ for all $\gamma \in \Gamma$. Again we consider a system of centered Gaussian random variables X_n^γ with covariance

$$\mathbb{E}(X_n^\gamma X_n^\delta) = \phi_n(\gamma\delta^{-1}), \quad \forall \gamma, \delta \in \Gamma$$

with (Ω_n, μ_n) the probability space on which the X_n^γ are defined and generate the σ-algebra. By Theorem 3.59 the induced Γ action $(\Omega_n, \mu_n, \Gamma)$ for such a Gauss system is weakly mixing iff the closed subspace of $L^2(\Omega_n, \mu_n)$ spanned linearly by the X_n^γ's (the first chaos) has no finite dimensional Γ-invariant subspaces. Since this first chaos is unitarily the same as the closed subspace generated by $\{\pi(\gamma)v_n\}_{\gamma\in\Gamma}$, this condition is satisfied and thus we conclude that $(\Omega_n, \mu_n, \Gamma)$ is a weakly mixing action.

Form now

$$(\Omega, \mu) = \prod_{n=1}^{\infty} (\Omega_n, \mu_n)$$

which is weakly mixing as the product of weakly mixing actions. The proof will be completed by Proposition 13.14 as soon as we exhibit in Ω a non trivial sequence of asymptotically invariant sets and for that we just repeat, verbatim, the proof of part 2, Case 1 of Theorem 13.3. □

5. A characterization of the Haagerup property

In a recent book [**41**] Cherix, Cowling, Jolissaint, Julg and Valette, study various aspects of a property called the "Haagerup property". For our purposes it will be convenient to consider the negation of the Haagerup property.

13.16. DEFINITION. We call a group Γ *weakly Kazhdan*, or say that it has property T_w, if Γ admits no non-zero mixing unitary representation with almost invariant vectors.

Clearly every Kazhdan group is also weak Kazhdan. Refer to [**41**] for examples of T_w but not T groups (one such example is the semidirect product $SL_2(\mathbb{Z}) \ltimes \mathbb{Z}$). As in the case of property T, and with an easier proof, we get the following analogue of Theorem 13.3.

13.17. THEOREM. *The group Γ has property T_w iff every mixing Γ-system is strongly ergodic.*

13.18. REMARK. Note that by the theorem of Bekka and Valette [**16**], it follows that Γ has property T iff Γ admits no non-zero *weakly mixing* unitary representation with almost invariant vectors.

In [**41**] many of the known characterizations of property T are modified to analogous characterizations of property T_w. At the end of the book the authors ask whether an analogue of the Glasner-Weiss characterization of property T exists for the weak Kazhdan property. In fact such a characterization exists and it can be obtained by appropriate modifications of the proofs of Theorem 13.12 and Proposition 13.13 as follows.

13.19. THEOREM. *If Γ has the weak Kazhdan property then for any topological system (X, Γ), with $M_\Gamma(X) \neq \emptyset$, the closure of the set of mixing measures in $M_\Gamma(X)$ is contained in $M_\Gamma^{\mathrm{erg}}(X)$.*

PROOF. Just repeat the proof of Theorem 13.12 and note that we now assume that all the measures μ_n are mixing, so that the measure $\nu = \lambda$ is automatically mixing hence ergodic. □

13.20. THEOREM. *If in $M_\Gamma(\Omega)$ the closure of the set of mixing measures is contained in the set $M_\Gamma^{\rm erg}(\Omega)$ then Γ has the weak Kazhdan property. Thus Γ has property T_w iff this condition is satisfied.*

PROOF. Again repeat the proof of Proposition 13.13 and note that the image of a mixing measure under a homomorphism is mixing. □

Finally one also has the analogue of Theorem 13.15.

13.21. THEOREM. *The group Γ does not have the weak Kazhdan property (i.e. it is a Haagerup group) iff in the simplex $M_\Gamma(\Omega)$the closure of the set of mixing measures is convex.*

PROOF. If Γ is T_w then the closure of the collection of mixing measures is contained in the set of ergodic measures (Theorem 13.19) and is therefore evidently not convex. For the converse we assume that Γ is not T_w and all we have to show is that, given mixing measures ν_1, ν_2 in $M_\Gamma(\Omega)$, the measure $\frac{1}{2}(\nu_1 + \nu_2)$ is a limit of a sequence of mixing measures in $M_\Gamma(\Omega)$. We apply the proof of Proposition 13.14 together with the following observations. First, when ν_1, ν_2 are mixing measures then so is their product $\lambda = \nu_1 \times \nu_2$. Moreover, by assumption there is a mixing system $\mathbf{X} = (X, \mathfrak{X}, \mu, \Gamma)$ which is not strongly ergodic and we take $\sigma = \mu \times \nu_1 \times \nu_2$; again a mixing measure on $X \times \Omega \times \Omega$. Finally the measures $\eta_n = \theta_n(\sigma)$, constructed in the proof of Proposition 13.14, form a sequence of mixing measures in $M_\Gamma(\Omega)$converging to $\frac{1}{2}(\nu_1 + \nu_2)$ as required. □

6. Notes

Sources: Mainly Connes and Weiss [**42**] and Glasner and Weiss [**102**].

Strong ergodicity and property T: Lemma 13.2 is taken from Jones and Schmidt [**130**]. The 'only if' part of Theorem 13.3 is due to Schmidt [**227**], the 'if' part to Connes and Weiss [**42**].

A theorem of Bekka and Valette: Sources: mostly the book [**116**] by de la Harpe and Valette and the paper [**16**] by Bekka and Valette. I learned the proof of Proposition 13.7 from A. Valette.

A characterization of not having property T: In [**102**] this theory and in particular the dichotomy Bauer-Poulsen for $M_\Gamma(\Omega)$ is extended to locally compact second countable topological groups G. The role of the space of configurations $\Omega = \{0,1\}^\Gamma$ is taken by the space Σ, of closed subsets of the one point compactification G^* of G which contain the point at infinity, and the space $M_G(\Sigma)$ replaces $M_\Gamma(\Omega)$.

A characterization of the Haagerup property: As in [**102**] these characterizations are valid for any locally compact second countable topological group.

Part 2

Entropy Theory for $\mathbb{Z}$-systems

CHAPTER 14

Entropy

A single natural number, its dimension, suffices to completely describe a vector space over a given field. Can one hope to achieve such a nice description of a dynamical system? The answer is certainly not. We know that the problem of finding isomorphism invariants (i.e. some numerical or algebraic entities attached to a dynamical system that are invariant under isomorphisms) — not to speak of obtaining a complete set of invariants — for all ergodic dynamical systems is as complicated as such a problem can be (see Hjorth [**122**] and Foreman and Weiss [**68**]). Nonetheless much of ergodic theory is concerned with finding invariants for ergodic systems, and even a complete set of invariants within some restricted classes of ergodic systems.

The crudest sort of invariant for a dynamical system is its Koopman representation. For a $\mathbb{Z}$-dynamical system this is just the unitary operator $U_T : L^2(X, \mu) \to L^2(X, \mu)$ (which by the spectral theorem is completely described by a measure class on $\mathbb{T}$ and a multiplicity function).

As we have seen some of our basic definitions, like ergodicity and various mixing properties, are spectral in nature; that is, they are stated in terms of the Koopman representation and therefore are shared by all systems having unitarily isomorphic Koopman representations. Thus for example Theorem 5.13 tells us that all K-systems (hence by Proposition 3.51 all Bernoulli systems) are spectrally indistinguishable.

A related issue is the isomorphism problem. Two isomorphic dynamical systems may be presented to us in a way that makes them look different. Let us consider the following well known example due to Meshalkin. The Bernoulli systems $B(\frac{1}{4}, \frac{1}{4}, \frac{1}{4}, \frac{1}{4})$ and $B(\frac{1}{2}, \frac{1}{8}, \frac{1}{8}, \frac{1}{8}, \frac{1}{8})$ are isomorphic as measure-preserving transformations. In fact Meshalkin produced a *finitary isomorphism* between the two systems; i.e. an injective map $\phi : \{1, 2, 3, 4\}^{\mathbb{Z}} \to \{1, 2, 3, 4, 5\}^{\mathbb{Z}}$ which maps the measure $\mu = (\frac{1}{4}, \frac{1}{4}, \frac{1}{4}, \frac{1}{4})^{\mathbb{Z}}$ onto the measure $\nu = (\frac{1}{2}, \frac{1}{8}, \frac{1}{8}, \frac{1}{8}, \frac{1}{8})^{\mathbb{Z}}$, intertwines the shifts on these spaces and has the property that for μ almost every x the value $(\phi(x))_0$ depends on the finite word $x_{-n}^n = (x_n, \ldots, x_0, \ldots, x_n)$ for some $n = n(x)$ (see e.g. [**260**]).

Entropy was introduced as an important isomorphism invariant in ergodic theory by Kolmogorov in 1958 (following its introduction to information theory some ten years earlier by C. Shannon). The entropy of a probability vector $p = (p_1, \ldots, p_\ell)$ is defined as $H(p) = -\sum_{j=1}^{\ell} p_j \log(p_j)$. For a dynamical system $(X, \mathcal{X}, \mu, T)$ and a finite partition $\alpha = \{A_1, \ldots, A_\ell\}$ the information function $I(\alpha)(x)$ is defined as $-\log(\mu(A_j))$ for $x \in A_j$ and we set

$$H(\alpha) = \int I(\alpha)\, d\mu = -\sum_{j=1}^{\ell} \mu(A_j) \log(\mu(A_j)).$$

The entropy $h(\alpha) = h(\alpha, T)$ (or $h_\mu(\alpha)$) of the partition α with respect to T is then defined as the limit

$$h(\alpha, T) = \lim_{n\to\infty} \frac{1}{n} H(\alpha_0^{n-1}),$$

where α_0^{n-1} is the partition of X obtained by the mutual refinement of the partitions $\alpha, T^{-1}\alpha, \ldots, T^{-n+1}\alpha$. We interpret this as the average information per unit in the time evolution of α. We sometimes think of α as a random variable $\alpha : (X, \mu) \to (\{1, \ldots, \ell\}, p)$ and then talk about the corresponding stationary sequence of random variables $T^{-n}\alpha$ as a "process". (It is stationary in the sense that for any k the joint distribution of the random variables $(T^{-(n+1)}\alpha, T^{-(n+2)}\alpha, \ldots, T^{-(n+k)}\alpha)$ is independent of n.) The entropy of the dynamical system $h(\mathbf{X}) = h_\mu$ is then defined as the supremum of the numbers $h_\mu(\alpha)$ over all possible finite partitions. It is sometimes necessary to work with countable rather than finite partitions. The corresponding definitions are obtained in the obvious way.

Kolmogorov and Sinai have shown that the entropy of a dynamical system is attained by the entropy of any of its generators (i.e. countable (or finite) partitions that, together with their translations, $T^n\alpha, n \in \mathbb{Z}$, generate the σ-algebra $\mathcal{X}$). In particular, for a Bernoulli system, presented in its canonical form as $B(p_1, \ldots, p_\ell) = (\Omega(\ell), \mu_p)$ — where $p = (p_1, \ldots, p_\ell)$ is a probability vector and μ_p is the corresponding product measure — the partition $\alpha = \{A_1, \ldots, A_\ell\}$ with $A_j = \{x \in \Omega(\ell) : x_0 = j\}$ (and $\mu(A_j) = p_j$) is a generator and therefore

$$h(\Omega(\ell), \mu_p, S) = -\sum_{j=1}^{\ell} p_j \log(p_j).$$

As a result we immediately see that, e.g. the Bernoulli systems $B(\frac{1}{2}, \frac{1}{2})$ with entropy $\log 2$ and $B(\frac{1}{3}, \frac{1}{3}, \frac{1}{3})$ with entropy $\log 3$ are not isomorphic. On the other hand for the Meshalkin example the two Bernoulli systems indeed have the same entropy $\log 4$.

Perhaps the most far reaching achievement of ergodic theory today is D. Ornstein's isomorphism theorem: entropy, a single number $0 < h = h(\mathbf{X}) \leq \infty$, is a complete isomorphism invariant for Bernoulli systems. The methods and ideas introduced by Ornstein go much beyond the proof of the isomorphism theorem. For example, the various "Bernoullicity" criteria are of great importance in deciding whether a given dynamical system is Bernoulli (see for example Ornstein and Weiss [**198**], [**194**]). Many examples (arising from various branches of mathematics such as Lie group theory, number theory and mathematical physics) were proved to be Bernoulli systems using these methods. Ornstein himself has shown that mixing Markov systems are Bernoulli and produced many important examples such as the first example of a K-system which is not Bernoulli. In 1979 Keane and Smorodinsky succeeded in showing that in fact "Bernoulli schemes of the same entropy are finitarily isomorphic" [**151**]. We shall present a proof of Ornstein's theorem in Chapter 18.

Another interpretation of the entropy $h(\alpha, T)$ is obtained by the formula

$$h(\alpha, T) = H(\alpha|\alpha_{-\infty}^{-1}) = \lim_{n\to\infty} H(\alpha|\alpha_{-n}^{-1}).$$

Here $H(\alpha|\alpha_{-n}^{-1})$ measures the information conveyed by α when the past evolution from time $-n$ up to time -1, α_{-n}^{-1} (or the whole past $\alpha_{-\infty}^{-1}$ in the case of $H(\alpha|\alpha_{-\infty}^{-1})$) is completely known. Thus, the process $\{T^n\alpha\}$ is "deterministic" when $h(\alpha, T) =$

$H(\alpha|\alpha_{-\infty}^{-1}) = 0$. In fact in such a case we can find a measurable function f on $\Omega(\ell)$ such that $x_0 = f(\ldots, x_{-2}, x_{-1})$, where as usual $x_n = j$ iff $T^n x \in A_j$ iff $x \in T^{-n}A_j$. On the other hand we have $H(\alpha|\alpha_{-\infty}^{-1}) = H(\alpha)$, that is α is independent of its "past", iff the process corresponding to the partition α is an independent process (i.e. a Bernoulli process).

A third interpretation of the entropy of a process α is given by the "ergodic theorem of information theory" or as it is usually called in ergodic theory the Shannon-McMillan-Breiman theorem (SMB). It says that for an ergodic system the functions $\frac{1}{n}I(\alpha_0^{n-1})$ converge almost surely and in $L^1(X, \mu)$ to the entropy of the process $h(\alpha, T)$. As we shall see this shows that the number $h(\alpha, T)$ measures the "typical" size of an atom in α_0^{n-1}.

The interested reader is referred to Smorodinsky's article [**239**] concerning the origin of entropy theory, Ornstein isomorphism theory and related developments

Topological entropy was defined in 1964 by Adler, Konheim and McAndrew [4] and a new and constructive approach was introduced by Bowen [**32**]. The "variational principle" (due to Goodwyn, Dinaburg and Goodman) states that the topological entropy $h_{\text{top}}(X, T)$ of a topological system (X, T) is the supremum over all the numbers $h(X, \mu)$ with μ a T-invariant probability measure on X. After several stages its proof was completed in 1971.

We introduce topological entropy in Section 1, then measure entropy in Section 2. In Section 3 we prove the theorem of Kolmogorov and Sinai. In Section 5 the SMB theorem is proved. Note that our log is always the natural logarithm.

1. Topological entropy

14.1. DEFINITION. Let (X, T) be a topological system. For a finite open cover $\mathcal{U} = \{U_1, \ldots, U_n\}$ let $N(\mathcal{U})$ be the minimal cardinality of a subcover of $\mathcal{U}$. Set $H(\mathcal{U}) = \log N(\mathcal{U})$.

If $\mathcal{V} = \{V_1, \ldots, V_m\}$ is another open cover we let $\mathcal{U} \vee \mathcal{V}$ denote the *join* of $\mathcal{U}$ and $\mathcal{V}$, which is the cover consisting of all non-empty sets $U_i \cap V_j$. The join of a finite collection of covers is defined similarly. The cover $T\mathcal{U}$ is defined by $T\mathcal{U} = \{TU_1, \ldots, TU_n\}$. Write $\mathcal{U}_m^n$ for the cover $\bigvee_{j=m}^n T^{-j}\mathcal{U}$. In particular

$$\mathcal{U}_0^{n-1} = \mathcal{U} \vee T^{-1}\mathcal{U} \vee \cdots \vee T^{-n+1}\mathcal{U}.$$

We say that $\mathcal{V}$ *refines* $\mathcal{U}$ and write $\mathcal{V} \succ \mathcal{U}$ if for every $V \in \mathcal{V}$ there is a $U \in \mathcal{U}$ with $V \subset U$. In particular if $\mathcal{V} \subset \mathcal{U}$ is a subcover of $\mathcal{U}$ then $\mathcal{V} \succ \mathcal{U}$. Clearly $H(T\mathcal{U}) = H(\mathcal{U})$ and $\mathcal{V} \succ \mathcal{U}$ implies $H(\mathcal{V}) \geq H(\mathcal{U})$.

14.2. LEMMA. *Given a topological system and a finite open cover $\mathcal{U}$, the sequence $a_n = H(\mathcal{U}_0^{n-1})$ satisfies: $a_{m+n} \leq a_m + a_n$, for $m, n \geq 1$. Hence the limit*

$$h(\mathcal{U}, T) = h(\mathcal{U}) = \lim_{n\to\infty} \frac{1}{n} H(\mathcal{U}_0^{n-1}) = \inf \frac{1}{n} H(\mathcal{U}_0^{n-1})$$

exists.

PROOF. It is easy to see that for any two open covers $\mathfrak{U}$ and $\mathcal{V}$ we have $N(\mathfrak{U} \vee \mathcal{V}) \leq N(\mathfrak{U})N(\mathcal{V})$ hence $H(\mathfrak{U} \vee \mathcal{V}) \leq H(\mathfrak{U}) + H(\mathcal{V})$. Hence

$$\begin{aligned} a_{m+n} &= H(\mathfrak{U}_0^{n+m-1}) = H(\mathfrak{U}_0^{n-1} \vee T^{-n}\mathfrak{U}_0^{m-1}) \\ &\leq H(\mathfrak{U}_0^{n-1}) + H(T^{-n}\mathfrak{U}_0^{m-1}) \\ &= a_n + a_m. \end{aligned}$$

Now apply the subadditivity lemma (Theorem A.1). □

14.3. DEFINITION. Let

$$h_{\text{top}}(X,T) = \sup\{h(\mathfrak{U}) : \mathfrak{U} \text{ a finite open cover of } X\}.$$

The non-negative number (or $+\infty$) $h_{\text{top}}(X,T)$ is called the *topological entropy* of the system (X,T).

The proof of the following proposition is straightforward.

14.4. PROPOSITION. 1. *If (Y,T) is a subsystem of (X,T), then we have $h_{\text{top}}(X,T) \geq h_{\text{top}}(Y,T)$.*
2. *If $(X,T) \to (Y,T)$ is a factor map, then $h_{\text{top}}(X,T) \geq h_{\text{top}}(Y,T)$.*
3. *$h_{\text{top}}(X,T) = h_{\text{top}}(X,T^{-1})$.*

The next lemma provides a more practical way of computing the entropy of a dynamical system.

14.5. LEMMA. *Let $\mathfrak{U}_1 \prec \mathfrak{U}_2 \prec \cdots$ be a sequence of open covers with* diam $(\mathfrak{U}_n) = \sup\{\text{diam}\,(U) : U \in \mathfrak{U}_n\} \to 0$ *(with respect to some compatible metric d), then*

$$\lim_{n\to\infty} h(\mathfrak{U}_n) = h_{\text{top}}(X,T).$$

PROOF. Let $\mathcal{V}$ be any open cover and let δ be a Lebesgue number for $\mathcal{V}$ (i.e. for every subset $A \subset X$ with diam $(A) < \delta$ there exists an element $V \in \mathcal{V}$ such that $A \subset V$), then $\mathfrak{U}_n \succ \mathcal{V}$ for every n with diam $(\mathfrak{U}_n) < \delta$. This clearly proves our lemma. □

14.6. LEMMA. *For every open cover $\mathcal{V}$ and $n \in \mathbb{N}$*

$$h(\mathcal{V}_{-n}^n) = h(\mathcal{V}).$$

PROOF.

$$\begin{aligned} h(\mathcal{V}) \leq h(\mathcal{V}_{-n}^n) &= \lim_{k\to\infty} \frac{1}{k} H((\mathcal{V}_{-n}^n)_0^{k-1}) = \lim_{k\to\infty} \frac{1}{k} H(\mathcal{V}_{-n}^{n+k-1}) \\ &\leq \lim_{k\to\infty} \frac{1}{k} (H(\mathcal{V}_{-n}^{-1}) + H(\mathcal{V}_0^{k-1}) + H(T^{-k}\mathcal{V}_0^{n-1})) \\ &= \lim_{k\to\infty} \frac{1}{k} H(\mathcal{V}_0^{k-1}) = h(\mathcal{V}). \end{aligned}$$

□

Recall that for the Bernoulli system $\Omega = \Omega(\ell) = \mathfrak{L}^{\mathbb{Z}}$, with $\mathfrak{L} = \{1, 2, \ldots, \ell\}$, we denote the basic cylinder sets by $\mathbf{a}(j) = \{x \in \Omega : x_0 = j\}$ and the corresponding partition of Ω into clopen sets by $P_0 = \{\mathbf{a}(j) : 1 \le j \le \ell\}$. For integers $m < n$, $P_m^n = \bigvee_{i=m}^n S^{-i} P_0$ is the partition corresponding to the cylinders on the coordinates $[m, n]$, and we write $P_n = P_{-n}^n$.

Let (X, S) be a subshift. Let $\mathfrak{U}_m^n = P_m^n \cap X$, and let u_n be the number of distinct n blocks that appear in elements of X; i.e. $u_n = \operatorname{card}(\mathfrak{U}_0^{n-1})$.

14.7. COROLLARY. *For a subshift* (X, S),

$$h_{\mathrm{top}}(X, S) = \lim_{n \to \infty} \frac{1}{n} u_n.$$

In particular for $X = \Omega$ *we have* $h_{\mathrm{top}}(\Omega, S) = \log \ell$.

PROOF. Clearly the sequence $\mathfrak{U}_n = \mathfrak{U}_{-n}^n$ satisfies the conditions of Lemma 14.5 and therefore, by this lemma and in view of Lemma 14.6:

$$\begin{aligned} h_{\mathrm{top}}(X, T) &= \lim_{n \to \infty} h(\mathfrak{U}_n) = h(\mathfrak{U}_0) \\ &= \lim_{k \to \infty} \frac{1}{k} H(\mathfrak{U}_0^{k-1}) = \lim_{k \to \infty} \frac{1}{k} \log N(\mathfrak{U}_0^{k-1}) \\ &= \lim_{k \to \infty} \frac{1}{k} \log u_k. \end{aligned}$$

For $X = \Omega$, $u_k = \ell^k$, we get $h_{\mathrm{top}}(\Omega, S) = \lim_{k \to \infty} \frac{1}{k} \log \ell^k = \log \ell$. $\square$

14.8. EXERCISES.

1. If $X \subset \Omega(\ell)$ is a subshift and $h_{\mathrm{top}}(X) = \log \ell$ then $X = \Omega(\ell)$.
2. Show that $h_{\mathrm{top}}(X, S) = 0$ for the Morse dynamical system, 1.19.1. [[Every block that appears in X is a subblock of blocks of the form $B_n B_n$, $B_n \bar{B}_n$, $\bar{B}_n B_n$, or $\bar{B}_n \bar{B}_n$. Hence $h_{\mathrm{top}}(X, S) \le \lim_{k \to \infty} \frac{1}{2^k} \log(4 \cdot 2^k) = 0$.]]
3. $h_{\mathrm{top}}(X, S) = 0$ for every substitution system, Examples 1.19.5.
4. Use Exercise 1.3.3 to show that for the golden mean subshift (X, S) we have $h_{\mathrm{top}}(X, S) = \log \lambda$ with $\lambda = \frac{1}{2}(1 + \sqrt{5})$.

14.9. DEFINITION (Bowen's definition of topological entropy). Let (X, d) be a metric space and $0 < \epsilon \in \mathbb{R}$. A subset $F \subset X$ is called *ϵ-spanning* if for every $x \in X$ there exists $x' \in F$ with $d(x, x') < \epsilon$. A subset $F \subset X$ is called *ϵ-separated* if for every distinct points $x, x' \in F$, $d(x, x') \ge \epsilon$.

When (X, T) is a compact dynamical system, d a compatible metric on X and $1 \le n \in \mathbb{N}$, we set

$$d_n(x, x') = \max_{0 \le k \le n-1} d(T^k x, T^k x').$$

It is easy to see that d_n is also a compatible metric on X.

A subset $F \subset X$ is called *(n, ϵ)-spanning* (*(n, ϵ)-separated*) if it is ϵ-spanning (ϵ-separated) with respect to the metric d_n. Write $sp_n(\epsilon)$ for the minimal cardinality of an (n, ϵ)-spanning set and $sr_n(\epsilon)$ for the maximal cardinality of an (n, ϵ)-separated set.

Now define:

$$sp\,(\epsilon, T) = \limsup_{n\to\infty} \frac{1}{n} \log sp_n(\epsilon), \quad \text{and,}$$

$$sr\,(\epsilon, T) = \limsup_{n\to\infty} \frac{1}{n} \log sr_n(\epsilon).$$

One can easily check that both $sp\,(\epsilon, T)$ and $sr\,(\epsilon, T)$ increase when $\epsilon \searrow 0$, so that the limits

$$sp\,(T, d) = \lim_{\epsilon\searrow 0} sp\,(\epsilon, T), \quad \text{and} \quad sr\,(T, d) = \lim_{\epsilon\searrow 0} sr\,(\epsilon, T)$$

exist.

14.10. LEMMA. *For every* $n \in \mathbb{N}$

1. $sp_n(\epsilon) \le sr_n(\epsilon) \le sp_n(\epsilon/2)$,
2. $sp\,(\epsilon, T) \le sr\,(\epsilon, T) \le sp\,(\epsilon/2, T)$, *hence*
3. $sp\,(T,d) = sr\,(T,d)$.

PROOF. It is clear that an (n, ϵ)-separated set of maximal cardinality is also an (n, ϵ)-spanning set; thus $sp_n(\epsilon) \le sr_n(\epsilon)$. Conversely, if F is an $(n, \epsilon/2)$-spanning set of minimal cardinality and E is any (n, ϵ)-separated set, then for every $x \in E$ we can choose a point $\phi(x) \in F$ with $d_n(x, \phi(x)) \le \epsilon/2$. As E is (n, ϵ)-separated, it follows that ϕ is 1-1, whence $|F| \ge |E|$ so that $sr_n(\epsilon) \le sp_n(\epsilon/2)$. The remaining assertions of the lemma now follow readily. ☐

14.11. PROPOSITION. *For a topological system* (X, T) *we have*

$$h_{\text{top}}(X, T) = sp\,(T, d) = sr\,(T, d).$$

In particular the number $sp\,(T, d) = sr\,(T, d)$ *does not depend on the choice of a compatible metric* d.

PROOF. 1. First we fix an $\epsilon > 0$ and an $n \in \mathbb{N}$. Let $\mathcal{U}$ be an open cover of X with Lebesgue number ϵ and let F be an (n, ϵ)-spanning set of minimal cardinality; i.e. $|F| = sp_n(\epsilon)$. By definition of $sp_n(\epsilon)$ we have

$$\bigcup_{x\in F} \bigcap_{k=0}^{n-1} T^{-k} \bar{B}_\epsilon(T^k x) = X.$$

Now for every $x \in F$ and $0 \le k \le n-1$, $B_\epsilon(T^k x)$ is contained in an element of the cover $\mathcal{U}$ and we conclude that $N(\mathcal{U}_0^{n-1}) \le sp_n(\epsilon)$.

2. If $\mathcal{V}$ is an open cover with $\operatorname{diam}(\mathcal{V}) \le \epsilon$, let E be an (n, ϵ)-separated set of maximal cardinality; i.e. $|E| = sr_n(\epsilon)$. Now in every element of $\mathcal{V}_0^{n-1}$ there is at most one point of E and therefore $sr_n(\epsilon) \le N(\mathcal{V}_0^{n-1})$.

3. By 1 and 2

$$N(\mathcal{U}_0^{n-1}) \le sp_n(\epsilon) \le sr_n(\epsilon) \le N(\mathcal{V}_0^{n-1}), \quad \text{hence}$$

$$\frac{1}{n} H(\mathcal{U}_0^{n-1}) \le \frac{1}{n} \log sp_n(\epsilon) \le \frac{1}{n} \log sr_n(\epsilon) \le \frac{1}{n} H(\mathcal{V}_0^{n-1}) \tag{1.1}$$

4. Let $n \to \infty$ to get

$$h(\mathcal{U}) \le sp\,(\epsilon) \le sr\,(\epsilon) \le h(\mathcal{V}).$$

Finally letting diam $(\mathcal{U}) \searrow 0$ we get (by Lemma 14.5)

$$h_{\text{top}}(X,T) = sp\,(T,d) = sr\,(T,d).$$

□

From (1.1) we also obtain the following corollary.

14.12. COROLLARY. *Set*

$$\underline{sp}\,(\epsilon,T) = \liminf_{n\to\infty} \frac{1}{n} \log sp_n(\epsilon),$$

$$\underline{sr}\,(\epsilon,T) = \liminf_{n\to\infty} \frac{1}{n} \log sr_n(\epsilon),$$

$$\underline{sp}\,(T,d) = \lim_{\epsilon\searrow 0} \underline{sp}\,(\epsilon,T), \quad \textit{and} \quad \underline{sr}\,(T,d) = \lim_{\epsilon\searrow 0} \underline{sr}\,(\epsilon,T),$$

then also

$$h_{\text{top}}(X,T) = \underline{sp}\,(T,d) = \underline{sr}\,(T,d).$$

14.13. PROPOSITION. *For a topological dynamical system* (X,T) *and for every* $k \in \mathbb{Z}$, $h_{\text{top}}(X,T^k) = |k| h_{\text{top}}(X,T)$.

PROOF. The equality $h_{\text{top}}(X,T^{-1}) = h_{\text{top}}(X,T)$ follows from the fact that for every open cover $\mathcal{U}$ we have

$$\begin{aligned} h(\mathcal{U},T) &= \lim_{n\to\infty} \frac{1}{n} H(\mathcal{U}_0^{n-1}) = \lim_{n\to\infty} \frac{1}{n} H(T^{n-1}\mathcal{U}_0^{n-1}) \\ &= \lim_{n\to\infty} \frac{1}{n} H(\mathcal{U} \vee T\mathcal{U} \vee \cdots T^{n-1} \vee \mathcal{U}) = h(\mathcal{U},T^{-1}). \end{aligned}$$

Now fix a positive integer k. If $E \subset X$ is an (n,ϵ)-separated set for the homeomorphism T^k then clearly E is also a (kn,ϵ)-separated set for T. It follows that $sr_n(\epsilon,T^k) \le sr_{kn}(\epsilon,T)$, hence $h(X,T^k) \le kh(X,T)$. On the other hand, given $\epsilon > 0$ there exists, by uniform continuity of the family T^j, $0 \le j \le k$, a positive δ such that $d(x,x') < \delta$ implies $\max_{0\le j\le k} d(T^j x, T^j x') < \epsilon$. From this we deduce that every (n,δ)-spanning subset for T^k is also a (kn,ϵ)-spanning set for T. Thus $sp_{kn}(\epsilon,T) \le sp_n(\delta,T^k)$, hence

$$\begin{aligned} k\,\underline{sp}(\epsilon,T) &= \liminf_{n\to\infty} \frac{k}{n} \log sp_n(\epsilon,T) \le \liminf_{n\to\infty} \frac{k}{kn} \log sp_{kn}(\epsilon,T) \\ &\le \limsup_{n\to\infty} \frac{1}{n} \log sp_n(\delta,T^k) = sp(\delta,T^k). \end{aligned}$$

In view of Corollary 14.12 this yields $kh(X,T) \le h(X,T^k)$. □

14.14. THEOREM. *Let* (X,T) *and* (Y,T) *be two dynamical systems, then*

$$h(X \times Y, T \times T) = h(X,T) + h(Y,T).$$

PROOF. Observe that with respect to the metric

$$d((x,y),(x',y')) = \max\{d((x,x'),d(y,y')\}$$

the set $E \times F$ is an (n,ϵ)-spanning set whenever E and F are (n,ϵ)-spanning subsets of X and Y respectively. This implies $sp_n(\epsilon, X\times Y) \le sp_n(\epsilon,X)\cdot sp_n(\epsilon,Y)$, whence

$h(X \times Y, T \times T) \leq h(X,T) + h(Y,T)$. On the other hand if E and F are (n,ϵ)-separated subsets of X and Y respectively, then $E \times F$ is an (n,ϵ)-separated subset of $X \times Y$, whence $sr_n(\epsilon, X \times Y) \geq sr_n(\epsilon, X) \cdot sr_n(\epsilon, Y)$. Now

$$\begin{aligned} sr(\epsilon, X \times Y) &\geq \limsup_{n\to\infty} \frac{1}{n} \left(\log sr_n(\epsilon, X) + \log sr_n(\epsilon, Y)\right) \\ &\geq \liminf_{n\to\infty} \frac{1}{n} \log sr_n(\epsilon, X) + \liminf_{n\to\infty} \frac{1}{n} \log sr_n(\epsilon, Y) \\ &= \underline{sr}(\epsilon, X) + \underline{sr}(\epsilon, Y). \end{aligned}$$

This implies $h(X \times Y, T \times T) \geq h(X,T) + h(Y,T)$ and the proof is complete. □

2. Measure entropy

In the sequel we shall repeatedly use the function $\phi : [0,1] \to \mathbb{R}^+$ defined by

$$\phi(t) = \begin{cases} -t \cdot \log t & t \neq 0 \\ 0 & t = 0. \end{cases}$$

14.15. EXERCISE. The function ϕ is strictly concave: for every $t_i \in [0,1], p_i \geq 0$ with $\sum_{i=1}^k p_i = 1$,

$$\phi\left(\sum_{i=1}^k p_i t_i\right) \geq \sum_{i=1}^k p_i \phi(t_i),$$

and equality holds iff all the t_i with $p_i \neq 0$ are equal.

Let $(X, \mathcal{X}, \mu)$ be a measure space, $\alpha = \{A_j\}_{j\in J}$ a measurable partition, where $J = J_\alpha$ is a finite or countable set. For $x \in X$ let $\alpha(x) = j$ when $x \in A_j$. The *information function* corresponding to α is defined by:

$$I(\alpha)(x) = -\sum_{j\in J} \mathbf{1}_{A_j}(x) \log \mu(A_j)(x),$$

and the *entropy* of α is the number $H(\alpha) \in [0,\infty]$:

$$\begin{aligned} H(\alpha) &= \sum_{j\in J} \phi(\mu(A_j)) = -\sum_{j\in J} \mu(A_j) \log \mu(A_j) \\ &= \int I(\alpha)(x)\, d\mu(x). \end{aligned}$$

More generally, for a σ-algebra $\mathcal{F} \subset \mathcal{X}$, the *conditional information function* and the *conditional entropy of α with respect to $\mathcal{F}$* are defined by:

$$I^{\mathcal{F}}(\alpha)(x) = -\sum_{j\in J} \mathbf{1}_{A_j}(x) \log \mu(A_j|\mathcal{F})(x),$$

and

$$\begin{aligned} H(\alpha|\mathcal{F}) &:= \int I^{\mathcal{F}}(\alpha)(x)\, d\mu(x) \\ &= -\sum_{j\in J} \int \mathbf{1}_{A_j}(x) \log \mu(A_j|\mathcal{F})(x) d\,\mu(x) \end{aligned}$$

where $\mu(A|\mathcal{F}) = \mathbb{E}^{\mathcal{F}}(\mathbf{1}_A)$.

Note that the function $I^{\mathcal{F}}(\alpha)(x)$ is not necessarily $\mathcal{F}$-measurable. However we have

$$\mathbb{E}^{\mathcal{F}}(I^{\mathcal{F}}(\alpha)) = -\sum_{j\in J} \mu(A_j|\mathcal{F}) \log \mu(A_j|\mathcal{F}) = \sum_{j\in J} \phi(\mu(A_j|\mathcal{F})),$$

whence

$$\begin{aligned} H(\alpha|\mathcal{F}) &= \int I^{\mathcal{F}}(\alpha)\, d\mu = \mathbb{E}(\mathbb{E}^{\mathcal{F}} I^{\mathcal{F}}(\alpha)) \\ &= \sum_{j\in J} \int \phi(\mu(A_j|\mathcal{F}))\, d\mu. \end{aligned} \tag{2.1}$$

Let $\mathcal{N} = \{\emptyset, X\}$ be the trivial σ-algebra, then clearly: $I^{\mathcal{N}}(\alpha) = I(\alpha)$, and $H(\alpha|\mathcal{N}) = H(\alpha)$. For $\mathcal{F} = \mathcal{X}$ we get: $I^{\mathcal{X}}(\alpha) = 0$ and $H(\alpha|\mathcal{X}) = 0$.

For a partition β let $\mathcal{F}(\beta)$ be the σ-algebra generated by β. We say that β is finer than α, denoted $\beta \geq \alpha$, if $\alpha \subset \mathcal{F}(\beta)$. When convenient we shall use β to denote $\mathcal{F}(\beta)$, thus we write $I^{\beta}(\alpha)$ and $H(\alpha|\beta)$ for $I^{\mathcal{F}(\beta)}(\alpha)$ and $H(\alpha|\mathcal{F}(\beta))$ respectively; or $\alpha \vee \mathcal{F}$ for $\mathcal{F}(\alpha) \vee \mathcal{F}$, where $\mathcal{F}$ is a σ-algebra, etc.

When α and β are countable partitions, denoting as usual for $A \in \alpha, B \in \beta$ $\mu(A|B) = \frac{\mu(A\cap B)}{\mu(B)}$, we have

$$I^{\beta}(\alpha)(x) = -\sum_{i\in J_\alpha} \sum_{j\in J_\beta} \mathbf{1}_{A_i\cap B_j}(x) \log \mu(A_i|B_j)(x),$$

and

$$H_\mu(\alpha|\beta) = -\sum_{i\in J_\alpha} \sum_{j\in J_\beta} \mu(A_i \cap B_j) \log \mu(A_i|B_j).$$

For partitions α and β, denote their *join*, by $\alpha \vee \beta = \{A_i \cap B_j : i \in J_\alpha, j \in J_\beta\}$. Given three countable partitions α, β and γ, an easy calculation gives the formula

$$I^{\gamma}(\alpha \vee \beta) = I^{\gamma}(\alpha) + I^{\alpha\vee\gamma}(\beta).$$

We prove the following, more general formula, for an arbitrary σ-field $\mathcal{F} \subset \mathcal{X}$:

14.16. PROPOSITION (The information and entropy cocycle equations). *For countable partitions α and β and a σ-algebra $\mathcal{F} \subset \mathcal{X}$:*

$$I^{\mathcal{F}}(\alpha \vee \beta) = I^{\mathcal{F}}(\alpha) + I^{\alpha\vee\mathcal{F}}(\beta) \quad \text{a.e.}$$

In particular for $\mathcal{F} = \mathcal{N}$, we get $I(\alpha \vee \beta) = I(\alpha) + I^{\alpha}(\beta)$. The corresponding equations for the entropy are:

$$\begin{aligned} H(\alpha \vee \beta|\mathcal{F}) &= H(\alpha|\mathcal{F}) + H(\beta|\alpha \vee \mathcal{F}) \quad \text{and} \\ H(\alpha \vee \beta) &= H(\alpha) + H(\beta|\alpha). \end{aligned}$$

PROOF. We shall first show that for $B \in \beta$

$$\mu(B|\alpha \vee \mathcal{F}) = \sum_{A\in\alpha} \mathbf{1}_A \frac{\mu(B \cap A|\mathcal{F})}{\mu(A|\mathcal{F})}, \tag{2.2}$$

as elements of $L^1(\alpha \vee \mathcal{F})$. To see this we integrate both sides against a function of the form $\mathbf{1}_{A'\cap F}$ with $A' \in \alpha, F \in \mathcal{F}$.

$$\begin{aligned}
\int \mathbf{1}_{A'\cap F}\Big(\sum_{A\in\alpha} \mathbf{1}_A \frac{\mu(B\cap A|\mathcal{F})}{\mu(A|\mathcal{F})}\Big)d\mu &= \int_F \mathbf{1}_{A'}\frac{\mu(B\cap A'|\mathcal{F})}{\mu(A'|\mathcal{F})}d\mu \\
&= \int_F \mathbb{E}^{\mathcal{F}}\big(\mathbf{1}_{A'}\frac{\mu(B\cap A'|\mathcal{F})}{\mu(A'|\mathcal{F})}\big)d\mu \\
&= \int_F \mu(A'|\mathcal{F})\frac{\mu(B\cap A'|\mathcal{F})}{\mu(A'|\mathcal{F})}d\mu \\
&= \int \mathbf{1}_F\mu(B\cap A'|\mathcal{F})d\mu = \mathbb{E}(\mathbf{1}_F\mathbb{E}^{\mathcal{F}}(\mathbf{1}_{B\cap A'})) \\
&= \mathbb{E}(\mathbb{E}^{\mathcal{F}}(\mathbf{1}_{B\cap A'\cap F})) = \mu(B\cap A'\cap F) \\
&= \int_{A'\cap F} \mathbf{1}_B = \int_{A'\cap F} \mathbb{E}^{\alpha\vee\mathcal{F}}(\mathbf{1}_B)d,\mu \\
&= \int \mathbf{1}_{A'\cap F}\mu(B|\alpha\vee\mathcal{F})d\mu.
\end{aligned}$$

Taking log on both sides of (2.2) we get

$$\log\mu(B|\alpha\vee\mathcal{F}) = \sum_{A\in\alpha}\mathbf{1}_A\big(\log\mu(A\cap B|\mathcal{F}) - \log\mu(A|\mathcal{F})\big),$$

and finally:

$$\begin{aligned}
I^{\mathcal{F}}(\alpha\vee\beta) &= -\sum_{A\in\alpha}\sum_{B\in\beta}\mathbf{1}_{A\cap B}\log\mu(A\cap B|\mathcal{F}) \\
&= -\sum_{A\in\alpha}\sum_{B\in\beta}\mathbf{1}_A\mathbf{1}_B\log\mu(A|\mathcal{F}) - \sum_{B\in\beta}\mathbf{1}_B\log\mu(B|\alpha\vee\mathcal{F}) \\
&= -\sum_{A\in\alpha}\mathbf{1}_A\log\mu(A|\mathcal{F}) - \sum_{B\in\beta}\mathbf{1}_B\log\mu(B|\alpha\vee\mathcal{F}) \\
&= I^{\mathcal{F}}(\alpha) + I^{\alpha\vee\mathcal{F}}(\beta).
\end{aligned}$$

□

14.17. COROLLARY. *For countable partitions $\beta_1, \beta_2, \ldots, \beta_N$*

$$I\Big(\bigvee_{j=1}^{N}(\beta_j)\Big) = I(\beta_1) + \sum_{j=1}^{N-1} I^{\beta_1\vee\cdots\vee\beta_j}(\beta_{j+1}).$$

PROOF. Use induction. □

14.18. PROPOSITION. *For countable partitions α and β and σ-algebras $\mathcal{F}, \mathcal{F}_1, \mathcal{F}_2 \subset \mathcal{X}$, we have*

1. *$H(\alpha|\mathcal{F}) \geq 0$ with equality iff α is $\mathcal{F}$-measurable, and $H(\alpha|\beta) \geq 0$ with equality iff $\alpha \leq \beta$.*
2. *$\mathcal{F}_2 \subset \mathcal{F}_1 \Rightarrow H(\alpha|\mathcal{F}_1) \leq H(\alpha|\mathcal{F}_2)$.*
3. *$\alpha \leq \beta \Rightarrow H(\alpha|\mathcal{F}) \leq H(\beta|\mathcal{F})$, hence $H(\alpha) \leq H(\beta)$.*
4. *$H(\alpha\vee\beta|\mathcal{F}) \leq H(\alpha|\mathcal{F}) + H(\beta|\mathcal{F})$, hence $H(\alpha\vee\beta) \leq H(\alpha) + H(\beta)$.*

5. $H(T^{-1}\alpha|T^{-1}\mathcal{F}) = H(\alpha|\mathcal{F})$ *hence* $H(T^{-1}\alpha) = H(\alpha)$.

PROOF. 1. The inequality $H(\alpha|\mathcal{F}) \geq 0$ is clear. When α is $\mathcal{F}$-measurable $\mathbb{E}^{\mathcal{F}}(\mathbf{1}_A) = \mathbf{1}_A$, whence

$$H(\alpha|\mathcal{F}) = \sum_{A\in\alpha} \int \mu(A|\mathcal{F}) \log \mu(A|\mathcal{F})\, d\mu = 0.$$

Suppose $H(\alpha|\mathcal{F}) = 0$; then for every $A \in \alpha$, $\mu(A|\mathcal{F}) \log \mu(A|\mathcal{F}) = 0$ hence $\mu(A|\mathcal{F}) = \mathbf{1}_F$ for some $F \in \mathcal{F}$. We then have $\mu(F) = \mathbb{E}(\mu(A|\mathcal{F})) = \mu(A)$, but also $\mu(F \cap A) = \mathbb{E}(\mathbf{1}_F \mathbf{1}_A) = \mathbb{E}(\mathbf{1}_F \mathbb{E}^{\mathcal{F}} \mathbf{1}_A) = \mathbb{E}(\mathbf{1}_F) = \mu(F)$. Thus $F = A \pmod{\mu}$ so that α is indeed $\mathcal{F}$-measurable.

2. By Jensen's inequality we have

$$\mathbb{E}^{\mathcal{F}_2}(\phi(\mu(A|\mathcal{F}_1))) \leq \phi(\mathbb{E}^{\mathcal{F}_2}\mu(A|\mathcal{F}_1)) = \phi(\mu(A|\mathcal{F}_2)),$$

hence

$$\int \phi(\mu(A|\mathcal{F}_1))\, d\mu \leq \int \phi(\mu(A|\mathcal{F}_2))\, d\mu.$$

Summation over the elements $A \in \alpha$ yields the required inequality by (2.1).

The proof of the remaining statements is an easy exercise. □

14.19. PROPOSITION. *For countable partitions α, β the following are equivalent*

1. *α is independent of β (i.e. $\mu(A \cap B) = \mu(A)\mu(B)$ for every $A \in \alpha$ and $B \in \beta$; we write $\alpha \perp \beta$).*
2. *$H(\alpha \vee \beta) = H(\alpha) + H(\beta)$.*
3. *$H(\alpha|\beta) = H(\alpha)$.*

PROOF. The equivalence of 2 and 3 follows from the identity $H(\alpha \vee \beta) = H(\alpha) + H(\beta|\alpha)$. Suppose $\alpha \perp \beta$, then

$$\mu(A|B) = \frac{\mu(A \cap B)}{\mu(B)} = \frac{\mu(A)\mu(B)}{\mu(B)} = \mu(A),$$

hence for $x \in A \cap B$, $I^\beta(\alpha)(x) = -\log \mu(A|B)(x) = -\log \mu(A) = I(\alpha)(x)$. Thus $I^\beta(\alpha) = I(\alpha)$, hence $H(\alpha|\beta) = \int I^\beta(\alpha) = \int I(\alpha) = H(\alpha)$. Conversely, if $H(\alpha|\beta) = H(\alpha)$ then, for the concave function $\phi(t) = -t \log t$ we have, by (2.1),

$$0 = \sum_{A\in\alpha} \{\phi(\mu(A)) - \int \phi(\mu(A|\beta))\, d\mu\}. \tag{2.3}$$

By Jensen's inequality

$$\begin{aligned}\phi(\mu(A)) &= \phi(\mathbb{E}(\mathbf{1}_A)) = \phi(\mathbb{E}(\mathbb{E}^\beta(\mathbf{1}_A))) \\ &\geq \mathbb{E}(\phi \circ \mathbb{E}^\beta(\mathbf{1}_A)) = \int \phi(\mu(A|\beta))\, d\mu.\end{aligned}$$

It follows that each summand in (2.3) is non-negative hence vanishes. Thus, for every $A \in \alpha$ we get

$$\begin{aligned}\int \phi(\mu(A|\beta)) d\mu &= \phi(\mu(A)) = -\mu(A) \log \mu(A) \\ &= -\sum_{B\in\beta} \mu(B) \frac{\mu(A \cap B)}{\mu(B)} \log \frac{\mu(A \cap B)}{\mu(B)}\end{aligned}$$

$$= -\sum_{B\in\beta} \mu(B)\mu(A|B)\log\mu(A|B).$$

Denoting $\lambda_j = \mu(B_j)$ and $x_j = \mu(A|B_j)$ we have

$$(\mu(A), \phi(\mu(A))) = \sum \lambda_j(x_j, \phi(x_j)) \tag{2.4}$$

But the point on the left hand side of (2.4) is on the graph of the concave function ϕ and therefore can be expressed as a convex combination of points on that graph only when all these points coincide $x_j = \mu(A) = \mu(A|B_j)$; i.e. $\alpha \perp \beta$. □

In the sequel we shall adopt the following notation for a finite or countable partition α and $-\infty < m < n < \infty$:

$$\alpha_m^n = T^{-m}\alpha \vee T^{-m-1}\alpha \vee \cdots \vee T^{-n}\alpha.$$

In particular

$$\alpha_0^{n-1} = \alpha \vee T^{-1}\alpha \vee \cdots \vee T^{-n+1}\alpha.$$

It will sometimes be convenient to use the shorthand notation

$$\alpha^- = \alpha_{-\infty}^{-1} = \bigvee_{n=1}^{\infty} T^n\alpha, \quad \text{and} \quad \alpha^T = \alpha_{-\infty}^{\infty} = \bigvee_{n=-\infty}^{\infty} T^n\alpha.$$

Here e.g. $\bigvee_{n=1}^{\infty} T^n\alpha$ is the σ-algebra generated by the union of the finite algebras $\mathcal{F}(\bigvee_{j=1}^{n} T^j\alpha)$.

14.20. PROPOSITION. *For a countable partition α the sequence*

$$H(\alpha \vee T^{-1}\alpha \vee \cdots \vee T^{-n+1}\alpha) = H(\alpha_0^{n-1}),$$

is subadditive, hence the following limit exists and equals the infimum:

$$\lim_{n\to\infty} \frac{1}{n} H(\alpha_0^{n-1}) = \inf_{n\in\mathbb{N}} \frac{1}{n} H(\alpha_0^{n-1}).$$

PROOF. The calculation

$$\begin{aligned}
&H(\alpha_0^{n+m-1}) = H(\alpha \vee T^{-1}\alpha \vee \cdots \vee T^{-(n+m)+1}\alpha)\\
=&H(\alpha \vee T^{-1}\alpha \vee \cdots \vee T^{-n+1}\alpha + T^{-n}(\alpha \vee T^{-1}\alpha \vee \cdots \vee T^{-m+1}\alpha))\\
\le&H(\alpha_0^{n-1}) + H(T^{-n}\alpha_0^{m-1}) = H(\alpha_0^{n-1}) + H(\alpha_0^{m-1})
\end{aligned}$$

implies subadditivity. The rest follows from the well known properties of such sequences (Theorem A.1). □

14.21. DEFINITION. We denote the limit $\lim_{n\to\infty} \frac{1}{n} H(\alpha_0^{n-1})$, whose existence was proved in Proposition 14.20, by $h(\alpha) = h_\mu(\alpha, T)$ and call it *the entropy of the partition α with respect to T*. Note that we have $h(\alpha) \le H(\alpha)$.

14.22. PROPOSITION. *For two countable partitions α and β*

$$h(\alpha) \le h(\beta) + H(\alpha|\beta).$$

PROOF. Using the entropy cocycle equation, Proposition 14.16, we get

$$\begin{aligned} H(\alpha_0^n) &\le H(\alpha_0^n \vee \beta_0^n) \\ &= H(\beta_0^n) + H(\alpha_0^n|\beta_0^n) \\ &\le H(\beta_0^n) + \sum_{j=0}^{n} H(T^{-j}\alpha|\beta_0^n) \\ &\le H(\beta_0^n) + \sum_{j=0}^{n} H(T^{-j}\alpha|T^{-j}\beta) \\ &= H(\beta_0^n) + (n+1)H(\alpha|\beta), \end{aligned}$$

and the proposition follows when the limit in Definition 14.21 is computed. □

When $\mathcal{F} \subset \mathcal{X}$ is a T-invariant σ-algebra then we can condition the various entropies on $\mathcal{F}$. All the calculations above go through and we then define $h_\mu(\alpha|\mathcal{F})$, the *conditional entropy of the partition* α *given* $\mathcal{F}$ *with respect to* T.

14.23. DEFINITION. For a dynamical system $\mathbf{X}$ the number

$$h_\mu(T) = h(\mathbf{X}) = \sup\{h_\mu(\alpha) : \alpha \text{ a finite measurable partition of } X\},$$

is called the *entropy* of the system $\mathbf{X}$. If $\pi : \mathbf{X} \to \mathbf{Y}$ is a factor then the *conditional entropy* of $\mathbf{X}$ over $\mathbf{Y}$ is defined as

$$h_\mu(\mathbf{X}|\mathcal{Y}) = h(\mathbf{X}|\mathbf{Y}) = \sup\{h_\mu(\alpha|\mathcal{Y}) : \alpha \text{ a finite measurable partition of } X\}.$$

14.24. PROPOSITION. *For a dynamical system* $\mathbf{X} = (X, \mathcal{X}, \mu, T)$ *and for every* $k \in \mathbb{Z}$ *we have* $h(T^k) = |k|h(T)$. *In particular* $h(T) = h(T^{-1})$.

PROOF. For $k \ge 0$ we have

$$\begin{aligned} h(\alpha_0^{k-1}, T^k) &= \lim_{n\to\infty} \frac{1}{n} H(\bigvee_{j=0}^{n-1} T^{-kj}(\bigvee_{i=0}^{k-1} T^{-i}\alpha)) = \lim_{n\to\infty} \frac{k}{nk} H(\bigvee_{j=0}^{nk-1} T^{-i}\alpha) \\ &= k \lim_{n\to\infty} \frac{1}{nk} H(\alpha_{j=0}^{nk-1}) = kh(\alpha, T), \end{aligned}$$

hence

$$\begin{aligned} kh(T) &= k\mathrm{sup}_\alpha h(\alpha, T) = \mathrm{sup}_\alpha h(\alpha_0^{k-1}, T^k) \\ &\le \mathrm{sup}_\beta h(\beta, T^k) = h(T^k). \end{aligned}$$

On the other hand

$$h(\alpha, T^k) \le h(\alpha_0^{k-1}, T^k) = kh(\alpha, T) \le kh(T),$$

hence $h(T^k) \le kh(T)$. For negative k it suffices to show that $h(\alpha, T) = h(\alpha, T^{-1})$. Now

$$\begin{aligned} H(\alpha_{-n+1}^0) &= H(\bigvee_{i=0}^{n-1} T^i\alpha) = H(T^{n-1} \bigvee_{i=0}^{n-1} T^{-i}\alpha) \\ &= H(\bigvee_{i=0}^{n-1} T^{-i}\alpha) = H(\alpha_0^{n-1}), \end{aligned}$$

and thus

$$h(\alpha, T^{-1}) = \lim_{n\to\infty} \frac{1}{n} H(\alpha_{-n+1}^{0}) = \lim_{n\to\infty} \frac{1}{n} H(\alpha_0^{n-1}) = h(\alpha, T).$$

□

14.25. THEOREM. *Let (X,T) be a topological system and α a finite or countable Borel partition of X.*

1. *The functions $\mu \mapsto h_\mu(\alpha)$ and $\mu \mapsto h_\mu(T)$ from $M_T(X)$ into $\mathbb{R} \cup \{\infty\}$ are affine functions.*
2. *For $\mu \in M_T(X)$, with ergodic decomposition, $\mu = \int_\Omega \mu_\omega \, dm(\omega)$, and a finite measurable partition α we have*

$$h_\mu(\alpha) = \int h_{\mu_\omega}(\alpha)\, dm(\omega) \quad \text{and} \quad h_\mu = \int h_{\mu_\omega}\, dm(\omega).$$

PROOF. Let $\mu = a\nu + (1-a)\eta$, where ν and η are in $M_T(X)$ and $0 < a < 1$. Using the concavity of $\phi(t) = -t\log t$ and the fact that $\log(t)$ is an increasing function we have for every $A \in \mathcal{X}$

$$\begin{aligned}
0 \leq& \phi(\mu(A)) - a\phi(\nu(A)) - (1-a)\phi(\eta(A)) \\
=& - a\nu(A)\big(\log \mu(A) - \log(a\nu(A))\big) \\
&- (1-a)\eta(A)\big(\log \mu(A) - \log((1-a)\eta(A))\big) \\
&- \nu(A) a \log a - \eta(A)(1-a)\log(1-a) \\
\leq& - \nu(A) a \log a - \eta(A)(1-a)\log(1-a) \\
=& \nu(A)\phi(a) + \eta(A)\phi(1-a),
\end{aligned}$$

hence

$$\begin{aligned}
0 &\leq H_\mu(\alpha_0^n) - aH_\nu(\alpha_0^n) - (1-a)H_\eta(\alpha_0^n) \\
&\leq \phi(a) + \phi(1-a).
\end{aligned}$$

Thus $h_\mu(\alpha) = ah_\nu(\alpha) + (1-a)h_\eta(\alpha)$, so that $\mu \mapsto h_\mu(\alpha)$ is indeed affine. The proof for $\mu \mapsto h_\mu(T)$ follows by choosing a partition α such that the entropies $h_\mu(\alpha)$, $h_\nu(\alpha)$ and $h_\eta(\alpha)$ approximate the corresponding suprema $h_\mu(T)$, $h_\nu(T)$ and $h_\eta(T)$. We leave the details of this argument as an exercise. We shall prove part 2 later as Theorem 15.12. □

3. Applications of the martingale convergence theorem

In the sequel we shall use the martingale convergence theorem; we refer to Doob [**48**] or Breiman [**35**] for the proof.

Recall that when $(\Omega, \mathcal{F}, \mu)$ is a probability space and $\mathcal{F}_1 \subset \mathcal{F}_2 \subset \cdots$ an increasing sequence of sub-σ-algebras, a sequence of L^1 functions $\{f_n\}_{n=1}^\infty$ is called a *martingale with respect to the filtration* $\{\mathcal{F}_n\}_{n=1}^\infty$ if for every n the function f_n is $\mathcal{F}_n$-measurable and

(3.1) $$\mathbb{E}^{\mathcal{F}_n}(f_{n+1}) = f_n \quad \mu - \text{a.e.}$$

When we say that the sequence $\{f_n\}_{n=1}^{\infty}$ is a martingale this means that it is a martingale with respect to the sequence $\mathcal{F}_n = \mathcal{F}\{f_1, f_2, \ldots, f_n\}$, the σ-algebra generated by the functions $f_1, f_2, \ldots, f_n$.

The martingale $\{f_n\}_{n=1}^{\infty}$ is *L^1-bounded* if $\sup_n \mathbb{E}(|f_n|) < \infty$.

The martingale convergence theorem ensures the almost everywhere convergence of a L^1-bounded martingale.

14.26. THEOREM (Martingale convergence theorem). 1. *Each L^1-bounded martingale converges a.e.*

2. *If $\{f_n\}_{n=1}^{\infty}$ is a L^1-bounded martingale and $\sup_n |f_n| = f^*$ is an L^1 function then the sequence $\{f_n\}_{n=1}^{\infty}$ converges in $L^1(\mu)$ norm to an L^1 function.*
3. *If $\{\mathcal{F}_n\}_{n=1}^{\infty}$ is an increasing sequence of sub-σ-algebras of $\mathcal{F}$ and $f \in L^1(\mu)$ then the sequence $f_n = \mathbb{E}^{\mathcal{F}_n}(f)$ is a martingale with respect to the filtration $\{\mathcal{F}_n\}_{n=1}^{\infty}$ and denoting $\mathcal{F}_\infty = \vee_{n=1}^{\infty}\mathcal{F}_n$ and $f_\infty = \mathbb{E}^{\mathcal{F}_\infty}(f)$ we have $f_n \to f_\infty$, both a.e. and in $L^1(\mu)$.*

14.27. LEMMA (Chung). *Let $(X, \mathcal{X}, \mu)$ be a probability space and α a countable partition with $H(\alpha) < \infty$.*

1. *If $\mathcal{F}_1 \subset \mathcal{F}_2 \subset \cdots$ is an increasing sequence of σ-algebras with $\mathcal{F}_n \nearrow \mathcal{F}$ (i.e. $\mathcal{F}$ is the σ-algebra generated by $\cup \mathcal{F}_n$), then: the function $f = \sup_n I^{\mathcal{F}_n}(\alpha)$ is an $L^1(\mu)$ function and*

$$\int f \, d\mu \leq H(\alpha) + 1,$$

2. *The same conclusion holds when $\mathcal{F}_1 \supset \mathcal{F}_2 \supset \cdots$ is a decreasing sequence of σ-algebras with $\mathcal{F}_n \searrow \mathcal{F}$ (i.e. $\mathcal{F} = \bigcap \mathcal{F}_n$).*

PROOF. 1. We would like to have an estimate on the sets $C = \{x : f(x) > t\}$. For a fixed element $A \in \alpha$ set $g_n = \mu(A|\mathcal{F}_n)$; then, since $I^{\mathcal{F}_n}(\alpha)(x) = -\log g_n(x)$ for $x \in A$, we see that $f(x) > t$ implies $g_n(x) < e^{-t}$ for some n. Define

$$C_n = \{x \in X : g_k(x) \geq e^{-t} \text{ for } k < n \text{ but } g_n(x) < e^{-t}\},$$

then clearly $C_n \in \mathcal{F}_n$, the sets C_n are disjoint and our previous observation implies that $C \cap A \subset \cup_n C_n$. As C_n is $\mathcal{F}_n$-measurable

$$\mu(A \cap C_n) = \int_{C_n} \mathbf{1}_A \, d\mu = \int_{C_n} g_n \, d\mu < e^{-t}\mu(C_n).$$

Summing over n we get

$$\mu(\{x \in A : f(x) > t\}) \leq e^{-t},$$

hence

$$\mu(\{x \in X : f(x) > t\}) \leq \sum_{A \in \alpha} \min(\mu(A), e^{-t}).$$

Finally, denoting $F(t) = \mu(\{x \in X : f(x) > t\})$, we have

$$\begin{aligned}
\int f \, d\mu &= -\int_0^\infty t \, dF(t) \\
&= \left[-tF(t)\right]_0^\infty + \int_0^\infty F(t) \, dt \\
&\leq \int_0^\infty \mu(\{x \in X : f(x) > t\}) \, dt
\end{aligned}$$

$$\leq \int_0^\infty \sum_{A\in\alpha} \min(\mu(A), e^{-t})\, dt$$

$$= \sum_{A\in\alpha} \Big(\int_0^{-\log\mu(A)} \mu(A)\, dt + \int_{-\log\mu(A)}^\infty e^{-t}\, dt\Big)$$

$$= H(\alpha) + \sum_{A\in\alpha} \mu(A) = H(\alpha) + 1.$$

2. For every $n \geq 1$ the finite decreasing sequence $\mathfrak{F}_1 \supset \mathfrak{F}_2 \supset \cdots \supset \mathfrak{F}_n$ can be regarded as an *increasing* sequence and part 1 then yields the conclusion that the function $f_n = \sup_{1\leq j\leq n} I^{\mathfrak{F}_j}(\alpha)$ is an $L^1(\mu)$ function with

$$\int f_n\, d\mu \leq H(\alpha) + 1.$$

Now, the sequence f_n is an increasing sequence of non-negative functions with $f_n \nearrow f = \sup_{n\geq 1} I^{\mathfrak{F}_n}(\alpha)$, and the monotone convergence theorem implies that $\int f_n\, d\mu \to \int f\, d\mu \leq H(\alpha) + 1$. □

14.28. THEOREM. *If α is a countable partition with $H(\alpha) < \infty$ and if $\mathfrak{F}_1 \subset \mathfrak{F}_2 \subset \cdots$ is an increasing sequence of sub-σ-algebras with $\mathfrak{F}_n \nearrow \mathfrak{F}$ ($\mathfrak{F}_1 \supset \mathfrak{F}_2 \supset \cdots$ is a decreasing sequence of sub-σ-algebras with $\mathfrak{F}_n \searrow \mathfrak{F}$) then $I^{\mathfrak{F}_n}(\alpha) \to I^{\mathfrak{F}}(\alpha)$ a.e. and in $L^1(\mu)$. Moreover $H(\alpha|\mathfrak{F}_n) \searrow H(\alpha|\mathfrak{F})$ ($H(\alpha|\mathfrak{F}_n) \nearrow H(\alpha|\mathfrak{F})$).*

PROOF. The convergence a.e. of the sequence $I^{\mathfrak{F}_n}(\alpha)$ to $I^{\mathfrak{F}}(\alpha)$ follows from the martingale convergence theorem 14.26. By Chung's lemma (Lemma 14.27) the sequence $I^{\mathfrak{F}_n}(\alpha)$ is dominated by an $L^1(\mu)$ function. This implies uniform integrability, hence $L^1(\mu)$ convergence. Integration yields $H(\alpha|\mathfrak{F}_n) \to H(\alpha|\mathfrak{F})$ and this is a monotone convergence since $H(\alpha|\mathfrak{F}_n)$ is decreasing (increasing). □

14.29. THEOREM. *Let α be a countable partition with $H(\alpha) < \infty$, then*

$$h(\alpha) = \lim_{n\to\infty} H(\alpha|\alpha_{-n}^{-1}) = H(\alpha|\alpha^-).$$

PROOF. We have

$$\begin{aligned}
h(\alpha) &= \lim_{n\to\infty} \frac{1}{n} H(\alpha \vee T^{-1}\alpha \vee \cdots \vee T^{-n+1}\alpha) \\
&= \lim_{n\to\infty} \frac{1}{n} \left(H(T^{-1}\alpha \vee \cdots \vee T^{-n+1}\alpha) + H(\alpha|T^{-1}\alpha \vee \cdots \vee T^{-n+1}\alpha)\right) \\
&= \lim_{n\to\infty} \frac{1}{n} \left(H(\alpha \vee \cdots \vee T^{-n+2}\alpha) + H(\alpha|\alpha_1^{n-1})\right) \\
&= \cdots \\
&= \lim_{n\to\infty} \frac{1}{n} \left(H(\alpha) + H(\alpha|T^{-1}\alpha) + \cdots + H(\alpha|\alpha_1^{n-1})\right).
\end{aligned}$$

Since, by Theorem 14.28, $H(\alpha|\alpha_1^{n-1}) \searrow H(\alpha|\alpha_1^\infty) = H(\alpha|\alpha^-)$, we get $h(\alpha) = H(\alpha|\alpha^-)$ as claimed. □

4. Kolmogorov-Sinai theorem

As in the case of topological entropy (see Lemma 14.5) one needs a more practical way of computing the entropy. This is provided by the following theorem and by the theorem of Kolmogorov and Sinai, 14.33.

14.30. THEOREM. *Let α_n be a sequence of finite partitions such that the corresponding σ-algebras satisfy $\mathcal{F}(\alpha_n) \nearrow \mathcal{X}$ then*

$$h(\mathbf{X}) = \lim_{n\to\infty} h(\alpha_n).$$

PROOF. For any finite partition β we have $H(\beta|\alpha_n) \to H(\beta|\mathcal{X}) = 0$, since by Proposition 14.22, $h(\beta) \leq h(\alpha_n) + H(\beta|\alpha_n)$ we get $h(\beta) \leq \lim_{n\to\infty} h(\alpha_n)$, hence

$$h(\mathbf{X}) = \sup_\beta h(\beta) = \lim_{n\to\infty} h(\alpha_n).$$

□

14.31. THEOREM. *If $\mathbf{X}$ and $\mathbf{Y}$ are two dynamical systems then*

$$h(\mathbf{X} \times \mathbf{Y}) = h(\mathbf{X}) + h(\mathbf{Y}).$$

PROOF. Since for finite partitions α and β of X and Y respectively the partitions $\alpha \times Y$ and $X \times \beta$ are independent partitions with respect to the joining $\mu \times \nu$, it follows easily that $h(\alpha \times \beta) = h(\alpha \times Y) + h(X \times \beta)$. Now choose sequences $\alpha_n \nearrow \mathcal{X}$ and $\beta_n \nearrow \mathcal{Y}$, then $\alpha_n \times \beta_n \nearrow \mathcal{X} \otimes \mathcal{Y}$ and by Theorem 14.30 we have

$$\begin{aligned} h(\mathbf{X} \times \mathbf{Y}) &= \lim_{n\to\infty} h(\alpha_n \times \beta_n) \\ &= \lim_{n\to\infty} (h(\alpha_n) + h(\beta_n)) \\ &= h(\mathbf{X}) + h(\mathbf{Y}). \end{aligned}$$

□

14.32. DEFINITION. A finite or countable partition α of X is called a *generator* for the dynamical system $\mathbf{X}$ if $\alpha^T = \alpha_{-\infty}^{\infty} = \mathcal{X}$. It is a *strong generator* if $\alpha^- = \alpha_{-\infty}^{-1} = \mathcal{X}$.

In Chapter 18 we shall prove a theorem of W. Krieger according to which every dynamical system with finite entropy has a finite generator (**[160]**). Here we prove the following useful theorem.

14.33. THEOREM (Kolmogorov-Sinai). *If α is a generator for the dynamical system $\mathbf{X}$ then*

$$h(\mathbf{X}) = h(\alpha).$$

PROOF. For every k we have

$$\begin{aligned} h(\alpha_0^k, T) &= \lim_{n\to\infty} \frac{1}{n} H\Big(\bigvee_{j=0}^{n-1} T^{-j}\Big(\bigvee_{i=0}^{k} T^{-i}\alpha\Big)\Big) = \lim_{n\to\infty} \frac{1}{n} H\Big(\bigvee_{i=0}^{k+n-1} T^{-i}\alpha\Big) \\ &= \lim_{n\to\infty} \frac{k+n}{n} \frac{1}{k+n} H(\alpha_0^{k+n-1}) = h(\alpha, T), \end{aligned}$$

and therefore also $h(\alpha_{-n}^{n}, T) = h(\alpha_0^{2n}, T) = h(\alpha, T)$. Finally by Proposition 14.22, for every finite partition β

$$\begin{aligned} h(\beta, T) &\leq h(\alpha_{-n}^{n}, T) + H(\beta|\alpha_{-n}^{n}) = h(\alpha, T) + H(\beta|\alpha_{-n}^{n}) \\ &\to h(\alpha, T) + H(\beta|\alpha_{-\infty}^{\infty}) = h(\alpha, T). \end{aligned}$$

□

5. Shannon-McMillan-Breiman theorem

14.34. LEMMA (Breiman). *Let* $\mathbf{X}$ *be a dynamical system,* $g_n, g \in L^1(\mu)$ *with* $g_n \to g$ *a.e. and in* $L^1(\mu)$. *If*

$$\int \sup\nolimits_n |g_n| \, d\mu < \infty,$$

then

$$\frac{1}{N} \sum_{n=1}^{N} g_n(T^n x) \to \mathbb{E}^{\mathcal{J}}(g), \qquad \text{a.e. and in } L^1(\mu)\ .$$

PROOF. By the ergodic theorem

$$\lim_{n \to \infty} \frac{1}{n} \sum_{j=0}^{n-1} g(T^j x) = g^* = \mathbb{E}^{\mathcal{J}}(g),$$

μ a.e. and in $L^1(\mu)$. Writing

$$\frac{1}{n} \sum_{j=0}^{n-1} g_j(T^j x) = \frac{1}{n} \sum_{j=0}^{n-1} g(T^j x) + \frac{1}{n} \sum_{j=0}^{n-1} g_j(T^j x) - g(T^j x),$$

we see that it suffices to show that

$$\lim_{n \to \infty} \frac{1}{n} \sum_{j=0}^{n-1} |g_j(T^j x) - g(T^j x)| = 0.$$

Set

$$G_N = \sup\nolimits_{j \geq N} |g_j(x) - g(x)| \leq \sup\nolimits_{j \geq N} \{|g_j(x)| + |g(x)|\},$$

and observe that by Lebesgue's theorem $\mathbb{E}^{\mathcal{J}}(G_N) \to 0$. Now fix N and for $n > N+1$ write

$$\begin{aligned} \frac{1}{n} \sum_{j=0}^{n-1} |g_j(T^j x) - g(T^j x)| &\leq \frac{1}{n} \sum_{j=0}^{N-1} |g_j(T^j x) - g(T^j x)| \\ &\quad + \frac{n-N-1}{n} \left(\frac{1}{n-N-1} \sum_{j=0}^{n-N-1} G_N(T^j(T^N x)) \right). \end{aligned}$$

The first summand tends to zero, the second converges to $\mathbb{E}^{\mathcal{J}}(G_N)$ (both μ a.e. and in $L^1(\mu)$). □

14.35. THEOREM (Shannon-McMillan-Breiman). *Let* $\mathbf{X}$ *be a dynamical system and* α *a countable partition with* $H(\alpha) < \infty$. *Denoting* $g = I^{\alpha_{-\infty}^{-1}}(\alpha) = I^{\alpha^-}(\alpha)$, *we have*

$$\frac{1}{N} I(\alpha_0^{N-1}) \to \mathbb{E}^{\mathcal{I}}(g), \text{ a.e. and in } L^1(\mu)$$

and

$$\frac{1}{N} H(\alpha_0^{N-1}) \to H(\alpha|\alpha^-) = h(\alpha, T).$$

When $\mathbf{X}$ *is ergodic,*

$$\frac{1}{N} I(\alpha_0^{N-1}) \to H(\alpha|\alpha^-) = h(\alpha, T), \quad \text{a.e. and in } L^1(\mu) .$$

PROOF. Set, for every k, $g_k(x) = I^{\alpha_{-k}^{-1}}(\alpha)(x)$. We have, by Corollary 14.17,

$$\begin{aligned} I(\alpha_0^{N-1}) &= I(T^{-N+1}\alpha \vee T^{-N+2}\alpha \vee \cdots \vee T^{-1}\alpha \vee \alpha) \\ &= \sum_{i=0}^{N-1} I^{\alpha_0^{i-1}}(T^{-i}\alpha) = \sum_{i=0}^{N-1} I^{T^{-i}\alpha_{-i}^{-1}}(T^{-i}\alpha) \\ &= \sum_{i=0}^{N-1} g_i(T^i x). \end{aligned}$$

The last equality follows from the identity

$$I^{T^{-1}\beta}(T^{-1}\alpha)(x) = I^{\beta}(\alpha)(Tx)$$

which is easily verified. By the corollary of Chung's lemma (Theorem 14.28),

$$g_k(x) = I^{\alpha_{-k}^{-1}}(\alpha)(x) \to I^{\alpha^-}(\alpha)(x) = g(x).$$

Also by Chung's lemma (Lemma 14.27)

$$\int \sup g_k = \int h < H(\alpha) + 1 < \infty.$$

Applying Breiman's lemma (Lemma 14.34), we deduce that a.e. as well as in $L^1(\mu)$

$$\lim \frac{1}{N} \sum_{i=1}^{N} g_i(T^i x) = \mathbb{E}^{\mathcal{I}}(g).$$

We have $H(\alpha|\alpha^-) = h(\alpha, T)$ by Theorem 14.29. Finally when $\mathbf{X}$ is ergodic $\mathbb{E}^{\mathcal{I}}(g) = \int g \, d\mu$ and $\frac{1}{N} I(\alpha_0^{N-1}) \to H(\alpha|\alpha^-) = h(\alpha, T)$. □

14.36. COROLLARY. *If* $\mathbf{X}$ *is an ergodic system and* α *a countable partition with* $H(\alpha) < \infty$ *then, given* $\epsilon > 0$ *and* $\delta > 0$, *there exists an* N *such that for* $n \geq N$

1. $$\exp(-n(h(\alpha) + \epsilon)) \leq \mu(A) \leq \exp(-n(h(\alpha) - \epsilon))$$
   for all the elements of α_0^{n-1} *with the exception of a subset of elements whose union has measure less than* δ.
2. *There exists an* α_0^{n-1}*-measurable set* E *of measure less than* δ, *such that the number* M *of elements of* α_0^{n-1} *which are contained in* $X \setminus E$ *satisfies*
   $$\exp(n(h(\alpha) - \epsilon)) \leq M \leq \exp(n(h(\alpha) + \epsilon)).$$

PROOF. By the SMB theorem $\frac{1}{n}I(\alpha_0^{n-1}) \to h(\alpha)$ a.e. and hence also in measure. Thus given $\epsilon, \delta > 0$ there exists an N such that for $n \geq N$

$$\mu\{x : |\frac{1}{n}I(\alpha_0^{n-1})(x) - h(\alpha)| \geq \epsilon\} < \delta. \tag{5.1}$$

Since $I(\alpha_0^{n-1})(x) = -\log \mu(A)$ for $x \in A \in \alpha_0^{n-1}$ we see that 1. holds for elements A of α_0^{n-1} which satisfy the condition in (5.1). This proves the first assertion and the second follows easily. □

6. Examples

1. Clearly the natural partition $P_0 = \{\mathbf{a}(j) : 1 \leq j \leq \ell\}$, where $\mathbf{a}(j) = \{x \in \Omega : x_0 = j\}$, induces a generator $\alpha = P_0 \cap X$ on every subshift $\mathbf{X}$ (where $\mu \in M_S(\Omega(\ell))$ and $X = \operatorname{supp}(\mu) \subset \Omega(\ell) = \{1, \ldots, \ell\}^{\mathbb{Z}}$; see Examples 3.40). Thus the Kolmogorov-Sinai theorem (Theorem 14.33) implies $h(\mathbf{X}) = h(\alpha)$. In particular for the Bernoulli shift $B(p) = B(p_1, \ldots, p_\ell)$ we have, by Proposition 14.19,

$$H(\alpha_0^{n-1}) = H(\alpha \vee S^{-1}\alpha \vee \cdots \vee S^{-n+1}\alpha) = nH(\alpha) = -n\sum_{j=1}^{\ell} p_j \log p_j,$$

hence

$$h_p(\Omega, S) = h(\alpha, S) = \lim_{n\to\infty} \frac{1}{n}H(\alpha_0^{n-1}) = H(\alpha) = -\sum_{j=1}^{\ell} p_j \log p_j.$$

We have already observed that this was the way by which Kolmogorov and Sinai have shown, e.g., that the Bernoulli systems $B(\frac{1}{2}, \frac{1}{2})$ with entropy $\log 2$ and $B(\frac{1}{3}, \frac{1}{3}, \frac{1}{3})$ with entropy $\log 3$ are not isomorphic. On the other hand Meshalkin's examples $B(\frac{1}{4}, \frac{1}{4}, \frac{1}{4}, \frac{1}{4})$ and $B(\frac{1}{2}, \frac{1}{8}, \frac{1}{8}, \frac{1}{8}, \frac{1}{8})$ demonstrate the fact that two Bernoulli systems corresponding to different probability vectors have the same entropy $\log 4$ and the entropy in this case can not decide the isomorphism question.

2. For a Markov system $M(p, P)$ we get (with $\phi(t) = -t \log t$ and $\mu = \mu_{(p,P)}$)

$$\begin{aligned}
H(\alpha_0^n) &= \sum \{\phi(\mu(A)) : A \in \alpha_0^n\} \\
&= \sum \phi\left(\mu(A_{i_0} \cap S^{-1}A_{i_1} \cap \cdots \cap S^{-n}A_{i_n})\right) \\
&= - \sum_{i_0, \ldots, i_n} \phi(p_{i_0} p_{i_0 i_1} \cdots p_{i_{n-1} i_n}) \\
&= - \sum p_{i_0} p_{i_0 i_1} \cdots p_{i_{n-1} i_n} \log(p_{i_0} p_{i_0 i_1} \cdots p_{i_{n-1} i_n}) \\
&= - \sum p_{i_0} p_{i_0 i_1} \cdots p_{i_{n-2} i_{n-1}} (p_{i_{n-1} i_n} \log(p_{i_0} p_{i_0 i_1} \cdots p_{i_{n-2} i_{n-1}}) \\
&\quad + p_{i_{n-1} i_n} \log(p_{i_{n-1} i_n})) \\
&= \sum_{i_0, \ldots, i_{n-1}} \left(\sum_{i_n} p_{i_{n-1} i_n}\right) \phi(p_{i_0} p_{i_0 i_1} \cdots p_{i_{n-2} i_{n-1}})
\end{aligned}$$

$$
\begin{aligned}
&+\sum_{i_{n-1}i_n}\left(\sum_{i_0,\dots,i_{n-2}} p_{i_0}p_{i_0i_1}\cdots p_{i_{n-2}i_{n-1}}\right)\phi(p_{i_{n-1}i_n})\\
&=\sum_{i_0,\dots,i_{n-1}}\phi(p_{i_0}p_{i_0i_1}\cdots p_{i_{n-2}i_{n-1}})+\sum_{i_{n-1}i_n} p_{i_{n-1}}\phi(p_{i_{n-1}i_n})\\
&=\quad\cdots\\
&=-\sum_i p_i\log p_i - n\sum_{i,j} p_ip_{ij}\log p_{ij}.
\end{aligned}
$$

Therefore we finally get for the entropy of the Markov shift $M(p,P)$:

$$h_\mu = \lim_{n\to\infty}\frac{1}{n+1}H(\alpha_0^n) = -\sum_{i,j} p_ip_{ij}\log p_{ij}.$$

3. For the dynamical system $\mathbf{X} = (\mathbb{T},\lambda,R_\beta)$, where $R_\beta(x) = x+\beta$ is an irrational rotation and λ is Lebesgue measure on $\mathbb{T}$, we take $\alpha = \{A, A^c\}$ with $A = [0,\frac{1}{2})$. Clearly $\alpha^- = \alpha_{-\infty}^{-1} = \mathcal{X}$ (the Borel σ-algebra of $\mathbb{T}$); i.e. α is a strong generator. In particular α is α^--measurable and (again using Theorem 14.33) we have

$$h(\mathbf{X}) = h_\lambda(\alpha) = H(\alpha|\alpha^-) = 0.$$

4. As we shall see (Exercise 18.18) every dynamical system with singular spectrum as well as every measure distal system (Theorem 18.19) has zero entropy.

7. Notes

In this chapter I followed, often very closely, the books by B. Weiss [**260**], M. Denker, C. Grillenberger and K. Sigmund [**46**], W. Parry [**203**], and K. Petersen [**205**].

CHAPTER 15

Symbolic Representations

Symbolic dynamics was used in the investigation of smooth dynamical systems already by J. Hadamard. The fundamental work of R. L. Adler and B. Weiss [**5**], [**6**] introduced, for the first time, *Markov Partitions* as a systematic tool for studying smooth systems by means of symbolic systems (see also K. R. Berg [**18**] and refer to R. Bowen [**33**] for the definition of Markov partitions). For a general measure-preserving $\mathbb{Z}$ dynamical system the symbolic representation that is associated to a finite partition of the system is an indispensable tool.

The present chapter prepares the combinatorial ground needed for the three "heavy" theorems that we present in the second part of the book: the Jewett-Krieger 'unique ergodicity representation theorem', which we prove at the end of the chapter, Krieger's 'finite generator theorem', and Ornstein's 'isomorphism theorem', which will be presented and proved in the last chapter. In Section 1 we look once more at symbolic (or subshift) systems and define two basic tools for dealing with the combinatorics of such systems: (ϵ, N)-genericity and the Hamming distance. Section 2 introduces combinatorial tools constructed for a general dynamical system: the Rohlin and Kakutani towers. In Section 3 we study the symbolic representation that is associated with a finite partition of a dynamical system, and in Section 4 the notion of an (ϵ, N)-generic point is pulled back from a symbolic representation to the original system. This is used in the next two sections (Sections 5 and 6) to obtain "tower versions" of the ergodic theorem and the SMB theorem respectively. In Section 7 Ornstein's $\bar{d}$-metric is introduced and its close relationship with entropy and the Hamming distance is investigated. In the last section (Section 8) we describe in details Weiss' method of construction of uniform partitions which is then used to achieve a relative version of the Jewett-Krieger theorem.

1. Symbolic systems

For an integer $\ell \geq 2$, let $\mathfrak{L} = \{1, 2, \ldots, \ell\}$ (the alphabet or the set of *states* or *symbols*) and set $\Omega = \Omega(\ell) = \mathfrak{L}^{\mathbb{Z}}$. Let $S : \Omega \to \Omega$ be the shift transformation: $(Sx)_n = x_{n+1}$. The topological dynamical system (Ω, S), the topological Bernoulli shift, and its topological subsystems, the subshifts, were studied in earlier chapters. In the present section we are interested in the *symbolic* (measure theoretical) systems (Ω, μ, S) obtained by singling out an S-invariant probability measure μ in $M_S(\Omega)$.

For integers $m < n$ and a (finite or infinite) sequence $(a_j)_{j\in\mathbb{Z}}$ with $a_j \in \mathfrak{L}$, we shall use the notation $\mathbf{a}_m^n$ for both the sequence $(a_m, a_{m+1}, \ldots, a_n)$ and the *cylinder set* $\{x \in \Omega : x_j = a_j,\ m \leq j \leq n\} = [a_m, a_{m+1}, \ldots, a_n]_m^n$. Denote the partition of Ω into $(n+1-m)$-cylinders $\mathbf{a}_m^n$ by P_m^n and let $P_n = P_{-n}^n$. An S-invariant Borel probability measure μ on Ω is uniquely defined by the values $\mu(\mathbf{a}_m^n)$ and conversely every assignment $\mathbf{a}_m^n \mapsto \mu(\mathbf{a}_m^n)$ which satisfies:

1. $0 \leq \mu(\mathbf{a}_1^n) \leq 1$,
2. $\mu(\Omega) = 1$,
3. $\mu(\mathbf{a}_1^k) = \sum\{\mu(\mathbf{a}_1^{k+1}) : \mathbf{a}_1^{k+1}$ is an extension of $\mathbf{a}_1^k\}$, and
4. $\mu(\mathbf{a}_m^n) = \mu(\mathbf{b}_{m+1}^{n+1})$ when $b_{j+1} = a_j$, $j = m, m+1, \ldots, n$,

defines such a probability measure on Ω.

The collection of all S-invariant Borel probability measures is denoted, as usual, by $M_S(\Omega)$ and we use the metric

$$d_{\text{meas}}(\mu, \nu) = \sum_{n=1}^{\infty} 2^{-n} d_n(\mu, \nu), \tag{1.1}$$

where

$$d_n(\mu, \nu) = \frac{1}{2} \sum_{\mathbf{a}_{-n}^n \in P_n} |\mu(\mathbf{a}_{-n}^n) - \nu(\mathbf{a}_{-n}^n)|. \tag{1.2}$$

The corresponding topology is the weak* topology so that with the metric d_{meas}, $M_S(\Omega)$ is a compact convex metric space. Recall that the set of extreme points of $M_S(\Omega)$ coincides with the set $M_S^{\text{erg}}(\Omega)$ of ergodic S-invariant probability measures.

15.1. LEMMA. *The function $h : \mu \mapsto h_\mu$, $h : M_S(\Omega) \to \mathbb{R}$ is upper semi-continuous. In other words: if μ_n is a sequence of measures in $M_S(\Omega)$ converging to $\mu \in M_S(\Omega)$ then*

$$\limsup_{n \to \infty} h_{\mu_n} \leq h_\mu.$$

PROOF. By the Kolmogorov-Sinai theorem (Theorem 14.33) we have $h_\nu = h_\nu(P_0)$ for every $\nu \in M_S(\Omega)$. By Theorem 14.29, $h_\mu = h_\mu(P_0)$ is the limit of the decreasing sequence $H_\mu(P_0|P_{-k}^{-1})$. Given $\epsilon > 0$ there exists therefore a k with $H_\mu(P_0|P_{-k}^{-1}) \leq h_\mu + \epsilon$. This implies that for sufficiently large n

$$h_{\mu_n} \leq H_{\mu_n}(P_0|P_{-k}^{-1}) \leq H_\mu(P_0|P_{-k}^{-1}) + \epsilon \leq h(\mu) + 2\epsilon.$$

□

15.2. DEFINITION. 1. For positive integers $N \geq 2n+1$ and sequences $\mathbf{x}_1^N$ and $\mathbf{a}_1^{2n+1}$ set

$$p(\mathbf{a}_1^{2n+1}|\mathbf{x}_1^N) = \frac{\operatorname{card}\{i \in [n+1, N-n] : \mathbf{x}_{i-n}^{i+n} = \mathbf{a}_1^{2n+1}\}}{N - 2n},$$

the *density* or *empirical distribution* of occurrences of $\mathbf{a}_1^{2n+1}$ in $\mathbf{x}_1^N$.

2. For a measure $\nu \in M_S(\Omega)$ an $\epsilon > 0$ and integers $N \geq 2n+1$ let

$$G(\nu, \epsilon, N, 2n+1) = \\ \{\mathbf{x}_1^N \in \mathfrak{L}^N : |p(\mathbf{a}_1^{2n+1}|\mathbf{x}_1^N) - \nu(\mathbf{a}_1^{2n+1})| < \epsilon \ \forall\ \mathbf{a}_1^{2n+1} \in \mathfrak{L}^{2n+1}\}.$$

We call the sequences in $G(\nu, \epsilon, N, 2n+1)$ the $(\nu, \epsilon, N, 2n+1)$-*typical* or $(\epsilon, N, 2n+1)$-*typical for* ν.

3. Denote $L = L(\epsilon) = [\log(1/\epsilon)] + 1$ and let

$$M = \min\{\nu(\mathbf{a}_1^{2L+1}) : \nu(\mathbf{a}_1^{2L+1}) > 0\}.$$

We say that a sequence $\mathbf{x}_1^N \in \mathfrak{L}^N$ is (ν, ϵ, N)*-generic* or (ϵ, N)*-generic for* ν if for every $0 \le n \le L$ and every extension $\mathbf{x}_{-n+1}^{N+n}$ of $\mathbf{x}_1^N$, $\mathbf{x}_{-n+1}^{N+n}$ is $(\nu, \epsilon M, N+n, 2n+1)$-typical; i.e.

$$\mathbf{x}_{-n+1}^{N+n} \in G(\nu, \epsilon M, N+n, 2n+1)$$

or

$$|p(\mathbf{a}_1^{2n+1} | \mathbf{x}_{-n+1}^{N+n}) - \nu(\mathbf{a}_1^{2n+1})| < \epsilon M, \qquad \forall \mathbf{a}_1^{2n+1} \in \mathfrak{L}^{2n+1}.$$

15.3. Definition. The *Hamming distance* between two sequences $\mathbf{a}, \mathbf{b} \in \mathfrak{L}^N$ is defined by

$$d_{\text{ham}}(\mathbf{a}, \mathbf{b}) = \frac{1}{N} \text{card}\,\{n : a_n \ne b_n\}.$$

Clearly d_{ham} is a metric on $\mathfrak{L}^N$ and it follows that

$$d_{\text{ham}}(x, y) = \limsup_{n \to \infty} d_{\text{ham}}(x_{-n}^n, y_{-n}^n)$$

is a pseudometric on $\Omega(\ell)$.

2. Kakutani, Rohlin and K-R towers

Let $\mathbf{X}$ be a dynamical system and $B \subset X$; an array $\mathfrak{c} = \{B, TB, \dots, T^{N-1}B\}$ with T^jB, $0 \le j < N$ pairwise disjoint is called a *Rohlin tower* or a *column over* B *of height* N. The set B is called the *base* of the tower and $T^{N-1}B$ is its *roof*. Let $|\mathfrak{c}| = \cup_{j=0}^{N-1} T^jB$, the *carrier* of the tower $\mathfrak{t}$. A collection (finite or countable) $\mathfrak{t}$ of disjoint columns $\mathfrak{c}_k$ (with bases B_k and heights N_k) is called a *tower* and we let $|\mathfrak{t}| = \cup_{j=0}^{\infty} |\mathfrak{c}_k|$. The union of the bases $B = \cup_k B_k$ is the *base* of $\mathfrak{t}$, and the union of the roofs is the *roof* of $\mathfrak{t}$. We sometimes write (a bit imprecisely) $\mathfrak{t} = \{\mathfrak{c}_k : k = 1, 2, \dots\}$. The sets $\{T^ix : \ 0 \le i < N_k(x)\}$ for $x \in B$ are called the *fibers* of $\mathfrak{t}$.

Given a tower $\mathfrak{t}$ with columns $\{\mathfrak{c}_k : k = 1, 2, \dots\}$, base $B = \cup_k B_k$, and a finite (or countable) partition $\alpha = \{A_1, \dots, A_t\}$, we define an equivalence relation on B as follows: $x \sim y$ iff x and y are in the same B_k and for every $0 \le j < N_k$, T^jx and T^jy are in the same element of α; i.e. x and y have the same (α, N_k)*-name*. We now consider each equivalence class $B_{k,\mathbf{a}}$, with $\mathbf{a}$ a name in $\alpha_0^{N_k-1}$, as a basis of a column $\mathfrak{c}_{k,\mathbf{a}} = \{B_{k,\mathbf{a}}, TB_{k,\mathbf{a}}, \dots, T^{N_k-1}B_{k,\mathbf{a}}\}$ and say that the resulting tower $\mathfrak{t}_\alpha = \{\mathfrak{c}_{k,\mathbf{a}} : \mathbf{a} \in \alpha_0^{N_k-1}, \ k = 1, \dots\}$ is the tower $\mathfrak{t}$ *refined* according to α. When $\alpha = \{C, X \setminus C\}$, C a subset of X, we write $\mathfrak{t}_\alpha = \mathfrak{t}_C$ and say that $\mathfrak{t}$ is refined according to C.

A subset B of X is called a *sweeping set* if $\cup_{n \ge 0} T^nB = X$. For an ergodic system every set of positive measure is sweeping. Given any set B of positive measure, define the *return time function* $r_B : B \to \mathbb{N} \cup \{\infty\}$ by

$$r_B(x) = \min\{n \ge 1 : T^nx \in B\},$$

when this minimum is finite and $r_B(x) = \infty$ otherwise. Let $B_k = \{x \in B : r_B(x) = k\}$ and note that by Poincaré's recurrence theorem B_∞ is a null set. We let $\mathfrak{c}_k$ be the column $\{B_k, TB_k, \dots, T^{k-1}B_k\}$ and we call the tower $\mathfrak{t} = \mathfrak{t}(B) = \{\mathfrak{c}_k : k = 1, 2, \dots\}$, the *Kakutani tower over* B. Thus the Kakutani tower $\mathfrak{t}$ is a partition of the set $|\mathfrak{t}|$. For a sweeping set B, $|\mathfrak{t}| = X$ and the Kakutani tower over B is then a partition of the whole space. If the Kakutani tower over B has finitely many columns (i.e. the function r_B is bounded) we say that B has a *finite height* and we call the Kakutani tower over B a *K-R tower*. The number $\max r_B$ is called the *height* of B or the *height* of the K-R tower.

Note that for an ergodic $\mathbb{Z}$-system $\mathbf{X}$, either the space X consists of a finite set of points on which μ is equidistributed, or the measure μ is atom-less. In the first case the system is called *periodic*, and it is called *non-periodic* in the latter.

15.4. THEOREM (Rohlin's lemma). *Let* $\mathbf{X}$ *be a non-periodic ergodic system,* N *a positive integer and* $\epsilon > 0$*, then there exists a subset* B *such that the sets* $B, TB, \dots, T^{N-1}B$ *are pairwise disjoint and* $\mu(\cup_{j=0}^{N-1} T^j B) > 1 - \epsilon$.

PROOF. Let $C \subset X$ be a set with measure $0 < \mu(C) < \epsilon/N$. Consider the Kakutani tower $\mathfrak{t}(C)$ over C. For every $k \geq N$ divide the column $\mathfrak{c}_k = \{T^i C_k : 0 \leq i < k\}$, starting from its basis $C_k = \{x \in C : r_C(x) = k\}$, into blocks of size N. Mark the first level of each of these blocks as belonging to B. Taking the union of these marked levels over the columns $\mathfrak{c}_k, k = N, N+1, \dots,$ gives us a set B with $r_B \geq N$. Clearly $\mathfrak{q} = \{B, TB, \dots, T^{N-1}B\}$ is a Rohlin tower; i.e. the sets $T^j B$ for $0 \leq j < N$ are disjoint. Now $|\mathfrak{t}| \setminus |\mathfrak{q}| = X \setminus |\mathfrak{q}|$ is composed of the first N columns $\mathfrak{c}_k$ of height $k < N$ and some top levels from the other columns, with a contribution of at most $N - 1$ levels from each. A simple calculation shows therefore that

$$\mu(X \setminus |\mathfrak{q}|) < N\mu(C) < \epsilon.$$

□

15.5. THEOREM. *Let* $\mathbf{X}$ *be a non-periodic ergodic system.*

1. *For any positive integer* N *and* $\epsilon > 0$ *there exists a set* C *of finite height such that the K-R tower* $\mathfrak{t}(C)$ *satisfies* range $r_C \subset \{N, N+1\}$ *and* $\mu(|\mathfrak{t}(C)|) < \epsilon$.
2. *Given a K-R tower with base* C *and height* N*, for any* $n \geq 1$ *there is a bounded K-R tower with base* D *contained in* C *whose column heights are all at least* n *and at most* $n + 4N$.

PROOF. 1. Let $n > 10N^2$ and use Rohlin's lemma to construct a Rohlin tower $\mathfrak{q} = \{B, TB, \dots, T^{n-1}B\}$ with $\mu(|\mathfrak{t}(B)|) < \epsilon$. Thus the return time function $r_B(x)$ is greater than $10 \cdot N^2$ on B. Let

$$B_k = \{x \in B : r_B(x) = k\}.$$

When B_k is non-empty one can therefore write k as a positive combination of N and $N + 1$, say

$$k = Nu_k + (N+1)v_k.$$

Now divide the column $\mathfrak{c}_k = \{T^i B_k : 0 \leq i < k\}$ into u_k blocks of size N and v_k blocks of size $N+1$. The set C is now defined as the union over the various columns of the first levels of these blocks . Clearly the function r_C takes only two values, either N or $N + 1$ as required. If n is sufficiently large we also have $\mu(|\mathfrak{t}(C)|) < \epsilon$.

2. Start with a Rohlin tower $\mathfrak{q} = \{B, TB, \dots, T^{M-1}B\}$ with $M = 10(n+2N)^2$, and look at the unbounded (in general) Kakutani tower over B. Since, by ergodicity, C is a sweeping set we can choose $B \subset C$. Refine this tower according to C

and call the refined tower $\mathfrak{t}$. For each $m \geq 10\ (n+2N)^2$, the column $\mathfrak{q}_m$ over $B_m = \{x \in B : r_B = m\}$, is now split into a finite number of columns so that each level is either a subset of C or of C^c.

As in the first part of the proof we partition the tower $\mathfrak{t}$ into blocks of sizes $n+2N$ and $n+2N+1$. Since we want the base D of the K-R tower we construct to belong to C we move the base level of each block to the nearest level that belongs to C. Since the height of C is N we do not move these levels more than $N-1$ steps. The new blocks, with bases in C, are of size between n and $n+4N$, and we let D be the union of these bases. □

3. Partitions and symbolic representations

Let $\mathbf{X}$ be a dynamical system and $\alpha = \{A_j\}_{1\leq j\leq \ell}$ a finite partition (we usually assume $\mu(A_j) > 0$ for all j). We sometimes think of the partition α as a function $\xi_0 : X \to \mathfrak{L} = \{1, 2, \ldots, \ell\}$ defined by $\xi_0(x) = j$ for $x \in A_j$. The pair $(\mathbf{X}, \alpha)$ is traditionally called a *process*. The reason for this terminology is that with this pair one associates the sequence of random variables or the *stationary stochastic process* $\{\xi_n = \xi_0 \circ T^n : n \in \mathbb{Z}\}$. This sequence defines a homomorphism ϕ_α : $\mathbf{X} \to (\Omega, (\phi_\alpha)_*(\mu), S)$ of $\mathbf{X}$ into the symbolic system $(\Omega, (\phi_\alpha)_*(\mu), S)$ $(\Omega = \Omega(\ell) = \{1, 2, \ldots, \ell\}^{\mathbb{Z}}$ and $S =$ the shift), given by $\phi_\alpha(x) = \omega \in \Omega$, where

$$\omega_n = \xi_n(x) = \xi_0(T^n x),$$

so that $\omega_n(x) = j$ iff $T^n x \in A_j$. This homomorphism is sometimes called the *name map*; we shall call it the *symbolic representation* of the process $(\mathbf{X}, \alpha)$. When there is no room for confusion we may write, for a point $x \in X$ and integers $m < n$, x_m^n for the sequence or "word"

$$\phi_\alpha(x)_m^n = (\phi_\alpha(x)_m, \phi_\alpha(x)_{m+1}, \ldots, \phi_\alpha(x)_n) = \omega_m\omega_{m+1}\cdots\omega_n.$$

We denote the distribution of the stochastic process, $(\phi_\alpha)_*(\mu)$, by $\rho = \rho(\mathbf{X}, \alpha)$ and call it the *symbolic representation measure* of $(\mathbf{X}, \alpha)$. When it is clear which process is being referred to we may, given a name $\mathbf{a} \in \mathfrak{L}^{n-m+1}$, write $\mu(\mathbf{a}_m^n)$ instead of $\rho(\mathbf{a}_m^n)$ so that here $\mathbf{a}_m^n$ also denotes the set $\{x \in X : \phi_\alpha(x)_m^n = \mathbf{a}_m^n\}$.

15.6. LEMMA. *For a dynamical system* $\mathbf{X}$ *and a finite partition* $\alpha = \{A_j\}_{1\leq j\leq \ell}$ *we have*

$$h_\mu(\alpha) = h(\Omega(\ell), \rho(\mathbf{X}, \alpha), S).$$

PROOF. By Kolmogorov-Sinai's theorem (Theorem 14.33) $h(\Omega(\ell), \rho(\mathbf{X}, \alpha), S)$ $= h(P_0, \rho(\mathbf{X}, \alpha))$ and the equality

$$h_\mu(\alpha) = h(P_0, \rho(\mathbf{X}, \alpha)) = \lim_{n\to\infty} \frac{1}{n} \sum \{-\rho(\mathbf{a}_0^{n-1}) \log \rho(\mathbf{a}_0^{n-1}) : \mathbf{a}_0^{n-1} \in P_n\}$$

is obvious especially when one uses the above notation. □

A convenient way of producing partitions, and thus also symbolic representations, is by *copying* or *painting* names on towers . If $\mathfrak{t} = \{B, TB, \ldots, T^{N-1}B\}$ is a tower and $\mathbf{a}_0^{N-1} \in \mathfrak{L}^N$ then *copying the name* $\mathbf{a}_0^{N-1}$ *on the tower* $\mathfrak{t}$ means that on $|\mathfrak{t}| = \cup_{j=0}^{N-1} T^j B$ we define a partition by letting

$$A_k = \cup\{T^j B : a_j = k\}, \qquad k \in \mathfrak{L} = \{1, 2, \ldots, \ell\}.$$

In other words, thinking on a partition as a coloring, we paint the level T^jB with the color k if and only if $a_j = k$. If there are t towers $\mathfrak{t}_i = \{B_i, TB_i, \dots, T^{N-1}B_i\}$ and t names, $\mathbf{a}(i)_0^{N-1}$ $i = 1, 2, \dots, t$, then copying these names on the corresponding towers means that we define a partition of $\cup_{i=1}^t |\mathfrak{t}_i|$ by

$$A_k = \cup\{T^jB_i : a(i)_j = k,\ i = 1, 2, \dots, t\}, \qquad k \in \mathfrak{L} = \{1, 2, \dots, \ell\}.$$

These partitions can be extended to a partition $\alpha = \{A_1, A_2, \dots, A_\ell\}$ of the whole space by assigning, for example, the value 1 to the rest of the space.

Note that for a point $x \in B_i$, an integer n with $2n + 1 \le N$ and any element $C \in \alpha_{-n}^n$ with name $\mathbf{c}_{-n}^n$,

$$C = C(\mathbf{c}_{-n}^n) = T^n A_{c_n} \cap \dots \cap A_{c_0} \cap T^{-1} A_{c_{-1}} \cap \dots \cap T^{-n} A_{c_{-n}},$$

we have $x_0^{N-1} = a(i)_0^{N-1}$, $\mu(C) = \rho(\mathbf{c}_{-n}^n)$ and

$$p(\mathbf{c}_{-n}^n | x_0^{N-1}) = \frac{\mu(C \cap |\mathfrak{t}_i|)}{\mu(|\mathfrak{t}_i|)}. \tag{3.1}$$

15.7. Definition. Given a dynamical system $\mathbf{X}$ and $\ell \ge 2$ we define a metric on the collection of ℓ-set partitions of X as follows:

$$\begin{aligned} d_{\text{part}}(\alpha, \beta) &= \frac{1}{2}\|\alpha - \beta\|_1 = \frac{1}{2}\sum_{j=1}^{\ell} \int |\mathbf{1}_{A_j} - \mathbf{1}_{B_j}| \, d\mu \\ &= \frac{1}{2}\sum_{j=1}^{\ell} \mu(A_j \triangle B_j) = \mu\{x \in X : \alpha(x) \ne \beta(x)\}. \end{aligned}$$

Clearly the space $(\mathcal{P}_\ell, d_{\text{part}})$ of ℓ-set partitions is a complete metric space. Note that this metric takes into account the order of the partition's elements. If we need to compare two partitions of different cardinality we can take some of the sets to be the empty set. We shall often want to measure the distance between two partitions α and β on a measure space $(X, \mathcal{X})$ with respect to various probability measures μ on X. To emphasize the dependence of the distance on the measure we shall then write

$$d_{\text{part}}^{\mu}(\alpha, \beta) = \mu(\alpha \triangle \beta).$$

15.8. Lemma. *Let $\mathbf{X}$ be an ergodic system and $\alpha = \{A_1, \dots, A_\ell\}$ and $\beta = \{B_1, \dots, B_\ell\}$ two ℓ-set partitions.*

1. *For every integer $N \ge 1$*

$$d_{\text{part}}^{\mu}(\alpha_{-N}^N, \beta_{-N}^N) = \mu(\alpha_{-N}^N \triangle \beta_{-N}^N) \le (2N+1)\mu(\alpha \triangle \beta) = (2N+1) d_{\text{part}}^{\mu}(\alpha, \beta).$$

2. *For every $\epsilon > 0$ there exists a $\delta > 0$ such that $d_{\text{part}}^{\mu}(\alpha, \beta) < \delta$ implies $d_{\text{meas}}(\rho(\mu, \alpha), \rho(\mu, \beta)) < \epsilon$.*
3. *For any partition γ we have*

$$d_{\text{part}}^{\mu}(\alpha \vee \gamma, \beta \vee \gamma) = d_{\text{part}}^{\mu}(\alpha, \beta).$$

Proof. 1. This is a tedious computation. As a hint let us observe that e.g. $(A_1 \cap TA_1) \setminus (B_1 \cap TB_1) \subset (A_1 \setminus B_1) \cup T(A_1 \setminus B_1)$.

2. Denoting $\rho_1 = \rho(\mu, \alpha)$ and $\rho_2 = \rho(\mu, \beta)$ we have

$$\begin{aligned} d_{\text{meas}}(\rho_1, \rho_2) &= \sum_{n=1}^{\infty} 2^{-n-1}\Big(\sum_{\mathbf{a}\in P_n} |\rho_1(\mathbf{a}) - \rho_2(\mathbf{a})|\Big) \\ &= \sum_{n=1}^{\infty} 2^{-n-1}\Big(\sum_{\mathbf{a}\in P_n} |\mu(A_{\mathbf{a}}) - \mu(B_{\mathbf{a}})|\Big) \\ &\le \sum_{n=1}^{\infty} 2^{-n-1}\Big(\sum_{\mathbf{a}\in P_n} \mu(A_{\mathbf{a}} \triangle B_{\mathbf{a}})\Big). \end{aligned}$$

Since by part 1,

$$\begin{aligned} \sum_{\mathbf{a}\in P_n} \mu(A_{\mathbf{a}} \triangle B_{\mathbf{a}}) &\le (2N+1) \sum_{1\le j\le \ell} \mu(A_j \triangle B_j) \\ &= 2(2N+1) d^{\mu}_{\text{part}}(\alpha, \beta), \end{aligned}$$

we can choose an N with $\sum_{n=N}^{\infty} 2^{-n} < \epsilon/2$ and then $\delta = \epsilon/(2N+1)$ to get

$$d_{\text{meas}}(\rho_1, \rho_2) < \sum_{n<N} 2^{-n-1}\Big(\sum_{\mathbf{a}\in P_n} \mu(A_{\mathbf{a}} \triangle B_{\mathbf{a}})\Big) + \sum_{n=N}^{\infty} 2^{-n} < \epsilon.$$

3. This is easy to check. □

We want next to prove the continuity of the entropy $h_\mu(\alpha)$ as a function of the ℓ-set partition α. For this purpose it will be convenient to introduce a new metric on the space of (finite or countable) partitions with finite entropy. For two such partitions α and β set

$$d_{\text{ent}}(\alpha, \beta) = d^{\mu}_{\text{ent}}(\alpha, \beta) = H(\alpha|\beta) + H(\beta|\alpha).$$

15.9. LEMMA. 1. *The function $d_{\text{ent}}(\cdot, \cdot)$ is a metric on the space of all (finite and countable) partitions with finite entropy.*

2. *For a system $\mathbf{X}$ and fixed $\ell \ge 2$, given $\epsilon > 0$ there exists a $\delta = \delta(\epsilon, \ell) > 0$ such that for two ℓ-set partitions α and β we have*

$$d_{\text{part}}(\alpha, \beta) < \delta \Rightarrow d_{\text{ent}}(\alpha, \beta) < \epsilon.$$

3.

$$|H(\alpha) - H(\beta)| < d_{\text{ent}}(\alpha, \beta).$$

4.

$$|h_\mu(\alpha) - h_\mu(\beta)| < d_{\text{ent}}(\alpha, \beta).$$

5. *The entropy $h_\mu(\alpha)$ is a continuous function of the ℓ-set partition α with respect to the metric d_{part}: for every $\epsilon > 0$ there exists a $\delta > 0$ such that*

$$d_{\text{part}}(\alpha, \beta) < \delta \Rightarrow |h_\mu(\alpha) - h_\mu(\beta)| < \epsilon.$$

PROOF. 1. An easy exercise.

2. An easy calculation.

3.

$$\begin{aligned} |H(\alpha) - H(\beta)| &= |(H(\alpha) - H(\alpha \vee \beta)) - (H(\beta) - H(\alpha \vee \beta))| \\ &\le H(\alpha|\beta) + H(\beta|\alpha) = d_{\text{ent}}(\alpha, \beta). \end{aligned}$$

4. We first prove, by induction the inequality

$$d_{\text{ent}}(\alpha_0^{n-1}, \beta_0^{n-1}) \le n d_{\text{ent}}(\alpha, \beta). \tag{3.2}$$

For $n = 1$ this is an equality. Assume the validity of (3.2) for n. Now

$$\begin{aligned} d_{\text{ent}}(\alpha_0^n, \beta_0^n) =& H(\alpha_0^{n-1} \vee T^{-n}\alpha | \beta_0^{n-1} \vee T^{-n}\beta) + H(\beta_0^{n-1} \vee T^{-n}\beta | \alpha_0^{n-1} \vee T^{-n}\alpha) \\ \le& H(\alpha_0^{n-1} | \beta_0^{n-1} \vee T^{-n}\beta) + H(T^{-n}\alpha | \beta_0^{n-1} \vee T^{-n}\beta) \\ &+ H(\beta_0^{n-1} | \alpha_0^{n-1} \vee T^{-n}\alpha) + H(T^{-n}\beta | \alpha_0^{n-1} \vee T^{-n}\alpha) \\ \le& H(\alpha_0^{n-1} | \beta_0^{n-1}) + H(T^{-n}\alpha | \vee T^{-n}\beta) \\ &+ H(\beta_0^{n-1} | \alpha_0^{n-1}) + H(T^{-n}\beta | \vee T^{-n}\alpha) \\ =& d_{\text{ent}}(\alpha_0^{n-1}, \beta_0^{n-1}) + d_{\text{ent}}(T^{-n}\alpha, T^{-n}\beta) \\ \le& n d_{\text{ent}}(\alpha, \beta) + d_{\text{ent}}(\alpha, \beta) = (n+1) d_{\text{ent}}(\alpha, \beta). \end{aligned}$$

This establishes (3.2) and by 3 we get

$$|H(\alpha_0^{n-1}) - H(\beta_0^{n-1})| \le d_{\text{ent}}(\alpha_0^{n-1}, \beta_0^{n-1}) \le n d_{\text{ent}}(\alpha, \beta).$$

Hence

$$|h_\mu(\alpha) - h_\mu(\beta)| = \lim_{n\to\infty} \left| \frac{1}{n} H(\alpha_0^{n-1}) - \frac{1}{n} H(\beta_0^{n-1}) \right| \le d_{\text{ent}}(\alpha, \beta).$$

5. Follows from parts 2 and 4. ☐

15.10. REMARK. In fact it can be shown that the metric space $(\mathcal{P}, d_{\text{ent}})$, where $\mathcal{P}$ is the collection of countable partitions α with $H(\alpha) < \infty$ is a complete metric space (see [**201**]).

15.11. THEOREM. *Let* $\mathbf{Y}$ *be an ergodic system with* $h_\nu = h(\mathbf{Y}) > 0$. *For every* $0 \le s \le h_\nu$ *there exists a factor* $\pi : \mathbf{Y} \to \mathbf{Z}$ *with* $h(\mathbf{Z}) = s$.

PROOF. By the definition of $h(\mathbf{Y})$ as the supremum over the entropies of finite partitions, we can assume that $h(\mathbf{Y}) = h_\nu(\beta)$ for some finite partition β. Now in the metric space $(\mathcal{P}_\ell, d_{\text{part}})$ there is a path β_t, $0 \le t \le 1$ with $\beta_1 = \beta$ and $\beta_0 = \{X, \emptyset, \dots, \emptyset\}$. By Theorem 15.9.5, the function $f(t) = h_\nu(\beta_t)$ is a continuous function. It follows that range $f \supset [0, h_\nu(\beta)]$. If $f(t) = h_\nu(\beta_t) = s$ then the σ-algebra $(\beta_t)^T$ defines the required factor $\mathbf{Z}$. ☐

We conclude this section with the following important theorem for whose proof we shall use the apparatus of symbolic representations.

15.12. THEOREM. *Let* $\mathbf{X}$ *be a dynamical system and* α *a finite partition of* X. *If*

$$\mu = \int_Y \mu_y \, d\nu(y)$$

is the ergodic decomposition of μ *then*

1. $h_\mu(\alpha) = \int_Y h_{\mu_y}(\alpha) \, d\nu(y)$ *and*
2. $h_\mu(X,T) = \int_Y h_{\mu_y}(X,T) \, d\nu(y)$

PROOF. Consider the symbolic representation $\phi_\alpha : \mathbf{X} \to (\Omega(\ell), \rho(\mathbf{X}, \alpha), S)$ associated with α. It is easy to see that the ergodic decomposition of $\rho = \rho(\mathbf{X}, \alpha)$ is obtained as the image under ϕ_α of μ:

$$\rho = \int_Y \rho_y \, d\nu(y), \qquad \text{with } \rho_y = \phi_\alpha(\mu_y).$$

Now for the symbolic topological system $(\Omega(\ell), S)$ the function $\theta \mapsto h_\theta$ is affine (Theorem 14.25) and upper semi-continuous (Lemma 15.1) hence by Choquet's theorem (Theorem A.13) $h_\rho = \int_Y h_{\rho_y} \, d\nu(y)$ and thus, by Lemma 15.6 applied to the measures ρ and ρ_y,

$$h_\mu(\alpha) = \int_Y h_{\mu_y}(\alpha) \, d\nu(y).$$

This proves part 1 and part 2 is obtained by choosing a sequence of finite partitions $\alpha_n \nearrow \mathfrak{X}$, since Theorem 14.30 implies that $h_\mu(\alpha_n) \nearrow h_\mu(X, T)$ and for ν-a.e. y $h_{\mu_y}(\alpha_n) \nearrow h_{\mu_y}(X, T)$. □

4. (α, ϵ, N)-generic points

15.13. DEFINITION. Let $(\mathbf{X}, \alpha)$ be a process, $\rho = \rho(\mathbf{X}, \alpha)$ the corresponding symbolic representation measure, N a positive integer and $\epsilon > 0$. We say that x is

1. $(\alpha, \epsilon, N, 2n+1)$-*typical* if $x_{-n}^{N-1+n} = \phi_\alpha(x)_{-n}^{N-1+n}$ is $(\rho, \epsilon, N, 2n+1)$-typical; i.e.

$$|p(\mathbf{a}_{-n}^n | x_{-n}^{N-1+n}) - \mu(A)| = |p(\mathbf{a}_{-n}^n | x_{-n}^{N-1+n}) - \rho(\mathbf{a}_{-n}^n)| < \epsilon, \qquad \forall\, A \in \alpha_{-n}^n,$$

   where $\mathbf{a}_{-n}^n$ is the cylinder set in Ω which corresponds to A.
2. (α, ϵ, N)-*generic* if $x_{-n}^{N-1+n} = \phi_\alpha(x)_{-n}^{N-1+n}$ is (ρ, ϵ, N)-generic; i.e. for every $0 \le n \le L(\epsilon) = [\log(1/\epsilon)] + 1$ and for every $A \in \alpha_{-n}^n$ with $\mu(A) > 0$:

$$\left| \frac{1}{N} \sum_{j=0}^{N-1} \mathbf{1}_A(T^j x) - \mu(A) \right| < \epsilon M,$$

   where

$$\begin{aligned} M &= \min\{\mu(A) : A \in \alpha_{-L(\epsilon)}^{L(\epsilon)},\ \mu(A) > 0\} \\ &= \min\{\rho(\mathbf{a}_1^{2n+1}) : \rho(\mathbf{a}_1^{2n+1}) > 0,\ 0 \le n \le L\}. \end{aligned}$$

   Equivalently, $x \in X$ is (α, ϵ, N)-generic if for every $0 \le n \le L$, x_{-n}^{N-1+n} is $(\rho, \epsilon M, N+n, 2n+1)$-typical; i.e.

$$\text{(4.1)} \qquad \begin{aligned} &|p(\mathbf{a}_{-n}^n | x_{-n}^{N-1+n}) - \rho(\mathbf{a}_{-n}^n)| < \epsilon M, \\ &\text{for all cylinder sets } \mathbf{a}_{-n}^n \text{ corresponding to sets } A \in \alpha_{-n}^n. \end{aligned}$$

15.14. THEOREM. *Let $(\mathbf{X}, \alpha)$ be an ergodic process and $\epsilon > 0$. There exists N_0 such that for every $N \ge N_0$ the set of (ϵ, N)-generic points in X has measure $> 1 - \epsilon$.*

PROOF. The ergodic theorem implies the existence of N_0 such that for every $B \in \bigcup_{n=0}^{L(\epsilon)} \alpha_{-n}^{n}$ with $\mu(B) > 0$, $N > N_0$ and x in a subset of X of measure $> 1 - \epsilon$,

$$\left| \frac{1}{N} \sum_{j=0}^{N-1} 1_B(T^j x) - \mu(B) \right| < \epsilon M.$$

□

We illustrate these ideas with the following theorem. It provides a more concrete version of the assertion that the simplex $M_S(\Omega(\ell))$ is the Poulsen simplex, Theorem 13.15.

15.15. THEOREM (Density of ergodic measures). 1. *Let* $\mathbf{X}$ *be a system,* $\nu = \sum_{i=1}^{t} p_i \nu_i$ *a convex combination where* $\nu_i \in M_S^{\text{erg}}(\Omega(\ell))$*, and* $\epsilon > 0$*. Let* $\mathbf{a}(1)_0^{N-1}, \mathbf{a}(2)_0^{N-1}, \ldots, \mathbf{a}(t)_0^{N-1}$ *be* t *elements of* $\mathfrak{L}^N = \{1, 2, \ldots, \ell\}^N$ *such that every extension* $\mathbf{a}(i)_{-L}^{N-1+L}$, $1 \le i \le t$ *is* (ν_i, ϵ, N)*-generic. Suppose further that in* $\mathbf{X}$ *we are given a collection* $\mathfrak{t}_i$, $i = 1, 2, \ldots, t$, *of* t *disjoint towers of height* N *whose union is a set of* μ *measure* $> 1 - \epsilon$*. Copying the names* $\mathbf{a}(i)_0^{N-1}$ *on the towers* $\mathfrak{t}_i$, $i = 1, 2, \ldots, t$, *we obtain a partition* $\alpha = \{A_1, A_2, \ldots, A_\ell\}$ *of* X *such that*

$$d_{\text{meas}}(\rho, \nu) < 3\epsilon,$$

where $\rho = (\phi_\alpha)_*(\mu)$ *is the symbolic representation of the process* $(\mathbf{X}, \alpha)$*.*

2. *Let* $\mathbf{X}$ *be a non-periodic ergodic system,* $\nu \in M_S(\Omega(\ell))$ *and* $\epsilon > 0$*. Then there exists a partition* α *of* X *such that*

$$d_{\text{meas}}(\rho, \nu) < 4\epsilon,$$

where $\rho = (\phi_\alpha)_*(\mu)$ *is the symbolic representation of the process* $(\mathbf{X}, \alpha)$*.*

3. *Let* $\mathbf{X}$ *be a non-periodic ergodic system and* $\ell \ge 2$*. The symbolic representations of processes* $(\mathbf{X}, \alpha)$*, with* α *a partition of* X *into* ℓ *subsets, are dense in* $M_S(\Omega(\ell))$*. In particular the ergodic measures are dense in* $M_S(\Omega(\ell))$*.*

PROOF. 1. As $d_{\text{meas}}(\rho, \nu) = \sum_{n=1}^{\infty} 2^{-n} d_n(\rho, \nu)$ and $\sum_{n=L+1}^{\infty} 2^{-n} < \epsilon$ it is enough to deal with the terms $d_n(\rho, \nu) = \frac{1}{2} \sum_{\mathbf{c}_{-n}^{n} \in P_n} |\rho(\mathbf{c}_{-n}^{n}) - \nu(\mathbf{c}_{-n}^{n})|$ for $n \le L = [\log(1/\epsilon)] + 1$. Also, since $\mu(\cup_{i=1}^{t} |\mathfrak{t}_i|) > 1 - \epsilon$, we ignore the points which lie outside that union.

Now for any $C = C(\mathbf{c}_{-n}^{n}) \in \alpha_{-n}^{n}$ with $\mu(C) > 0$, and any point x in the base of the tower $\mathfrak{t}_i$, we have by (3.1) and (4.1):

$$\left| \frac{\mu(C \cap |\mathfrak{t}_i|)}{\mu(|\mathfrak{t}_i|)} - \nu(\mathbf{c}_{-n}^{n}) \right| = |p(\mathbf{c}_{-n}^{n} | x_0^{N-1}) - \nu(\mathbf{c}_{-n}^{n})| < \epsilon \nu(\mathbf{c}_{-n}^{n}).$$

By taking union over all towers and summing over all $C \in \alpha_{-n}^{n}$ we get

$$\sum_{C \in \alpha_{-n}^{n}} \left| \frac{\mu(C \cap (\cup_{i=1}^{t} |\mathfrak{t}_i|))}{\mu(\cup_{i=1}^{t} |\mathfrak{t}_i|)} - \nu(\mathbf{c}_{-n}^{n}) \right| < 2\epsilon.$$

Taking into account the points outside the union of the towers we obtain

$$d_n(\rho, \nu) = \frac{1}{2} \sum_{C \in \alpha_{-n}^{n}} |\mu(C(\mathbf{c}_{-n}^{n})) - \nu(\mathbf{c}_{-n}^{n})| < 2\epsilon.$$

So that finally

$$d(\rho, \nu) = \sum_{n=1}^{\infty} 2^{-n} d_n(\rho, \nu) < 3\epsilon.$$

2. By the Krein-Milman theorem ([**53**], page 440) there exists a convex combination $\lambda = \sum_{i=1}^{t} p_i \nu_i$ with $\nu_i \in M_S^{\text{erg}}(\Omega(\ell))$, such that $d(\lambda, \nu) < \epsilon$. By Theorem 15.14 there exist N and t sequences in $\mathfrak{L}^N$, $\mathbf{a}(1)_0^{N-1}, \mathbf{a}(2)_0^{N-1}, \dots, \mathbf{a}(t)_0^{N-1}$ such that for every i, every extension $\mathbf{a}(i)_{-L}^{N-1+L}$ is (ν_i, ϵ, N)-generic. Next construct a Rohlin tower $\mathfrak{t} = \{B, TB, \dots, T^{N-1}B\}$ in X with $\mu(|\mathfrak{t}|) > 1 - \epsilon$, and then partition B as $B = \cup_{i=1}^{t} B_i$ with

$$\frac{\mu(B_i)}{\mu(B)} = p_i$$

to create t towers $\mathfrak{t}_i = \{B_i, TB_i, \dots, T^{N-1}B_i\}$. On the tower $\mathfrak{t}_i$ copy the name $a(i)_0^{N-1}$ thus defining a partition α of X. By part 1 conclude that $d(\rho, \lambda) < 3\epsilon$, hence $d(\rho, \nu) < 4\epsilon$.

3. This is a direct consequence of part 2. □

5. An ergodic theorem for towers

The next lemma — a direct consequence of the ergodic theorem — will usually be applied to the elements C of a finite partition α in order to get good α empirical distributions. It will be useful in proving the Jewett-Krieger theorem and Ornstein's fundamental lemma.

15.16. LEMMA. *Let* $\mathbf{X}$ *be an ergodic system and* $\epsilon > 0$.

1. *Given* $C \subset X$ *there is an* N_0 *with the following property: for every tower* $\mathfrak{t} = \{\mathfrak{c}_k : k = 1, 2, \dots\}$ *(with* B_k *the base of the column* $\mathfrak{c}_k$, N_k *its height and* $B = \cup_k B_k$ *the base of* $\mathfrak{t}$*) such that* $\min_k N_k > N_0$ *and* $\mu(|\mathfrak{t}|) > 1 - \epsilon/2$, *those fibers* $\{x, Tx, \dots, T^{N_k(x)-1}x\}$, $x \in B$ *that satisfy*
$$\left| \frac{1}{N_k(x)} \sum_{i=0}^{N_k(x)-1} \mathbf{1}_C(T^i x) - \mu(C) \right| < \epsilon,$$
fill up at least $1 - \epsilon$ *of* X.
2. *Given a finite partition* α *of* X *there is an* N_0 *with the following property: for every tower* $\mathfrak{t} = \{\mathfrak{c}_k : k = 1, 2, \dots\}$ *with* $\min_k N_k > N_0$ *and* $\mu(|\mathfrak{t}|) > 1 - \epsilon/2$, *those fibers* $\{x, Tx, \dots, T^{N_k(x)-1}x\}$, $x \in B$ *such that* x *is* $(\alpha, \epsilon, N_k(x), 1)$*-typical fill up at least* $1 - \epsilon$ *of* X. *When* $\mathfrak{t} = \{B, TB, \dots, T^{N-1}B\}$ *has just one column we conclude that the set of those* $x \in B$ *which are* $(\alpha, \epsilon, N, 1)$*-typical has measure at least* $(1 - \epsilon)\mu(B)$.
3. *Given a finite partition* α *of* X *there is an* N_0 *with the following property: for every tower* $\mathfrak{t} = \{B, TB, \dots, T^{N-1}B\}$ *with* $N > N_0$ *and* $\mu(|\mathfrak{t}|) > 1 - \epsilon/2$ *the set of* $x \in B$ *which are* (α, ϵ, N)*-generic has measure at least* $(1-\epsilon)\mu(B)$.

PROOF. 1. Let $\delta = \epsilon^2/2$ and use the ergodic theorem to find an N such that the set of $x \in X$ which satisfy

$$\left| \frac{1}{N} \sum_{i=0}^{N-1} \mathbf{1}_C(T^i x) - \mu(C) \right| < \delta \tag{5.1}$$

has measure at least $1 - \delta$. Let F denote the set of x that satisfy (5.1). We take N_0 large so that N/N_0 is negligible.

Now consider a tower $\mathfrak{t} = \{\mathfrak{c}_k : k = 1, 2, \dots\}$, with B_k the base of the column $\mathfrak{c}_k$ and N_k its height, such that $\min_k N_k > N_0$ and $\mu(|\mathfrak{t}|) > 1 - \epsilon/2$. Let A be the union of all those fibers $\{x, Tx, \dots, T^{N_k(x)-1}x\}$, $x \in B$ with more than a $(1 - \epsilon)$ fraction of its points lying in F (i.e. $\sum_{i=0}^{N_k-1} \mathbf{1}_F(T^i x) \geq (1 - \epsilon)N_k$). By Fubini's theorem

$$\begin{aligned} \mu(|\mathfrak{t}|) - \delta \leq \mu(F \cap |\mathfrak{t}|) &= \int_{|\mathfrak{t}| \setminus A} \mathbf{1}_F \, d\mu + \int_A \mathbf{1}_F \, d\mu \\ &\leq \mu(|\mathfrak{t}| \setminus A)(1 - \epsilon) + \mu(A) \\ &= \mu(|\mathfrak{t}| \setminus A)(1 - \epsilon) + \mu(|\mathfrak{t}|) - \mu(|\mathfrak{t}| \setminus A), \end{aligned}$$

hence $\mu(|\mathfrak{t}| \setminus A) < \epsilon/2$ and $\mu(A) = \mu(|\mathfrak{t}|) - \mu(|\mathfrak{t}| \setminus A) > 1 - \epsilon$.

Fix such a good fiber $\{x, Tx, \dots, T^{N_k(x)-1}x\} \subset A$ and divide it into disjoint blocks of size N that will cover all points that lie in F — except for at most N points at the top — as follows. Move up the fiber and mark the first point in F as the base of the first block of height N. Starting from the top of the first block chose the next point up the fiber that belongs to F as the base of the second block. Continue in this way to (at most N levels before) the top. We now get that on most of the fiber the averages over those N-blocks are close to $\mu(C)$. The values outside these N-blocks can contribute to the sum $\sum_{i=0}^{N_k-1} \mathbf{1}_C(T^i x)$ at most $N_k\epsilon + N$. This completes the proof of part 1.

2. Apply the proof of part 1 to each of the elements $C \in \alpha$, and then take A as the intersection of the good sets. The last assertion follows since $\mu(|\mathfrak{t}|) = N\mu(B)$.

3. Again apply the proof of part 1, this time to each $C \in \alpha_{-L}^{L}$ and with $\delta = M\epsilon^2(2 \operatorname{card} \alpha_{-L}^{L})^{-1}$, where $L = L(\epsilon) = [-\log(\epsilon)] + 1$ and $M = \min\{\mu(A) : A \in \alpha_{-L}^{L}, \ \mu(A) > 0\}$. □

We shall later need the following fact, in whose proof Theorem 15.15 and Lemma 15.16 are applied: given a process $(\mathbf{Y}, \beta)$, there exists an approximation $(\mathbf{Y}, \gamma)$ with a slightly smaller entropy.

15.17. THEOREM. *Let $(\mathbf{Y}, \beta)$ be a non-periodic ergodic process with $h_\nu(\beta) > 0$. Given $\delta > 0$, there exists a partition γ of Y with the properties*

1. $d_{\text{part}}^{\nu}(\beta, \gamma) \leq \delta$ *and,*
2. $h_\nu(\beta) - \delta < h_\nu(\gamma) < h_\nu(\beta)$.

PROOF. Fix $0 < \theta < \delta/2$. We construct a sequence $\beta_N \in \mathcal{P}_\ell$ of modifications of β with $d_{\text{part}}(\beta_N, \beta) < 2\theta$ and such that the sequence $\rho_N = \rho(\mathbf{Y}, \beta_N)$ converges to the measure $\theta\delta_1 + (1 - \theta)\rho(\mathbf{Y}, \beta)$, where δ_1 is the point mass at $\mathbf{1} = (\dots, 1, 1, 1, \dots) \in \Omega(\ell)$. For every positive integer N construct a Rohlin tower $\mathfrak{t}_N = \{B_N, TB_N, \dots, T^{N-1}B_N\}$ with $\nu(|\mathfrak{t}_N|) > 1 - 1/N$. Let $B_N = B_N' \cup B_N''$ a disjoint union with $\nu(|\mathfrak{t}_N'|) = \theta$, where $\mathfrak{t}_N'$ is the column of height N over B_N'.

Now copy the name $111\dots 1$ over $\mathfrak{t}'_N$ and leave β unchanged on the rest of the space. This defines a new partition β_N with $d_{\text{part}}(\beta, \beta_N) \leq \theta + 1/N$. Next refine the tower $\mathfrak{t}''_N$ over B''_N according to the various (N, β) names. By Theorem 15.16, for every $\epsilon > 0$ for sufficiently large N the columns in $\mathfrak{t}''_N$ with (β, ϵ, N)-generic names fill up most of $\mathfrak{t}''_N$ and Theorem 15.15.1 implies:

$$\lim_{N\to\infty} d_{\text{meas}}(\rho_N, \theta\delta_{\mathbf{1}} + (1-\theta)\rho(\mathbf{Y}, \beta)) = 0,$$

where $\rho_N = \rho(\beta_N, \nu)$. Lemma 15.1 implies

$$\limsup_{N\to\infty} h_{\rho_N} \leq h_{\theta\delta_{\mathbf{1}}+(1-\theta)\rho(\mathbf{Y},\beta)} = (1-\theta)h(\mathbf{Y}, \beta).$$

On the other hand for sufficiently small θ and sufficiently large N, $d_{\text{part}}(\beta, \beta_N) \leq \theta + 1/N$ is sufficiently small to imply (Theorem 15.9)

$$h_\nu(\beta) - \delta < h_\nu(\beta_N) < h(\mathbf{Y}, \beta).$$

For such θ and N we set $\gamma = \beta_N$. □

6. A SMB theorem for towers

15.18. LEMMA. *Let $(\mathbf{X}, \alpha)$ be an ergodic process and $\delta, \epsilon > 0$.*

1. *For sufficiently large N, if $\mu(A) > \delta$ then there exists a subset $A_0 \subset A$ with $\mu(A_0) > (1-\delta)\mu(A)$ and*

$$\left|-\frac{1}{N}\log\big(\mu(\mathbf{a}_N(x) \cap A)/\mu(A)\big) - h_\mu(\alpha)\right| < \epsilon$$

for every $x \in A_0$.

2. *For sufficiently large N for every tower $\mathfrak{t} = \{B, TB, \dots, T^{N-1}B\}$ such that $\mu(|\mathfrak{t}|) > 1 - \delta$, the set I of (α, N)-names $\mathbf{a}$ for which*

$$\left|-\frac{1}{N}\log\big(\mu(B_{\mathbf{a}})/\mu(B)\big) - h_\mu(\alpha)\right| < \epsilon,$$

satisfies $\mu\big(\cup_{\mathbf{a}\in I} B_{\mathbf{a}}\big) > (1-\delta)\mu(B)$.

PROOF. 1. Since

$$-\frac{1}{N}\log\big(\mu(\mathbf{a}_N(x) \cap A)/\mu(A)\big) = -\frac{1}{N}\log(\mu(\mathbf{a}_N(x) \cap A) - \frac{1}{N}\log(\mu(A)),$$

it suffices to show that

$$\left|-\frac{1}{N}\log(\mu(\mathbf{a}_N(x) \cap A)) - h_\mu(\alpha)\right| < \epsilon$$

The SMB theorem 14.35 yields, for sufficiently large N,

$$\left|-\frac{1}{N}\log(\mu(\mathbf{a}_N(x))) - h_\mu(\alpha)\right| < \epsilon$$

for x in a subset $X_0 \subset X$ with measure $\mu(X_0) \geq 1 - \delta^2$. This implies

$$-\frac{1}{N}\log(\mu(\mathbf{a}_N(x) \cap A)) < h_\mu(\alpha) + \epsilon \tag{6.1}$$

for $x \in A_0 = A \cap X_0$, and as $\mu(A) > \delta$, we have $\mu(A_0) > \mu(A) - \delta^2 > \mu(A)(1-\delta)$.

For the other direction, let $D \subset A$ be the set of "bad" points where $h_\mu(\alpha) - \epsilon \geq -\frac{1}{N}\log(\mu(\mathbf{a}_N(x) \cap A))$; i.e. where $\mu(\mathbf{a}_N(x) \cap A) \geq \exp(-N(h_\mu(\alpha) - \epsilon))$. Note that for sufficiently large N there exists, by Corollary 14.36, a collection of at most $\exp(N(h_\mu(\alpha) + \epsilon))$ elements $\mathbf{a}_N$ of α_0^{N-1} which covers X up to a set of measure δ^2, each of which satisfies $\mu(\mathbf{a}_N) < \exp(-N(h_\mu(\alpha) - \epsilon))$. This collection will therefore cover A up to a set of measure $\delta\mu(A)$. Thus $\mu(D) < \delta\mu(A)$ and

$$h_\mu(\alpha) - \epsilon < -\frac{1}{N}\log(\mu(\mathbf{a}_N(x) \cap A)) \tag{6.2}$$

holds for $x \in A \setminus D$. Together the inequalities (6.1) and (6.2) complete the proof of the first part of the lemma.

2. Again, as in the proof of the first part, we see that it suffices to show

$$\left|-\frac{1}{N}\log(\mu(B_{\mathbf{a}})) - h_\mu(\alpha)\right| < \epsilon.$$

In order to be able to use part 1 of the lemma we let $M = [N\frac{\epsilon}{4h'}]$, where $h' = \max\{1, h_\mu(\alpha)\}$, and consider the thickened sets

$$A^+ = \bigcup_{j=0}^{M} T^j B \qquad \text{and} \qquad A^- = \bigcup_{j=-M}^{0} T^j B,$$

both of measure at least $\frac{\epsilon}{8h'}$ (for $\delta < 1/2$), independently of N.

Applying part 1 of the lemma to A^+ we know that, for sufficiently large N, a subset $A_0^+ \subset A^+$ with $\mu(A_0^+) > (1-\delta)\mu(A^+)$ is covered by $(N-M, \alpha)$-names $\mathbf{a}_{N-M}$ with $\mu(\mathbf{a}_{N-M} \cap A^+) < \exp(-(N-M)(h_\mu - \frac{\epsilon}{4}))$. Now for any $x \in B$ and $0 \le j \le M$,

$$T^j(\mathbf{a}_N(x) \cap B) \subset \mathbf{a}_{N-M}(T^j x) \cap A^+,$$

and therefore if $T^j x \in A_0^+$ then

$$\mu(B_{\mathbf{a}_N(x)}) = \mu(\mathbf{a}_N(x) \cap B) \le \exp(-N(h_\mu(\alpha) - \frac{\epsilon}{2})). \tag{6.3}$$

Note that the condition $x \in B, T^j x \in A_0^+$ for some $0 \le j \le M$ is satisfied by a subset of B of measure $(1-\delta)\mu(B)$.

For the other direction we need to be a little more careful. We use the first part of the lemma, with $\epsilon/8$ rather than $\epsilon/4$, to deduce that a subset $A_0^- \subset A^-$ with $\mu(A_0^-) > (1-\delta)\mu(A^-)$ is covered by $(N+M)$-names $\mathbf{a}_{N+M}$ with $\mu(\mathbf{a}_{N+M} \cap A^-) > \mu(A^-)\exp(-(N+M)(h_\mu(\alpha) + \frac{\epsilon}{8}))$. Thus $\exp((N+M)(h + \frac{\epsilon}{8}\epsilon))$ $(N+M)$-names suffice to cover A_0^-.

Let $B_0 \subset B$ be the set of points $x \in B$ such that for some $0 \le j \le M$, $T^{-j}x \in A_0^-$; then $\mu(B_0) > (1-\delta)\mu(B)$. Each $(N+M)$-name in A_0^- can give rise to M different N-names in B_0, one for each $0 \le j \le M$, and therefore B_0 can be covered by no more than $M\exp((N+M)(h + \frac{\epsilon}{8}\epsilon))$ N-names and for large enough N we get $M\exp((N+M)(h + \frac{\epsilon}{8}\epsilon)) < \exp(N(h_\mu(\alpha) + \frac{\epsilon}{4}))$.

Now the "bad" names are those names $\mathbf{a}_N(x)$ with $x \in B_0$ for which $\mu(\mathbf{a}_N(x) \cap B) \le \exp(-N(h_\mu(\alpha) + \frac{\epsilon}{2}))$, and by the previous discussion their total mass is bounded by

$$\exp(N(h_\mu(\alpha) + \frac{\epsilon}{4}))\exp(-N(h_\mu(\alpha) + \frac{\epsilon}{2})) = \exp(-N\frac{\epsilon}{4}).$$

Since $\mu(B) > \frac{1-\delta}{N}$, we see that for sufficiently large N the measure of the set of bad points in B_0 is less than $\exp(-N\frac{\epsilon}{4}) < c\mu(B)$ (for any given $c > 0$). Thus the measure of the set of bad points in B is less than $\delta\mu(B)$ and

$$\exp(-N(h_\mu(\alpha) - \frac{\epsilon}{2})) \le \mu(B_{\mathbf{a}_N(x)}) = \mu(\mathbf{a}_N(x) \cap B) \tag{6.4}$$

outside the bad set. This concludes the proof of the lemma. □

7. The $\bar{d}$-metric

In this section we introduce the $\bar{d}$ metric on $M_S(\Omega)$. It is much stronger than d_{meas} in the sense that a $\bar{d}$ convergent sequence is also d_{meas} convergent but not vice versa. E.g. it can be shown that the collections of ergodic measures and of mixing measures are both $\bar{d}$ closed in $M_S(\Omega)$ (see e.g. [**235**], Theorems I.9.15 and I.9.17), whereas they are both dense in $M_S(\Omega)$ with respect to the weak* topology which is the topology induced by d_{meas}. One can approach the $\bar{d}$ distance from various angles. Faithful to our basic theme we define it via joinings.

15.19. DEFINITION. Fix ℓ and let $\Omega = \Omega(\ell) = \mathfrak{L}^{\mathbb{Z}}$, where as usual $\mathfrak{L} = \{1, \dots, \ell\}$. For $\rho_1, \rho_2 \in M_S(\Omega)$ let

$$\bar{d}(\rho_1, \rho_2) = \inf\{\lambda(P_0 \times \Omega \triangle \Omega \times P_0) : \lambda \in J(\rho_1, \rho_2)\},$$

where P_0 is the partition $A_j = \{x \in \Omega : x_0 = j\}$, $j = 1, \dots, \ell$ (see Definition 15.7). It will often be convenient to work with processes rather than their representing measures. Accordingly we define for two processes $(\mathbf{X}_i, \alpha_i)$, $i = 1, 2$ with the same number of sets:

$$\bar{d}((\mathbf{X}_1, \alpha_1), (\mathbf{X}_2, \alpha_2)) = \bar{d}(\alpha_1, \alpha_2) = \inf\{\lambda(\alpha_1 \times X_2 \triangle X_1 \times \alpha_2) : \lambda \in J(\mu_1, \mu_2)\}.$$

It is easy to verify that in fact

$$\bar{d}((\mathbf{X}_1, \alpha_1), (\mathbf{X}_2, \alpha_2)) = \bar{d}(\rho_1, \rho_2),$$

where $\rho_i = \rho_i(\mathbf{X}_i, \alpha_i)$ are the representing measures.

15.20. PROPOSITION (Properties of the $\bar{d}$-metric). 1. *The infimum in the definition of the $\bar{d}$ metric is in fact a minimum; i.e. there exists a joining $\lambda \in J(\rho_1, \rho_2)$ such that $\bar{d}(\rho_1, \rho_2) = \lambda(P_0 \times \Omega \triangle \Omega \times P_0)$; moreover if ρ_1 and ρ_2 are ergodic then there exists an ergodic joining λ which realizes the minimum.*

2. *$(M_S(\Omega), \bar{d})$ is a metric space.*
3. *For every $\epsilon > 0$ there is a $\delta > 0$ such that*

$$\bar{d}(\rho_1, \rho_2) < \delta \quad \Rightarrow \quad d_{\text{meas}}(\rho_1, \rho_2) < \epsilon.$$

4. *The entropy h_ρ is $\bar{d}$ continuous.*
5. *For a fixed dynamical system $\mathbf{X}$ and $\ell \geq 2$, we have on the space $\mathcal{P}_\ell$ of ℓ-set partitions:*

$$\bar{d}((\mathbf{X}, \alpha), (\mathbf{X}, \beta)) \leq d^{\mu}_{\text{part}}(\alpha, \beta).$$

PROOF. 1. Considered as a function of λ, $f(\lambda) = \lambda(P_0 \times \Omega \triangle \Omega \times P_0)$ is a continuous function on the compact convex set $J(\rho_1, \rho_2)$ and therefore achieves its infimum, say at the joining λ_0. Moreover f is an affine function so that if $\lambda_0 = t\lambda_1 + (1 - t)\lambda_2$ with $\lambda_1, \lambda_2 \in J(\rho_1, \rho_2)$ and $0 \leq t \leq 1$, then it follows that $f(\lambda_i) = f(\lambda_0)$, $i = 1, 2$. This means that the set $\{\lambda \in J(\rho_1, \rho_2) : f(\lambda) = f(\lambda_0)\}$ is a *face* of the convex set $J(\rho_1, \rho_2)$ and therefore contains an extreme point of $J(\rho_1, \rho_2)$. By Theorem 6.2, when ρ_1 and ρ_2 are ergodic this extreme joining is also ergodic.

2. We get $\bar{d}(\rho,\rho)=0$ via the diagonal joining $\lambda = \mathrm{gr}\,(\rho,\mathrm{id})$. Conversely by part 1, $\bar{d}(\rho_1,\rho_2)=0$ implies $\lambda(P_0\times\Omega\bigtriangleup\Omega\times P_0)=0$ for some $\lambda\in J(\rho_1,\rho_2)$. A simple calculation (Lemma 15.8.1) shows that for every n

$$\lambda(P^n_{-n}\times\Omega\bigtriangleup\Omega\times P^n_{-n})\le(2n+1)\lambda(P_0\times\Omega\bigtriangleup\Omega\times P_0)=0,$$

hence

$$\mathcal{B}\times\Omega=\Omega\times\mathcal{B}\qquad(\mathrm{mod}\ \lambda).$$

By Theorem 6.5 we conclude that $\lambda=\mathrm{gr}\,(\rho,\mathrm{id})$ and it follows that $\rho_1=\rho_2$.

If

$$\bar{d}(\rho_1,\rho_2)=\lambda_{1,2}(P_0\times\Omega\bigtriangleup\Omega\times P_0)\quad\text{and}\quad\bar{d}(\rho_2,\rho_3)=\lambda_{2,3}(P_0\times\Omega\bigtriangleup\Omega\times P_0),$$

we form the relative independent joining

$$\lambda_{1,2,3}=\lambda_{1,2}\underset{\rho_2}{\times}\lambda_{2,3}\in J(\lambda_{1,2},\lambda_{2,3}),$$

and observe that

$$\begin{aligned}&\lambda_{1,2,3}(P_0\times\Omega\times\Omega\bigtriangleup\Omega\times\Omega\times P_0)\\&\le\lambda_{1,2,3}(P_0\times\Omega\times\Omega\bigtriangleup\Omega\times P_0\times\Omega)+\lambda_{1,2,3}(\Omega\times P_0\times\Omega\bigtriangleup\Omega\times\Omega\times P_0)\\&=\lambda_{1,2}(P_0\times\Omega\bigtriangleup\Omega\times P_0)+\lambda_{2,3}(P_0\times\Omega\bigtriangleup\Omega\times P_0).\end{aligned}$$

Since $\lambda:=\lambda_{1,2,3}\upharpoonright\mathcal{B}\times\Omega\times\mathcal{B}$ is in $J(\rho_1,\rho_3)$, we get

$$\bar{d}(\rho_1,\rho_3)\le\bar{d}(\rho_1,\rho_2)+\bar{d}(\rho_2,\rho_3).$$

3. This is the same as the proof of Lemma 15.8.2; here is the proof anyhow. Again let

$$\bar{d}(\rho_1,\rho_2)=\lambda(P_0\times\Omega\bigtriangleup\Omega\times P_0)=d^{\lambda}_{\mathrm{part}}(P_0\times\Omega,\Omega\times P_0)<\delta,$$

then as in the proof of part 2

$$\lambda(P^n_{-n}\times\Omega\bigtriangleup\Omega\times P^n_{-n})\le(2n+1)\lambda(P_0\times\Omega\bigtriangleup\Omega\times P_0)=(2n+1)\delta.$$

Since

$$|\lambda(A\times\Omega)-\lambda(\Omega\times A)|=|\lambda(A\times\Omega\setminus\Omega\times A)-\lambda(\Omega\times A\setminus A\times\Omega)|\le\lambda(A\times\Omega\Delta\Omega\times A),$$

we get

$$\frac{1}{2}\sum_{A\in P^n_{-n}}|\rho_1(A)-\rho_2(A)|\le(2n+1)\delta.$$

Choose N with $\sum_{n=N}^{\infty}2^{-n}<\epsilon/2$ and then $\delta=\epsilon/(2(2N+1))$ to get

$$d_{\mathrm{meas}}(\rho_1,\rho_2)<\sum_{n<N}\Big(\frac{1}{2}\sum_{A\in P^n_{-n}}|\rho_1(A)-\rho_2(A)|\Big)+\sum_{n=N}^{\infty}2^{-n}<\epsilon.$$

4. Once more we let

$$\bar{d}(\rho_1,\rho_2)=\lambda(P_0\times\Omega\bigtriangleup\Omega\times P_0)$$

for $\lambda\in J(\rho_1,\rho_2)$. By Lemma 15.9.2, given $\epsilon>0$ there is $\delta>0$ such that

$$d^{\lambda}_{\mathrm{part}}(P_0\times\Omega,\Omega\times P_0)=\lambda(P_0\times\Omega\bigtriangleup\Omega\times P_0)<\delta,$$

implies

$$d^{\lambda}_{\mathrm{ent}}(P_0\times\Omega,\Omega\times P_0)=|H(P_0\times\Omega)-H(\Omega\times P_0)|<\epsilon.$$

Since $\rho(\lambda, P_0 \times \Omega) = \rho_1$ and $\rho(\lambda, \Omega \times P_0) = \rho_2$, Lemma 15.9.4 yields

$$|h_{\rho_1} - h_{\rho_2}| < d^{\lambda}_{\text{ent}}(P_0 \times \Omega, \Omega \times P_0) < \epsilon.$$

5. The measure μ on X defines, for every $\alpha, \beta \in \mathcal{P}_\ell$, a joining of the processes $(\mathbf{X}, \alpha)$ and $(\mathbf{X}, \beta)$, hence the inequality

$$\bar{d}((\mathbf{X}, \alpha), (\mathbf{X}, \beta)) \leq d^{\mu}_{\text{part}}(\alpha, \beta) = \mu(\alpha \triangle \beta)$$

follows from the definition of the $\bar{d}$ metric. □

Proposition 15.20.3 tells us that the metric d_{meas} is $\bar{d}$ continuous (the converse is far from being true). It will be convenient to have this estimation in a more explicit way. Define for processes $(\mathbf{X}, \alpha)$ and $(\mathbf{Y}, \beta)$ with $\operatorname{card} \alpha = \operatorname{card} \beta = \ell$ and an integer $N \geq 0$,

$$\bar{d}^N(\alpha, \beta) = \bar{d}(\alpha^N_{-N}, \beta^N_{-N}) = \bar{d}(\rho_1, \rho_2),$$

where $\rho_i \in M_S(\Omega(\ell^{2N+1}))$ are the representing measures of the processes $(\mathbf{X}, \alpha^N_{-N})$ and $(\mathbf{Y}, \beta^N_{-N})$ respectively.

15.21. PROPOSITION. 1. $\bar{d}^N(\alpha, \beta) \leq (2N+1)\bar{d}(\alpha, \beta)$.
2. *For every* $\epsilon > 0$ *there exists an* N_0 *such that for* $N \geq N_0$

$$d_{\text{meas}}(\rho_1, \rho_2) \leq \bar{d}^N(\alpha, \beta) + \epsilon.$$

PROOF. 1. Let $\lambda \in J(\mu, \nu)$ achieve the $\bar{d}$ distance between α and β, then

$$\begin{aligned}\bar{d}^N(\alpha, \beta) &\leq \lambda(\alpha^N_{-N} \times Y \triangle X \times \beta^N_{-N}) \\ &\leq (2N+1)\lambda(\alpha \times Y \triangle X \times \beta) \\ &= (2N+1)\bar{d}(\alpha, \beta).\end{aligned}$$

2. This is just the proof of Proposition 15.20.3. □

Given $\rho \in M_S(\Omega(\ell))$ with $h_\rho < \log(\ell)$ (i.e. ρ is not the Bernoulli measure $(1/\ell, \ldots, 1/\ell)^{\mathbb{Z}}$) we shall show next that one can always approximate ρ in $\bar{d}^N$ with a measure $\eta \in M_S(\Omega(\ell))$ with strictly higher entropy.

15.22. THEOREM. *For every* $\rho \in M_S(\Omega(\ell))$, *an integer* $N \geq 1$ *and* $0 < \epsilon < 1$, *there exists* $\eta \in M_S(\Omega(\ell))$ *such that*

1. $\bar{d}^N(\rho, \eta) < 2\epsilon$.
2. $h_\eta \geq h_\rho + \epsilon(\log(\ell) - h_\rho)$.

Moreover when ρ *is ergodic we can choose* η *to be ergodic.*

PROOF. Let $\mathbf{X}$ be an arbitrary weakly mixing system. Choose an integer M such that $2N/M < \epsilon/2$ and use Rohlin's lemma to construct a tower $\mathfrak{t} = \{B, TB, \ldots, T^M B\}$ with $\mu(X \setminus |\mathfrak{t}|) < \epsilon/2$. Let B_1 be any subset of B with $\mu(B_1)/\mu(B) > 1 - \epsilon/2$; then the partition $\alpha_0 = \{A, A^c\}$ where $A = \cup_{n=0}^{M-1} T^n B_1$, satisfies

$$\mu(\cap^N_{-N} T^{-n}(A)) > (1 - \epsilon)\mu(A),$$

and $\mu(A^c) < \epsilon$. Changing B_1 slightly, we make $\mu(A^c) = \epsilon$.

Let $\eta_0 \in M_S(\Omega(\ell))$ be the Bernoulli measure $\eta_0 = (\ell^{-1}, \ell^{-1}, \ldots, \ell^{-1})^{\mathbb{Z}}$ and construct the product system $\mathbf{Z} = (\Omega \times X \times \Omega, \rho \times \mu \times \eta_0, S \times T \times S)$. Since $\mathbf{X}$ as well as (Ω, η_0, S) are weakly mixing systems it follows that, when ρ is ergodic so is $\mathbf{Z}$. Next define an ℓ set partition α on $Z = \Omega \times X \times \Omega$ by

$$\alpha(y_1, x, y_2) = \begin{cases} P_0(y_1) & \text{if } x \in A \\ P_0(y_2) & \text{if } x \in A^c, \end{cases}$$

and set $\eta = \rho(\mathbf{Z}, \alpha)$, the representing measure of the process $(\mathbf{Z}, \alpha)$.

Now consider the measure $\lambda = \rho \times \mu \times \eta_0$ on Z as a joining of the processes $(\Omega, P_0(y_1))$ and $(Z, \alpha(y_1, x, y_2))$ and observe that for $x \in \cap_{-N}^{N} T^{-n}(A)$ the points (y_1, x, y_2) and y_1 belong to the corresponding elements of the partitions α_{-N}^{N} and P_{-N}^{N} respectively. Thus

$$\bar{d}^N(\rho, \eta) < 1 - \mu(\cap_{-N}^{N} T^{-n}(A)) < 2\epsilon,$$

which proves part 1.

To prove part 2 we first observe that $\alpha_{-\infty}^{-1}$ is $P_{-\infty}^{-1} \times (\alpha_0)_{-\infty}^{0} \times P_{-\infty}^{-1}$-measurable and then make the following calculation.

$$\begin{aligned} h_\eta &= h_\lambda(\alpha) = \int I^{\alpha_{-\infty}^{-1}}(\alpha) d\lambda \\ &\geq \int I^{P_{-\infty}^{-1} \times (\alpha_0)_{-\infty}^{0} \times P_{-\infty}^{-1}}(\alpha) d\lambda \\ &= \int_{\Omega \times A^c \times \Omega} I^{P_{-\infty}^{-1} \times (\alpha_0)_{-\infty}^{0} \times P_{-\infty}^{-1}}(\Omega \times X \times P_0) d\lambda \\ &\quad + \int_{\Omega \times A \times \Omega} I^{P_{-\infty}^{-1} \times (\alpha_0)_{-\infty}^{0} \times P_{-\infty}^{-1}}(P_0 \times X \times \Omega) d\lambda \\ &= \int_{A^c \times \Omega} I^{P_{-\infty}^{-1}}(P_0) d(\mu \times \eta_0) + \int_{\Omega \times A} I^{P_{-\infty}^{-1}}(P_0) d(\rho \times \mu) \\ &= \epsilon \log(\ell) + (1 - \epsilon) h_\rho. \end{aligned}$$

□

The $\bar{d}$ distance of two processes is dominated by the asymptotic Hamming distance between ϵ-generic names of these processes. More precisely we have

15.23. THEOREM. *Let $(\mathbf{X}, \alpha)$ and $(\mathbf{Y}, \beta)$ be ergodic processes. Let $N_k \nearrow \infty$ and and $\epsilon_k \searrow 0$. If $\mathbf{a}_k$ and $\mathbf{b}_k$ are ϵ_k-generic names in $\mathfrak{L}^{N_k}$ for $(\mathbf{X}, \alpha)$ and $(\mathbf{Y}, \beta)$ respectively (with $\mathfrak{L} = \{1, \dots, \ell\}$), then*

$$\bar{d}(\rho(\mathbf{X}, \alpha), \rho(\mathbf{Y}, \beta)) \leq \liminf_k d_{\text{ham}}(\mathbf{a}_k, \mathbf{b}_k).$$

PROOF. Let $\mathbf{Z} = (Z, \mathcal{Z}, \eta, T)$ be an arbitrary ergodic non-periodic system. For each k construct a Rohlin tower $\mathfrak{t}_k = \{B_k, TB_k, \dots, T^{N_k - 1} B_k\}$ with $\mu(|\mathfrak{t}_k|) > 1 - \epsilon_k$. On the tower $\mathfrak{t}_k$ copy the double name $\mathbf{a}_k \times \mathbf{b}_k \in (\mathfrak{L} \times \mathfrak{L})^{N_k}$. This operation defines three partitions of Z, γ_k according to the $\mathfrak{L} \times \mathfrak{L}$ value, and γ_k^1 and γ_k^2 with $\gamma_k = \gamma_k^1 \vee \gamma_k^2$, according to the $\mathfrak{L}$-value of the first and second marginals respectively. Set

$$\rho_k = \rho(\mathbf{Z}, \gamma_k) \in M_S(\Omega(\ell \times \ell)),$$

and let $\lambda = \lim_{i \to \infty} \rho_{k_i}$ be a weak* limit obtained by passing to a subsequence for which also

$$\lim_{i \to \infty} d_{\text{ham}}(\mathbf{a}_{k_i}, \mathbf{b}_{k_i}) = \liminf_k d_{\text{ham}}(\mathbf{a}_k, \mathbf{b}_k).$$

Now, by Theorem 15.15

$$d_{\text{meas}}(\rho(\mathbf{X},\alpha),\rho(\mathbf{Z},\gamma_k^1)) < 2\epsilon_k \quad \text{and}$$
$$d_{\text{meas}}(\rho(\mathbf{Y},\beta),\rho(\mathbf{Z},\gamma_k^2)) < 2\epsilon_k.$$

Also, it follows directly from the definition of the d_{ham} metric and the fact that $|\mathbf{t}_k|$ fills up all but ϵ_k of the space, that

$$\eta(\gamma_k^1 \triangle \gamma_k^2) \leq d_{\text{ham}}(\mathbf{a}_k,\mathbf{b}_k) + \epsilon_k.$$

Denoting by $P_0 = P_0^{(1)} \times P_0^{(2)}$ the partition of $\Omega(\ell \times \ell)$ according to the 0 coordinate we have

$$\lim_{i\to\infty} \rho(\mathbf{Z},\gamma_{k_i}^1) = \rho(\Omega(\ell\times\ell),\lambda,P_0^{(1)}) = \rho(\mathbf{X},\alpha) \quad \text{and}$$
$$\lim_{i\to\infty} \rho(\mathbf{Z},\gamma_{k_i}^2) = \rho(\Omega(\ell\times\ell),\lambda,P_0^{(2)}) = \rho(\mathbf{Y},\beta).$$

In other words λ is a joining of $\rho(\mathbf{X},\alpha)$ and $\rho(\mathbf{Y},\beta)$. Therefore

$$\begin{aligned}\bar{d}(\rho(\mathbf{X},\alpha),\rho(\mathbf{Y},\beta) &\leq \lambda(P_0^{(1)} \triangle P_0^{(2)}) = \lim_{i\to\infty} \rho_{k_i}(P_0^{(1)} \triangle P_0^{(2)}) \\ &= \lim_{i\to\infty} \eta(\gamma_{k_i}^1 \triangle \gamma_{k_i}^2) = \lim_{i\to\infty} d_{\text{ham}}(\mathbf{a}_{k_i},\mathbf{b}_{k_i}) \\ &= \liminf_k d_{\text{ham}}(\mathbf{a}_k,\mathbf{b}_k).\end{aligned}$$

□

It will be important for us later to have a way of computing the $\bar{d}$-distance of two stationary measures in $M_S(\Omega)$ by means of the $\bar{d}$-distance of their finite dimensional projections, which in turn is computed as the infimum, over *all* the joinings (of these finite dimensional projections), of average of Hamming distance with respect to these joinings. More precisely, for measures $\mu,\nu \in M(\mathfrak{L}^k)$ let

$$\begin{aligned}\bar{d}(\mu,\nu) &= \inf_{\lambda\in J(\mu,\nu)} \int_{\mathfrak{L}^k\times\mathfrak{L}^k} d_{\text{ham}}(x,y)\,d\lambda(x,y) \\ &= \inf_{\lambda\in J(\mu,\nu)} \mathbb{E}_\lambda(d_{\text{ham}}).\end{aligned}$$

Write μ_n for the finite dimensional projection of a measure $\mu \in \Omega(\ell)$ onto the coordinates $(-n,-n+1,\ldots,0,\ldots n)$; i.e. its restriction to the subalgebra generated by the partition

$$\mathfrak{L}^{\{\ldots,-n-1\}} \times P_{-n}^n \times \mathfrak{L}^{\{n+1,\ldots\}}.$$

A similar notation is used for $\lambda \in J(\mu,\nu)$. Now for measures $\mu,\nu \in M_S(\Omega(\ell))$ define

$$\hat{d}(\mu,\nu) = \limsup_{n\to\infty} \bar{d}(\mu_n,\nu_n).$$

15.24. LEMMA. 1. *For $\mu,\nu \in M(\mathfrak{L}^k)$*

$$\bar{d}(\mu,\nu) = \min_{\lambda\in J(\mu,\nu)} \int d_{\text{ham}}(x,y)\,d\lambda(x,y).$$

2. *$\bar{d}$ is a metric on $M(\mathfrak{L}^k)$.*

3. *For $\mu, \nu \in M_S(\Omega(\ell))$, $\lambda \in J(\mu, \nu)$ and $k \geq 1$,*

$$\int_{\mathfrak{L}^{2k+1} \times \mathfrak{L}^{2k+1}} d_{\text{ham}}(x_{-k}^k, y_{-k}^k)\, d\lambda_k(x_{-k}^k, y_{-k}^k) = \int_{\Omega \times \Omega} d_{\text{ham}}(x_0, y_0)\, d\lambda(x, y)$$
$$= \lambda(P_0 \times \Omega \,\triangle\, \Omega \times P_0)$$

(where $d_{\text{ham}}(x_0, y_0)$ is 0 or 1 according to whether $x_0 = y_0$ or not).

4. *For $\mu, \nu \in M_S(\Omega(\ell))$*

$$\hat{d}(\mu, \nu) = \lim_{n \to \infty} \bar{d}(\mu_n, \nu_n) = \sup_n \bar{d}(\mu_n, \nu_n).$$

PROOF. 1. Considered as a function of λ, $f(\lambda) = \int d_{\text{ham}}(x, y) d\lambda(x, y)$ is a continuous function on the compact convex set $J(\mu, \nu)$ and therefore it achieves its infimum.

2. We get $\bar{d}(\mu, \mu) = 0$ via the diagonal joining $\lambda = \text{gr}\,(\mu, \text{id})$. Conversely by part 1, $\bar{d}(\mu, \nu) = 0$ implies $\int d_{\text{ham}}(x, y) d\lambda(x, y)$ for some $\lambda \in J(\mu, \nu)$. Therefore, $\text{supp}\,(\lambda) \subset \Delta$, hence $\lambda = \text{gr}\,(\mu, \text{id})$ and it follows that $\nu = \mu$. If

$$\bar{d}(\mu, \nu) = \int d_{\text{ham}}(x, y)\, d\lambda_{1,2}(x, y) \quad \text{and} \quad \bar{d}(\nu, \eta) = \int d_{\text{ham}}(x, y)\, d\lambda_{2,3}(x, y),$$

we form the relative independent joining

$$\lambda_{1,2,3} = \lambda_{1,2} \underset{\rho_2}{\times} \lambda_{2,3} \in J(\lambda_{1,2}, \lambda_{2,3}),$$

let $\lambda_{1,3}$ denote the corresponding projection and observe that

$$\begin{aligned}
\bar{d}(\mu, \eta) &\leq \int d_{\text{ham}}(x, z)\, d\lambda_{1,3}(x, z) = \int d_{\text{ham}}(x, z)\, d\lambda_{1,2,3}(x, y, z) \\
&= \int d_{\text{ham}}(x, z)\, d(\lambda_{1,2})_y(x)\, d(\lambda_{2,3})_y(z)\, d\nu(y) \\
&\leq \int \big(d_{\text{ham}}(x, y) + d_{\text{ham}}(x, y)\big)\, d(\lambda_{1,2})_y(x)\, d(\lambda_{2,3})_y(z)\, d\nu(y) \\
&= \int d_{\text{ham}}(x, y)\, d\lambda_{1,2}(x, y) + \int d_{\text{ham}}(y, z)\, d\lambda_{2,3}(y, z) \\
&= \bar{d}(\mu, \nu) + \bar{d}(\nu, \eta).
\end{aligned}$$

3. We have by stationarity

$$\begin{aligned}
\int d_{\text{ham}}(x_{-k}^k, y_{-k}^k)\, d\lambda_k(x_{-k}^k, y_{-k}^k) &= \int \frac{1}{2k+1} \sum_{|j| \leq k} d_{\text{ham}}(x_j, y_j)\, d\lambda_k(x_{-k}^k, y_{-k}^k) \\
&= \int_{\Omega \times \Omega} d_{\text{ham}}(x_0, y_0)\, d\lambda(x, y) \\
&= \lambda(P_0 \times \Omega \,\triangle\, \Omega \times P_0).
\end{aligned}$$

4. For $\mu, \nu \in M_S(\Omega)$ one can easily check the relation

$$m\bar{d}(\mu_1^m, \nu_1^m) + n\bar{d}(\mu_{m+1}^{m+n}, \nu_{m+1}^{m+n}) \leq (m+n)\bar{d}(\mu_1^{m+n}, \nu_1^{m+n}).$$

As μ and ν are S-invariant, $d(\mu_{m+1}^{m+n}, \nu_{m+1}^{m+n}) = d(\mu_1^n, \nu_1^n)$ and we obtain

$$m\bar{d}(\mu_1^m, \nu_1^m) + n\bar{d}(\mu_1^n, \nu_1^n) \leq (m+n)\bar{d}(\mu_1^{m+n}, \nu_1^{m+n}).$$

This subadditivity implies that the lim sup in the definition of $\hat{d}$ is both a limit and a supremum (Theorem A.1). □

15.25. THEOREM. *For $\mu, \nu \in M_S(\Omega(\ell))$*

$$\hat{d}(\mu, \nu) = \bar{d}(\mu, \nu).$$

PROOF. Let $\lambda \in J(\mu, \nu)$ achieve the $\bar{d}$ distance

$$\bar{d}(\mu, \nu) = \lambda(P_0 \times \Omega \triangle \Omega \times P_0),$$

then, for each n, $\lambda_n \in J(\mu_n, \nu_n)$ and therefore by Lemma 15.24,

$$\begin{aligned}\hat{d}(\mu, \nu) &= \lim_{n\to\infty} \bar{d}(\mu_n, \nu_n)\\ &\le \lim_{n\to\infty} \int d_{\text{ham}}(x_{-n}^n, y_{-n}^n)\, d\lambda_n(x_{-n}^n, y_{-n}^n)\\ &= \bar{d}(\mu, \nu).\end{aligned}$$

For the other direction, for each n let $\lambda_n \in J(\mu_n, \nu_n)$ achieve the $\bar{d}$ distance

$$\bar{d}(\mu_n, \nu_n) = \int d_{\text{ham}}(x, y)\, d\lambda_n(x, y).$$

Let λ^n be the product measure $\cdots \times \lambda_n \times \lambda_n \times \cdots$ on

$$\Omega \times \Omega = \cdots \times (\mathfrak{L}^{2n+1} \times \mathfrak{L}^{2n+1}) \times (\mathfrak{L}^{2n+1} \times \mathfrak{L}^{2n+1}) \times \cdots$$

and let $\hat{\lambda}^n = \frac{1}{2n+1} \sum_{|j|\le n} (S \times S)^j \lambda^n \in M_{S\times S}(\Omega \times \Omega)$.

The S-invariance of μ implies that for a fixed m and any subset $A \subset \mathfrak{L}^{2m+1}$, for sufficiently large n and any $|i| \le n - m$,

$$\begin{aligned}\lambda^n\big((S \times S)^i((\mathfrak{L} \times \mathfrak{L})^{\{\dots,-m-1\}} \times (A \times \mathfrak{L}^{\{-m,\dots,m\}}) \times (\mathfrak{L} \times \mathfrak{L})^{\{m+1,\dots\}})\big)\\ = \mu_n(S^i(\mathfrak{L}^{\{\dots,-m-1\}} \times A \times \mathfrak{L}^{\{m+1,\dots\}}))\\ = \mu(S^i(\mathfrak{L}^{\{\dots,-m-1\}} \times A \times \mathfrak{L}^{\{m+1,\dots\}}))\\ = \mu(\mathfrak{L}^{\{\dots,-m-1\}} \times A \times \mathfrak{L}^{\{m+1,\dots\}}).\end{aligned}$$

This clearly implies that for every $B \in \mathfrak{B}$

$$\lim_{n\to\infty} \hat{\lambda}^n(B \times \Omega) = \mu(B).$$

A similar argument shows that also

$$\lim_{n\to\infty} \hat{\lambda}^n(\Omega \times B) = \nu(B);$$

i.e. $\hat{\lambda}^n \in J(\mu, \nu)$.

Our next goal is to show that

$$\bar{d}(\mu_n, \nu_n) = \int d_{\text{ham}}(x_0, y_0) d\hat{\lambda}^n(x, y). \tag{7.1}$$

Denoting by $\lambda_{n,j}$ the projection of λ_n on the j coordinate ($|j| \le n$), and writing for $a, b \in \mathfrak{L}$, $\hat{\lambda}^n\big((\mathfrak{L} \times \mathfrak{L})^{\{\dots,-1\}} \times \{(a, b)\} \times (\mathfrak{L} \times \mathfrak{L})^{\{1,\dots\}}\big) = \hat{\lambda}^n(a, b)$, we have

$$\hat{\lambda}^n(a, b) = \frac{1}{2n+1} \sum_{|j|\le n} \lambda_{n,j}(a, b),$$

hence

$$\begin{aligned}\int d_{\mathrm{ham}}(x_0, y_0) d\,\hat{\lambda}^n(x,y) &= \sum_{(a,b)\in\mathfrak{L}\times\mathfrak{L}} \hat{\lambda}^n(a,b)\\ &= \sum_{(a,b)\in\mathfrak{L}\times\mathfrak{L}} \frac{1}{2n+1}\sum_{|j|\le n} \lambda_{n,j}(a,b)\\ &= \frac{1}{2n+1}\sum_{|j|\le n}\sum_{(a,b)\in\mathfrak{L}\times\mathfrak{L}} \lambda_{n,j}(a,b)\\ &= \int d_{\mathrm{ham}}(x_{-n}^n, y_{-n}^n)\, d\,\lambda_n(x_{-n}^n, y_{-n}^n)\\ &= \bar{d}(\mu_n, \nu_n),\end{aligned}$$

Finally if we let $\lambda \in J(\mu,\nu)$ be any limit point of the sequence $\hat{\lambda}^n$, then

$$\begin{aligned}\bar{d}(\mu,\nu) \le \lambda(P_0\times\Omega\,\triangle\,\Omega\times P_0) &= \int d_{\mathrm{ham}}(x_0,y_0)\,d\,\lambda(x,y)\\ &= \lim_{k\to\infty}\int d_{\mathrm{ham}}(x_0,y_0)\,d\,\hat{\lambda}^{n_k}(x,y)\\ &= \lim_{k\to\infty}\bar{d}(\mu_{n_k},\nu_{n_k}) = \hat{d}(\mu,\nu).\end{aligned}$$

This completes the proof of the theorem. □

We shall also need the following lemma concerning the $\bar{d}$ distance of measures on finite spaces.

15.26. LEMMA. *For $\mu,\nu\in M(\mathfrak{L}^k)$.*

1. *There exists $\lambda\in J(\mu,\nu)$ satisfying*

$$\lambda(\mathbf{a},\mathbf{a}) = \min(\mu(\mathbf{a}),\nu(\mathbf{a})), \quad \forall\,\mathbf{a}\in P_1^k.$$

2.

$$\bar{d}(\mu,\nu)\le d_k(\mu,\nu) = \frac{1}{2}\sum_{\mathbf{a}\in P_1^k}|\mu(\mathbf{a})-\nu(\mathbf{a})|,$$

with equality when $k=1$.

3. *The metric space $(M(\mathfrak{L}^k),\bar{d})$ is complete.*

PROOF. 1. We shall actually prove the assertion, by induction on n, for every $\mu,\nu\in M(n)$, where $M(n) = M(\{1,2,\dots,n\})$. For $n=1$ this is trivial. For $n=2$ we may assume that $\mu(2)\le\nu(2)$ and then define $\lambda(2,2)=\mu(2), \lambda(2,1) = 0, \lambda(1,1)=\nu(1), \lambda(1,2)=\nu(2)-\mu(2)$. Assuming the claim holds for n consider $\mu,\nu\in M(n+1)$. Again we assume with no loss of generality that $\mu(n+1)\le\nu(n+1)$ and therefore define $\lambda(n+1,n+1)=\mu(n+1), \lambda(n+1,j)=0$ for $1\le j\le n$. There are now two cases to consider. The first is when $\nu(j)\le\mu(j)$ for all $1\le j\le n$. In this case we are forced to define $\lambda(k,k)=\nu(k)$, $\lambda(i,j)=0$ for $i\ne j$, $1\le i\le n+1$, $1\le j\le n$; and $\lambda(n+1,n+1)=\mu(n+1), \lambda(j,n+1)=\mu(j)-\nu(j)$ for $1\le j\le n$.

The second case is when there exists at least one $1\le j\le n$ with $\nu(j)>\mu(j)$. In this case let $c = 1-\mu(n+1)$ and consider the measures $\mu' = (1/c)(\mu(1),\dots,\mu(n))$ and $\nu' = (1/c)(\nu(1),\dots,\nu'(j),\nu(j+1),\dots,\nu(n))$, where $\nu'(j) = \nu(j)+\nu(n+1)-\mu(n+1)$. These measures are in $M(n)$ and the induction hypothesis yields an appropriate joining $\lambda' \in J(\mu', \nu')$. Using λ' it is now easy to construct an appropriate $\lambda \in J(\mu, \nu)$.

2. The identity $|x - y| = x + y - 2\min(x, y)$ implies

$$d_k(\mu, \nu) = 1 - \sum_{\mathbf{a} \in P_1^k} \min(\mu(\mathbf{a}), \nu(\mathbf{a})).$$

Now part 1 of the lemma provides a joining $\lambda \in J(\mu, \nu)$ with

$$\lambda(\mathbf{a}, \mathbf{a}) = \min(\mu(\mathbf{a}), \nu(\mathbf{a})), \quad \forall \mathbf{a} \in P_1^k,$$

hence

$$\begin{aligned} \bar{d}(\mu, \nu) &\le \int d_{\text{ham}}(\mathbf{a}, \mathbf{b})\, d\lambda(\mathbf{a}, \mathbf{b}) \le \lambda(\{(\mathbf{a}, \mathbf{b}) : \mathbf{a} \ne \mathbf{b}\}) \\ &= 1 - \sum_{\mathbf{a} \in P_1^k} \lambda(\mathbf{a}, \mathbf{a}) = d_k(\mu, \nu). \end{aligned}$$

For $k = 1$, $\bar{d}(\mu, \nu) = \int d_{\text{ham}}(a, b)\, d\lambda(a, b) = 1 - \sum_{a \in \mathfrak{L}} \lambda(a, a)$.

3. This follows from part 2 as $(M(\mathfrak{L}^k), d_k)$ is a compact metric space. □

8. The Jewett-Krieger theorem

In this section we shall prove the following generalization of the Jewett-Krieger theorem, which is due to B. Weiss. Recall that when $\mathbf{X} = (X, \mathcal{X}, \mu, T)$ is an ergodic dynamical system we say that the system $\hat{\mathbf{X}} = (\hat{X}, \hat{\mathcal{X}}, \hat{\mu}, T)$ is a *topological model* (or just a model) for $\mathbf{X}$ if $(\hat{X}, T)$ is a topological system, $\hat{\mu} \in M_T(\hat{X})$ and the systems $\mathbf{X}$ and $\hat{\mathbf{X}}$ are measure theoretically isomorphic. Similarly we say that $\hat{\pi} : \hat{\mathbf{X}} \to \hat{\mathbf{Y}}$ is a *topological model* for $\pi : \mathbf{X} \to \mathbf{Y}$ when $\hat{\pi}$ is a topological factor map and there exist measure theoretical isomorphisms ϕ and ψ such that the diagram

$$\begin{array}{ccc} \mathbf{X} & \xrightarrow{\phi} & \hat{\mathbf{X}} \\ {\scriptstyle \pi}\downarrow & & \downarrow{\scriptstyle \hat{\pi}} \\ \mathbf{Y} & \xrightarrow[\psi]{} & \hat{\mathbf{Y}} \end{array}$$

is commutative.

15.27. THEOREM (B. Weiss). *If* $\pi : \mathbf{X} = (X, \mathcal{X}, \mu, T) \to \mathbf{Y} = (Y, \mathcal{Y}, \nu, T)$ *is a factor map with* $\mathbf{X}$ *ergodic and* $\hat{\mathbf{Y}}$ *is a uniquely ergodic model for* $\mathbf{Y}$ *then there is a uniquely ergodic model* $\hat{\mathbf{X}}$ *for* $\mathbf{X}$ *and a factor map* $\hat{\pi} : \hat{\mathbf{X}} \to \hat{\mathbf{Y}}$ *which is a model for* $\pi : \mathbf{X} \to \mathbf{Y}$.

In particular, taking $\mathbf{Y}$ to be the trivial one point system we get:

15.28. THEOREM (Jewett-Krieger). *Every ergodic system has a uniquely ergodic model.*

The notion of a uniform partition that we introduce below is at the basis of our approach. Such a partition yields a uniquely ergodic symbolic representation (not necessarily faithful).

15.29. DEFINITION. A finite partition α is a *uniform* partition if the corresponding symbolic system $\phi_\alpha(\mathbf{X})$ is uniquely ergodic. A subset $A \subset X$ is called *uniform* if

$$\lim_{N\to\infty} \text{ ess sup}_x \left| \frac{1}{N} \sum_{0}^{N-1} \mathbf{1}_A(T^i x) - \mu(A) \right| = 0.$$

Clearly the partition α is uniform iff every set in $\cup_{n=0}^{\infty} \alpha_{-n}^{n}$ is uniform.

A brief consideration of the definitions will convince the reader of the validity of the following proposition.

15.30. PROPOSITION. *Let $\mathbf{X}$ be an ergodic system. A finite or countable partition α is uniform if and only if the corresponding symbolic representation $\mathbf{X}_\alpha = \phi_\alpha(\mathbf{X})$ is a strictly ergodic subshift.*

PROOF OF THEOREM 15.27. We begin with an outline of the proof. Our first object is to prove the existence of a non-trivial uniform partition β. Here we have to distinguish between two cases. The first is when $\mathbf{X}$ has an infinite factor with discrete spectrum, in which case this very factor can serve as a strictly ergodic factor, providing many uniform partitions. The other case is when $\mathbf{X}$ is totally ergodic (i.e. $(X, \mathfrak{X}, \mu, T^k)$ is ergodic for every $0 \neq k \in \mathbb{Z}$). Under this assumption in every K-R tower $\mathfrak{t}$ with columns $\{\mathfrak{c}_k : k = 1, 2, \dots n\}$ the g.c.d of the heights N_k is 1 and we use this fact in the construction of a uniform partition β in the neighborhood of an arbitrary partition α.

In the relative case when we assume that $\mathbf{Y}$ is a strictly ergodic system, there will be of course many $\mathbf{Y}$-measurable uniform partitions. Given a non-trivial $\mathbf{Y}$-measurable uniform partition β, an arbitrary partition $\alpha \succ \beta$ and $\delta > 0$, we show how to construct a partition $\gamma = \gamma(\alpha, \beta, \delta)$ with the properties (i) γ is uniform (ii) $\gamma \succ \beta$ and (iii) $d_{\text{part}}(\gamma, \alpha) < \delta$. Let $\mathbf{X}_\gamma$ denote the corresponding strictly ergodic symbolic representation factor. Now let $\alpha_n \succ \beta_n$ be a sequence of partitions, with β_n uniform and $\mathbf{Y}$-measurable, such that $\beta_n \nearrow \mathcal{Y}$ and $\alpha_n \nearrow \mathfrak{X}$. It then follows that $\hat{\mathbf{X}} = \underleftarrow{\lim}_n \mathbf{X}_{\gamma_n}$, with $\gamma_n = \gamma(\alpha_n, \beta_n, \frac{1}{n})$ is the required strictly ergodic model.

15.31. LEMMA. *Let $\mathbf{X}$ be a totally ergodic system, then in every K-R tower $\mathfrak{t}$ with columns $\{\mathfrak{c}_k : k = 1, 2, \dots, n\}$ the g.c.d of the heights N_k is 1.*

PROOF. Let d be the g.c.d of the heights N_k. Define a partition of X by copying the name $0, 1, 2, \dots, d-1$ periodically, starting at the base level of each column. Since $d|N_k$, the map $x \mapsto i$ when x belongs to a level copied by i, defines a factor of $\mathbf{X}$ with d points and our assumption implies that $d = 1$. □

15.32. PROPOSITION. *Given a totally ergodic system $\mathbf{X}$, a finite partition α_0 and $\delta > 0$ there is a uniform partition α with*

$$d_{\text{part}}(\alpha_0, \alpha) < \epsilon.$$

PROOF. Starting from the partition α_0 we shall construct inductively a sequence of partitions α_n, all within δ of α_0, and a triangular array of K-R towers

$\mathfrak{t}_k(n)$ with bases $B_k(n)$, $(k = 1, 2, \ldots;\ n \geq k)$ with $B_{k+1}(n) \subset B_k(n)$, so that $\mathfrak{t}_{k+1}(n)$ refines $\mathfrak{t}_k(n)$:

$$\begin{array}{cccc} \mathfrak{t}_1(1) & \mathfrak{t}_1(2) & \mathfrak{t}_1(3) & \ldots \\ & \mathfrak{t}_2(2) & \mathfrak{t}_2(3) & \ldots \\ & & \mathfrak{t}_3(3) & \ldots \\ & & & \ldots \end{array}$$

In the d_{part} metric we shall have $\lim_n B_k(n) = B_k$ and $\lim_n \alpha_n = \alpha$ with α a uniform partition.

Step 1: By lemma 15.16, applied to the indicator functions of the elements of α_0 and $\epsilon = 1/10$, there exists N_1' such that any bounded K-R tower with minimal height $> N_1'$ at least 9/10 of the fibers will have the empirical distribution of α_0 1-blocks within 1/10 of the α_0 distribution. Use Theorem 15.5 to build a bounded K-R tower $\mathfrak{t}_1(1)$ with base $B_1(1)$ and heights N_1, $N_1 + 1$ with $N_1 > N_1'$. Refine $\mathfrak{t}_1(1)$ according to α_0. Choosing an arbitrary good N_1 column, we now copy its name on the remaining bad N_1 columns, and similarly for the bad $N_1 + 1$ columns.

We have defined a new partition α_1 so that all the fibers of our K-R tower have an empirical distribution of α_1 names within 1/10 of the α_0 distributions. Again if N_1 is sufficiently large the distance $d_{\text{part}}(\alpha_0, \alpha_1)$ will be small and we shall get also that all fibers have an empirical distribution of α_1 names within 1/10 of the α_1 distributions. Further changes in α_1 will not change the N_1, $N_1 + 1$ blocks that we shall see on fibers of the tower over our ultimate B_1. Therefore, we shall get a uniformity (up to 1/10) on all blocks of size $\hat{N}_1 = 100N_1$. The 100 is to get rid of the edge effects since we only know the distribution across fibers over points in $B_1(1)$.

Step 2: Next apply Theorem 15.16 to indicator functions of elements of the partition $(\alpha_1)_0^1$ with $\epsilon = 1/100$. Choose N_2' so large that N_1/N_2' is negligible and so that any bounded K-R tower with minimal height at least N_2' has for at least 99/100 of its fibers an empirical distribution of α_1 2-blocks within 1/100 of the actual $(\alpha_1)_0^1$ distribution. Use Theorem 15.5 to build a bounded K-R tower $\mathfrak{t}_2(2)$ with base $B_2(2) \subset B_1(1)$ such that its column heights are between $N_2 > N_2'$ and $N_2 + 4N_1$. Refine $\mathfrak{t}_2(2)$ according to α_1. For columns with good α_1 2-block distribution we make no change. For the others, we would like to choose good names and copy them on the bad columns. However it may happen that no fiber with the precise height of the bad column is good, and we have to preserve the N_1, $N_1 + 1$ blocks.

Now the heights of columns in the K-R tower $\mathfrak{t}_1(1)$, N_1 and N_1+1, are of course relatively prime. This means that taking $M_2 = 10N_1^2$, both M_2 and $h - M_2$ — where h is the height of the bad column — can be written as a combination of N_1 and $N_1 + 1$. We can now copy on most of the bad column (except for the top M_2 levels) most of the name from one of the good columns, preserving the $B_1(1)$-name so that we preserve the N_1, $N_1 + 1$ blocks. The top M_2 levels are copied on with good N_1, $N_1 + 1$ blocks.

This correction of the bad names on the bad columns also changes the levels where the bases of the N_1, $N_1 + 1$ blocks occur. Thus it defines a new base for the first tower that we call $B_1(2)$, and therefore a new K-R tower $\mathfrak{t}_1(2)$. It also defines and a new partition α_2. The partition α_2 has the properties that all the fibers of the tower $\mathfrak{t}_1(2)$ over $B_1(2)$ have good (up to 1/10) α_2 1-block distributions, and all the fibers of the tower $\mathfrak{t}_1(2)$ over $B_2(2)$ have good (up to $1/100$) α_2 2-block distributions. These will not change in the subsequent steps of the construction. Therefore, we shall get a uniformity (up to $1/10$) on all blocks of size $\hat{N}_2 = 10^3 N_2$, where again the 10^3 is to get rid of the edge effects. Note that the change from $B_1(1)$, to $B_1(2)$ (hence also from α_1 to α_2) can be made arbitrarily small by choosing N_2 sufficiently large.

The construction will be complete when we describe the third step of the inductive procedure since that step is already completely general in that one can replace 2 by n and 3 by $n+1$.

Step 3: Next apply Theorem 15.16 to indicator functions of elements of the partition $(\alpha_2)_0^2$ with $\epsilon = 1/10^3$. Choose N_3' so large that N_2/N_3' is negligible and so that any bounded K-R tower with height at least N_3' has for at least $999/1000$ of its fibers an empirical distribution of α_2 3-blocks within $1/10^3$ of the actual $(\alpha_2)_0^2$ distribution. Use Theorem 15.5 to build a bounded K-R tower $\mathfrak{t}_3(3)$ with base $B_3(3) \subset B_2(2)$ such that its column heights are between $N_3 > N_3'$ and $N_3 + 4N_2$. Refine $\mathfrak{t}_3(3)$ according to α_2. For columns with good α_2 3-block distribution we make no change.

Now by Lemma 15.31 the heights of columns in the K-R tower $\mathfrak{t}_2(2)$ are relatively prime. This means that we can find a number M_3 such that M_3/N_3 is negligible and such that both M_3 and $h - M_3$, where h is the height of the bad column, can be written as a combination of heights of columns from $\mathfrak{t}_2(2)$. We can now copy on most of the bad column (except for the top M_3 levels) most of the name from one of the good columns, preserving the $B_2(2)$-name so that we preserve the N_2, $N_2 + 4N_1$ blocks. The final M_3 levels are copied with good N_2, $N_2 + 4N_1$ blocks.

This copying defines new bases for the previous towers that we call $B_1(3)$ and $B_2(3)$, and therefore new K-R towers $\mathfrak{t}_1(3)$ and $\mathfrak{t}_2(3)$. It also defines a new partition α_3. The partition α_3 has the properties that all the fibers of the towers $\mathfrak{t}_1(3)$ over $B_1(3)$ have good (up to $1/10$) α_3 1-block distributions, all the fibers of the tower $\mathfrak{t}_2(3)$ over $B_2(3)$ have good (up to $1/100$) α_3 2-block distributions and all the fibers of the tower $\mathfrak{t}_3(3)$ over $B_3(3)$ have good (up to $1/10^3$) α_3 3-block distributions. These will not change in the subsequent steps of the construction. Therefore, we shall get a uniformity (up to $1/100$) on all blocks of size $\hat{N}_3 = 10^4 N_3$. Again the changes in $B_1(2)$ and $B_2(2)$ to $B_1(3)$ and $B_2(3)$ respectively (hence also from α_2 to α_3), can be made arbitrarily small by choosing N_3 sufficiently large.

With this construction completed we end up with a sequence of partitions α_n that converges to a partition α and towers $\mathfrak{t}_k(n)$ with bases $B_k(n)$ converging to a tower $\mathfrak{t}_k$ with base B_k, such that the α-names of all fibers of the tower $\mathfrak{t}_k$ over points in B_k have a good (up to $1/10^k$) distribution of α k-blocks. This clearly gives the uniformity of the partition α. □

15.33. PROPOSITION. *Every ergodic non-periodic system* $\mathbf{X}$ *has a uniquely ergodic non-periodic factor.*

PROOF. When $\mathbf{X}$ is totally ergodic this follows from Proposition 15.32. If $\mathbf{X}$ admits an adding machine as a factor, then this adding machine is the required uniquely ergodic non-periodic factor (see Exercise 5.10). The only remaining case is when $\mathbf{X}$ admits a maximal finite factor $\mathbf{X} \to \{0, 1, \ldots, d-1\}$. It then follows that

$\mathbf{X}$ is isomorphic to $\mathbf{X}_0 \times \{0, 1, \dots, d-1\}$, where $\mathbf{X}_0$ is a totally ergodic T^d-system and for $x \in X_0, 0 \le j < d-1, T(x, j) = (x, j+1)$, while $T(x, d-1) = (T^d x, 0)$. Clearly a uniform factor for this system will also yield a uniform factor of $\mathbf{X}$. □

15.34. PROPOSITION. *Given a uniform partition β and an arbitrary partition α_0 that refines β, for any $\epsilon > 0$ there is a uniform partition α that also refines β and satisfies*

$$d_{\text{part}}(\alpha_0, \alpha) < \epsilon.$$

PROOF. We would like to proceed along the lines of the proof of Proposition 15.32; however we now face a severe restriction in that we are no longer free to change the basic partition β. As the various partitions α_n that will be constructed refine the partition β, we shall think of the alphabet of the α_n partitions as composed of pairs of symbols (i, j) where i denotes an index of an element of the partition β and j an index of an element of the partition α_n, a subset of the former. We shall refer to the i part of the name as the β part. When redefining names we are not allowed to change the β part of a name. In order to overcome this problem we change our strategy.

We no longer try to make all the columns of the K-R tower good, but we make sure that the bad ones are seen less and less frequently, uniformly over the whole space. What helps us in this strategy is the presence of the uniform σ-algebra $\mathcal{B} = \beta^T$ generated by β. We shall build the K-R towers so that they will be measurable with respect to $\mathcal{B}$. As we shall see this will mean that when refining the tower according to α_n, the union of the subcolumns with bad names represents a set in $\mathcal{B}$ of small measure and therefore for sufficiently large N, *uniformly over the whole space*, we shall not encounter this set too often.

To make this precise we shall now describe, given the partition α_2 and the $\mathcal{B}$ towers $\mathfrak{t}_2(1)$ and $\mathfrak{t}_2(2)$, the construction of the partition α_3 and the $\mathcal{B}$ towers $\mathfrak{t}_3(j)$, $1 \le j \le 3$.

Step 3: Apply Theorem 15.16 to indicator functions of elements of the partition $(\alpha_2)_0^2$ with $\epsilon = 1/10^3$. Choose N_3' so large that N_2/N_3' is negligible and so that any bounded K-R tower with height at least N_3' has for at least 999/1000 of its fibers an empirical distribution of α_2 3-blocks within $1/10^3$ of the actual $(\alpha_2)_0^2$ distribution. Use Theorem 15.5 to build an $\mathcal{B}$*-measurable* bounded K-R tower $\mathfrak{t}_3(3)$ with base $B_3(3) \subset B_2(2)$ such that its column heights are between $N_3 > N_3'$ and $N_3 + 4N_2$. Refine $\mathfrak{t}_3(3)$ first according to the β part of the name of a fiber. A subcolumn in this refined tower is still a $\mathcal{A}$-measurable set. If this subcolumn has at least one fiber with good α_2 3-block distribution we copy the name of this good fiber on the rest of the subcolumn. This copying will not change the partition β.

After accomplishing this procedure for all the subcolumns of the refined tower we are left with those β columns all of whose α_2 names have bad empirical distribution of α_2 3-blocks. If we denote by A the union of all these subcolumns, then A is $\mathcal{B}$-measurable and $\mu(A) < 1/10^3$. Thus for sufficiently large $\hat{N}_3$, *uniformly over the whole space*, on blocks of size $\hat{N}_3$ we shall encounter these bad columns with frequency less than $1/10^3$ and therefore shall see a good empirical distribution of α_2 3-blocks. □

We can now finish the proof of Theorem 15.27. according to the plan outlined at the beginning of the proof. □

One can describe Theorem 15.27 as asserting that every diagram of ergodic systems of the form $\mathbf{X} \to \mathbf{Y}$ has a strictly ergodic model. What can we say about more complicated commutative diagrams? A moments reflection will show that a repeated application of Theorem 15.27 proves the following theorem.

15.35. THEOREM. *Any commutative diagram in the category of ergodic* $\mathbb{Z}$ *dynamical systems with the structure of an inverted tree, i.e. no portion of it looks like*

$$\mathbf{X} \xleftarrow{\ \alpha\ } \mathbf{Z} \xrightarrow{\ \beta\ } \mathbf{Y} \tag{8.1}$$

has a strictly ergodic model. On the other hand there exists a diagram of the form (8.1) *that does not admit a strictly ergodic model.*

PROOF. We only need to prove the last assertion. Take $\mathbf{X} = \mathbf{Y}$ to be any nontrivial weakly mixing system, then $\mathbf{X} \times \mathbf{X}$ is ergodic and the diagram

$$\mathbf{X} \xleftarrow{\ p_1\ } \mathbf{X} \times \mathbf{X} \xrightarrow{\ p_2\ } \mathbf{X} \tag{8.2}$$

is our counter example. In fact if (8.1) is a uniquely ergodic model in this situation then it is easy to establish that the condition in Theorem 6.23 is satisfied and we apply this theorem to conclude that $\mathbf{X}$ is disjoint from itself. Since in a nontrivial system $\mu \times \mu$ and $\mathrm{gr}\,(\mu, \mathrm{id})$ are different ergodic joinings, this contradiction proves our assertion. □

9. Notes

Most of the chapter follows Rudolph [**220**]. In Sections 1,3 and 7 I used material and notations from P. Shields' book [**235**].

Kakutani, Rohlin and K-R towers: Rohlin's lemma (Theorem 15.4) holds for every aperiodic $\mathbb{Z}$-system $(X, \mathcal{X}, \mu, T)$ (that is a system satisfying $\mu(\{x : \exists\ n \neq 0,\ T^n x = x\}) = 0$), but the proof is considerably more intricate (see Halmos' book [**112**]).

Jewett-Krieger's theorem: The original papers of Jewett and Krieger are [**129**] and [**161**] respectively. I closely follow B. Weiss' exposition of his proof in [**264**]. In the paper Weiss [**261**] there is only an outline of a proof. The latter is also the source of the discussion of models for commutative diagrams. The notion of uniform partition was first used in this context by G. Hansel and J. P. Raoult [**114**].

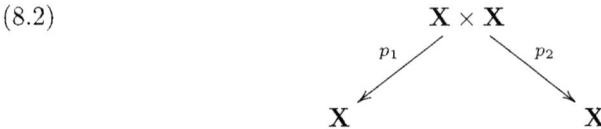

Let me mention two more striking representation results. The first is the following theorem which is due to B. Weiss and can be found in the papers [**236**] and [**263**]. A topological dynamical system (X, T) is called *prime* if it admits no non-trivial factors.

15.36. THEOREM. *An ergodic dynamical system has a topological, minimal, prime model iff it has zero entropy.*

For the second result one needs the notion of a UPE topological system. Suffices it to say here that the notion of UPE is one of several competing definitions of a "topological K-system" (see Chapter 19 for the theory of entropy pairs and UPE systems). The second theorem then (Glasner and Weiss [**98**]) asserts that an ergodic dynamical system has a strictly ergodic, UPE model iff it has positive entropy.

CHAPTER 16

Constructions

We have already seen (and shall further see in the next few chapters) the essential role that is played by Kakutani-Rohlin towers and in particular by Rohlin's lemma (Theorem 15.4) in ergodic theory. It is hardly surprising therefore to find that the same line of ideas serves also as a main source for construction of examples of dynamical systems. The idea is to consider on a Lebesgue space (say the unit interval with Lebesgue measure) a "geometric" Kakutani-Rohlin tower that exhausts most of the measure space (i.e. the geometric picture we associate with such a tower that arises from a dynamical system) and then *define*, by means of this picture — i.e. by mapping each interval onto the one above it — a transformation on all the levels of the tower except for the top ones. If we follow this construction successively on a sequence of nesting towers in a careful way we end up with a well defined measure-preserving transformation. This way of building a dynamical system is called "cutting and stacking"; it is extremely flexible and with clever combinatorics one can force the constructed transformation to have any one of a vast variety of properties. For instance the famous examples of D. Ornstein of a mixing transformation with no square root, [**190**], and of a K-automorphism which is not Bernoulli, [**191**], are both cutting and stacking constructions.

We shall treat in this short chapter only a very restricted sort of cutting and stacking, namely the "rank one" construction. For a systematic treatment of the method in general we refer to [**70**], [**220**] and [**235**]. For a survey of the theory and history of finite rank systems see [**66**]. In the first section we briefly examine rank one systems, and in the second we introduce the rank one Chacón system. In the third section we apply rank one cutting and stacking to construct α-weakly mixing systems.

1. Rank one systems

16.1. DEFINITION. A dynamical system $\mathbf{X}$ is a *rank 1* system if there exists a sequence of Rohlin towers

$$\mathfrak{t}_n = \{B_n, TB_n, \ldots, T^{h_n-1}B_n\},$$

with $h_n \to \infty$,

$$\mu\Big(\bigcup_{j=0}^{h_n-1} T^j B_n\Big) \to 1,$$

and such that $\mathcal{F}(\cup_n \mathfrak{t}_n) = \mathcal{X}$, where $\mathcal{F}(\cup_n \mathfrak{t}_n)$ is the σ-algebra generated by $\cup_n \mathfrak{t}_n$.

16.2. EXERCISE. Adding machines and the Chacón system are rank 1.

16.3. EXERCISE. If $\mathbf{X}$ is of rank 1, there is a sequence of Rohlin towers as in the definition above which in addition satisfies $\mathfrak{t}_n \prec \mathfrak{t}_{n+1}$ (i.e. $\mathbf{X}$ is obtained by a cutting and stacking process with one column).

16.4. THEOREM. *A rank 1 system is ergodic.*

PROOF. Given $A \in \mathcal{X}, \mu(A) > \delta > 0$, choose n and $C_n \in \mathcal{F}(\mathfrak{t}_n)$ satisfying $\mu(A \triangle C_n) < \eta$, where $2\eta = \delta(\mu(A) - \delta)$. If for every $0 \le j \le h_n - 1$, $\mu(A \cap T^j B_n) < (1 - \delta)\mu(B_n)$, then summing on those j's with $T^j B_n \subset C_n$, we get $\mu(A \cap C_n) < (1 - \delta)\mu(C_n)$, hence

$$\mu(A) - \eta < \mu(A \cap C_n) < (1 - \delta)\mu(C_n) < (1 - \delta)(\mu(A) + \eta).$$

This leads to $\eta < -\delta$, a contradiction.

Thus there exists $0 \le j_0 \le h_n - 1$ for which

$$\mu(A \cap T^{j_0} B_n) \ge (1 - \delta)\mu(B_n)$$

and summing over the column we get:

$$\sum_{j+j_0=1}^{h_n-1} \mu(T^j A \cap T^{j+j_0} B_n) \ge (1 - \delta)\mu(|\mathfrak{t}_n|),$$

hence

$$\mu((\bigcup_{j\in\mathbb{Z}} T^j A) \cap |\mathfrak{t}_n|) \ge (1 - \delta)\mu(|\mathfrak{t}_n|),$$

and we conclude that

$$\mu(\bigcup_{j\in\mathbb{Z}} T^j A) = 1.$$

□

16.5. THEOREM. *The spectral type of a rank 1 system is simple.*

PROOF. By Baire's category theorem it suffices to show that for any $f \in L_0^2(\mu)$ and $\delta > 0$, the set $V = \{g \in L_0^2(\mu) : d(f, Z(g)) < \delta\}$ is a dense open set, since then, for a dense sequence $f_n \in L_0^2(\mu)$, every element g of

$$U = \bigcap_n \bigcap_k \{g \in L_0^2(\mu) : d(f_n, Z(g)) < 1/k\},$$

satisfies $Z(g) = L_0^2(\mu)$.

Clearly V is open. Fix $h \in L_0^2(\mu)$ and $0 < \epsilon < \frac{1}{2}\delta$. Let $\alpha = \{A_1, A_2, \dots, A_\ell\}$ be a measurable partition of X such that there exist $\mathcal{F}(\alpha)$-measurable functions, $\tilde{f}$ and $\tilde{h}$, with $||f - \tilde{f}||$ and $||h - \tilde{h}||$ less than ϵ. By the rank 1 property, there exits a Rohlin tower $\mathfrak{t} = \{B, TB, \dots, T^{d-1}B\}$ and a partition $\alpha' = \{A'_1, A'_2, \dots, A'_\ell\}$, measurable with respect to the algebra generated by the partition $\mathfrak{t} \cup \{X \setminus |\mathfrak{t}|\}$ (which we denote by $\mathcal{F}(\mathfrak{t})$), such that

$$d(\alpha, \alpha') = \sum_{j=1}^{\ell} \mu(A_j \triangle A'_j) < \epsilon.$$

So we can assume that $\tilde{f}$ and $\tilde{h}$ are $\mathcal{F}(\mathfrak{t})$-measurable and that they both vanish on $X \setminus |\mathfrak{t}|$. By the nature of $\mathfrak{t}$ it follows that $\tilde{f}$ and $\tilde{h}$ are in $Z(\mathbf{1}_B)$. We now have

$$d(f, Z(\mathbf{1}_B)) < ||f - \tilde{f}|| < \epsilon.$$

Write $\tilde{h} = p(U_T)\mathbf{1}_B$ and $\tilde{f} = q(U_T)\mathbf{1}_B$, where $p(z)$ and $q(z)$ are polynomials. By a small perturbation (say, of the form $p(z) \mapsto p(z+\delta)$), we can find a polynomial $s(z)$ that does not vanish on $K = \{z : |z| = 1\}$ (so that $0 \notin s(K) \supset s(\mathrm{spec}\,(U_T)) = \mathrm{spec}\,(s(U_T))$; i.e. $s(U_T)$ is invertible), and such that

$$||(p(U_T) - s(U_T))\mathbf{1}_B|| < \epsilon.$$

Denoting $g = s(U_T)\mathbf{1}_B$, we have

$$||h - g|| \leq ||h - \tilde{h}|| + ||\tilde{h} - g|| < 2\epsilon.$$

Next chose a polynomial $r(z)$ so that $r(U_T)$ approximates $s(U_T)^{-1}$ in the sense that:

$$||r(U_T)g - \mathbf{1}_B|| = ||r(U_T)g - s(U_T)^{-1}g|| < \frac{\epsilon}{||q(U_T)||},$$

hence

$$\begin{aligned} d(f, Z(g)) &\leq ||f - q(U_T)r(U_T)g|| \\ &\leq ||f - \tilde{f}|| + ||q(U_T)\mathbf{1}_B - q(U_T)r(U_T)g|| \\ &\leq \epsilon + ||q(U_T)||||\mathbf{1}_B - r(U_T)g|| < 2\epsilon. \end{aligned}$$

We thus found in a 2ϵ-neighborhood of h a function g with $d(f, Z(g)) < 2\epsilon$. This shows that V is dense and completes the proof. □

2. Chacón's transformation

Chacón's dynamical system (or transformation), introduced in [**40**] as an example of a weakly mixing but not mixing system, has played a crucial role in both rank one theory and the theory of minimal self-joinings. In this section we briefly indicate its construction as a rank one system and consider some of its basic properties.

For this construction our Lebesgue space is the interval $[0, 3/2]$ with normalized Lebesgue measure.

Set $B_1 = [0, 1/3], TB_1 = [1/3, 2/3]$, then $T^2B_1 = S_1 = [1, 1 + 1/3]$ and $T^3B_1 = [2/3, 1]$. Thus $h_1 = 3 + 1 = 4$ and

$$\mathfrak{t}_1 = \{B_1, TB_1, T^2B_1, T^3B_1\}$$

is our first tower. Next set $B_2 = [0, 1/9], TB_2 = [1/3, 1/3 + 1/9], T^2B_2 = [1, 1 + 1/9]T^3B_2 = [2/3, 2/3+1/9], T^4B_2 = [1/9, 2/9], T^5B_2 = [1/3+1/9, 1/3+2/9], T^6B_2 = [1 + 1/9, 1 + 2/9], T^7B_2 = [2/3 + 1/9, 2/3 + 2/9]$ as dictated by the first tower $\mathfrak{t}_1$. Now we let $S_2 = [1 + 1/3, 1 + 1/3 + 1/9]$ and set $T^8B_2 = S_1, T^9B_2 = [2/9, 1/3], T^{10}B_2 = [1/3 + 2/9, 2/3], T^{11}B_2 = [1 + 2/9, 1 + 1/3], T^{12}B_2 = [2/3 + 2/9, 1]$. Thus $h_2 = 13 = 3 \cdot 4 + 1$ and

$$\mathfrak{t}_2 = \{B_2, TB_1, \ldots, T^{12}B_1\}.$$

At the $n + 1$ stage we cut the n-th tower $\mathfrak{t}_n$ vertically into three columns of equal width $3^{-(n+1)}$, stack the second column on top of the first, a single "spacer level"

S_{n+1} — an interval of length $3^{-(n+1)}$ — on top of the second column, and finally the third column on top of the spacer.

This procedure defines, with the exception of a countable set of points, a 1-1 transformation T of the interval $X = [0, 3/2]$ onto itself. Equipped with its Borel σ-algebra $\mathcal{X}$ and normalized Lebesgue measure μ, the dynamical system $\mathbf{X} = (X, \mathcal{X}, \mu, T)$ is the Chacón transformation.

We now state without proof the following theorem. Good references here are [**205**] and [**220**].

16.6. THEOREM. *The Chacón system* $\mathbf{X}$ *has the following properties.*

1. $\mathbf{X}$ *is measure theoretically isomorphic to the uniquely ergodic topological system given as a substitution minimal system in Examples 1.19 (see Theorem 4.11).*
2. *It is weakly mixing (in fact mildly mixing) but not mixing.*
3. $\mathbf{X}$ *has minimal self-joinings of all orders, hence* $\mathrm{Aut}(\mathbf{X}) = \{T^n : n \in \mathbb{Z}\}$ *and* $\mathbf{X}$ *is prime (see Corollary 12.5).*
4. *Let* σ *be the spectral type of* $\mathbf{X}$*, then the family of convolution powers* $\sigma^n = \sigma * \sigma * \cdots * \sigma$ *(n-times),* $n = 1, 2, \ldots$ *is pairwise singular (see also Theorem 7.12).*

3. A rank one α-weakly mixing system

As promised in Chapter 7 we describe in the present section a construction of α-weakly mixing systems. Our Lebesgue space is the unit interval $[0, 1]$ and all the partitions that will be considered during the construction will consist of subintervals. We set aside a certain special interval the subintervals of which will be called *spacers*.

Starting from the n tower (or column) $\mathfrak{t}_n$, we shall describe the construction of the $n+1$ tower by cutting and stacking columns from the n tower and adding spacers. The height of the n tower we denote by $h = h_n$ and its h levels — h intervals of equal length stacked on top of each other — will be labeled $0, 1, 2, \ldots, h-1$; thus $\mathbf{a} = 012 \cdots h - 1$ is the "name" of points at the base of $\mathfrak{t}_n$. The spacer will be denoted by x. Cut the n tower vertically into many columns of equal width and let $\mathbf{a} = 012 \cdots h - 1$ denote the name of each of these columns; just how many, will be determined by the construction. In the next paragraph we show how to construct the $n+1$ tower $\mathfrak{t}_{n+1}$, with name $\mathbf{b}$, from these columns $\mathbf{a}$ and the spacer intervals, which are new intervals, having the same length as the levels of $\mathbf{a}$ columns cut from $\mathfrak{t}_n$, each labeled as x.

First put $\mathbf{c} = \mathbf{a}\mathbf{a} \cdots \mathbf{a} = \mathbf{a}^C$, where C is a parameter to be specified later. (Of course we write $\mathbf{c}$ horizontally, but again we think of it as a stack or column of intervals of equal length stacked one on top of the other.) Next choose another parameter M and for each $0 \leq s \leq M$ let

$$\mathbf{c}_s = x^s \mathbf{c} x^{M-s}.$$

The common length (or height) of the columns $\mathbf{c}_s$,

$$w_n = w = |\mathbf{c}_s| = Ch + M,$$

will turn out to be the nth element of the α-weak mixing sequence. Next choose yet another parameter N, an even number, and let

$$\mathbf{s} = (s_0, s_1, s_2, \ldots, s_{N-1}) \in \{0, 1, 2, \ldots, M\}^N.$$

The main idea of the construction is to think of this vector as an i.i.d. sequence of random variables uniformly distributed over $\{0, 1, 2, \ldots, M\}$ and then, by the law of large numbers, to choose a suitable N and a suitable realization of $\mathbf{s}$. Our final step is to choose two more parameters L_1 and L_2, let $\mathbf{b}_1 = \mathbf{c}(0)^{NL_1}$ and

$$\mathbf{b}_2 = (\mathbf{c}(0)\mathbf{c}(1))^{L_2}(\mathbf{c}(2)\mathbf{c}(3))^{L_2} \cdots (\mathbf{c}(N-2)\mathbf{c}(N-1))^{L_2},$$

where we write $\mathbf{c}(j)$ for $\mathbf{c}_{s_j}$, and set

$$\mathbf{b} = \mathbf{b}_1\mathbf{b}_2.$$

This completes the construction of the $n+1$ tower $\mathfrak{t}_{n+1}$ with name $\mathbf{b}$ and hence also the formal description of our rank one system. We shall denote by $\mathfrak{t}^1$ and $\mathfrak{t}^2$ the parts of the tower $\mathfrak{t}_{n+1}$ corresponding to the names $\mathbf{b}_1$ and $\mathbf{b}_2$ respectively. Thus $|\mathfrak{t}^1|$ denotes the subset of X — our system's phase space — formed by the union of the levels of $\mathfrak{t}^1$, while $|\mathbf{b}_1| = NL_1w$ is the length of the name $\mathbf{b}_1$ or the height of the tower $\mathfrak{t}^1$.

We are now given h and $\epsilon > 0$ and our task is to determine the values of the various parameters, $C, M, N, L_k, k = 1, 2$, as well as the values of the random vector s, in such a way that — up to ϵ — for sets A, B in the finite algebra $\mathcal{F}(\mathfrak{t}_n)$; i.e. sets formed by union of levels of $\mathfrak{t}_n$,

$$\mu(T^w A \cap B) \overset{\epsilon}{\sim} \alpha\mu(A)\mu(B) + (1-\alpha)\mu(A \cap B), \tag{3.1}$$

and so that the total mass added by the new spacers is less than ϵ. It is enough to check this relation for A and B level sets of $\mathfrak{t}_n$; thus for sets A and B formed as the union of all levels of $\mathbf{b}$ labeled by l and l' respectively, where $0 \le l, l' \le h-1$. Of course, once C and M are chosen, the numbers L_k ought to be sufficiently large so that $w/|\mathbf{b}|$ is small compared with ϵ so that (3.1) makes sense. We write A_k for the part of A that lies in $\mathbf{b}_k$, $k = 1, 2$, and B_k is defined similarly.

In order to obtain the right proportion of mixing and rigidity we require that

$$\frac{L_2}{L_1 + L_2} \overset{\epsilon}{\sim} \alpha.$$

The rigidity resides in the $\mathbf{b}_1$ part of the tower $\mathbf{b} = \mathfrak{t}_{n_1}$. In fact this part is composed of N strings of L_1 periodic blocks, each of period w. It thus follows that $\mu(T^w A_1 \Delta A_1) \sim \epsilon$ and $\mu(T^{-w} B_1 \Delta B_1) \sim \epsilon$, hence

$$\mu(T^w A_1 \cap B_1) < \epsilon, \qquad \mu(T^w A_1 \cap B_2) < \epsilon,$$
$$\mu(T^w A_2 \cap B_1) = \mu(A_2 \cap T^{-w} B_1) < \epsilon$$

when $A \neq B$ (i.e. $l \neq l'$) and

$$\mu(T^w A_1 \cap B_1) \overset{\epsilon}{\sim} \mu(A_1) = (1-\alpha)\mu(A) = (1-\alpha)\mu(A \cap B),$$
$$\mu(T^w A_1 \cap B_2) < \epsilon, \quad \mu(T^w A_2 \cap B_1) = \mu(A_2 \cap T^{-w} B_1) < \epsilon$$

when $A = B$ ($l = l'$). In order to obtain (3.1) it therefore suffices to show that

$$\mu(T^w A_2 \cap B_2) \overset{\epsilon}{\sim} \alpha\mu(A)\mu(B). \tag{3.2}$$

Let us denote by ν the normalized — up to ϵ — probability measure obtained by restricting μ to $|\mathbf{b}_2|$; i.e. $\nu(D) = (1/\alpha)\mu(D \cap |\mathrm{t}^2|)$. Since $\mu(A_2) = \mu(A \cap |\mathrm{t}_2|) \stackrel{\epsilon}{=} \alpha\mu(A)$ and $\mu(B_2) = \mu(B \cap |\mathrm{t}^2|) \stackrel{\epsilon}{=} \alpha\mu(B)$, it follows that (3.2) is the same as

$$\nu(T^w A_2 \cap B_2) \stackrel{\epsilon}{\sim} \alpha\nu(A_2)\nu(B_2), \tag{3.3}$$

which we now proceed to establish.

As mentioned above the strategy of the proof is to consider the parameters $\mathbf{s} = (s_0, s_1, s_2, \dots, s_{N-1}) \in \{0, 1, 2, \dots, M\}^N$ as an i.i.d. process with each s_j a random variable uniformly distributed over $\{0, 1, 2, \dots, M\}$. For $l, l' \in \{0, 1, \dots, h-1\}$ let

(3.4)
$$\pi(l, l') = (|\mathbf{b}_2| - 1 - 2w)^{-1}\mathrm{card}\,\{t \in [0, |\mathbf{b}_2| - 1 - 2w] : \mathbf{b}_2(t) = l,\ \mathbf{b}_2(t+w) = l'\}.$$

π is (almost) a probability distribution on $[0, 1, \dots, h-1] \times [0, 1, \dots, h-1]$ and our goal is to show that for an appropriate choice of N and a good realization of the random vector $\mathbf{s}$, the distribution π is within ϵ of the uniform distribution on $[0, 1, \dots, h-1] \times [0, 1, \dots, h-1]$, which is just a reformulation of (3.3). For a subset $F \subset [0, |\mathbf{b}_2| - 1 - 2w]$ we shall use the notation π_F for the distribution

$$\pi(l, l') = |F|^{-1}\mathrm{card}\,\{t \in F : \mathbf{b}_2(t) = l,\ \mathbf{b}_2(t+w) = l'\}.$$

Let W be the interval $[0, w-1]$ in $\mathbb{Z}$ and $0 \le r \le 2L_2$, and set

$$E = E(r) = \bigcup_{q=0}^{\frac{N-2}{2}} (W + rw + 2qL_2).$$

We think of W as a "window" in $\mathbf{b}_2$ and then E is a choice of the rth window $W' = W + rw + 2qL_2$ in each periodic block of $\mathbf{b}_2$. Ignoring the "edges" what we see in the window W' is the block $\mathbf{c}_{s_{2q}}$ in $\mathbf{b}_2$ and the block $\mathbf{c}_{s_{2q+1}}$ in $T^w\mathbf{b}_2$ if r is even and $\mathbf{c}_{s_{2q+1}}$ in $\mathbf{b}_2$ and $\mathbf{c}_{s_{2q+2}}$ in $T^w\mathbf{b}_2$ if r is odd. So by the definition of $\mathbf{c}_s$ we see two $\mathbf{c}$ blocks separated by a shift of either $d = s_{2q+1} - s_{2q}$ or $d = s_{2q+2} - s_{2q+1}$. If M is sufficiently small compared to $|\mathbf{c}| = Cw$, the overlap of these two copies of $\mathbf{c}$ fills most of the window W'. Since $\mathbf{c}$ is a concatenation of many copies of $\mathbf{a}$, it follows that $\pi_{W'}$ is close to the uniform distribution on the subset $\{(l, l+d) : 0 \le l \le h-1\}$ of $[0, 1, \dots, h-1] \times [0, 1, \dots, h-1]$. Denote this uniform distribution by δ_d ($d \in [-M, M]$).

The distribution of the random variable $s_{2q+1} - s_{2q}$ (or $d = s_{2q+2} - s_{2q+1}$) is the unique tent shape distribution on $[-M, M]$, which we denote by σ_M. Now the sequence $s_{2q+1} - s_{2q}$, $q = 0, 1, \dots, 2N-2/2$ (or $s_{2q+2} - s_{2q+1}$, $q = 0, 1, \dots, 2N-2/2$ depending on the parity of r) is an i.i.d. sequence and the law of large numbers guarantees that if N is sufficiently large, then with high probability the empirical distribution of the sequence $s_{2q+1} - s_{2q}$, $q = 0, 1, \dots, (2N-2)/2$ is close to σ_M. We can therefore choose a realization of $\mathbf{s}$ such that

$$\pi_E \stackrel{\epsilon}{\sim} \sum_{d=-M}^{M} \sigma_M(d)\delta_d.$$

The distribution σ_M has a unique representation as $\sigma_M = \sum_{m=0}^{M} a_m u_m$, where u_m is the uniform distribution on $[-m, m]$ and $a_m > 0$ sum up to 1. Moreover, given $H > 0$, if M is sufficiently large the total weight of u_m with $m \le H$, $\sum_{m \le H} a_m$ is small compared with ϵ. Thus

$$\begin{aligned}\pi_E &\overset{\epsilon}{\sim} \sum_{d=-M}^{M} \sigma_M(d)\delta_d \\ &\overset{\epsilon}{\sim} \sum_{d=-M}^{M} \left(\sum_{m \geq H} a_m u_m(d) \right) \delta_d \\ &= \sum_{m \geq H} a_m \sum_{d=-M}^{M} u_m(d)\delta_d.\end{aligned}$$

If H is sufficiently large compared to h, all the inner sums above will be within ϵ of the uniform distribution on $[0, 1, \ldots, h-1] \times [0, 1, \ldots, h-1]$, so we conclude that π_E is ϵ close to this uniform distribution. Finally, since the distribution π is a convex combination of the distributions $\pi_E(r)$, $0 \leq r \leq 2L_2$, this is also true for π, proving (3.1).

To sum up we review the correct choice of parameters. At the n-th stage of the construction we are given $h = h_n$ and $\epsilon = \epsilon_n$. The first numbers to be chosen are H and M. Next the parameter C is picked to make $|\mathbf{c}|/w$ small compared to ϵ. The numbers L_k, $k = 1, 2$ and the number N we can choose independently; the first two so that $\frac{L_2}{L_1+L_2} \overset{\epsilon}{\sim} \alpha$, and then N in accordance with the law of large numbers as explained above. Now $h_{n+1} = |\mathbf{b}|$ is determined and we choose ϵ_{n+1} so that $\sum_i \epsilon_i$ is finite. This completes the construction of the rank 1 system $\mathbf{X}$ and the proof that this system is α-weakly mixing.

4. Notes

Rank one systems: For more examples see the books by Friedman [**70**], Rudolph [**220**] and Shields [**235**] and the papers by del Junco [**131**] and Ferenczi [**66**]. The proof of Theorem 16.5 is due to J.-P. Thouvenot.

A rank one α-weakly mixing system: The construction is taken almost verbatim from del Junco and Lemańczyk [**134**].

CHAPTER 17

The Relation Between Measure and Topological Entropy

The variational principle asserts that for a topological $\mathbb{Z}$-dynamical system (X,T) the topological entropy equals the supremum of the measure entropies computed over all the invariant probability measures on X. It was already conjectured in the original paper of Adler, Konheim and McAndrew [**4**] where topological entropy was introduced; and then, after many stages (mainly by Goodwyn, Bowen and Dinaburg; see for example [**46**]) matured into a theorem in Goodman's paper [**104**]. We present in Section 1 Misiurewicz' short and elegant proof [**182**] of that theorem. Sections 2 and 3 describe a strong version of the part $h_{\mathrm{top}} \le \sup h_\mu$ of the variational principle due to Blanchard, Glasner and Host, [**27**]. Namely, given an open cover $\mathcal{U}$ of X, there exists an invariant measure μ such that $h_\mu(\alpha) \ge h_{\mathrm{top}}(\mathcal{U})$ for all Borel partitions α finer than $\mathcal{U}$. As a first application of this result we show in Section 4 that an expansive dynamical system admits a measure of maximal entropy. In Section 5 we present another application of the BGH theorem to the notion of entropy capacity. Further applications will be described in Chapter 19.

1. The variational principle

17.1. THEOREM (The variational principle). *Let (X,T) be a topological dynamical system, then*

$$h_{\mathrm{top}}(X,T) = \sup\{h_\mu : \mu \in M_T(X)\} = \sup\{h_\mu : \mu \in M_T^{\mathrm{erg}}(X)\}.$$

PROOF. Theorem 15.12 expresses the entropy h_μ as an integral $\int_Y h_{\mu_y}\, d\nu(y)$ of the entropies of the ergodic components of $\mu \in M_T(X)$. Thus the two suprema in the statement of the theorem are equal.

Part I: We first show that for any $\mu \in M_T(X)$, $h_\mu \le h_{\mathrm{top}}(X,T)$. Recall that if $\alpha = \{A_1, A_2, \ldots, A_k\}$ and $\beta = \{B_1, B_2, \ldots, B_k\}$ are measurable partitions then

$$H_\mu(\alpha|\beta) = -\sum_{i=1}^{k}\sum_{j=1}^{k} \mu(A_i \cap B_j) \log \mu(A_i|B_j).$$

Therefore if $\sum_{i=1}^k \mu(A_i \triangle B_i)$ is sufficiently small then $H_\mu(\alpha|\beta) < 1$. Given a partition $\alpha = \{A_1, A_2, \ldots, A_k\}$ choose compact sets $B_i \subset A_i$, $i = 1, 2, \ldots, k$, such that $H_\mu(\alpha|\beta) < 1$ for the partition $\beta = \{B_1, B_2, \ldots, B_k, B_{k+1}\}$, where $B_{k+1} = (\cup_{i=1}^k B_i)^c$ (and $A_{k+1} = \emptyset$). Now for every $n > 0$

$$\begin{aligned} H_\mu(\alpha_0^{n-1}|\beta_0^{n-1}) &\le H_\mu(\alpha|\beta_0^{n-1}) + H_\mu(T^{-1}\alpha|\beta_0^{n-1}) + \cdots + H_\mu(T^{-n+1}\alpha|\beta_0^{n-1}) \\ &\le H_\mu(\alpha|\beta) + H_\mu(T^{-1}\alpha|T^{-1}\beta) + \cdots + H_\mu(T^{-n+1}\alpha|T^{-n+1}\beta) \\ &= nH_\mu(\alpha|\beta) \le n. \end{aligned}$$

Therefore

$$\begin{aligned} H_\mu(\alpha_0^{n-1}) &\le H_\mu(\alpha_0^{n-1} \vee \beta_0^{n-1}) \\ &= H_\mu(\beta_0^{n-1}) + H_\mu(\alpha_0^{n-1} | \beta_0^{n-1}) \\ &\le H_\mu(\beta_0^{n-1}) + n, \end{aligned}$$

and

$$h_\mu(\alpha) = \lim_{n\to\infty} \frac{1}{n} H_\mu(\alpha_0^{n-1}) \le \lim_{n\to\infty} \frac{1}{n} \left(H_\mu(\beta_0^{n-1}) + n \right) = h_\mu(\beta) + 1.$$

Now set

$$U_i = (\cup_{j\neq i} B_j)^c \quad \text{and} \quad \mathcal{U} = \{U_i\}_{i=1}^k,$$

then $\mathcal{U}$ is an open cover of X and each U_i intersects at most two elements of the cover β, B_i and B_{k+1}. Let $\mathcal{V}(n)$ be an open sub-cover of minimal cardinality of $\mathcal{U}_0^{n-1}$. If $x \in U_{i_1} \cap T^{-1}U_{i_2} \cap \cdots \cap T^{-n+1}U_{i_n} \in \mathcal{V}(n)$, then x can be an element of at most 2^n elements of β_0^{n-1}. In fact if $x \in B_{l_1} \cap T^{-1}B_{l_2} \cap \cdots \cap T^{-n+1}B_{l_n} \in \beta_0^{n-1}$ then $x \in B_{l_j} \cap U_{i_j} \neq \emptyset$ for $1 \le j \le n$, hence, either $B_{l_j} = B_{i_j}$ or $B_{l_j} = B_{k+1}$. In any other case (than these $2^n \operatorname{card} \mathcal{V}(n)$ cases) the intersection $B_{l_1} \cap T^{-1}B_{l_2} \cap \cdots \cap T^{-n+1}B_{l_n}$ is empty. It therefore follows that

$$b = \operatorname{card} \beta_0^{n-1} \le 2^n \operatorname{card} \mathcal{V}(n) = 2^n N(\mathcal{U}_0^{n-1}).$$

Since $H_\mu(\beta_0^{n-1}) \le \log b$ we get

$$H_\mu(\beta_0^{n-1}) \le \log b \le n \log 2 + \log N(\mathcal{U}_0^{n-1}).$$

This implies

$$\begin{aligned} h_\mu(\alpha) \le h_\mu(\beta) + 1 &= \lim_{n\to\infty} \frac{1}{n} H_\mu(\beta_0^{n-1}) + 1 \\ &\le \log 2 + h_{\text{top}}(\mathcal{U}) + 1. \end{aligned}$$

As this holds for every α we conclude that $h_\mu(T) \le \log 2 + h_{\text{top}}(X,T) + 1$. Observe that, in view of Propositions 14.13 and 14.24, the same argument yields, for every $k \ge 1$, the formula

$$kh_\mu(T) = h_\mu(T^k) \le \log 2 + h_{\text{top}}(X, T^k) + 1 = \log 2 + kh_{\text{top}}(X,T) + 1,$$

hence

$$h_\mu(T) \le h_{\text{top}}(X,T)$$

as required.

Part II: Let $\epsilon > 0$ be given; our goal is to produce a measure $\mu \in M_T(X)$ with $h_\mu \ge sr(\epsilon, T)$. For each $n \ge 1$, let E_n be an (n,ϵ)-separating subset of X of maximal cardinality; thus $\operatorname{card} E_n = e_n = sr_n(\epsilon)$. Define

$$\nu_n = \frac{1}{e_n} \sum_{x\in E_n} \delta_x \quad \text{and} \quad \mu_n = \frac{1}{n} \sum_{j=0}^{n-1} \nu_n \circ T^{-j},$$

and choose a subsequence n_k satisfying

$$(i)\ sr(\epsilon, T) = \lim_{k\to\infty} \frac{1}{n_k} \log sr_{n_k}(\epsilon) \quad \text{and} \quad (ii)\ \lim_{k\to\infty} \mu_{n_k} = \mu \in M_T(X).$$

Next choose a finite measurable partition α with the properties

$$(iii)\ \operatorname{diam}(\alpha) < \epsilon \quad \text{and} \quad (iv)\ \mu(\partial A) = 0,\ \forall A \in \alpha.$$

Since for every $x \in X$ only a countable number of the sets $\partial B_t(x)$ can have positive measure, we can find a finite cover of X by open balls $\{B_i\}_{i=1}^k$ of diameter $< \epsilon$ whose boundary has measure zero. Thus the partition $\{A_i\}_{i=1}^k$, where $A_1 = B_1$, $A_2 = B_2 \setminus B_1, \dots, A_n = B_n \setminus (\cup_{j<n} B_j)$, satisfies these requirements.

If $x, y \in A = A_{i_1} \cap T^{-1}A_{i_2} \cap \cdots \cap T^{-n+1}A_{i_n} \in \alpha_0^{n-1}$ then for every $0 \le j \le n-1$ $T^j x, T^j y \in A_{i_{j+1}}$, hence $d(T^j x, T^j y) < \epsilon$. We conclude that $A \cap E_n$ contains at most one point and therefore

$$\begin{aligned} H_{\nu_n}(\alpha_0^{n-1}) &= -\sum_{A \in \alpha_0^{n-1}} \nu_n(A) \log \nu_n(A) \\ &= -\sum_{A \in \alpha_0^{n-1},\ A \cap E_n \neq \emptyset} \nu_n(A) \log \nu_n(A) \\ &= -sr_n(\epsilon) \frac{1}{sr_n(\epsilon)} \log \frac{1}{sr_n(\epsilon)} = \log sr_n(\epsilon). \end{aligned} \tag{1.1}$$

Next fix a positive integer m and for $n > m$ write $n = dm + r$ with $0 \le r < m$. By concavity of $\phi(t) = -t \log t$ we get

$$H_{\mu_n}(\alpha_0^{m-1}) \ge \frac{1}{n} \sum_{j=0}^{n-1} H_{\nu_n}(T^{-j}\alpha_0^{m-1}) \ge \frac{1}{n} \sum_{i=0}^{m-1} \sum_{j=0}^{d-1} H_{\nu_n}(T^{-jm-i}\alpha_0^{m-1}). \tag{1.2}$$

Since for every $1 \le i < m$ we have

$$\begin{aligned} \alpha_0^{n-1} &= \bigvee_{j=0}^{d-1} T^{-jm}\alpha_0^{m-1} \vee \alpha_{dm}^{dm+r-1} \\ &\prec \alpha_0^{i-1} \vee \left(\bigvee_{j=0}^{d-1} T^{-jm-i}\alpha_0^{m-1} \right) \vee \alpha_{dm}^{dm+r-1}, \end{aligned}$$

it follows that

$$\begin{aligned} H_{\nu_n}(\alpha_0^{n-1}) &\le \sum_{j=0}^{d-1} H_{\nu_n}(T^{-jm-i}\alpha_0^{m-1}) + H_{\nu_n}(\alpha_0^{i-1} \vee \alpha_{dm}^{dm+r-1}) \\ &\le \sum_{j=0}^{d-1} H_{\nu_n}(T^{-jm-i}\alpha_0^{m-1}) + 2m \log \operatorname{card} \alpha. \end{aligned} \tag{1.3}$$

Put together, these inequalities yield the following estimation:

$$\begin{aligned} H_{\mu_n}(\alpha_0^{m-1}) &\ge \frac{1}{n} \sum_{i=0}^{m-1} \sum_{j=0}^{d-1} H_{\nu_n}(T^{-jm-i}\alpha_0^{m-1}) && \text{(by 1.2)} \\ &\ge \frac{1}{n} \sum_{i=0}^{m-1} (H_{\nu_n}(\alpha_0^{n-1}) - 2m \log \operatorname{card} \alpha) && \text{(by 1.3)} \\ &= \frac{m}{n} H_{\nu_n}(\alpha_0^{n-1}) - \frac{2m^2}{n} \log \operatorname{card} \alpha \\ &= \frac{m}{n} \log \operatorname{card} E_n - \frac{2m^2}{n} \log \operatorname{card} \alpha && \text{(by 1.1)} \\ &= \frac{m}{n} \log sr_n(\epsilon) - \frac{2m^2}{n} \log \operatorname{card} \alpha. \end{aligned}$$

Now set $n = n_k$ and then let $k \to \infty$ to get

$$\lim_{k\to\infty} H_{\mu_n}(\alpha_0^{m-1}) = H_\mu(\alpha_0^{m-1}) \geq m \cdot sr\,(\epsilon, T)$$

(the first equality follows from the w^* convergence of the sequence $\mu_{n_k} \to \mu$ and from the choice of the partition α; see Theorem A.5). Next let $m \to \infty$ to conclude that

$$h_\mu \geq h_\mu(\alpha, T) \geq sr\,(\epsilon, T).$$

Of course the measure $\mu = \mu(\epsilon)$ and the partition α satisfying this inequality depend on ϵ, but nonetheless, letting $\epsilon \searrow 0$ we finally get

$$\sup_{\mu\in M_T(X)} h_\mu(T) \geq h_{\text{top}}(X, T)$$

and the proof is complete. □

17.2. DEFINITION. Given a topological dynamical system (X, T) , a measure $\mu \in M_T(X)$ for which $h_\mu(X, T) = h_{\text{top}}(X, T)$ is called a *measure of maximal entropy*. The set of measures of maximal entropy is denoted by $M_{\text{max}}(X)$.

It turns out that a measure of maximal entropy need not always exist; even in a minimal system (see the discussion in [**46**], Chapter 19). However, for many classes of dynamical systems measures of maximal entropy do exist.

17.3. PROPOSITION. 1. *If the function $h : \mu \mapsto h_\mu$, $h : M_T(X) \to \mathbb{R}$ is upper semi-continuous (see Lemma 15.1), then $M_{\text{max}}(X) \neq \emptyset$.*
2. *$M_{\text{max}}(X) \neq \emptyset$ for any subshift system (X, S) (with $X \subset \Omega(\ell)$).*
3. *If $h_{\text{top}}(X, T) = \infty$ then $M_{\text{max}}(X) \neq \emptyset$.*

PROOF. The first claim is a direct consequence of Theorem 17.1. The second follows from Lemma 15.1. For the third claim, again we use Theorem 17.1 to choose a sequence $\mu_n \in M_T^{\text{erg}}(X)$ with $h_{\mu_n} > 2^n$ and then let $\mu = \sum_{n=1}^\infty 2^{-n}\mu_n$. The fact that the entropy function $h : \mu \mapsto h_\mu$ is affine (Theorem 14.25) implies that $h_\mu = \sum_{n=1}^\infty 2^{-n} h_{\mu_n} = \infty$, whence $\mu \in M_{\text{max}}(X)$. □

17.4. REMARK. When $M_{\text{max}}(X)$ is nonempty and contains a single measure, the topological dynamical system (X, T) is called *intrinsically ergodic*. In [**199**], Parry has shown that topologically transitive subshifts of finite type are intrinsically ergodic.

2. A combinatorial lemma

Throughout the rest of this chapter, $\phi : [0, 1] \to \mathbb{R}$ denotes the function

$$\phi(x) = -t \log t \quad \text{for } 0 < t \leq 1;\ \phi(0) = 0\ .$$

Let $\mathfrak{L} = \{1, 2, \ldots, \ell\}$ be a finite set, called the *alphabet*; sequences $\omega = \omega_1 \ldots \omega_n \in \mathfrak{L}^n$, for $n \geq 1$, are called *words of length n on the alphabet $\mathfrak{L}$* . Let n and k be two integers with $1 \leq k \leq n$.

For every word ω of length n and every word θ of length k on the same alphabet, we denote by $p(\theta|\omega)$ the frequency of appearances of θ in ω, i.e.

$$p(\theta|\omega) = \frac{1}{n-k+1} \text{card}\left\{i :\ 1 \leq i \leq n-k+1,\ \omega_i\omega_{i+1}\ldots\omega_{i+k-1} = \theta_1\theta_2\ldots\theta_k\right\}\ .$$

For every word ω of length n on the alphabet $\mathfrak{L}$, we let

$$H_k(\omega) = \sum_{\theta \in \mathfrak{L}^k} \phi\big(p(\theta|\omega)\big) \, .$$

17.5. LEMMA. *For every $h > 0$, $\epsilon > 0$, every integer $k \geq 1$ and every sufficiently large integer n,*

$$\operatorname{card}\big\{\omega \in \mathfrak{L}^n : H_k(\omega) \leq kh\big\} \leq \exp\big(n(h+\epsilon)\big) \, .$$

Remark. It is equally true that, if $h \leq \log(\operatorname{card}\mathfrak{L})$, for sufficiently large n,

$$\operatorname{card}\big\{\omega \in \mathfrak{L}^n : H_k(\omega) \leq kh\big\} \geq \exp\big(n(h-\epsilon)\big) \, .$$

We do not prove this inequality here, since we have no use for it in the sequel.

PROOF. **The case $k = 1$.**
We have

$$\operatorname{card}\big\{\omega \in \mathfrak{L}^n : \; H_1(\omega) \leq h\big\} = \sum_{q \in K} \frac{n!}{q_1! \ldots q_\ell!} \tag{2.1}$$

where K is the set of $q = (q_1, \ldots, q_\ell) \in \mathbb{N}^\ell$ such that

$$\sum_{i=1}^{\ell} q_i = n \;\; \text{and} \;\; \sum_{i=1}^{\ell} \phi(\frac{q_i}{n}) \leq h \, .$$

By Stirling's formula, there exist two universal constants c and c' such that

$$c(\frac{m}{e})^m \sqrt{m} \leq m! \leq c'(\frac{m}{e})^m \sqrt{m}$$

for every $m > 0$. From this we deduce the existence of a constant $C(\ell)$ such that for every $q \in K$,

$$\frac{n!}{q_1! \ldots q_\ell!} \leq C(\ell) \exp\big(n \sum_{i-1}^{\ell} \phi(\frac{q_i}{n})\big) \leq C(\ell) \exp(nh) \, .$$

Now the sum (2.1) contains at most $(n+1)^\ell$ terms; so that we have

$$\operatorname{card}\big\{\omega \in \mathfrak{L}^n : \; H_1(\omega) \leq h\big\} \leq (n+1)^\ell C(\ell) \exp(nh) \leq \exp\big(n(h+\epsilon)\big)$$

for all sufficiently large n, as was to be proved.

The case $k > 1$.

For every word ω of length $n \geq 2k$ on the alphabet $\mathfrak{L}$, and for $0 \leq j < k$, we let n_j be the integral part of $\frac{n-j}{k}$, and $\omega^{(j)}$ the word

$$(\omega_{j+1} \ldots \omega_{j+k}) \, (\omega_{j+k+1} \ldots \omega_{j+2k}) \, \ldots \, (\omega_{j+(n_j-1)k+1} \ldots \omega_{j+n_j k})$$

of length n_j on the alphabet $B = \mathfrak{L}^k$.

Let now θ be a word of length k on the alphabet $\mathfrak{L}$; we also consider θ as an element of B. One easily verifies that, for every word ω of length n on the alphabet $\mathfrak{L}$,

$$\big|p(\theta|\omega) - \frac{1}{k} \sum_{j=0}^{k-1} p(\theta|\omega^{(j)})\big| \leq \frac{k}{n-2k+1} \, .$$

The function ϕ being uniformly continuous, we see that for sufficiently large n, and for every word ω of length n on $\mathfrak{L}$,

$$\sum_{\theta \in B} \left| \phi(p(\theta|\omega)) - \phi\left(\frac{1}{k} \sum_{j=0}^{k-1} p(\theta|\omega^{(j)}) \right) \right| < \frac{\epsilon}{2}$$

and by convexity of ϕ,

$$\frac{1}{k} \sum_{j=0}^{k-1} H_1(\omega^{(j)}) = \frac{1}{k} \sum_{j=0}^{k-1} \sum_{\theta \in B} \phi(p(\theta|\omega^{(j)})) \leq \frac{\epsilon}{2} + \sum_{\theta \in \mathfrak{L}^k} \phi(p(\theta|\omega)) = \frac{\epsilon}{2} + H_k(\omega).$$

Thus, if $H_k(\omega) \leq kh$, there exists a j such that $H_1(\omega^{(j)}) \leq \frac{\epsilon}{2} + kh$.

Now, given j and a word u of length n_j on the alphabet B, there exist $\ell^{n-n_j k} \leq \ell^{2k-2}$ words ω of length n on $\mathfrak{L}$ such that $\omega^{(j)} = u$. Thus for sufficiently large n, by the first part of the proof,

$$\begin{aligned} \operatorname{card}\left\{\omega \in \mathfrak{L}^n : H_k(\omega) \leq kh\right\} &\leq \ell^{2k-2} \sum_{j=0}^{k-1} \operatorname{card}\left\{u \in B^{n_j} : H_1(u) \leq \frac{\epsilon}{2} + kh\right\} \\ &\leq \ell^{2k-2} \sum_{j=0}^{k-1} \exp\left(n_j(\epsilon + kh)\right) \\ &\leq \ell^{2k-2} k \exp\left(n(\frac{\epsilon}{k} + h)\right) \leq \exp\left(n(h+\epsilon)\right) . \end{aligned}$$

□

3. A variational principle for open covers

Let (X,T) be a compact dynamical system. As usual we denote by $M_T(X)$ the set of T-invariant probability measures on X, and by $M_T^{\text{erg}}(X)$ the subset of ergodic measures.

We say that a partition α is finer than a cover $\mathfrak{U}$ when every atom of α is contained in an element of $\mathfrak{U}$. If $\alpha = \{A_1, \ldots, A_\ell\}$ is a partition of X, $x \in X$ and $N \in \mathbb{N}$, we write $\omega(\alpha, N, x)$ for the word of length N on the alphabet $\mathfrak{L} = \{1, \ldots, \ell\}$ defined by

$$\omega(\alpha, N, x)_n = i \quad \text{if} \quad T^{n-1}x \in A_i, \qquad 1 \leq n \leq N .$$

17.6. Theorem (The variational principle for open covers). *Let (X,T) be a topological dynamical system, and $\mathfrak{U}$ an open cover of X, then there exists a measure $\mu \in M_T(X)$ such that $h_\mu(\alpha) \geq h_{\text{top}}(\mathfrak{U})$ for all Borel partitions α finer than $\mathfrak{U}$.*

17.7. Lemma. *Let $\mathfrak{U}$ be a cover of X, $h = h_{\text{top}}(\mathfrak{U})$, $K \geq 1$ an integer, and $\{\alpha_l : 1 \leq l \leq K\}$ a finite sequence of partitions of X, all finer than $\mathfrak{U}$. For every $\epsilon > 0$ and sufficiently large N, there exists an $x \in X$ such that*

$$H_k\big(\omega(\alpha_l, N, x)\big) \geq k(h - \epsilon) \textit{ for every } k, l \textit{ with } 1 \leq k,\ l \leq K.$$

Proof. One can assume that all the partitions α_l have the same number of elements ℓ and we let $\mathfrak{L} = \{1, \ldots, \ell\}$. For $1 \leq k \leq K$ and $N \geq K$, denote

$$\Omega(N,k) = \{\omega \in \mathfrak{L}^N : \ H_k(\omega) < k(h-\epsilon)\} .$$

By Lemma 2, for sufficiently large N

$$\operatorname{card}(\Omega(N,k)) \le \exp(N(h-\epsilon/2)) \quad \text{for all } k \le K.$$

Let us choose such an N which moreover satisfies $K^2 < \exp(N\epsilon/2)$. For $1 \le k, l \le K$, let

$$Z(k,l) = \{x \in X : \omega(\alpha_l, N, x) \in \Omega(N,k)\} .$$

The set $Z(k,l)$ is the union of $\operatorname{card}(\Omega(N,k))$ elements of $(\alpha_l)_0^{N-1}$. Now this partition is finer than the cover $\mathfrak{U}_0^{N-1}$, hence $Z(k,l)$ is covered by

$$\operatorname{card}(\Omega(N,k)) \le \exp(N(h-\epsilon/2))$$

elements of $\mathfrak{U}_0^{N-1}$. Finally,

$$\bigcup_{1\le k,l\le K} Z(k,l)$$

is covered by $K^2\exp(N(h-\epsilon/2)) < \exp(Nh)$ elements of $\mathfrak{U}_0^{N-1}$. As every subcover of $\mathfrak{U}_0^{N-1}$ has at least $\exp(Nh)$ elements,

$$\bigcup_{1\le k,l\le K} Z(k,l) \ne X.$$

This completes the proof of the lemma. □

PROOF OF THEOREM 17.6. Let $\mathfrak{U} = \{U_1, \dots, U_\ell\}$ be an open cover of X. It is clearly sufficient to consider Borel partitions α of X of the form

$$\alpha = \{A_1, \dots, A_\ell\} \quad \text{with } A_i \subset U_i \quad \text{for every } i. \tag{3.1}$$

Step 1: Assume first that X is 0-dimensional.

The family of partitions finer than $\mathfrak{U}$, consisting of clopen sets and satisfying (3.1) is countable; let $\{\alpha_l : l \ge 1\}$ be an enumeration of this family. According to the previous lemma, there exists a sequence of integers N_K tending to $+\infty$ and a sequence x_K of elements of X such that:

$$H_k\big(\omega(\alpha_l, N_K, x_K)\big) \ge k\Big(h - \frac{1}{K}\Big) \quad \text{for every } 1 \le k,\ l \le K. \tag{3.2}$$

Write

$$\mu_K = \frac{1}{N_K}\sum_{i=0}^{N_K-1} \delta_{T^i x_K} .$$

Replacing the sequence μ_K by a subsequence (this means replacing the sequence N_K by a subsequence, and the sequence x_K by the corresponding subsequence preserving the property (3.2)), one can assume that the sequence of measures μ_K converges weak* to a probability measure μ. This measure μ is clearly T-invariant. Fix $k, l \ge 1$, and let F be an atom of the partition $(\alpha_l)_0^{k-1}$, with name $\theta \in \{1, \dots, \ell\}^k$. For every K one has

$$\big|\mu_K(F) - p\big(\theta|\omega(\alpha_l, N_K, x_K)\big)\big| \le \frac{2k}{N_K}.$$

Now as F is clopen,

$$\mu(F) = \lim_{K\to\infty} \mu_K(F) = \lim_{K\to\infty} p\big(\theta|\omega(\alpha_l, N_K, x_K)\big) \text{ hence}$$
$$\phi(\mu(F)) = \lim_{K\to\infty} \phi\big(p\big(\theta|\omega(\alpha_l, N_K, x_K)\big)\big)$$

and, summing over $\theta \in \{1, \dots, \ell\}^k$, one gets

$$H_\mu\big((\alpha_l)_0^{k-1}\big) = \lim_{K\to\infty} H_k\big(\omega(\alpha_l, N_K, x_K)\big) \geq kh.$$

Finally, by sending k to infinity one obtains $h_\mu(\alpha_l) \geq h$.

Now, as X is 0-dimensional, the family of partitions $\{\alpha_l\}$ is dense in the collection of Borel partitions of X satisfying (3.1), with respect to the distance associated with $L^1(\mu)$. Thus, $h_\mu(\alpha) \geq h$ for every partition of this kind.

Step 2: The general case.

Let us recall a well known fact: there exists a topological system (Y, T), where Y is 0-dimensional, and a continuous surjective map $\pi : Y \to X$ with $\pi \circ T = T \circ \pi$.

(Proof : as X is a compact metric space, it is easy to construct a Cantor set K and a continuous surjective $f : K \to X$. Put

$$Y = \{y \in K^{\mathbb{Z}} : \ f(y_{n+1}) = Tf(y_n) \ \text{ for every } \ n \in \mathbb{Z}\}$$

and let $\pi : Y \to X$ be defined by $\pi(y) = f(y_0)$.

Y is a closed subset of $K^{\mathbb{Z}}$ — where the latter is equipped with the product topology — and is invariant under the shift T on $K^{\mathbb{Z}}$. It is easy to check that π satisfies the required conditions.)

Let $\mathcal{V} = \pi^{-1}(\mathcal{U}) = \{\pi^{-1}(U_1), \dots, \pi^{-1}(U_d)\}$ be the preimage of $\mathcal{U}$ under π ; one has $h_{\text{top}}(\mathcal{V}) = h_{\text{top}}(\mathcal{U}) = h$. By the above remark, there exists $\nu \in M(Y, T)$ such that $h_\nu(\mathcal{Q}) \geq h$ for every Borel partition $\mathcal{Q}$ of Y finer than $\mathcal{V}$. Let $\mu = \nu \circ \pi^{-1}$ the measure which is the image of ν under π. One has $\mu \in M_T(X)$ and, for every Borel partition α of X finer than $\mathcal{U}$, $\pi^{-1}(\alpha)$ is a Borel partition of Y which is finer than $\mathcal{V}$ with

$$h_\mu(\alpha) = h_\nu\big(\pi^{-1}(\alpha)\big) \geq h.$$

This completes the proof of the theorem. □

17.8. Corollary. *Let (X, T) be a topological system, $\mathcal{U}$ an open cover of X and α a Borel partition finer than $\mathcal{U}$, then, there exists a T-invariant ergodic measure μ on X such that $h_\mu(\alpha) \geq h_{\text{top}}(\mathcal{U})$.*

Proof. By Theorem 17.6 there exists $\mu \in M_T(X)$ with $h_\mu(\alpha) \geq h_{\text{top}}(\mathcal{U})$; let $\mu = \int_\omega \mu_\omega \, dm(\omega)$ be its ergodic decomposition. The corollary follows from the formula (Theorem 15.12)

$$\int h_{\mu_\omega}(\alpha) \, dm(\omega) = h_\mu(\alpha).$$

□

4. An application to expansive systems

17.9. DEFINITION. Let (X, T) be a dynamical system and let d be a compatible metric.

1. (X, T) is called *expansive* if there exists an $\epsilon > 0$ with the property that for every pair x, y of distinct points in X there exists an $n \in \mathbb{Z}$ with $d(T^n x, T^n y) \geq \epsilon$.
2. A finite open cover $\mathfrak{U} = \{U_1, \ldots, U_\ell\}$ of X is called *separating* if for every sequence $\{k_i\}_{i\in\mathbb{Z}} \in \{1, 2, \ldots, \ell\}^{\mathbb{Z}}$ the intersection
$$\bigcap_{i\in\mathbb{Z}} T^i \overline{U_{k_i}}$$
contains at most one point.

Note that the next lemma implies that expansiveness does not depend on the choice of a metric.

17.10. LEMMA. *The following conditions are equivalent:*

1. *(X, T) is expansive.*
2. *There exists a finite open cover $\mathfrak{U}$ with*
$$\operatorname{diam}(\mathfrak{U}_{-n}^{n}) = \sup\{\operatorname{diam}(U) : U \in \mathfrak{U}_{-n}^{n}\} \to 0.$$
3. *(X, T) has a separating open cover.*

PROOF. If $\epsilon > 0$ is an expansive constant (as in Definition 17.9) then it is easy to verify that every finite open cover $\mathfrak{U} = \{U_1, \ldots, U_\ell\}$ with $\operatorname{diam}(\mathfrak{U}) = \sup\{\operatorname{diam}(U) : U \in \mathfrak{U}\} < \epsilon$ satisfies $\operatorname{diam}(\mathfrak{U}_{-n}^{n}) \to 0$. The equivalence of conditions 2 and 3 is easy to verify ($2 \Rightarrow 3$ is clear and $3 \Rightarrow 2$ is proved by a compactness argument). Finally assume that 3 holds for a finite open cover $\mathfrak{U} = \{U_1, \ldots, U_\ell\}$ and let $\epsilon > 0$ be a Lebesgue number for $\mathfrak{U}$. We claim that ϵ is an expansive constant. In fact, if $x \neq y$ and $d(T^n x, T^n y) < \epsilon$ for every $n \in \mathbb{Z}$ then, choosing for every $i \in \mathbb{Z}$ an element $U_{k_i} \in \mathfrak{U}$ such that $B_\epsilon(T^{-i}x) \subset U_{k_i}$, we have $\{x, y\} \subset \bigcap_{i\in\mathbb{Z}} T^i \overline{U_{k_i}}$, contradicting condition 3. □

17.11. EXERCISE. A dynamical system (X, T) with X zero-dimensional is expansive iff it is isomorphic to a subshift $Y \subset \Omega(\ell)$ for some $\ell \in \mathbb{N}$.

17.12. THEOREM. *If (X, T) is an expansive system then $M_{\max}(X)$ is nonempty; i.e. there exists a measure $\mu \in M_T(X)$ with the property*
$$h_{\text{top}}(X, T) = h_\mu(X, T).$$

PROOF. By Lemma 17.10 there exists a finite open cover $\mathfrak{U}$ with $\operatorname{diam}(\mathfrak{U}_{-n}^{n}) \to 0$. By Lemma 14.6, for every $n \in \mathbb{N}$,
$$h(\mathfrak{U}_{-n}^{n}) = h(\mathfrak{U})$$
and by Lemma 14.5
$$h(\mathfrak{U}) = \lim_{n\to\infty} h(\mathfrak{U}_{-n}^{n}) = h_{\text{top}}(X, T).$$
Applying Theorem 17.6 to $\mathfrak{U}$ we get a measure $\mu \in M_T(X)$ such that $h_\mu(\alpha) \geq h(\mathfrak{U})$ for all Borel partitions α finer than $\mathfrak{U}$. In particular then $h_\mu(X, T) \geq h(\mathfrak{U}) =$

$h_{\text{top}}(X,T)$. Finally, the variational principle (Theorem 17.1) implies $h_\mu(X,T) = h_{\text{top}}(X,T)$. □

5. Entropy capacity

Let (X,T) be a topological system; for K a closed subset of X, let R_K be the collection of covers of K consisting of open subsets of X, and let P_K be the collection of finite Borel partitions of K.

Given a closed subset K of X, it is easy to establish necessary and sufficient conditions so that there exists a measure $\mu \in M_T^{\text{erg}}(X)$ with $\mu(K) > 0$. For example, this property is equivalent to:

$$\exists \epsilon > 0,\ \forall N > 0\ \exists x \in X \ \text{ such that } \ \frac{1}{N}\text{card}\,\{n : 0 \le n < N,\ T^n x \in K\} \ \ge \ \epsilon.$$

It is more difficult to decide whether there exists such a measure with some additional properties, for example a measure of positive entropy.

If $\mathfrak{U}$ is a cover of K, denote by $(\mathfrak{U}, K^c)$ the cover of X formed by the elements of $\mathfrak{U}$ and the set K^c ; we use the same convention for the partitions of K. Let

$$c(K) = \sup\{h_{\text{top}}(\mathfrak{U}, K^c) : \mathfrak{U} \in R_K\};$$

and, for $\mu \in M_T(X)$,

$$c_\mu(K) = \sup\{h_\mu(\alpha, K^c) : \alpha \in P_K\}.$$

One can easily verify that the function c is monotone and subadditive ($K \subset L \Rightarrow c(K) \le c(L)$ and $c(K \cup L) \le c(K) + c(L)$). This suggests that it can serve as a kind of "capacity" function on the closed subsets of a dynamical system. The following theorem gives an explicit formula for c in terms of the invariant measures of the system.

17.13. THEOREM. *Let (X,T) be a topological dynamical system and K a closed subset of X.*

1. *For every $\mu \in M_T^{\text{erg}}(X)$ one has*

$$c_\mu(K) = \begin{cases} 0 & \textit{if } \mu(K) = 0 \\ h_\mu(T) & \textit{if } \mu(K) > 0. \end{cases}$$

2. $$c(K) = \sup\{h_\mu(T) : \mu \in M_T^{\text{erg}}(X),\ \mu(K) > 0\}.$$

Thus in order that there exists $\mu \in M_T^{\text{erg}}(X)$ with $h_\mu(T) > 0$ and $\mu(K) > 0$ it is necessary and sufficient that there exists an open cover $\mathfrak{U}$ of K with $h_{\text{top}}(\mathfrak{U}, K^c) > 0$ and in fact one can find such a cover of K consisting of two open sets.

PROOF. 1. By Corollary 17.8 we get

$$\sup\{c_\mu(K) : \mu \in M_T^{\text{erg}}(X)\} \ge c(K).$$

We clearly have $c_\mu(K) \le h_\mu(T)$; the case when $\mu(K) = 0$ being trivial, we now suppose that $\mu(K) > 0$. Let $\epsilon > 0$; since the measure μ is ergodic, one can choose n sufficiently large so that $\mu\big(\bigcup_{i=0}^n T^{-i}K\big) > 1 - \epsilon$. Choose $\delta > 0$ such that

$$d(x,y) < \delta \Longrightarrow d(T^{-i}x, T^{-i}y) < \epsilon \ \text{ for } \ i = 0, \dots, n$$

and α_ϵ a partition of K each element of which has a diameter $< \delta$.

All the atoms of $\beta_\epsilon = (\alpha_\epsilon, K^c)_0^n$ have a diameter $< \epsilon$, with the exception of at most one of them which has measure $< \epsilon$. We therefore have

$$h_\mu(T) = \sup_\epsilon h_\mu(\beta_\epsilon) = \sup_\epsilon h_\mu(\alpha_\epsilon, K^c) \leq c_\mu(K).$$

2. We now show that for every $\mu \in M_T(X)$, $c_\mu(K) \leq c(K)$. Let $\{K_1, \ldots, K_d\}$ be a Borel partition of K, $\alpha = \{K_1, \ldots, K_d, K^c\}$, and $\epsilon > 0$.

For $1 \leq i \leq d$, let F_i be a closed subset of K_i, and put $F = \bigcup_i F_i$; we assume that these closed subsets have been chosen so that the partition $\beta = \{F_1, \ldots, F_d, F^c\}$ satisfies $|h_\mu(\beta) - h_\mu(\alpha)| < \epsilon$, and so that $h_\mu(\{K \setminus F, F \cup K^c\}) < \epsilon$.

Let V be a neighborhood of K with $h_\mu(\{V \setminus K, K \cup V^c\}) < \epsilon$; let γ denote the partition $\{V \setminus F, F \cup V^c\}$ and ζ the partition $\{V \setminus F, F_1, \ldots, F_d, V^c\}$.

As

$$\{K \setminus F, F \cup K^c\} \vee \{V \setminus K, K \cup V^c\} \succ \gamma$$

one has $h_\mu(\gamma) \leq 2\epsilon$.

Let also $\mathfrak{U} = \{U_1, \ldots, U_d\}$ be an open cover of K, with $F_i \subset U_i \subset V$ for all i, and $U_i \cap F_j = \emptyset$ for $i \neq j$.

For every element A of the partition γ and every element B of the cover $(\mathfrak{U}, K^c)$, $A \cap B$ is a subset of an element of the partition ζ. This immediately implies the same property for γ_0^n, $(\mathfrak{U}, K^c)_0^n$ and ζ_0^n. We therefore conclude that

$$\forall A \in \gamma_0^n, \ \operatorname{card}\{B \in \zeta_0^n : B \subset A\} \leq N\big((\mathfrak{U}, K^c)_0^n\big),$$

therefore $H_\mu(\zeta_0^n) \leq H_\mu(\gamma_0^n) + \log\big(N\big((\mathfrak{U}, K^c)_0^n\big)\big)$ and $h_\mu(\zeta) \leq h_\mu(\gamma) + h_{\mathrm{top}}(\mathfrak{U}, K^c) \leq 2\epsilon + h_{\mathrm{top}}(\mathfrak{U}, K^c) \leq 2\epsilon + c(K)$. Finally, $h_\mu(\alpha) \leq \epsilon + h_\mu(\beta) \leq \epsilon + h_\mu(\zeta) \leq 3\epsilon + c(K)$, and the result follows. This completes the proof of the theorem. □

We have, in fact, shown that

$$\begin{aligned} c(K) &= \sup\{c_\mu(K) : \mu \in M_T(X)\} = \sup\{c_\mu(K) : \mu \in M_T^{\mathrm{erg}}(X)\} \\ &= \sup\{h_\mu(T) : \mu \in M_T^{\mathrm{erg}}(X),\ \mu(K) > 0\}. \end{aligned}$$

17.14. COROLLARY. *A universally null set K (i.e. $\mu(K) = 0$ for every $\mu \in M_T(X)$) has zero capacity. When (X, T) is uniquely ergodic and has positive entropy, the converse is also true.*

17.15. REMARK. In a uniquely ergodic system (X, T) , for every closed subset $K \subset X$ either $c(K) = 0$ or $c(K) = h_{\mathrm{top}}(X, T)$. For an example of a minimal system having a closed subset K with $0 < c(K) = \log 2 < h_{\mathrm{top}}(T) = \log 3$, see [**101**].

6. Notes

The variational principle: This proof of the variational principle is due to Misiurewicz [**182**]. I followed its exposition in Petersen's book [**205**].

A variational principle for open covers: Sections 2, 3 and 5 of this chapter are based on the paper by Blanchard, Glasner and Host [**27**]. The main application of the variational principle for open covers in [**27**] is to the theory of entropy pairs; this theory is developed in Chapter 19 below.

An application to expansive systems: See the book by Denker, Grillenberger and Sigmund [**46**] for an extensive treatment of various kinds of expansive properties and their connection to the existence of measures of maximal entropy.

CHAPTER 18

The Pinsker Algebra, CPE and Zero Entropy Systems

Many questions in ergodic theory fall naturally into two distinct kinds, those concerning zero entropy systems and those dealing with positive entropy. The existence of the Pinsker algebra of a dynamical system makes this distinction easy to handle. For example one can show, that a system is mixing of all orders iff this is the case for its Pinsker factor, thus reducing the question whether mixing implies mixing of all orders to the case of zero-entropy systems. In Section 1 the Pinsker algebra $\Pi(\mathbf{X})$ of a dynamical system is introduced as the maximal zero entropy factor of the system. In Section 2 we prove the Rohlin-Sinai theorem which, for example, identifies the Pinsker algebra as the "remote past"

$$\bigcap_{m=1}^{\infty} \bigvee_{j=m}^{\infty} T^j \alpha,$$

of a generating partition α. In particular this theorem shows that the K-property is characterized by the property of "complete positive entropy" (CPE); i.e. the property that all non-trivial factors of the system have positive entropy. In Section 3 we investigate the class of zero entropy dynamical systems. We observe that zero entropy systems are disjoint from CPE systems and show that quasifactors of zero entropy systems have zero entropy. Finally we show that measure dynamical systems arising from invariant probability measures on a topologically distal system have zero entropy.

1. The Pinsker algebra

For a dynamical system $\mathbf{X}$ set

$$\Pi(\mathbf{X}) = \{A \in \mathcal{X} : h(\{A, A^c\}) = 0\}.$$

We leave the proof of the following theorem as an exercise.

18.1. THEOREM. 1. *$\Pi(\mathbf{X})$ is a T-invariant sub-σ-algebra*

2. *A finite partition α is $\Pi(\mathbf{X})$-measurable iff $h(\alpha) = H(\alpha|\alpha^-) = 0$ iff α is $\alpha^- = \alpha_{-\infty}^{-1}$-measurable.*
3. *For any $0 \neq k \in \mathbb{Z}$ we have $\Pi(X, \mu, T) = \Pi(X, \mu, T^k)$*

[[For the σ-additivity use Proposition 14.22 and the fact that the entropy $H(\alpha|\beta)$ is continuous in both arguments (see Lemma 15.9). To see that $h(\alpha) = 0$ for $\Pi(\mathbf{X})$-measurable $\alpha = \{A_1, A_2, \dots, A_k\}$ note that $\alpha \prec \vee_{j=1}^{k}\{A_j, A_j^c\}$.]]

This σ-algebra is called the *Pinsker algebra* and the corresponding factor (also denoted $\Pi(\mathbf{X})$) is called the *Pinsker factor* of the system $\mathbf{X}$. The condition "α is α^--measurable" is interpreted in probabilistic terms as saying that the process

α is *deterministic*; it is determined by its past α^-. Accordingly $\Pi(\mathbf{X})$, which is the largest zero entropy factor, is also called the largest deterministic factor of the system $\mathbf{X}$. We say that a system is *deterministic* when $\mathcal{X} = \Pi(\mathbf{X})$, which is just another way of saying that it has zero entropy.

We shall need several entropy formulas which we put together in the following lemma.

18.2. LEMMA. *Let α, β, γ be finite partitions.*
1. *If $\beta \prec \alpha$ then $\lim_{n\to\infty} \frac{1}{n} H(\alpha_0^n | \beta_{-\infty}^{-1}) = H(\alpha|\alpha_{-\infty}^{-1})$.*
2. *If $\alpha \prec \beta$ then $\lim_{n\to\infty} \frac{1}{n} H(\alpha_0^n | \beta_{-\infty}^{-1}) = H(\alpha|\alpha_{-\infty}^{-1})$.*
3. *If $\alpha \prec \beta$ then $\lim_{n\to\infty} H(\alpha|\beta_{-\infty}^{-1} \vee T^n\gamma_{-\infty}^{-1}) = H(\alpha|\beta_{-\infty}^{-1})$.*

PROOF. 1. Since $\beta \prec \alpha$ we have $T^n(\beta_{-\infty}^{-1} \vee \alpha_0^{n-1}) \nearrow \alpha_{-\infty}^{-1}$, hence

$$H(T^{-n}\alpha|\alpha_0^{n-1} \vee \beta_{-\infty}^{-1}) = H(\alpha|T^n(\alpha_0^{n-1} \vee \beta_{-\infty}^{-1})) \to H(\alpha|\alpha_{-\infty}^{-1}).$$

Now

$$H(\alpha_0^n|\beta_{-\infty}^{-1}) = H(\alpha|\beta_{-\infty}^{-1}) + H(T^{-1}\alpha|\alpha \vee \beta_{-\infty}^{-1}) + \cdots + H(T^{-n}\alpha|\alpha_0^{n-1} \vee \beta_{-\infty}^{-1}),$$

hence $\lim_{n\to\infty} \frac{1}{n} H(\alpha_0^n|\beta_{-\infty}^{-1}) = H(\alpha|\alpha_{-\infty}^{-1})$.

2. If $\alpha \prec \beta$ then by part 1,

$$\frac{1}{n} H(\alpha_0^n|\beta_{-\infty}^{-1}) \le \frac{1}{n} H(\alpha_0^n|\alpha_{-\infty}^{-1}) \to H(\alpha|\alpha_{-\infty}^{-1}).$$

For the other direction we have

$$\begin{aligned}\frac{1}{n} H(\alpha_0^n|\beta_{-\infty}^{-1}) &= \frac{1}{n} H(\beta_0^n|\beta_{-\infty}^{-1}) - \frac{1}{n} H(\beta_0^n|\alpha_0^n \vee \beta_{-\infty}^{-1}) \\ &\ge \frac{1}{n} H(\beta_0^n|\beta_{-\infty}^{-1}) - \frac{1}{n} H(\beta_0^n|\alpha_0^n \vee \alpha_{-\infty}^{-1}).\end{aligned}$$

Hence (again using part 1 twice, first with $\alpha = \beta$ and then exchanging their roles)

$$\begin{aligned}\lim_{n\to\infty} \frac{1}{n} H(\alpha_0^n|\beta_{-\infty}^{-1}) &\ge H(\beta|\beta_{-\infty}^{-1}) - \lim_{n\to\infty} \frac{1}{n} H(\beta_0^n|\alpha_0^n \vee \beta_{-\infty}^{-1}) \\ &= \lim_{n\to\infty} \frac{1}{n} \left(H(\beta_0^n|\alpha_{-\infty}^{-1}) - H(\beta_0^n|\alpha_0^n \vee \alpha_{-\infty}^{-1})\right) \\ &= \lim_{n\to\infty} \frac{1}{n} H(\alpha_0^n|\alpha_{-\infty}^{-1}) = H(\alpha|\alpha_{-\infty}^{-1}),\end{aligned}$$

where we use $\alpha \prec \beta$ and the entropy cocycle equation (Proposition 14.16) to deduce $H(\beta_0^n|\alpha_{-\infty}^{-1}) - H(\beta_0^n|\alpha_0^n \vee \alpha_{-\infty}^{-1}) = H(\alpha_0^n|\alpha_{-\infty}^{-1})$.

3. By the entropy cocycle equation

$$\begin{aligned}&H(\alpha|\beta_{-\infty}^{-1} \vee T^n\gamma_{-\infty}^{-1}) = \\ &H(\beta|\beta_{-\infty}^{-1} \vee T^n\gamma_{-\infty}^{-1}) - H(\beta|\alpha \vee \beta_{-\infty}^{-1} \vee T^n\gamma_{-\infty}^{-1}),\end{aligned} \tag{1.1}$$

and

$$\begin{aligned}&H(\beta_0^n|\beta_{-\infty}^{-1} \vee \gamma_{-\infty}^{-1}) \\ &= H(\beta|\beta_{-\infty}^{-1} \vee \gamma_{-\infty}^{-1}) + H(T^{-1}\beta|T^{-1}\beta_{-\infty}^{-1} \vee \gamma_{-\infty}^{-1}) \\ &\quad + \cdots + H(T^{-n}\beta|T^{-n}\beta_{-\infty}^{-1} \vee \gamma_{-\infty}^{-1}) \\ &= H(\beta|\beta_{-\infty}^{-1} \vee \gamma_{-\infty}^{-1}) + H(\beta|\beta_{-\infty}^{-1} \vee T\gamma_{-\infty}^{-1}) + \cdots + H(\beta|\beta_{-\infty}^{-1} \vee T^n\gamma_{-\infty}^{-1}).\end{aligned}$$

This together with part 2 implies

$$H(\beta|\beta_{-\infty}^{-1}) = \lim_{n\to\infty} \frac{1}{n} H(\beta_0^n|\beta_{-\infty}^{-1} \vee \gamma_{-\infty}^{-1}) = \lim_{n\to\infty} H(\beta|\beta_{-\infty}^{-1} \vee T^n\gamma_{-\infty}^{-1}).$$

Applying this to (1.1) we get

$$\lim_{n\to\infty} H(\alpha|\beta_{-\infty}^{-1} \vee T^n\gamma_{-\infty}^{-1}) = H(\beta|\beta_{-\infty}^{-1}) - \lim_{n\to\infty} H(\beta|\alpha \vee \beta_{-\infty}^{-1} \vee T^n\gamma_{-\infty}^{-1}).$$

And therefore

$$\begin{aligned}\lim_{n\to\infty} H(\alpha|\beta_{-\infty}^{-1} \vee T^n\gamma_{-\infty}^{-1}) &\geq H(\beta|\beta_{-\infty}^{-1}) - H(\beta|\alpha \vee \beta_{-\infty}^{-1}) \\ &= H(\beta \vee \alpha|\beta_{-\infty}^{-1}) - H(\beta|\alpha \vee \beta_{-\infty}^{-1}) \\ &= H(\alpha|\beta_{-\infty}^{-1}).\end{aligned}$$

The other direction is clear. □

We can now prove the following very useful formula.

18.3. PROPOSITION (Pinsker's formula). *Let α and β be finite partitions, then*

$$h(\alpha \vee \beta) = h(\beta) + H(\alpha|\alpha^- \vee \beta^T).$$

PROOF. We have

$$\begin{aligned}\frac{1}{n} H((\alpha \vee \beta)_0^n|(\alpha \vee \beta)_{-\infty}^{-1}) &= \frac{1}{n} H(\alpha_0^n \vee \beta_0^n|\alpha_{-\infty}^{-1} \vee \beta_{-\infty}^{-1}) \\ &= \frac{1}{n} H(\beta_0^n|\alpha_{-\infty}^{-1} \vee \beta_{-\infty}^{-1}) + \frac{1}{n} H(\alpha_0^n|\alpha_{-\infty}^{-1} \vee \beta_{-\infty}^{-1} \vee \beta_0^n) \\ &= \frac{1}{n} H(\beta_0^n|\alpha_{-\infty}^{-1} \vee \beta_{-\infty}^{-1}) + \frac{1}{n} \sum_{k=0}^{n} H(\alpha|\alpha_{-\infty}^{-1} \vee \beta_{-\infty}^{-1} \vee \beta_0^k),\end{aligned}$$

and applying Lemma 18.2 we get $h(\alpha \vee \beta) = h(\beta) + H(\alpha|\alpha_{-\infty}^{-1} \vee \beta_{-\infty}^{\infty})$. □

18.4. COROLLARY. *Let α and β be finite partitions, then*

$$h(\alpha \vee \beta) \leq h(\alpha) + h(\beta).$$

PROOF. By Pinsker's formula

$$\begin{aligned}h(\alpha \vee \beta) &= h(\beta) + H(\alpha|\alpha_{-\infty}^{-1} \vee \beta_{-\infty}^{\infty}) \\ &\leq h(\beta) + H(\alpha|\alpha_{-\infty}^{-1}) = h(\beta) + h(\alpha).\end{aligned}$$

□

Lemma 18.2 and Proposition 18.3 have "relative" analogues which are proved similarly; e.g. we have

18.5. PROPOSITION. *Let α and β be finite partitions and $\mathcal{Y} \subset \mathcal{X}$ a T-invariant sub-σ-algebra, then*

$$h(\alpha \vee \beta|\mathcal{Y}) = h(\beta|\mathcal{Y}) + H(\alpha|\alpha^- \vee \beta^T \vee \mathcal{Y}).$$

Our first application of Proposition 18.3 is to show that the Pinsker algebra is the union of all the "remote pasts" or "tails" of processes corresponding to finite partitions.

18.6. THEOREM. *Let* $\mathbf{X}$ *be a dynamical system, then*

$$\Pi(\mathbf{X}) = \bigvee\{\bigcap_{n=0}^{\infty} \alpha_{-\infty}^{-n} : \alpha \text{ a finite partition}\}$$
$$= \bigvee\{\bigcap_{n=0}^{\infty} T^n\alpha_{-\infty}^{-1} : \alpha \text{ a finite partition}\}.$$

PROOF. By Theorem 18.1 for every $\Pi(\mathbf{X})$-measurable finite partition α we have $\alpha \subset \alpha_{-\infty}^{-1} = \bigcap_{n=0}^{\infty} T^n\alpha_{-\infty}^{-1}$. Thus

$$\Pi(\mathbf{X}) \subset \bigvee\{\bigcap_{n=0}^{\infty} T^n\alpha_{-\infty}^{-1} : \alpha \text{ a finite partition}\}.$$

For the other direction let β be a finite partition with $\beta \subset \bigcap_{n=0}^{\infty} T^n\alpha_{-\infty}^{-1}$ for some finite partition α, then $\beta \subset \beta_{-\infty}^{\infty} \subset \alpha_{-\infty}^{-1}$ and by Pinsker's formula (18.3)

$$\begin{aligned} H(\alpha \vee \beta|\alpha_{-\infty}^{-1} \vee \beta_{-\infty}^{-1}) &= H(\alpha \vee \beta|\alpha_{-\infty}^{-1}) = H(\alpha|\alpha_{-\infty}^{-1}) \\ &= H(\alpha|\alpha_{-\infty}^{-1} \vee \beta_{-\infty}^{\infty}) + H(\beta|\beta_{-\infty}^{-1}) \\ &= H(\alpha|\alpha_{-\infty}^{-1}) + H(\beta|\beta_{-\infty}^{-1}). \end{aligned}$$

So $h(\beta) = H(\beta|\beta_{-\infty}^{-1}) = 0$ and $\beta \subset \Pi(\mathbf{X})$. □

2. The Rohlin-Sinai theorem

The Rohlin-Sinai theorem theorem deals with a sub-σ-algebra $\mathcal{F} \subset \mathcal{X}$ with the property that $T^{-1}\mathcal{F} \subset \mathcal{F}$. In preparation for the proof of this theorem we establish the following lemma.

18.7. LEMMA. *Let* α *be a finite partition and* $\mathcal{F} \subset \mathcal{X}$ *with* $T^{-1}\mathcal{F} \subset \mathcal{F}$.

1. $H(\alpha|\alpha_{-\infty}^{-1} \vee \mathcal{F} \vee \Pi(\mathbf{X})) = H(\alpha|\alpha_{-\infty}^{-1} \vee \mathcal{F})$.
2. *Denote* $T^n\mathcal{F} \nearrow \bigvee_{n=0}^{\infty} T^n\mathcal{F} = \mathcal{F}_{\infty}$ *and* $T^{-n}\mathcal{F} \searrow \bigcap_{n=0}^{\infty} T^{-n}\mathcal{F} = \mathcal{F}_{-\infty}$ *and assume* $\alpha \subset \mathcal{F}_{\infty}$, *then* $H(\alpha|\mathcal{F}_{-\infty} \vee \Pi(\mathbf{X})) = H(\alpha|\mathcal{F}_{-\infty})$.
3. *If* $\mathcal{F}_{\infty} = \mathcal{X}$ *then* $\Pi(\mathbf{X}) \subset \mathcal{F}_{-\infty}$.

PROOF. 1. For any finite $\Pi(\mathbf{X})$-measurable partition β we have by Pinsker's formula (18.5)

$$H(\alpha \vee \beta|\alpha_{-\infty}^{-1} \vee \beta_{-\infty}^{-1} \vee \mathcal{F}) = H(\beta|\alpha_{-\infty}^{\infty} \vee \beta_{-\infty}^{-1} \vee \mathcal{F}) + H(\alpha|\alpha_{-\infty}^{-1} \vee \mathcal{F}).$$

Since $\beta \subset \beta_{-\infty}^{-1}$ this implies

$$H(\alpha|\alpha_{-\infty}^{-1} \vee \beta_{-\infty}^{-1} \vee \mathcal{F}) = H(\alpha|\alpha_{-\infty}^{-1} \vee \mathcal{F}).$$

Choosing a sequence of finite partitions β_n with $\beta_n \nearrow \Pi(\mathbf{X})$ we deduce (by Theorem 14.28)

$$H(\alpha|\alpha_{-\infty}^{-1} \vee \mathcal{F} \vee \Pi(\mathbf{X})) = H(\alpha|\alpha_{-\infty}^{-1} \vee \mathcal{F}).$$

2. For a finite partition α let $\alpha_{-\infty}^{-1}(p)$ denote the σ-algebra $\bigvee_{-\infty}^{-1} T^{-jp}\alpha$. Suppose first that α is $\mathcal{F}$-measurable, then using part 1 and then Theorem 14.28 we

have

$$\begin{aligned} H(\alpha|\mathcal{F}_{-\infty} \vee \Pi(\mathbf{X})) &\geq \lim_{p\to\infty} H(\alpha|\alpha_{-\infty}^{-1}(p) \vee \mathcal{F}_{-\infty} \vee \Pi(\mathbf{X})) \\ &= \lim_{p\to\infty} H(\alpha|\alpha_{-\infty}^{-1}(p) \vee \mathcal{F}_{-\infty}) \\ &= H(\alpha|\mathcal{F}_{-\infty}), \end{aligned}$$

and therefore

$$H(\alpha|\mathcal{F}_{-\infty} \vee \Pi(\mathbf{X})) = H(\alpha|\mathcal{F}_{-\infty}).$$

The same argument works for α which is $T^n\mathcal{F}$-measurable for any integer n. Assume α is $\mathcal{F}_\infty$-measurable. Given $\epsilon > 0$ there exist an integer n and a finite partition β which is $T^n\mathcal{F}$-measurable such that $d_{\text{ent}}(\alpha, \beta) = H(\alpha|\beta) + H(\beta|\alpha) < \epsilon$. Now

$$\begin{aligned} H(\alpha|\mathcal{F}_{-\infty}) &\leq H(\alpha \vee \beta|\mathcal{F}_{-\infty}) \leq H(\beta|\mathcal{F}_{-\infty}) + H(\alpha|\beta) \\ &\leq H(\beta|\mathcal{F}_{-\infty} \vee \Pi(\mathbf{X})) + \epsilon \leq H(\alpha \vee \beta|\mathcal{F}_{-\infty} \vee \Pi(\mathbf{X})) + \epsilon \\ &\leq H(\alpha|\mathcal{F}_{-\infty} \vee \Pi(\mathbf{X})) + H(\beta|\alpha) + \epsilon \leq H(\alpha|\mathcal{F}_{-\infty} \vee \Pi(\mathbf{X})) + 2\epsilon. \end{aligned}$$

Thus

$$H(\alpha|\mathcal{F}_{-\infty}) \leq H(\alpha|\mathcal{F}_{-\infty} \vee \Pi(\mathbf{X}))$$

and the proof is complete as the reverse inequality is clear.

3. By part 2 we have

$$H(\alpha|\mathcal{F}_{-\infty}) \leq H(\alpha|\mathcal{F}_{-\infty} \vee \Pi(\mathbf{X}))$$

for every α and therefore the assumption $\alpha \subset \mathcal{F}_{-\infty} \vee \Pi(\mathbf{X})$ implies also $\alpha \subset \mathcal{F}_{-\infty}$, hence $\Pi(\mathbf{X}) \subset \mathcal{F}_{-\infty}$. □

18.8. DEFINITION. A dynamical system is called a *completely positive entropy system* (or a CPE-system) if $\Pi(\mathbf{X}) = \{\emptyset, X\}$. Equivalently, if every non-trivial factor system has positive entropy.

When $\mathbf{X}$ is a K-system there exists, by definition, a σ-algebra $\mathcal{K}$ with $\mathcal{K}_{-\infty} = \{\emptyset, X\}$ and $\mathcal{K}_\infty = \mathcal{X}$. Thus part 3 of Lemma 18.7 implies that every K-system is CPE. As we shall see next the converse is also true.

18.9. THEOREM (Rohlin-Sinai's theorem). *Let $\mathbf{X}$ be a dynamical system, then there exists a sub-σ-algebra $\mathcal{F} \subset \mathcal{X}$ such that $T^{-1}\mathcal{F} \subset \mathcal{F}$, $\bigvee_{n=0}^\infty T^n\mathcal{F} = \mathcal{F}_\infty = \mathcal{X}$ and $\bigcap_{n=0}^\infty T^n\mathcal{F} = \mathcal{F}_{-\infty} = \Pi(\mathbf{X})$. In particular the system $\mathbf{X}$ is a K-system iff it is CPE.*

PROOF. Begin by choosing an increasing sequence α_n of finite partitions with $\alpha_n \nearrow \mathcal{X}$. Next define inductively a second sequence $\beta_p = \beta_{p-1} \vee T^{n_p}\alpha_p$, $(p \geq 2, \beta_1 = \{X\})$ with n_p satisfying the following conditions. For every fixed $q \geq 1$ and $1 \leq p \leq q-1$,

$$H(\beta_p|\beta_{q-1}^-) - H(\beta_p|\beta_{q-1}^-) < \frac{1}{p}\frac{1}{2^{q-p}}.$$

Such a choice of n_p is possible since for $1 \leq j \leq q-1$, by Lemma 18.2.3,

$$\lim_{n\to\infty} H(\beta_j|\beta_{q-1}^- \vee T^n\alpha_q^-) = \lim_{n\to\infty} H(\beta_j|\beta_{q-1}^- \vee T^n\alpha_p) = H(\beta_j|\beta_{q-1}^-).$$

For a fixed p and any $n > p$, summing over $q = p+1, p+2, \dots, n$ we get:

$$H(\beta_p|\beta_p^-) - H(\beta_p|\beta_n^-) < \frac{1}{p}. \tag{2.1}$$

Define $\mathcal{C}$ as the limit of the increasing sequence $\beta_n \nearrow \mathcal{C}$, then $\beta_n^- \nearrow \mathcal{C}^-$ and we let $\mathcal{F} = \mathcal{C}^-$.

Next check that $\mathcal{F}$ satisfies the three properties required in the theorem. The inclusion $T^{-1}\mathcal{F} \subset \mathcal{F}$ is clear. Since $T^{n_p+1}\alpha_p \subset \beta_p^- \subset \mathcal{F}$ and since $T^n\mathcal{F}$ is increasing, it follows that $T^n\mathcal{F} \nearrow \mathcal{X}$. By 18.7.3 $\Pi(\mathbf{X}) \subset \mathcal{F}_{-\infty}$ and it only remains to show the reversed inclusion. Let $\alpha \subset \mathcal{F}_{-\infty} = \bigcap_{j=1}^{\infty} T^{-j}\mathcal{F}$, then $\alpha^T \subset \mathcal{F}_{-\infty} \subset \mathcal{F}$ and by the Pinsker formula (Proposition 18.5) we have

$$\begin{aligned} h(\alpha) &= H(\alpha|\alpha^-) \\ &= H(\beta_p \vee \alpha|\beta_p^- \vee \alpha_p^-) - H(\beta_p|\beta_p^- \vee \alpha^T) \\ &\le H(\beta|\beta_p^-) + H(\alpha|\beta_p^-) - H(\beta_p|\mathcal{F}). \end{aligned}$$

Since $\beta_n^- \nearrow \mathcal{F}$ we have, by (2.1), $H(\beta_p|\beta_p^-) - H(\beta_p|\mathcal{F}) < \frac{1}{p}$, hence $\lim_{p\to\infty} H(\beta_p|\beta_p^-) - H(\beta_p|\mathcal{F}) = 0$. Also $\lim_{p\to\infty} H(\alpha|\beta_p^-) = 0$ and we conclude that $h(\alpha) = 0$, hence that $\alpha \subset \Pi(\mathbf{X})$. Thus $\mathcal{F}_{-\infty} \subset \Pi(\mathbf{X})$ and the proof of the theorem is complete. □

18.10. DEFINITION. We call any sub-σ-algebra $\mathcal{F}$ satisfying the conditions of the theorem an *extremal algebra* (see [**201**]). Thus an extremal algebra for a K-system satisfies $T^{-1}\mathcal{F} \subset \mathcal{F}$, and for $n \nearrow \infty$, $T^n\mathcal{F} \nearrow \mathcal{F}_\infty = \mathcal{X}$ and $T^n\mathcal{F} \searrow \mathcal{F}_{-\infty} = \{\emptyset, X\}$.

The characterization of a K-system as a CPE system immediately yields the following corollary.

18.11. COROLLARY. 1. *If $\mathbf{X} = (X, \mathcal{X}, \mu, T)$ is a K-system then so are the systems $(X, \mathcal{X}, \mu, T^n)$, $(k \neq 0)$.*
2. *The K property is inherited by (non-trivial) factors.*

As another application of the Rohlin-Sinai theorem we next complete the proof of Theorem 3.53, on the equivalence of uniform mixing and the K property, whose proof was left incomplete.

COMPLETION OF THE PROOF OF THEOREM 3.53. We have seen already that in a uniformly mixing system $\mathbf{X}$, for every $B_1, \dots, B_k \in \mathcal{X}$, the sequence of σ-algebras $\mathcal{F}(B_1, \dots, B_k, n)$ decreases to the trivial σ-algebra $\mathcal{E}_0 = \{\emptyset, X\}$. This clearly implies that every finite partition α has a trivial remote past $\bigcap_{n=0}^{\infty} \alpha_{-\infty}^{-n}$ and by Theorem 18.6 we conclude that the Pinsker algebra $\Pi(\mathbf{X})$ is trivial. The Rohlin-Sinai theorem now completes the proof of the theorem. □

The K-property is not preserved by joinings. For example we cite (without a proof) the following surprising theorem of Smorodinsky and Thouvenot, [**240**].

18.12. THEOREM. *Let $\mathbf{X}$ be any ergodic system of positive entropy, then there exist three Bernoulli factors $\mathcal{B}_j \subset \mathcal{X}$, $j = 1, 2, 3$ such that $\mathcal{B}_1 \vee \mathcal{B}_2 \vee \mathcal{B}_3 = \mathcal{X}$. Thus $\mathbf{X}$ is represented as a joining of three Bernoulli systems.*

However, the K property is preserved under products. This is a special case of the following theorem.

18.13. THEOREM. *Let* $\mathbf{X}$ *and* $\mathbf{Y}$ *be ergodic systems, then*

$$\Pi(\mathbf{X} \times \mathbf{Y}) = \Pi(\mathbf{X}) \otimes \Pi(\mathbf{Y}).$$

In particular if $\mathbf{X}$ *and* $\mathbf{Y}$ *are* K*-systems then so is the product system* $\mathbf{X} \times \mathbf{Y}$.

PROOF. Let $\mathcal{F}(\mathbf{X})$ and $\mathcal{F}(\mathbf{Y})$ be extremal algebras for the K-systems $\mathbf{X}$ and $\mathbf{Y}$ respectively. We shall show that $\mathcal{F}(\mathbf{X}) \times \mathcal{F}(\mathbf{Y})$ is an extremal algebra for $\mathbf{X} \times \mathbf{Y}$. We have

$$(T \times T)^n(\mathcal{F}(\mathbf{X}) \times \mathcal{F}(\mathbf{Y})) \nearrow \mathcal{X} \otimes \mathcal{Y}$$

hence, by 18.7.3,

$$\begin{aligned} \Pi(\mathbf{X} \times \mathbf{Y}) &\subset \bigcap_{n=1}^{\infty} (T \times T)^{-n}(\mathcal{F}(\mathbf{X}) \times \mathcal{F}(\mathbf{Y})) \\ &\subset \bigcap_{n=1}^{\infty} (T^{-n}\mathcal{F}(\mathbf{X}) \times \mathcal{Y}) = \Pi(\mathbf{X}) \otimes \mathcal{Y}. \end{aligned}$$

Likewise

$$\Pi(\mathbf{X} \times \mathbf{Y}) \subset \mathcal{X} \otimes \Pi(\mathbf{Y}),$$

hence

$$\Pi(\mathbf{X} \times \mathbf{Y}) \subset (\Pi(\mathbf{X}) \otimes \mathcal{Y}) \cap (\mathcal{X} \otimes \Pi(\mathbf{Y})) = \Pi(\mathbf{X}) \otimes \Pi(\mathbf{Y}).$$

Since clearly $\Pi(\mathbf{X}) \times \{\emptyset, Y\}$ and $\{\emptyset, X\} \times \Pi(\mathbf{Y})$ are subalgebras of $\Pi(\mathbf{X} \times \mathbf{Y})$ we also have

$$\Pi(\mathbf{X}) \otimes \Pi(\mathbf{Y}) \subset \Pi(\mathbf{X} \times \mathbf{Y}).$$

□

We conclude this section with an open problem. A system $\mathbf{X}$ satisfies the *weak Pinsker property*, if for every $\delta > 0$ there exist a Bernoulli factor $\mathbf{X} \to \mathbf{Z}$ and an independent factor $\mathbf{X} \to \mathbf{Y}$ with $h_\mu(\mathbf{Y}) = \delta$ such that $\mathbf{X} = \mathbf{Z} \times \mathbf{Y}$, see [**245**] and [**67**]. One of the central open problems in ergodic theory is the weak Pinsker conjecture:

18.14. PROBLEM. Does every positive entropy ergodic system have the weak Pinsker property?

3. Zero entropy

The class of zero entropy dynamical systems behaves well with respect to the standard operations of ergodic theory. Nevertheless, some of the big mysteries of the theory are encompassed within the study of this class. For example both Rohlin's and Banach's problems (Problems 5.12 and 3.48) are about zero entropy systems. We begin our study with the following observation.

18.15. LEMMA. 1. *If* $\mathbf{X}$ *and* $\mathbf{Y}$ *are zero entropy systems and* $\lambda \in J(\mathbf{X}, \mathbf{Y})$ *is a joining, then* $(X \times Y, \lambda)$ *is also a zero entropy system.*
2. *If* $\mathbf{Y}_i$ *is an inverse system of zero entropy systems then so is the inverse limit* $\mathbf{Y} = \bigvee \mathbf{Y}_i$.

PROOF. 1. Observe that $\Pi(X \times Y, \lambda)$ contains both $\mathcal{X}$ and $\mathcal{Y}$.

2. Again the assertion follows from Corollary 18.4 and the observation that the collection of finite partitions measurable with respect to some $\mathcal{Y}_i$ is dense in the set of all finite measurable partitions of $\mathbf{Y}$. □

We can now add the following theorem to the list of disjointness theorems 6.26, 6.27 and 8.16.

18.16. THEOREM. *An ergodic system is a K-system iff it is disjoint from every ergodic zero entropy system.*

PROOF. Disjointness from every zero entropy system implies no nontrivial zero entropy factor i.e. it implies CPE which we know is equivalent to K. Now let $\mathbf{X}$ be a zero entropy system and $\mathbf{Y}$ a CPE system. Since we have shown that the property zero entropy is preserved under countable joinings and factors, applying Theorem 6.25 we see that the common factor $\mathbf{Z}$ in that theorem is both zero entropy and CPE, hence trivial. Thus $\mathcal{Y}$ and $\mathcal{X}^{\mathbb{Z}}$ are independent with respect to the joining λ_∞ and a fortiori $\mathcal{Y}$ and $\mathcal{X}$ are independent with respect to λ; i.e. $\lambda = \mu \times \nu$. (An alternative proof is provided by Corollary 8.13.) □

Also we can now apply Corollary 8.13 to deduce the following.

18.17. THEOREM. *A quasifactor of a zero entropy system is a zero entropy system.*

Note that together with Theorem 8.4 this provides an alternative proof of Theorem 18.16.

18.18. EXERCISE. Show that if $\mathbf{X}$ be a dynamical system of positive entropy then the Koopman operator U_T has countable Lebesgue spectrum on the orthogonal complement of $L^2(\Pi(\mathbf{X}))$. Thus if $\mathbf{X}$ has singular spectrum or a spectrum of finite multiplicity then $\mathbf{X}$ has zero entropy. [[Follow the proof of Kolmogorov's theorem 5.13. Another way of proving the second assertion is as follows. If $\mathbf{X}$ has a singular spectral type then Theorem 6.28 implies that $\mathbf{X}$ is disjoint from every Bernoulli system; however, as we shall see (Sinai's factor theorem 20.13), every ergodic system of positive entropy admits a Bernoulli factor.]]

Our next goal is to show that for distal extensions the conditional entropy (Definition 14.23) is zero. In particular, distal systems have zero entropy.

18.19. THEOREM. *Let $\pi : \mathbf{X} \to \mathbf{Y}$ be a distal extension, then $h(\mathbf{X}|\mathbf{Y}) = 0$. In particular if $\mathbf{X}$ is a distal system, then $h(\mathbf{X}) = 0$.*

PROOF. We prove the theorem for a distal system $\mathbf{X}$; the relative version is obtained by conditioning throughout on the σ-algebra $\mathcal{Y}$. Theorem 15.12 shows that it suffices to prove our claim for measure distal ergodic systems; we therefore assume that $\mathbf{X}$ is ergodic. By definition either X is finite or there exists a separating sieve for $\mathbf{X}$ (see Definition 10.4). In the first case our claim is obvious and we therefore assume that there exists a sequence $A_1 \supset A_2 \supset \cdots$ of sets in $\mathcal{X}$ with $\mu(A_n) > 0$ and $\mu(A_n) \to 0$ and a subset $X_0 \subset X$ with $\mu(X_0) = 1$ such that for every $x, x' \in X_0$ the condition "for every $n \in \mathbb{Z}$ there exists $k_n \in \mathbb{Z}$ with $T^{k_n}x, T^{k_n}x' \in A_n$" implies

$x = x'$. If we now set $A_0 = X, A_{-1} = \emptyset$ and for $n \geq 0$, $D_n = A_n \setminus A_{n-1}$, then $\alpha = \{D_0, D_1, D_2, \dots\}$ is a partition of X. With no loss of generality we can assume that $H(\alpha) = \sum_{n=1}^{\infty} -\mu(D_n)\log(D_n) < \infty$.

We claim that the σ-algebra $\mathcal{A} = \alpha_{-\infty}^{-1}$ separates points in X. In fact, if for $x, x' \in X_0$, $T^n x$ and $T^n x'$ belong to the same atom of α for every $n \geq 0$, then for every k, $T^n x \in A_k$ iff $T^n x' \in A_k$. By ergodicity almost every $z \in X_0$ satisfies $\lim_{m\to\infty} \mathbb{A}_m(\mathbf{1}_{A_n}(z)) = \mu(A_n)$ for every $n \geq 1$. If x is such a point then there is a sequence m_j such that for every $j \geq 1$, $T^{m_j}x \in A_j$ and therefore also $T^{m_j}x' \in A_j$. Since $x, x' \in X_0$, this implies $x = x'$.

Now Theorem 2.8.3 implies that $\mathcal{A} = \mathfrak{X}$; i.e. that α is a strong generator and in particular α is α^--measurable. Since $H(\alpha) < \infty$, Theorem 14.29 applies and we conclude that $h_\mu(\alpha) = H(\alpha|\alpha_{-\infty}^{-1}) = 0$. □

18.20. COROLLARY. *Let* $\mathbf{X}$ *be a dynamical system and* $\pi : \mathbf{X} \to \mathbf{Z}$ *its Pinsker factor (i.e.* $\pi^{-1}(\mathcal{Z}) = \Pi(\mathbf{X})$*), then the extension* π *is a weakly mixing extension. In particular, when* $\mathbf{X}$ *is ergodic and has positive entropy then — in the corresponding Rohlin decomposition (Theorem 3.18)* $\mathbf{X} \cong \mathbf{Z} \underset{\alpha}{\times} (U, \rho) = (Y \times U, \mathcal{Z} \otimes \mathcal{U}, \eta \times \rho, T)$, *where* $(U, \mathcal{U}, \rho)$ *is a standard probability space,* $\alpha : Y \to \mathrm{Aut}\,(U, \rho)$ *is a measurable cocycle and* $T(z, u) = (Tz, \alpha(z)u)$ *— the measure* ρ *has no atoms. Equivalently, the relative product measure*

$$\lambda = \int_Z (\mu_z \times \mu_z)\, d\eta(z) \cong \eta \times \rho \times \rho,$$

considered as a measure on $X \times X$, *satisfies* $\lambda(\Delta_X) = 0$.

PROOF. The first assertion follows from Theorem 18.19 and the Furstenberg-Zimmer structure theorem (Theorem 10.15) applied to the extension π. The proof of the second claim follows from Exercise 10.16. □

18.21. COROLLARY. *If* (X, T) *is a minimal distal topological system then* h_{top} $(X, T) = 0$.

PROOF. As (X, T) is minimal, it is either finite or X is a perfect compact space. Since the finite case is clear we can assume that X has no isolated points. By the variational principle (Theorem 17.1) it suffices to show that for any T-invariant ergodic probability measure μ on X the measure entropy $h_\mu(X, T) = 0$. Now for every $\mu \in M_T^{\mathrm{erg}}(X)$ the measure-preserving dynamical system $\mathbf{X} = (X, \mathfrak{X}, \mu, T)$ is ergodic and measure distal (see the remark following the definition of a separating sieve (Definition 10.4)), and by Theorem 18.19 has zero entropy. □

4. Notes

Sources: mainly Parry's book [**203**]. See also [**200**] and [**268**].

CHAPTER 19

Entropy Pairs

As we have noted in the introduction, the theories of measurable dynamics (ergodic theory) and topological dynamics exhibit a remarkable parallelism. Usually one translates 'ergodicity' as 'topological transitivity','weak mixing' as 'topological weak mixing', 'mixing' as 'topological mixing' and 'measure distal' as 'topologically distal'. One often obtains this way parallel theorems in both theories, though the methods of proof may be very different.

What is then the topological analogue of being a K-system? In [**25**] and [**26**] F. Blanchard introduced a notion of 'topological K' for $\mathbb{Z}$-systems which he called UPE (uniformly positive entropy). This is defined as follows: a topological dynamical system (X,T) is called a UPE system if every open cover of X by two non-dense open sets U and V has positive topological entropy. A local version of this definition led to the concept of an entropy pair. A pair $(x,x') \in X \times X$, $x \neq x'$ is an entropy pair if for every open cover $\mathcal{U} = \{U,V\}$ of X, with $x \in \mathrm{int}\,(U^c)$ and $x' \in \mathrm{int}\,(V^c)$, the topological entropy $h(\mathcal{U})$ is positive. The set of entropy pairs is denoted by $E_X = E_{(X,T)}$ and it follows that the system (X,T) is UPE iff $E_X = (X \times X)\setminus\Delta$. In general $E^* = E_X \cup \Delta$ is a $T\times T$-invariant closed symmetric and reflexive relation. Is it also transitive? When the answer to this question is affirmative then the quotient system X/E^*_X is the topological analogue of the Pinsker factor. Unfortunately this need not always be true even when (X,T) is a minimal system (see [**101**] for a counter example). In Glasner and Weiss [**98**] it was shown that if X supports an invariant measure μ for which the corresponding measure theoretical system $(X,\mathcal{X},\mu,T)$ is a K-system, then (X,T) is UPE. Using the Jewett-Krieger theorem it is now possible to obtain a great variety of strictly ergodic UPE systems.

Given a T-invariant probability measure μ on X, a pair $(x,x') \in X\times X$, $x \neq x'$ is called a μ-entropy pair if for every Borel partition $\alpha = \{Q,Q^c\}$ of X with $x \in \mathrm{int}\,(Q)$ and $x' \in \mathrm{int}\,(Q^c)$ the measure entropy $h_\mu(\alpha)$ is positive. This definition was introduced by Blanchard, Host, Maass, Martínez and Rudolph in [**28**] and it was shown there that for every invariant probability measure μ the set E_μ of μ-entropy pairs is contained in E_X. Since for a K-measure μ clearly every pair of distinct points is in E_μ, the result of [**98**] about K-measures follows. It was also shown in [**28**] that when (X,T) is uniquely ergodic the converse is also true: $E_X = E_\mu$ for the unique invariant measure μ on X. In order to gain a better understanding of the relationship between measure entropy pairs and topological entropy pairs the variational principle for open covers (Theorem 17.6) was proved in Blanchard, Glasner and Host [**27**]. Two important applications of this principle were: (i) the construction, for a general system (X,T), of a measure $\mu \in M_T(X)$ with $E_X = E_\mu$, and (ii) the proof that under a homomorphism $\pi : (X,\mu,T) \to (Y,\nu,T)$ every entropy pair in E_ν is the image of an entropy pair in E_μ.

Given a topological dynamical system (X, T) and a measure $\mu \in M_T(X)$, let $\pi : (X, \mathcal{X}, \mu, T) \to (Z, \mathcal{Z}, \eta, T)$ be the *measure-theoretical* Pinsker factor of (X, μ, T), and let $\mu = \int_Z \mu_z \, d\eta(z)$ be the disintegration of μ over (Z, η). Set

$$\lambda = \int_Z (\mu_z \times \mu_z) \, d\eta(z),$$

the independent joining of μ with itself over η. Finally let $\Lambda_\mu = \operatorname{supp}(\lambda)$ be the topological support of λ in $X \times X$. It was shown in Glasner [**87**] that $E_\mu = \Lambda_\mu \setminus \Delta$. One consequence of this characterization of the set of μ-entropy pairs is a description of the set of entropy pairs of a product system; in particular it follows that the product of two UPE systems is UPE. Another is the fact that the set $P \cap \bar{E}_X$ of proximal topological entropy pairs is residual in E_X.

Many of the main results of the papers [**25**], [**26**], [**98**], [**28**], [**27**] and [**87**] are reproduced in this chapter. In light of the above discussion, the table of contents gives a good description of its various sections.

1. Topological entropy pairs

19.1. Definition. Let (X, T) be a topological dynamical system and x, x' two distinct points of X. Call $(x, x') \in X \times X$ an *entropy pair of* (X, T) if, for every open cover $\mathcal{U} = \{U, V\}$ of X with $x \in \operatorname{int}(U^c)$, $x' \in \operatorname{int}(V^c)$ we have $h(\{U, V\}) > 0$. It will be convenient to state the same definition in terms of the complements $A = U^c$ and $B = V^c$. Thus for two distinct points a, $b \in X$ the pair (x, x') is an entropy pair of (X, T) if, for every closed neighborhood A of x and every closed neighborhood B of x' with $A \cap B = \emptyset$, one has $h(\{A^c, B^c\}) > 0$. Call an open cover $\mathcal{U} = \{U, V\}$ of X with U, V non-dense open sets a *standard cover*. We say that a standard cover $\mathcal{U}$ separates the points $x, x' \in X$ if $x \in \operatorname{int}(U^c), x' \in \operatorname{int}(V^c)$. We denote by $E_X = E_{(X,T)}$ the set of entropy pairs of (X, T). The system (X, T) is called a *UPE system* if every standard cover of X has positive topological entropy, or equivalently when $E_X = (X \times X) \setminus \Delta$. It is called a (topological) *CPE system* if every non trivial factor of (X, T) has positive entropy.

19.2. Proposition. *Let (X, T) be a topological dynamical system and let Δ denote the diagonal in $X \times X$.*

1. *The set $E_{(X,T)} \cup \Delta$ is a closed invariant subset of $X \times X$.*
2. *$E_X = \emptyset$ if and only if $h_{\mathrm{top}}(X, T) = 0$.*
3. *Let A and B be two disjoint closed subset of X. If $h(\{A^c, B^c\}) > 0$, then $A \times B$ contains an entropy pair.*

Proof. Part 1 follows directly from the definition of entropy pairs.

2. Evidently the existence of an entropy pair implies positive topological entropy. Assume to the contrary that $h(X, T) > 0$ and $E_X = \emptyset$. For $\delta > 0$ set

$$K_\delta = \{(x, x') \in X \times X : d(x, x') \geq \delta\}.$$

Given $(x, x') \in K_\delta$ — since by assumption (x, x') is not an entropy pair — there exists a standard cover $\mathcal{U} = \{U, V\}$ with $x \in \operatorname{int}(U^c)$, $x' \in \operatorname{int}(V^c)$ and $h(\mathcal{U}) = 0$. Choose $\epsilon = \epsilon(x, x') > 0$ with $B_\epsilon(x) \subset \operatorname{int}(U^c)$, $B_\epsilon(x') \subset \operatorname{int}(V^c)$ and let

$$U(x, x') = (\bar{B}_\epsilon(x))^c \qquad V(x, x') = (\bar{B}_\epsilon(x'))^c.$$

Clearly then $\mathcal{U}(x,x') = \{U(x,x'), V(x,x')\} \prec \mathcal{U}$ and therefore

$$h(\mathcal{U}(x,x')) \leq h(\mathcal{U}) = 0.$$

Let $\{(x_i,x_i')\}_{i=1}^k$ be a finite subset of K_δ such that $\{B_i, B_i'\}_{i=1}^k$ is an open cover of K_δ, where $B_i = B_{\epsilon(x_i,x_i')}(x_i)$ and $B_i' = B_{\epsilon(x_i,x_i')}(x_i')$. Let $U_i = (\bar{B}_i)^c, V_i = (\bar{B}_i')^c$, $\mathcal{U}_i = \{U_i, V_i\}$ and $\mathcal{V} = \bigvee_{i=1}^k \mathcal{U}_i$. We then have

$$h(\mathcal{V}) \leq \sum_{i=1}^k h(\mathcal{U}_i) = 0.$$

We show next that $\max\{\operatorname{diam}(W) : W \in \mathcal{V}\} \leq \delta$. Suppose $W = W_1 \cap W_2 \cap \cdots \cap W_k \in \bigvee_{i=1}^k \mathcal{U}_i$ is non-empty, where for each i, $W_i = U_i$ or $W_i = V_i$. Let $x \in W$; if $(x,x') \in K_\delta$ then there exists an i with $(x,x') \in B_i \times B_i'$. Hence $x \notin U_i$ so that $x \in W_i = V_i$. However $x' \in B_i'$ implies $x_i' \notin V_i$, hence $x' \notin W$. This proves our claim and since $\delta > 0$ is arbitrary, we deduce that $h_{\text{top}}(X,T) = 0$ (Lemma 14.5). This contradicts our assumption and the proof of part 2 is complete.

3. Given $\delta_1 > 0$ we can write $A = \cup_{i=1}^k A_i$ as a union of a finite number of closed sets A_i with $\operatorname{diam}(A_i) < \delta_1$ for every i. The cover $\bigvee_{i=1}^k \{A_i^c, B^c\}$ is clearly finer than $\{A^c, B^c\}$ and therefore

$$0 < h(\{A^c, B^c\}) \leq h(\bigvee_{i=1}^k \{A_i^c, B^c\}) \leq \sum_{i=1}^k h(\{A_i^c, B^c\}),$$

so that at least one of the summands on the right is positive; i.e. for some i the open cover $\{A_i^c, B^c\}$ has positive entropy. Repeating the same process with the cover $\{A_i^c, B^c\}$, but now decomposing B as a union $B = \cup_{j=1}^l B_j$ of closed sets with diameter smaller than δ_1, we get two closed sets $A_i = A^{(1)} \subset A$ and $B^{(1)} \subset B$ with diameter less than δ_1 and with $h(\{(A^{(1)})^c, (B^{(1)})^c\}) > 0$. By induction construct two decreasing sequences $A^{(n)}$, $B^{(n)}$ of closed sets with diameter less than δ_n, for a sequence $\delta_n \searrow 0$, and such that for every n, $h(\{(A^{(n)})^c, (B^{(n)})^c\}) > 0$. Let $\{x\} = \cap_{n=1}^\infty A^{(n)}$ and $\{x'\} = \cap_{n=1}^\infty B^{(n)}$. Clearly $x \in A$, $x' \in B$ and we claim that (x,x') is an entropy pair. In fact, if $\{U,V\}$ is a standard cover with $x \in \operatorname{int}(U^c)$ and $x' \in \operatorname{int}(V^c)$ then for some n, $x \in A^{(n)} \subset \operatorname{int}(U^c)$, $x' \in B^{(n)} \subset \operatorname{int}(V^c)$ and thus

$$h(\{U,V\}) \geq h(\{(A^{(n)})^c, (B^{(n)})^c\}) > 0.$$

□

19.3. PROPOSITION. *Let $\pi : (X,T) \to (Y,T)$ be a homomorphism of dynamical systems, then $(\pi \times \pi)(E_X \cup \Delta_X) = E_Y \cup \Delta_Y$.*

PROOF. Let $(x,x') \in E_X$ and suppose $y = \pi(x) \neq \pi(x') = y'$. For any open cover $\hat{\mathcal{U}} = \{\hat{U}, \hat{V}\}$ of Y with $y \in \operatorname{int}(\hat{U}^c)$, $y' \in \operatorname{int}(\hat{V}^c)$ we have $x \in \operatorname{int}(U^c)$, $x' \in \operatorname{int}(V^c)$ for the cover $\mathcal{U} = \{U,V\}$ of X with $U = \pi^{-1}(\hat{U})$ and $V = \pi^{-1}(\hat{V})$. Therefore $h(\{\hat{U},\hat{V}\}) = h(\{U,V\}) > 0$ and it follows that $(y,y') \in E_Y$.

Conversely, suppose $(y,y') \in E_Y$ and let $\hat{A}, \hat{B}$ be disjoint closed neighborhoods of y and y' respectively in Y. Denoting $A = \pi^{-1}(\hat{A}), B = \pi^{-1}(\hat{B})$, we have $0 < h(\{\hat{A}^c, \hat{B}^c\}) = h(\{A^c, B^c\})$ and by Proposition 3 we conclude that there exists $(x,x') \in (A \times B) \cap E_X$. If we let now $\hat{A}_n \searrow \{y\}, \hat{B}_n \searrow \{y'\}$ as above then the corresponding sequence $(x_n, x'_n) \in (A_n \times B_n) \cap E_X$ has a convergent subsequence $(x_{n_k}, x'_{n_k}) \to (x, x') \in E_X \cup \Delta_X$. Since $\pi(x) = y \neq y' = \pi(x')$, $x \neq x'$ hence $(x, x') \in E_X$. □

19.4. COROLLARY. *Every UPE system is CPE. There exist CPE systems which are not UPE .*

PROOF. The first statement is a direct corollary of the definitions and Proposition 19.3. For the counterexamples refer to [**25**] and [**101**]. □

It is very likely that there are also minimal CPE systems which are not UPE, but no such example is known.

2. Measure entropy pairs

19.5. DEFINITION. Let (X,T) be a topological dynamical system, and $\mu \in M_T(X)$. If x, x' are two distinct points in X, we say that the pair (x, x') is a *μ-entropy pair* if $h_\mu(\{Q, Q^c\}) > 0$ for every $Q \in \mathfrak{X}$ such that $x \in \operatorname{int}(Q)$ and $x' \in \operatorname{int}(Q^c)$. A measurable partition $\alpha = \{Q, Q^c\}$ with both $\operatorname{int}(Q)$ and $\operatorname{int}(Q^c)$ non-empty will be called a *replete partition*. If A and B are two disjoint subsets in X we say that the partition $\alpha = \{Q, Q^c\}$ *separates* A and B if $A \subset Q$ and $B \subset Q^c$. Write $E_\mu = E_\mu(X,T)$ for the set of μ-entropy pairs in (X,T). A pair (x, x') is called a *measure entropy pair* if it is in E_μ for some $\mu \in M_T(X)$. We denote by Π_μ the Pinsker σ-algebra of the measure-theoretical dynamical system $(X, \mathfrak{X}, T, \mu)$.

19.6. PROPOSITION. *Let (X,T) be a topological dynamical system, and $\mu \in M_T(X)$.*

1. *$E_\mu(X,T) \cup \Delta$ is a closed and invariant subset of $X \times X$.*
2. *Let A and B be two disjoint closed subsets. If $h_\mu(\{Q, Q^c\}) > 0$ for every Borel partition $\alpha = \{Q, Q^c\}$ separating A and B, then there exists a μ-entropy pair (a, b) with $a \in A$ and $b \in B$.*
3. *$E_\mu(X,T) = \emptyset$ if and only if $h_\mu(X,T) = 0$.*

PROOF. Part 1 follows directly from the definitions.

2. Assume to the contrary that $A \times B$ contains no μ-entropy pairs. Fix $x \in A$ and choose for every $x' \in B$ a partition $\alpha = \{Q_{x'}, Q^c_{x'}\}$ with $x \in \operatorname{int}(Q_{x'})$, $x' \in \operatorname{int}(Q^c_{x'})$ and $h_\mu(\{Q, Q^c\}) = 0$. Note that this means that the set $Q_{x'}$ belongs to the Pinsker σ-algebra Π_μ. By compactness of B the collection $\{Q^c_{x'}\}_{x' \in B}$ has a finite subcover $\{Q^c_1, \dots, Q^c_n\}$ and we let $P_x = \cap_{j=1}^n Q_j$. It follows that P_x has the following properties: (i) it is a neighborhood of x, (ii) it is an element of Π_μ and (iii) $P_x \cap B = \emptyset$.

By compactness of A the collection $\{P_x\}_{x \in A}$ has a finite subcover $\{P_1, \dots, P_m\}$. The set $P = \cup_{j=1}^m P_j$ satisfies $A \subset P \subset B^c$ and $P \in \Pi_\mu$; i.e. $\{P, P^c\}$ separates A and B, but $h_\mu(\{P, P^c\}) = 0$. This contradiction completes the proof.

3. If $h_\mu(X,T) = 0$, then $\Pi_\mu = \mathfrak{X}$, $h_\mu(\{Q, Q^c\}) = 0$ for every $Q \in \mathfrak{X}$ and in particular $E_\mu = \emptyset$. Conversely, assuming $E_\mu = \emptyset$ we conclude by part 2 that for every pair of disjoint closed subsets A and B there exists a separating partition $\{Q, Q^c\}$ with $h_\mu(\{Q, Q^c\}) = 0$. Now fix a closed subset A and let $B_n \nearrow A^c$ be an increasing sequence of closed subset whose union is A^c. Let $\{Q_n, Q^c_n\}$ be the

corresponding partitions separating A and B_n; we then have $\cap_{n=1}^{\infty} Q_n = A \in \Pi_\mu$ and therefore $\Pi_\mu = \mathfrak{X}$. □

19.7. PROPOSITION. *Let $\pi : (X,T) \to (Y,T)$ be a homomorphism of dynamical systems, and $\mu \in M_T(X)$ with $\nu = \pi(\mu) \in M_T(Y)$, then for $(x,x') \in E_\mu$ with $y = \pi(x) \neq \pi(x') = y'$ we have $(y,y') \in E_\nu$.*

PROOF. If $\{\hat{Q}, \hat{Q}^c\}$ is a replete partition with $y \in \mathrm{int}\,(\hat{Q}), y' \in \mathrm{int}\,(\hat{Q}^c)$ then, with $Q = \pi^{-1}(\hat{Q})$, the partition $\{Q, Q^c\}$ of X is replete with $x \in \mathrm{int}\,(Q), x' \in \mathrm{int}\,(Q^c)$ and $h_\nu(\{\hat{Q}, \hat{Q}^c\}) = h_\mu(\{Q, Q^c\}) > 0$. □

As we shall see below (Theorem 19.21) the converse is also true: every ν-entropy pair is the image of a μ-entropy pair. However, this result requires a considerably more elaborate machinery for its proof.

The next lemma is an important tool in understanding the relationship between covers and partitions. For its formulation we need the following definition.

19.8. DEFINITION. Let $\mu \in M_T(X)$ and A, B be two disjoint Borel subsets of X. We say that the pair (A, B) (or equivalently, the Borel cover $\{A^c, B^c\}$) satisfies the property $S(\mu)$ if $h_\mu(\{Q, Q^c\}) > 0$ for every Borel partition $\{Q, Q^c\}$ separating A and B.

Thus, by Proposition 19.6:

- If a, b are two distinct points in X, $(a,b) \in E_\mu(X,T)$ if and only if the property $S(\mu)$ holds for every pair of disjoint neighborhoods A of a and B of b.
- If A and B are two closed disjoint subsets satisfying the property $S(\mu)$, then $(A \times B) \cap E_\mu(X,T) \neq \emptyset$.

19.9. LEMMA. *Let A, B be two disjoint Borel subsets of X, then (A, B) satisfies the property $S(\mu)$ if and only if*

$$\mathbb{E}^{\Pi_\mu}(\mathbf{1}_A)(x)\,\mathbb{E}^{\Pi_\mu}(\mathbf{1}_B)(x) \quad \textit{is not} \quad \mu \ \textit{a.e.equaltozero}\ . \tag{2.1}$$

PROOF. If $h_\mu(\{Q, Q^c\}) = 0$ for some Borel partition $\{Q, Q^c\}$ separating A and B, then $Q \in \Pi_\mu$ and

$$\begin{aligned}\int \mathbb{E}^{\Pi_\mu}(\mathbf{1}_A)(x)\,\mathbb{E}^{\Pi_\mu}(\mathbf{1}_B)(x)\,d\mu(x) &\leq \int \mathbb{E}^{\Pi_\mu}(\mathbf{1}_Q)(x)\,\mathbb{E}^{\Pi_\mu}(\mathbf{1}_{Q^c})(x)\,d\mu(x) \\ &= \int \mathbf{1}_Q(x) \cdot \mathbf{1}_{Q^c}(x)\,d\mu(x) = 0.\end{aligned}$$

Conversely, suppose

$$\int \mathbb{E}^{\Pi_\mu}(\mathbf{1}_A)(x)\,\mathbb{E}^{\Pi_\mu}(\mathbf{1}_B)(x)\,d\mu(x) = 0.$$

If $\mu(A) = 0$ then $A \in \Pi_\mu$ and we let $Q = A$. Otherwise let

$$F = \{x : \mathbb{E}^{\Pi_\mu}(\mathbf{1}_A)(x) > 0\}.$$

Then F is Π_μ-measurable and for μ almost every $x \in F$ our assumption implies that $\mathbb{E}^{\Pi_\mu}(\mathbf{1}_B)(x) = 0$. Thus

$$\mu(F \cap B) = \mathbb{E}(\mathbf{1}_F \mathbb{E}^{\Pi_\mu}(\mathbf{1}_B)) = 0,$$

and

$$\mu(A\backslash F) = \mathbb{E}(\mathbf{1}_{F^c}\mathbb{E}^{\Pi_\mu}(\mathbf{1}_A)) = 0.$$

It follows that the set $Q = A\cup(F\backslash B)$ is in Π_μ, so that $h_\mu(\{Q,Q^c\}) = 0$ and $\{Q,Q^c\}$ separates A and B. □

3. A measure entropy pair is an entropy pair

Let (X,T) be a topological dynamical system, and $\mu \in M_T(X)$. It will be convenient in the sequel to deal with finite Borel (rather than open) covers $\mathfrak{U} = \{U_1, U_2, \ldots, U_n\}$. Given such a cover we define its *combinatorial entropy* $h_c(\mathfrak{U})$ in the same way that the topological entropy of an open cover is defined. Thus if $N(\mathfrak{U})$ is the minimal cardinality of a subcover of $\mathfrak{U}$ then $H(\mathfrak{U}) = \log N(\mathfrak{U})$ and

$$h_c(\mathfrak{U}) = \lim_{n\to\infty} \frac{1}{n} H(\mathfrak{U}_0^{n-1}).$$

Given a two set Borel cover $\mathfrak{U} = \{U,V\}$ of X, set $A = V^c, B = U^c$, $C = U\cap V$ and define two measurable partitions:

$$\alpha = \{A,B,C\}, \qquad \text{and} \qquad \beta = \{A\cup B, C\}.$$

19.10. LEMMA. *For a Borel cover $\mathfrak{U} = \{U,V\}$ as above*

$$0 \le h_\mu(\alpha) - h_\mu(\beta) \le h_c(\mathfrak{U}).$$

PROOF. For every $n \in \mathbb{N}$ we have (Proposition 14.16)

$$0 \le H_\mu(\alpha_0^{n-1}) - H_\mu(\beta_0^{n-1}) = H_\mu(\alpha_0^{n-1}|\beta_0^{n-1}) = \sum_{L\in\beta_0^{n-1}} \mu(L)K(L),$$

where

$$K(L) = -\sum\{\mu(M|L)\log\mu(M|L) : M \in \alpha_0^{n-1}, M \subset L\}.$$

Let $c_n(L) = \text{card}\,\{M : M \in \alpha_0^{n-1}, M \subset L\}$ then, since the uniform partition of L into $c_n(L)$ sets of equal measure maximizes the number $K(L)$ (Exercise 14.15), by replacing each $\mu(M|L)$ by $c_n(L)^{-1}$ we get

$$H_\mu(\alpha_0^{n-1}|\beta_0^{n-1}) \le \sum_{L\in\beta_0^{n-1}} \mu(L)\log c_n(L).$$

Let $N = N(\mathfrak{U}_0^{n-1})$ and let $\{W_1, W_2, \ldots, W_N\}$ be a subcover of $\mathfrak{U}_0^{n-1}$ of minimal cardinality. Fix $L \in \beta_0^{n-1}$. If $L\cap W_j$ is non-empty then it represents a joint $(\mathfrak{U},\beta), n$-name and, since $\alpha = \mathfrak{U}\vee\beta$, this name determines a unique α, n-name; i.e. an atom $M \in \alpha_0^{n-1}$ such that $M = L\cap W_j$. This clearly implies that, for each $L \in \beta_0^{n-1}$, $N(\mathfrak{U}_0^{n-1}) \ge c_n(L)$.

This estimation yields:

$$H_\mu(\alpha_0^{n-1}|\beta_0^{n-1}) \le \sum_{L\in\beta_0^{n-1}} \mu(L)\log c_n(L) \le \log N(\mathfrak{U}_0^{n-1}).$$

The required inequality is now obtained upon dividing by n and taking the limit as $n \to \infty$. □

We now formulate the main result of this section:

19.11. THEOREM. *Every measure entropy pair is a topological entropy pair.*

Obviously Theorem 19.11 will follow from the following proposition:

19.12. PROPOSITION. *Let* $\mathbf{X} = (X, \mathcal{X}, \mu, T)$ *be a measure dynamical system. Suppose* $\mathcal{U} = \{U, V\}$ *is a measurable cover such that every measurable two-set partition* $\gamma = \{H, H^c\}$ *which (as a cover) is finer than* $\mathcal{U}$ *satisfies* $h_\mu(\gamma) > 0$*; then* $h_c(\mathcal{U}) > 0$.

PROOF. As above, set $A = V^c, B = U^c$, $C = U \cap V$ and define the two measurable partitions:

$$\alpha = \{A, B, C\}, \qquad \text{and} \qquad \beta = \{A \cup B, C\}.$$

Applying Lemma 19.10 to the T^m action, for any $m \in \mathbb{N}$, we get

$$0 \le h_\mu(\alpha, T^m) - h_\mu(\beta, T^m) \le h_c(\mathcal{U}, T^m). \tag{3.1}$$

On the other hand, by Theorem 14.29 and Lemma 18.7.1, we have

$$\begin{aligned} h_\mu(\alpha, T^m) - h_\mu(\beta, T^m) &= H_\mu(\alpha|\alpha^-_{T^m}) - H_\mu(\beta|\beta^-_{T^m}) \\ &= H_\mu(\alpha|\alpha^-_{T^m} \vee \Pi_\mu) - H_\mu(\beta|\beta^-_{T^m} \vee \Pi_\mu), \end{aligned}$$

and by Theorem 14.28 we get

$$\lim_{m\to\infty} (h_\mu(\alpha, T^m) - h_\mu(\beta, T^m)) = H_\mu(\alpha|\Pi_\mu) - H_\mu(\beta|\Pi_\mu). \tag{3.2}$$

By definition — and since the set C belongs to both α and β — we have

$$\begin{aligned} &H_\mu(\alpha|\Pi_\mu) - H_\mu(\beta|\Pi_\mu) = \\ &-\int \Big(\sum_{D\in\alpha} \mathbb{E}^{\Pi_\mu}(\mathbf{1}_D)\log(\mathbb{E}^{\Pi_\mu}(\mathbf{1}_D)) - \sum_{D\in\beta} \mathbb{E}^{\Pi_\mu}(\mathbf{1}_D)\log(\mathbb{E}^{\Pi_\mu}(\mathbf{1}_D))\Big)\, d\mu = \\ &-\int \Big(\mathbb{E}^{\Pi_\mu}(\mathbf{1}_A)\log(\mathbb{E}^{\Pi_\mu}(\mathbf{1}_A)) + \mathbb{E}^{\Pi_\mu}(\mathbf{1}_A)\log(\mathbb{E}^{\Pi_\mu}(\mathbf{1}_A)) \\ &\quad - \mathbb{E}^{\Pi_\mu}(\mathbf{1}_{A\cup B})\log(\mathbb{E}^{\Pi_\mu}(\mathbf{1}_{A\cup B}))\Big)\, d\mu. \end{aligned} \tag{3.3}$$

By Lemma 19.9 there exists a subset $H \in \mathcal{X}$ with $\mu(H) > 0$ such that for every $x \in H$ both $\mathbb{E}^{\Pi_\mu}(\mathbf{1}_A)(x)$ and $\mathbb{E}^{\Pi_\mu}(\mathbf{1}_B)(x)$ are positive. By the convexity of the function $-x \cdot \log x$ the expression under the integral sign of the last integral in (3.3) is positive on H and it follows that

$$H_\mu(\alpha|\Pi_\mu) - H_\mu(\beta|\Pi_\mu) > 0.$$

From (3.1) and (3.2) we conclude that for some m, $h_c(\mathcal{U}, T^m) > 0$ and therefore also $h_c(\mathcal{U}, T) = h_c(\mathcal{U}) > 0$. □

19.13. COROLLARY. *If* (X, T) *is a topological dynamical system for which there exists a full measure* $\mu \in M_T(X)$ *such that the corresponding measure dynamical system* $\mathbf{X} = (X, \mathcal{X}, \mu, T)$ *is a* K*-system, then* (X, T) *is a UPE system.*

19.14. COROLLARY. *Every* K *system has a uniquely ergodic UPE model. More precisely, let* $\mathbf{X} = (X, \mathcal{X}, \mu, T)$ *be a measure dynamical* K*-system, then there exists a strictly ergodic topological system* (Y, T) *with a unique invariant measure* ν *such that the measure dynamical systems* $\mathbf{X} = (X, \mathcal{X}, \mu, T)$ *and* $\mathbf{Y} = (Y, \mathcal{Y}, \nu, T)$ *are isomorphic.*

PROOF. Follows directly from Corollary 19.13 and the Jewett-Krieger theorem (Theorem 15.28). □

Almost the same proof (as that of Proposition 19.12) yields the following lemma which we shall need later on.

19.15. LEMMA. *Let A, B be two disjoint Borel subsets of X satisfying the property $S(\mu)$, then*

$$\inf\{h_\mu(\{Q,Q^c\}) : \{Q,Q^c\} \text{ separates } A \text{ and } B\} > 0.$$

PROOF. Notation as in Proposition 19.12 we see that — as in the proof of this proposition — for sufficiently large m, $h_\mu(\alpha, T^m) > h_\mu(\beta, T^m)$. Let now $\gamma = \{Q, Q^c\}$ be a partition separating A and B; we then have $\gamma \vee \beta \succ \alpha$, hence $h_\mu(\gamma, T^m) \geq h_\mu(\alpha, T^m) - h_\mu(\beta, T^m)$, and therefore

$$h_\mu(\gamma) \geq \frac{1}{m} h_\mu(\gamma, T^m) \geq \frac{1}{m}(h_\mu(\alpha, T^m) - h_\mu(\beta, T^m)) > 0.$$

□

4. A characterization of E_μ

Let (X,T) be a dynamical system. For $\mu \in M_T(X)$ denote $\mathrm{s}(\mu) = \mathrm{supp}\,(\mu)$, $\mathrm{s}^2(\mu) = \{(x,x) : x \in \mathrm{s}(\mu)\}$ and let $X_m = \mathrm{cls} \bigcup\{\mathrm{s}(\mu) : \mu \in M_T(X)\}$. There is always a measure $\mu \in M_T(X)$ for which $\mathrm{s}(\mu) = X_m$. Let $\pi : (X, \mathfrak{X}, \mu, T) \to (Z, \mathfrak{Z}, \eta, T)$ be the measure theoretical Pinsker factor of $(X, \mathfrak{X}, \mu, T)$, and let $\mu = \int_Z \mu_z \, d\eta(z)$ be the disintegration of μ over (Z, η). Let

$$\lambda = \int_Z (\mu_z \times \mu_z) \, d\eta(z)$$

be the independent product of μ with itself over η. Finally let $\Lambda_\mu = \mathrm{supp}\,(\lambda)$ be the topological support of λ. Although the Pinsker factor is, in general, only defined measure theoretically, the measure λ is a well defined element of $M_{T\times T}(X \times X)$. Notice that for a measure $\mu \in M_T(X)$ with zero entropy $\Lambda_\mu = \mathrm{s}^2(\mu)$. Also note that $\Lambda_\mu \cap \Delta_X = \mathrm{s}^2(\mu)$. In fact, if $(x,x) \in \Lambda_\mu \cap \Delta$ and U is a neighborhood of x then

$$0 < \lambda(U \times U) = \int_Z \mu_z(U) \cdot \mu_z(U) \, d\eta(z) = \int_Z \mu_z(U)^2 \, d\eta(z),$$

hence also

$$\mu(U) = \int_Z \mu_z(U) \, d\eta(z) > 0.$$

Thus $x \in \mathrm{s}(\mu)$, whence $(x,x) \in \mathrm{s}^2(\mu)$.

19.16. THEOREM. *Let (X,T) be a topological dynamical system and let $\mu \in M_T(X)$.*

1. $E_\mu = \Lambda_\mu \setminus \Delta$ *and* $\Lambda_\mu = E_\mu \cup \mathrm{s}^2(\mu)$.
2. $\mathrm{cls}\, E_\mu \subset \Lambda_\mu$.
3. *If μ is ergodic with positive entropy then* $\mathrm{cls}\, E_\mu = \Lambda_\mu$.

PROOF. 1. If μ has zero entropy $\Lambda_\mu = \mathrm{s}^2(\mu)$ and $E_\mu = \Lambda_\mu \setminus \Delta = \emptyset$. Let $\mu \in M_T(X)$ have positive entropy and suppose $(x,y) \notin E_\mu \cup \Delta$, then there exists a replete Borel partition $\alpha = \{Q, Q^c\}$ with $x \in \operatorname{int}(Q)$, $y \in \operatorname{int}(Q^c)$ and such that $h_\mu(\alpha) = 0$. This implies that Q is in the Pinsker algebra Π_μ, and we have

$$\lambda(Q \times Q^c) = \int \mu_z(Q)\mu_z(Q^c)\, d\eta(z) = \int \mathbb{E}^{\Pi_\mu}(\mathbf{1}_Q)(x)\mathbb{E}^{\Pi_\mu}(\mathbf{1}_{Q^c})(x)\, d\mu(x)$$
$$= \int \mathbf{1}_Q(x) \cdot \mathbf{1}_{Q^c}(x)\, d\mu(x) = 0.$$

Thus $(x, y) \notin \Lambda$.

Conversely, suppose $(x, y) \notin \Lambda \cup \Delta$. Then there exist disjoint closed neighborhoods A and B of x and y respectively with

$$0 = \lambda(A \times B) = \int \mu_z(A)\mu_z(B)\, d\eta(z)$$
$$= \int \mathbb{E}^{\Pi_\mu}(\mathbf{1}_A)(x)\, \mathbb{E}^{\Pi_\mu}(\mathbf{1}_B)(x)\, d\mu(x).$$

By Lemma 19.9 this implies the existence of a Borel subset Q of X such that $A \subset Q$, $B \subset Q^c$ and $h_\mu(\alpha) = 0$, where $\alpha = \{Q, Q^c\}$. Thus $(x, y) \notin E_\mu$.

We have shown that $E_\mu = \Lambda_\mu \setminus \Delta$ and since $\Lambda_\mu \cap \Delta_X = \mathbf{s}^2(\mu)$ we conclude that $\Lambda_\mu = E_\mu \cup \mathbf{s}^2(\mu)$.

2. By part 1, $E_\mu \subset \Lambda_\mu$ and taking closure on both sides we get $\operatorname{cls} E_\mu \subset \Lambda_\mu$.

3. Now assume that μ is ergodic and has positive entropy. By Corollary 18.20 we see that λ has the form $\lambda = \eta \times \rho \times \rho$, where $\mathbf{X} \cong \mathbf{Z} \underset{\alpha}{\times} (U, \rho)$ is the Rohlin decomposition corresponding to the extension π (Theorem 3.18), and ρ has no atoms. It then follows that $\lambda(\Delta) = 0$. Now let $(x, x) \in \Lambda_\mu \cap \Delta$ and let V be a neighborhood of x; then $\lambda((V \times V) \setminus \Delta) > 0$ and, by part 1, $(V \times V) \cap E_\mu \neq \emptyset$, hence $\operatorname{cls} E_\mu = \Lambda_\mu$. □

5. A measure μ with $E_\mu = E_X$

In Section 3 it was shown that for every measure $\mu \in M_T(X)$ one has $E_\mu \subset E_{(X,T)}$. We ask next whether for every $(a, b) \in E_{(X,T)}$ there exists a measure $\mu \in M_T(X)$ such that $(a, b) \in E_\mu$? The following result answers that question in the affirmative and shows, moreover that it is possible to find a single measure which is good for all entropy pairs. As can be expected the crucial tool in the proof of this statement is the variational principle for open covers (Theorem 17.6).

19.17. THEOREM. *Let (X, T) be a topological dynamical system. There exists a measure $\mu \in M_T(X)$ such that $E_X = E_\mu = \Lambda_\mu \setminus \Delta$.*

PROOF. Let $\{(a_n, b_n) : n \geq 1\}$ be a dense sequence of points in E_X. For every $n \geq 1$ and $r \geq 1$, let $C_{n,r}$ and $D_{n,r}$ be the closed balls with centers a_n and b_n respectively and radii $d(a_n, b_n)/3r$. For every n and every r, $h(\{C^c_{n,r}, D^c_{n,r}\}) > 0$, since (a_n, b_n) is an entropy pair. Thus, by Theorem 17.6, there exists a measure $\mu_{n,r} \in M_T(X)$ with $h_{\mu_{n,r}}(\{Q, Q^c\}) > 0$ for every Borel partition $\{Q, Q^c\}$ separating $C_{n,r}$ and $D_{n,r}$. Let

$$\mu = \sum_{n=1}^\infty \sum_{r=1}^\infty 2^{-n-r} \mu_{n,r}.$$

For $(a, b) \in E_X$, we now show that $(a, b) \in E_\mu(X, T)$.

Let U and V be open neighborhoods of a and b respectively, with $U \cap V = \emptyset$. There exists an n such that $a_n \in U$ and $b_n \in V$, and there exists an r with $C_{n,r} \subset U$ and $D_{n,r} \subset V$. Every partition $\alpha = \{Q, Q^c\}$ which separates U and V separates also $C_{n,r}$ and $D_{n,r}$, thus $h_{\mu_{n,r}}(\alpha) > 0$, whence

$$h_\mu(\alpha) \geq 2^{-n-r} h_{\mu_{n,r}}(\alpha) > 0,$$

which proves our assertion. □

One cannot always choose the measure whose existence is guaranteed in Theorem 19.17 to be ergodic:

19.18. PROPOSITION. *There exist a topological dynamical system and an entropy pair for that system, which is not a measure entropy pair for any ergodic measure.*

PROOF. Let (X, T) be a UPE system and (Y, T) the symbolic system constructed by Weiss in [**259**]. This system has the following properties:

(i) it has zero topological entropy,
(ii) it is topologically transitive,
(iii) it contains a dense subset of periodic points and
(iv) no ergodic measure has all of Y as its topological support.

The properties (ii) and (iii) are just the conditions defining the property of being a P-system (Chapter 1, Section 1), and in particular it follows that (Y, T) is an E-system (Definition 4.25), so that $Y_m = Y$. Consider the product system $(X \times Y, T \times T)$. Since (X, T) is a UPE system and since (Y, T) satisfies (i), it follows from Theorem 19.24.5 (below) that

$$E(X \times Y, T \times T) = \big\{\big((x, y), (x', y)\big) : x,\, x' \in X,\, x \neq x',\, y \in Y\big\}.$$

In particular, if y_0 is a transitive point of Y, and x, x' are two distinct points in X, $\big((x, y_0), (x', y_0)\big)$ is an entropy pair for $(X \times Y, T \times T)$.

Let now $\mu \in M^{\text{erg}}_{T\times T}(X \times Y, T \times T)$ and let ν be its projection on Y. By (iii) – since ν is ergodic and as y_0 is transitive – it follows that y_0 does not belong to the support of ν, and there exists a neighborhood U of y_0 with $\nu(U) = 0$. Let A be a closed neighborhood of x which does not contain x', and let α be the partition $\{A \times U, (A \times U)^c\}$; α separates the points (x, y_0) and (x', y_0), and $h_\mu(\alpha) = 0$ since $\mu(A \times U) \leq \nu(U) = 0$. Thus $\big((x, y_0), (x', y_0)\big)$ is not a μ-entropy pair. □

6. Entropy pairs and the ergodic decomposition

19.19. THEOREM. *Let $\mu \in M_T(X)$,*

$$\mu = \int_\Omega \mu_\omega \, dm(\omega)$$

its ergodic decomposition.

1. *For m-almost every ω, $E_{\mu_\omega} \subset E_\mu$.*
2. *If $(a, b) \in E_\mu$, then, for every neighborhood V of (a, b),*

$$m\{\omega : V \cap E_{\mu_\omega} \neq \emptyset\} > 0.$$

Thus for an appropriate choice of Ω,

$$\mathrm{cls}(\bigcup\{E_{\mu_\omega} : \omega \in \Omega\}) = \bar{E}_\mu.$$

PROOF. 1. Let A, B be two open subsets with $\bar{A} \cap \bar{B} = \emptyset$ and $(\bar{A} \times \bar{B}) \cap E_\mu = \emptyset$. $\bar{A}$, $\bar{B}$ do not satisfy the property $S(\mu)$ (Definition 19.8), and therefore there exists a partition $\alpha = \{Q, Q^c\}$ separating $\bar{A}$ and $\bar{B}$ with $h_\mu(\alpha) = 0$. By Theorem 15.12

$$\int h_{\mu_\omega}(\alpha)\, dm(\omega) = h_\mu(\alpha) = 0$$

hence, for m-almost every ω, $h_{\mu_\omega}(\alpha) = 0$. It follows from the definition of entropy pairs that, $A \times B$ does not contain a μ_ω-entropy pair and $(A \times B) \cap E_{\mu_\omega} = \emptyset$.

Since $E_\mu \cup \Delta$ is closed in $X \times X$, its complement can be written as a countable union of sets of the form $A \times B$, where A and B are two open sets as above. By definition, $E_{\mu_\omega} \cap \Delta = \emptyset$ for all ω, and we conclude that for almost every ω,

$$E_{\mu_\omega} \cap (E_\mu)^c = \emptyset.$$

2. With no loss in generality, we can consider the case where $V = A \times B$, with A and B closed neighborhoods of a and b respectively, and $A \cap B = \emptyset$. Thus A, B satisfy the property $S(\mu)$.

Put

$$K = \{\omega \in \Omega : (A,\ B) \ \text{ satisfy } \ S(\mu_\omega)\} .$$

By Proposition 19.6,

$$K \subset \{\omega \in \Omega : (A \times B) \cap E_{\mu_\omega} \neq \emptyset\} ,$$

and it suffices to show that $m(K) > 0$. Suppose $m(K) = 0$. As the space $(X, \mathcal{X})$ is a Lebesgue space, there exists a countable family $(\alpha_i : i \in I)$ of partitions separating A and B which, for every probability measure ν on X, is $L^1(\nu)$-dense in the collection of partitions separating A and B. By Lemma 19.15

$$c = \inf\{h_\mu(\{Q, Q^c\}) : \{Q, Q^c\} \text{ separates } A \text{ and } B\} > 0$$

and we choose an $0 < \epsilon < c$.

For $\omega \notin K$, there exists a partition α separating A and B, with $h_{\mu_\omega}(\mathcal{P}) = 0$, hence by density and Lemma 15.9, there exists an $i \in I$ with $h_{\mu_\omega}(\alpha_i) \leq \epsilon$. It follows that there exist a countable partition $\{\Omega_n : n \in \mathbb{N}\}$ of $\Omega \setminus K$, with $m(\Omega_n) > 0$ for every n, and a sequence of partitions $\beta_n = \{Q_n, Q_n^c\}$ of X separating A and B, such that for every n

$$h_{\mu_\omega}(\beta_n) < \epsilon \text{ for every } \ \omega \in \Omega_n .$$

For every n write

$$t_n = m(\Omega_n) \ ; \quad \mu_n = \frac{1}{t_n}\int_{\Omega_n} \mu_\omega\, dm(\omega) .$$

We have

$$h_{\mu_n}(\beta_n) = \frac{1}{t_n}\int_{\Omega_n} h_{\mu_\omega}(\beta_n)\, dm(\omega) \leq \epsilon .$$

$$\sum t_n = 1 \text{ and } \mu = \sum t_n \mu_n$$

The measures μ_n are mutually singular, and there exists a sequence of Borel subsets X_n of X such that, for every n, $\mu_n(X_n) = 1$ and $\mu_n(X_k) = 0$ for every $k \neq n$.

Let $E = \cap_n Q_n$ and set $Q = E \cup \bigcup_n (Q_n \cap X_n)$; then $A \subset E \subset Q \subset B^c$ and for every k,

$$Q_k \triangle Q \subset \big(\bigcup_{n \neq k} X_n\big) \cup X_k^c.$$

Denoting $\gamma = \{Q, Q^c\}$, we see that γ is a Borel partition separating A and B and for every n, $\gamma = \beta_n \bmod \mu_n$. In particular, $h_{\mu_n}(\gamma) = h_{\mu_n}(\beta_n) \le \epsilon$ and by linearity, $h_\mu(\gamma) \le \epsilon$. This however, contradicts the way ϵ was chosen and the proof is complete. □

19.20. Corollary. *Let (X, T) be a topological dynamical system. Notation as in Theorem 19.16,*

$$\bar{E}_X = \operatorname{cls} \bigcup \{\Lambda_\nu : \nu \in M_T^{\mathrm{erg}}(X),\ h_\nu > 0\}.$$

Proof. By Theorem 19.17 there exists $\mu \in M_T(X)$ with $E_\mu = E_X$. Theorems 19.11, 19.16.3 and 19.19 imply

$$\begin{aligned} \bar{E} = \bar{E}_\mu &= \operatorname{cls} \bigcup \{E_\nu : \nu \text{ ergodic},\ h_\nu > 0\} \\ &= \operatorname{cls} \bigcup \{\Lambda_\nu : \nu \text{ ergodic},\ h_\nu > 0\}. \end{aligned}$$

□

7. Measure entropy pairs and factors

In section 2 we have seen that if the image of a measure entropy pair under a homomorphism is non-degenerate then it is an entropy pair (Proposition 19.7). As noted there the converse is also true but this result is much harder to prove.

19.21. Theorem. *Let $\pi : (X, T) \to (Y, S)$ be a homomorphism of topological dynamical systems, μ a T-invariant measure on X, and ν its image under π. Then, for every ν-entropy pair $(a, b) \in E_\nu(Y, S)$ there exists a μ-entropy pair $(a', b') \in E_\mu(X, T)$ with $\pi(a') = a$ and $\pi(b') = b$.*

Proof. Assume first that $(X, \mathcal{X}, \mu, T)$ is ergodic. Identify $\mathcal{Y}$, the Borel σ-algebra on Y, with the σ-algebra $\pi^{-1}(\mathcal{Y})$ and set $\mathcal{Z} = \pi^{-1}(\Pi_\nu)$, where Π_ν is the Pinsker subalgebra of $\mathcal{Y}$ corresponding to ν. We then have $\mathcal{Z} = \mathcal{Y} \cap \Pi_\mu$ and from Theorem 1.2 in [**100**] we know that the factors $\mathcal{Y}$ and Π_μ are relatively independent over $\mathcal{Z}$ (this is the relative version of Theorem 18.16).

This means that for every function $f \in L^1(\mu)$ which is $\mathcal{Y}$-measurable one has $\mathbb{E}^{\mathcal{Z}}(f) = \mathbb{E}^{\Pi_\mu}(f)$. Thus for every $g \in L^1(\nu)$, $g \circ \pi$ is $\mathcal{Y}$-measurable and

$$\mathbb{E}^{\Pi_\mu}(g \circ \pi) = \mathbb{E}^{\mathcal{Z}}(g \circ \pi) = \mathbb{E}^{\pi^{-1}(\Pi_\nu)}(g \circ \pi) = \mathbb{E}^{\Pi_\nu}(g) \circ \pi\ .$$

Let now (a, b) be a ν-entropy pair, A a closed neighborhood of a, B a closed neighborhood of b with $A \cap B = \emptyset$, $A' = \pi^{-1}(A)$ and $B' = \pi^{-1}(B)$. Applying the preceding remark to the functions $g = \mathbf{1}_A$ and $g = \mathbf{1}_B$, one obtains:

$$\begin{aligned} \int \mathbb{E}^{\Pi_\mu}(\mathbf{1}_{A'})(x)\, \mathbb{E}^{\Pi_\mu}(\mathbf{1}_{B'})(x)\, d\mu(x) &= \int \mathbb{E}^{\Pi_\nu}(\mathbf{1}_A)(\pi(x))\, \mathbb{E}^{\Pi_\nu}(\mathbf{1}_B)(\pi(x))\, d\mu(x) \\ &= \int \mathbb{E}^{\Pi_\nu}(\mathbf{1}_A)(y)\, \mathbb{E}^{\Pi_\nu}(\mathbf{1}_B)(y)\, d\nu(y). \end{aligned}$$

As (a,b) is a ν-entropy pair, (A,B) satisfy the property $S(\nu)$ and it follows from Lemma 19.9 that the last integral and therefore also the first one, do not vanish. Applying (the other direction) of Lemma 19.9, we conclude that $h_\mu(\alpha) > 0$ for every partition $\alpha = \{D', D'^c\}$ of X such that $A' \subset D'$ and $B' \subset D'^c$.

By Proposition 19.6 there exists a μ-entropy pair $(a', b') \in E_\mu$ with $a' \in A'$ and $b' \in B'$, i.e. $\pi(a') \in A$ and $\pi(b') \in B$.

As this is true for all closed neighborhoods A and B of a and b respectively, and as $E_\mu \cup \Delta$ is closed, there exists a μ-entropy pair (a', b') with $\pi(a') = a$ and $\pi(b') = b$, and our proof is complete.

In the general case, let (a,b) be a ν-entropy pair, and A, B neighborhoods of a and b respectively. Let

$$\int_\Omega \mu_\omega \, dm(\omega)$$

be the ergodic decomposition of μ, and let ν_ω be the image of μ_ω under π ; for m-almost every ω, ν_ω appears in the ergodic decomposition of ν.

By Theorem 19.19.2, for an m-positive set of $\omega \in \Omega$ there exists a ν_ω-entropy pair (c,d) with $c \in A$ and $d \in B$. By the above, there exists a μ_ω-entropy pair (c', d') with $\pi(c') = c$ and $\pi(d') = d$, hence $\pi(c') \in A$ and $\pi(d') \in B$. By Theorem 19.19.1, for almost every ω, (c', d') is a μ-entropy pair. Since $E_\mu \cup \Delta$ is closed, there exists a μ-entropy pair (a', b') with $\pi(a') = a$ and $\pi(b') = b$. □

8. Topological Pinsker factors

Given a dynamical system (X,T) and a nonempty T-invariant subset $A \subset X \times X$ we denote by $\langle A \rangle$ the smallest invariant closed equivalence relation (icer) containing $A \cup A^{-1} \cup \Delta$ (with $A^{-1} = \{(x,x') : (x',x) \in A\}$). An explicit description of $\langle A \rangle$ can be obtained as follows. Set $A_0 = \mathrm{cls}\,(A \cup A^{-1} \cup \Delta)$, then $A_\omega = \mathrm{cls}\,(\bigcup_{n<\omega} A_n)$, where $A_n = A_0 \circ \cdots \circ A_0$ (n times) and for $B, C \subset X$,

$$B \circ C = \{(x,x'') : \exists\, x' \in X, \text{ with } (x,x') \in B,\ (x',x'') \in C\}.$$

Use transfinite induction to define A_α for every ordinal and let $\langle A \rangle = A_\eta$ where η is the first ordinal with $A_{\eta+1} = A_\eta$.

Let now (X,T) be a topological dynamical system, $\mu \in M_T(X)$ a T-invariant probability measure. Let $R(\mu) \subset X \times X$ be the icer $\langle \Lambda_\mu \rangle$ generated by Λ_μ. Considering the dynamical system (Y,T), where $Y = \mathrm{s}(\mu) \subset X$ as a subsystem of (X,T) we note that for a point $x \in X \setminus Y$ the $R(\mu)$ equivalence class of x is the singleton $\{x\}$, so that (with a slightly loose notation) we have

$$X/R(\mu) = Y/R(\mu) \cup (X \setminus Y).$$

We denote the quotient dynamical system $X/R(\mu)$ by $X_P(\mu)$ and call it the μ *topological Pinsker factor* of (X,T). Note that for μ with zero entropy we have $X_P(\mu) = X$.

Next define X_P, the *topological Pinsker factor* of (X,T), to be the quotient system $X_P(\mu) = X/R(\mu)$ for some measure $\mu \in M_T(X)$ with $E_X = E_\mu$. Such a measure always exists when $E_X \neq \emptyset$ (Theorem 19.17) and when $E_X = \emptyset$ we set $X_P = X$.

19.22. PROPOSITION. *The system X_P is well defined; i.e. it does not depend on the choice of the measure μ.*

PROOF. By theorem 19.16, $\Lambda_\mu = E_\mu \cup \mathrm{s}(\mu) = E_X \cup \mathrm{s}(\mu)$, hence $\langle \Lambda_\mu \rangle = \langle \Lambda_\mu \cup \Delta_X \rangle = \langle E_X \cup \Delta_X \rangle$ is independent of μ. □

19.23. THEOREM. *Let (X,T) be a topological dynamical system, $\mu \in M_T(X)$.*

1. *$X_P(\mu)$ is the largest topological factor with zero μ-entropy.*
2. *X_P is the largest topological factor with zero topological entropy.*

PROOF. 1. If $h_\nu(X_P(\mu)) > 0$ — where ν is the image of μ under the quotient map $\pi : X \to X_P(\mu)$ — then, by Proposition 19.6, $E_\nu \neq \emptyset$ and Theorem 19.21 implies that, given $(y,y') \in E_\nu$, $\pi(x,x') = (y,y')$ for some $(x,x') \in E_\mu$. Since by definition $\pi(x) = \pi(x')$, we have $(y,y') \in \Delta_{X_P(\mu)} \cap E_\nu$, a contradiction.

Suppose now that $\phi : (X,T) \to (Y,T)$ is a factor with $h_\theta = 0$, where $\theta = \phi(\mu)$. Denoting $R_\phi = \{(x,x') : \phi(x) = \phi(x')\}$ we observe that, by Proposition 19.7, $E_\mu \subset R_\phi$, whence $R_\pi \subset R_\phi$.

2. Choosing $\mu \in M_T(X)$ with $E_X = E_\mu$ (Theorem 19.17) use the fact that $X_P = X_P(\mu)$ and apply the first part. □

9. The entropy pairs of a product system

The characterization of measure entropy pairs and the fact that the Pinsker algebra of the product of two measure dynamical systems coincides with the product of the Pinsker algebras of the components (see Theorem 18.13), yield the theorems of this section.

In the sequel, given two dynamical systems (X_1,T) and (X_2,T), we shall often identify the two product spaces $(X_1 \times X_2) \times (X_1 \times X_2)$ and $(X_1 \times X_1) \times (X_2 \times X_2)$ via the canonical isomorphism: $((x_1,x_2),(x_1',x_2')) \mapsto ((x_1,x_1'),(x_2,x_2'))$.

19.24. THEOREM. *Let (X_1,T) and (X_2,T) be two topological systems, $\mu_i \in M_T(X_i)$, $(i=1,2)$ invariant probabilities. Then:*

1. $\lambda_{\mu_1\times\mu_2} = \lambda_{\mu_1} \times \lambda_{\mu_2}$, *whence* $\Lambda_{\mu_1\times\mu_2} = \Lambda_{\mu_1} \times \Lambda_{\mu_2}$.
2. $E_{\mu_1\times\mu_2} = (E_{\mu_1} \times E_{\mu_2}) \cup (E_{\mu_1} \times \mathrm{s}^2(\mu_2)) \cup (\mathrm{s}^2(\mu_1) \times E_{\mu_2})$.
3. $\bar{E}_{\mu_1\times\mu_2} = (\bar{E}_{\mu_1} \times \bar{E}_{\mu_2}) \cup (\bar{E}_{\mu_1} \times \mathrm{s}^2(\mu_2)) \cup (\mathrm{s}^2(\mu_1) \times \bar{E}_{\mu_2})$.
4. *When μ_1 and μ_2 are ergodic with positive entropy then*
$$\bar{E}_{\mu_1\times\mu_2} = \Lambda_{\mu_1\times\mu_2} = \bar{E}_{\mu_1} \times \bar{E}_{\mu_2} = \Lambda_{\mu_1} \times \Lambda_{\mu_2}.$$
5. $E_{X_1\times X_2} = (E_{X_1} \times E_{X_2}) \cup (E_{X_1} \times \Delta_{(X_2)_m}) \cup (\Delta_{(X_1)_m} \times E_{X_2})$.
6. *The product of two UPE systems is UPE.*

PROOF. 1. Let $(X_i,\mu_i,T) \xrightarrow{\pi_i} (Z_i,\nu_i,T)$ be the measure-theoretical Pinsker factors of (X_i,μ_i,T), and let $\mu_i = \int_{Z_i} (\mu_i)_z \, d\nu_i(z)$ be the disintegration of μ_i over (Z_i,ν_i), $i=1,2$. Set
$$\lambda_{\mu_i} = \int_{Z_i} (\mu_i)_z \times (\mu_i)_z \, d\nu_i(z),$$
the independent product of μ_i with itself over ν_i. Finally let $\Lambda_{\mu_i} = \mathrm{supp}\,(\lambda_{\mu_i})$ be the topological support of λ_{μ_i}. By Theorem 18.13, the Pinsker factor of the product system is given by
$$(X_1 \times X_2, \mu_1 \times \mu_2, T \times T) \xrightarrow{\pi_1\times\pi_2} (Z_1 \times Z_2, \nu_1 \times \nu_2, T \times T).$$

Since clearly the disintegration of $\mu_1 \times \mu_2$ over $Z_1 \times Z_2$ is given by

$$\mu_1 \times \mu_2 = \int\int_{Z_1 \times Z_2} (\mu_1)_{z_1} \times (\mu_2)_{z_2} \, d\nu_1(z_1) d\nu_2(z_2),$$

we have

$$\lambda = \int\int_{Z_1 \times Z_2} ((\mu_1)_{z_1} \times (\mu_2)_{z_2}) \times ((\mu_1)_{z_1} \times (\mu_2)_{z_2}) \, d\nu_1(z_1) d\nu_2(z_2).$$

Integration in the last formula first with respect to ν_1 then with respect to ν_2 yields, via the identification map $((x_1, x_2), (x_1', x_2')) \mapsto ((x_1, x_1'), (x_2, x_2'))$ of $(X_1 \times X_2) \times (X_1 \times X_2)$ with $(X_1 \times X_1) \times (X_2 \times X_2)$,

$$\lambda_{\mu_1 \times \mu_2} = \lambda_{\mu_1} \times \lambda_{\mu_2} \quad \text{whence} \quad \Lambda_{\mu_1 \times \mu_2} = \Lambda_{\mu_1} \times \Lambda_{\mu_2}.$$

2. Applying part 1 and Theorem 19.16 we get

$$\begin{aligned} E_{\mu_1 \times \mu_2} &= \Lambda_{\mu_1 \times \mu_2} \setminus \mathbf{s}^2(\mu_1 \times \mu_2) = (\Lambda_{\mu_1} \times \Lambda_{\mu_2}) \setminus \mathbf{s}^2(\mu_1 \times \mu_2) \\ &= \big((E_{\mu_1} \cup \mathbf{s}^2(\mu_1)) \times (E_{\mu_2} \cup \mathbf{s}^2(\mu_2))\big) \setminus \mathbf{s}^2(\mu_1 \times \mu_2) \\ &= (E_{\mu_1} \times E_{\mu_2}) \cup (E_{\mu_1} \times \mathbf{s}^2(\mu_2)) \cup (\mathbf{s}^2(\mu_1) \times E_{\mu_2}). \end{aligned}$$

3. Take closure on both sides of the equality in part 2.

4. By part 3 and Theorem 19.16

$$\begin{aligned} \bar{E}_{\mu_1 \times \mu_2} &= \bar{E}_{\mu_1} \times \bar{E}_{\mu_2} \cup \bar{E}_{\mu_1} \times \mathbf{s}^2(\mu_2) \cup \mathbf{s}^2(\mu_1) \times \bar{E}_{\mu_2} \\ &= \bar{E}_{\mu_1} \times \bar{E}_{\mu_2} = \Lambda_{\mu_1} \times \Lambda_{\mu_2} = \Lambda_{\mu_1 \times \mu_2}. \end{aligned}$$

5. By Theorem 19.17 we can choose $\mu_i \in M_{X_i}$ with $E_{\mu_i} = E_{X_i}$ as well as $\mathbf{s}(\mu_i) = (X_i)_m$, $(i = 1, 2)$ (or just the second condition if μ_i has zero entropy). Now

$$\begin{aligned} &(E_{X_1} \cup \Delta_{(X_1)_m}) \times (E_{X_2} \cup \Delta_{(X_2)_m}) \setminus \Delta_{X_1 \times X_2} \\ &= (E_{X_1} \times E_{X_2}) \cup (E_{X_1} \times \Delta_{(X_2)_m}) \cup (\Delta_{(X_1)_m} \times E_{X_2}) \\ &= (E_{\mu_1} \times E_{\mu_2}) \cup (E_{\mu_1} \times \mathbf{s}^2(\mu_2)) \cup (\mathbf{s}^2(\mu_1) \times E_{\mu_2}) \\ &E_{\mu_1 \times \mu_2} \subset E_{X_1 \times X_2} \\ &\subset (E_{X_1} \cup \Delta_{(X_1)_m}) \times (E_{X_2} \cup \Delta_{(X_2)_m}) \setminus \Delta_{X_1 \times X_2}. \end{aligned}$$

6. This clearly follows from part 5. □

Our next goal is to determine the μ-Pinsker (Pinsker) factor of a product system. The straightforward proof of the following lemma is left to the reader.

19.25. LEMMA. *Let (X_i, T), $i = 1, 2$ be two dynamical systems and $A \subset X_1 \times X_1$, $B \subset X_2 \times X_2$ symmetric subsets containing the diagonal Δ, then with the identification $((x_1, x_2), (x_1', x_2')) \leftrightarrow ((x_1, x_1'), (x_2, x_2'))$ of $(X_1 \times X_2) \times (X_1 \times X_2)$ with $(X_1 \times X_1) \times (X_2 \times X_2)$ we have*

$$\langle A \times B \rangle = \langle A \rangle \times \langle B \rangle.$$

19.26. THEOREM. *Let (X_1, T) and (X_2, T) be two topological systems, $\mu_i \in M_T(X_i)$, $i = 1, 2$ invariant probability measures. Then denoting, $\mathbf{s}(\mu_i) = Y_i$, $i = 1, 2$ we have*

1.

$$(X_1 \times X_2)_P(\mu_1 \times \mu_2) \\ = \big((Y_1)_P(\mu_1) \times (Y_2)_P(\mu_2)\big) \cup \big((X_1 \times X_2) \setminus (Y_1 \times Y_2)\big).$$

In particular

$$(X_1 \times X_2)_P(\mu_1 \times \mu_2) = (X_1)_P(\mu_1) \times (X_2)_P(\mu_2)$$

when $X_1 = \mathbf{s}(\mu_1)$ *and* $X_2 = \mathbf{s}(\mu_2)$.

2.

$$(X_1 \times X_2)_P = \big((X_1)_P \times (X_2)_P\big) \cup \big((X_1 \times X_2) \setminus ((X_1)_m \times (X_2)_m)\big)$$

and

$$(X_1 \times X_2)_P = (X_1)_P \times (X_2)_P$$

when $X_1 = (X_1)_m$ *and* $X_2 = (X_2)_m$.

PROOF. 1. Apply Lemma 19.25 to the relations $A = \Lambda_{\mu_1}$ and $B = \Lambda_{\mu_2}$. Note however that when $\mathbf{s}(\mu_i) \neq X_i$, $i = 1, 2$,

$$(A \times B) \cup \Delta_{X_1 \times X_2} \subsetneq (A \cup \Delta_{X_1}) \times (B \cup \Delta_{X_2}).$$

This means that in order to apply Lemma 19.25 we have to work with the subsystems Y_1, Y_2 and $Y_1 \times Y_2$, and then extend the relations to $X_1 \times X_2$ in the trivial way.

2. As in part 1 this follows from Lemma 19.25 applied to the relations $A = \Lambda_{\mu_1}$ and $B = \Lambda_{\mu_2}$ for an appropriate choice of measures $\mu_i \in M_T(X_i)$ (see Proposition 19.22). □

10. An application to the proximal relation

19.27. THEOREM. *Let (X, T) be a topological dynamical system, P the proximal relation on X. Then:*

1. *For every T-invariant ergodic measure μ of positive entropy the dynamical system $(\bar{E}_\mu, T \times T)$ is topologically transitive.*
2. *For every T-invariant ergodic measure μ of positive entropy the set $P \cap E_\mu$ is residual in the G_δ set E_μ of μ entropy pairs.*
3. *When $E_X \neq \emptyset$ the set $P \cap E_X$ is residual in the G_δ set E_X of topological entropy pairs.*

PROOF. 1. Notation is as in Section 4. By Corollary 18.20 the extension $\pi : (X, \mathcal{X}, \mu, T) \to (Z, \mathcal{Z}, \eta, T)$, where $(Z, \mathcal{Z}, \eta, T)$ is the measure-theoretical Pinsker factor of the system $(X, \mathcal{X}, \mu, T)$, is weakly mixing and therefore the measure λ is ergodic for $T \times T$ and satisfies $\lambda(\Delta_X) = 0$. This implies that the topological dynamical system $(\Lambda, T \times T)$ is topologically transitive.

2. Since $\bar{E}_\mu \cap \Delta = \Lambda_\mu \cap \Delta = \mathbf{s}^2(\mu) \neq \emptyset$, it follows that $P \setminus \Delta$ is dense in E_μ. Since P is always a G_δ subset of $X \times X$ (Exercise 1.9) and since E_μ is a G_δ subset as well, our claim follows.

3. By Theorem 19.19, $\bigcup\{E_\mu : \mu \text{ ergodic}\}$ is dense in E_X. Thus we deduce from part 2 that $P \cap E_X$ is dense in E_X, and the proof is concluded as in part 2. □

11. Notes

As was explained in the introduction to this chapter, my main sources are the papers [**25**], [**26**], [**98**], [**28**], [**27**] and [**87**].

CHAPTER 20

Krieger's and Ornstein's Theorems

In the last chapter we put together the combinatorial machinery (that was built up in the second part of the book) and the theory of joinings (developed in the first part) to achieve a proof of the deep theorems of Krieger, Sinai and Ornstein. The idea of using joinings and the Baire category theorem to smooth and streamline the original proofs of these theorems is due to R. Burton and A. Rothstein. An outline of this exists in the form of a preliminary report ([**39**]) that was never published. This approach was then presented, in a complete and coherent way, in Rudolph's book [**220**]. (See also the more recent works of Serafin [**225**] and Burton, Keane and Serafin [**38**].) We shall follow Rudolph's exposition closely; hopefully in a more leisurely and detailed manner.

Although this was not clear at the beginning, it turns out that Krieger's theorem requires for its proof almost the same tools that were used by Ornstein in proving his theorem. These tools are developed in Section 1, and then applied in Section 2 to prove Krieger's finite generators theorem. Ornstein's key notion of "finitely determined process", which is defined with the aid of the $\bar{d}$ metric, is introduced in Section 3 and it is then shown that two finitely determined processes with the same entropy are isomorphic. In Section 4 we show that Bernoulli processes are finitely determined. Sinai's factor theorem — which again completely fits into the framework of Ornstein's proof — and finally Ornstein's isomorphism theorem are proven in the last section (Section 5).

As explained in [**220**] the proof of Ornstein's isomorphism theorem given here applies only to systems with finite entropy. The corresponding statement for Bernoulli systems with infinite entropy holds as well and we refer to [**192**]. The book book [**220**] presents also a discussion of "weakly Bernoulli processes" which leads to Ornstein's proof of the fact that such processes are finitely determined, and its corollary that mixing Markov systems with the same entropy are isomorphic. Finally we refer to J.-P. Thouvenot's fundamental work [**244**] which provides an extension of the isomorphism theorem to the relative case.

1. Ornstein's fundamental lemma

The following lemma is the combinatorial key to Ornstein's proof. Given two non-periodic ergodic processes $(\mathbf{X}, \mu, \alpha)$ and $(\mathbf{Y}, \nu, \beta)$ with $h_\mu(\alpha) > h_\nu(\beta)$ and an ergodic joining λ, the ergodic theorem and the SMB theorem are used to construct, via a marriage lemma (which is implicit in the proof), a 1-1 matching ϕ of good β names with good α names in such a way that each pair $(\phi(\mathbf{b}), \mathbf{b})$ is a λ-good $\alpha \times \beta$ name. Here $\alpha \times \beta$ is the partition $\{A \times B : A \in \alpha, B \in \beta\}$.

20.1. LEMMA. *Let $(\mathbf{X}, \alpha)$ and $(\mathbf{Y}, \beta)$ be ergodic non-periodic processes with $h_\mu(\alpha) > h_\nu(\beta)$ and let $\lambda \in J_e(\mu, \nu)$. Given $\epsilon > 0$ there exist an N and a tower*

$\mathfrak{c} = \{C, TC, \dots, T^{N-1}C\}$ in $\mathbf{Y}$ with $\nu(|\mathfrak{c}|) > 1 - \frac{1}{2}\epsilon$ such that, denoting by I and J the collections of (N, ϵ)-generic $(\mathbf{X}, \alpha)$ and $(\mathbf{Y}, \beta)$ names respectively, there exists a subset $J_0 \subset J$ and a 1-1 function $\phi : J_0 \to I$ with the following properties:

1. *for every $\mathbf{b} \in J_0$ the double name $(\phi(\mathbf{b}), \mathbf{b})$ is (N, ϵ)-generic for $(X \times Y, \alpha \times \beta, \lambda)$,*
2. *$\nu(\cup_{\mathbf{b} \in J_0} C_{\mathbf{b}}) > (1 - \epsilon)\nu(C)$.*

PROOF. Choose $0 < \theta \leq 1$ with $h_\mu(\alpha) - h_\nu(\beta) > \theta$. By Theorems 15.16 and 15.18 there exists N_0 such that for every $N > N_0$, *any* tower $\mathfrak{t} = \{B, (T \times T)B, \dots, (T \times T)^{N-1}B\}$ with $\lambda(|\mathfrak{t}|) > 1 - \frac{1}{2}\epsilon\theta$ satisfies the properties:

1. $\lambda(\cup\{B_{\mathbf{c}} : \mathbf{c}$ is $\frac{1}{2}\epsilon\theta$ generic for $\alpha \times \beta\}) > (1 - \frac{1}{2}\epsilon\theta)\lambda(B)$.
2. The set of points $(x, y) \in B$ such that
   (a) $\left|-\frac{1}{N}\log(\lambda(B_{\mathbf{a}_N(x)})/\lambda(B)) - h_\mu(\alpha)\right| < \frac{1}{2}\epsilon\theta$ and
   (b) $\left|-\frac{1}{N}\log(\lambda(B_{\mathbf{b}_N(y)})/\lambda(B)) - h_\nu(\beta)\right| < \frac{1}{2}\epsilon\theta$,
   has λ measure $> (1 - \frac{1}{2}\epsilon\theta)\lambda(B)$.

Next apply Rohlin's lemma to construct a tower $\mathfrak{c} = \{C, TC, \dots, T^{N-1}C\}$ in $\mathbf{Y}$ with $N > N_0$ and $\nu(|\mathfrak{c}|) > 1 - \frac{1}{2}\epsilon\theta$. Set $B = X \times C$ and $\mathfrak{t} = \{B, (T \times T)B, \dots, (T \times T)^{N-1}B\}$. Thus $\lambda(|\mathfrak{t}|) > 1 - \frac{1}{2}\epsilon\theta$ and $\lambda(B_0) > (1 - \frac{1}{2}\epsilon\theta)\lambda(B)$, where B_0 is the set of points $(x, y) \in B$ whose $N, \alpha \times \beta$ name $\mathbf{c}$ is $\frac{1}{2}\epsilon\theta$ generic and for which (a) and (b) above hold.

Set

$$I = \{\mathbf{a} = \mathbf{a}_N(x) : (x, y) \in B_0\} \quad \text{and}$$
$$J = \{\mathbf{b} = \mathbf{b}_N(y) : (x, y) \in B_0\}.$$

We now have, for any $\mathbf{a} \in I$ and $\mathbf{b} \in J$, by 2

$$\frac{\lambda(B_{\mathbf{a}})}{\lambda(B)} < e^{-N(h_\mu(\alpha) - \frac{1}{2}\epsilon\theta)} \quad \text{and}$$
$$\frac{\lambda(B_{\mathbf{b}})}{\lambda(B)} > e^{-N(h_\nu(\beta) + \frac{1}{2}\epsilon\theta)}.$$

Hence for sufficiently large N

$$\frac{\lambda(B_{\mathbf{a}})}{\lambda(B_{\mathbf{b}})} < e^{-N(h_\mu(\mu) - h_\nu(\beta))} e^{-N\epsilon\theta} < e^{-N\theta(1+\epsilon)} < \epsilon/2. \tag{1.1}$$

We say that the names $\mathbf{a} \in I$ and $\mathbf{b} \in J$ are *λ-related* if $\mathbf{a} = \mathbf{a}_N(x)$ and $\mathbf{b} = \mathbf{b}_N(y)$ for some $(x, y) \in B_0$; let $\Lambda \subset I \times J$ be the corresponding relation. We would like to match names in J in a 1-1 way with λ-related names in I. For this purpose let Φ be the collection of 1-1 maps $\phi : J_0 \to I$ with $J_0 \subset J$ and $\phi(\mathbf{b})$ λ-related to $\mathbf{b}$ for every $\mathbf{b} \in J_0$. On Φ define the natural partial order: $\phi \prec \psi$ when ψ is an extension of ϕ. We now fix a maximal element $\phi : J_0 \to I$ in Φ and our task is to show that this maximal ϕ satisfies the claims of the lemma. Clearly claim 1. is satisfied and we now proceed to prove claim 2.

Observe that

$$\text{range}(\phi) := I_0 \supset \cup\{\Lambda(\mathbf{b}) : \mathbf{b} \notin J_0\},$$

for otherwise there would be $(\mathbf{a}, \mathbf{b}) \in \Lambda$ with $\mathbf{b} \notin J_0$, $\mathbf{a} \notin I_0$, and by defining $\phi(\mathbf{b}) = \mathbf{a}$ we shall contradict the maximality of ϕ.

It therefore follows from (1.1) that

$$\begin{aligned}\lambda(\{(x,y)\in B_0 : \mathbf{b}_N(y)\notin J_0\}) &\le \lambda(\{(x,y)\in B_0 : \mathbf{a}_N(x)\in I_0\})\\ &\le \sum_{\mathbf{a}\in I_0}\lambda(B_{\mathbf{a}})\\ &< \frac{\epsilon}{2}\sum_{\mathbf{b}\in J_0}\lambda(B_{\mathbf{b}}) \le \frac{\epsilon}{2}\lambda(B) = \frac{\epsilon}{2}\nu(C).\end{aligned}$$

Since $B = X\times C$ we now have

$$\begin{aligned}\nu\big(\cup_{\mathbf{b}\in J_0} C_{\mathbf{b}}\big) &= \lambda\big(\cup_{\mathbf{b}\in J_0} B_{\mathbf{b}}\big)\\ &\ge \lambda(B_0) - \lambda\big(\{(x,y)\in B_0 : \mathbf{b}_N(y)\notin J_0\}\big)\\ &\ge \lambda(B_0) - \frac{\epsilon}{2}\mu(C) > (1-\epsilon)\nu(C).\end{aligned}$$

□

We need one more lemma that will enable us to introduce unmistakable "markers".

20.2. LEMMA. *Let $(\mathbf{X},\alpha)$ be ergodic non-periodic process and $\epsilon>0$. There exist an N and a partition β with $d_{\text{part}}(\alpha,\beta)<\epsilon$ and such that*

$$\mu\big(\{x\in X : \beta(T^j x) = 1,\quad \forall\, 0\le j<N\}\big) = 0.$$

PROOF. Since $\mathbf{X}$ is non-periodic μ has no atoms, hence $\lim_{N\to\infty}\mu\big(\{x\in X : \alpha(T^j x) = 1,\ \forall\, 0\le j<N\}\big) = 0$. Choose an N such that $\mu\big(\{x\in X : \alpha(T^j x) = 1,\ \forall\, 0\le j<N\}\big)<\epsilon$ and set

$$\beta(x) = \begin{cases} 2 & \text{if } \alpha(T^j x) = 1,\ \forall\, 0\le j<N,\\ \alpha(x) & \text{otherwise.}\end{cases}$$

□

Given two partitions α and β of a system $\mathbf{X}$ and $\epsilon>0$, we write $\alpha \overset{\epsilon}{\subset} \beta$ when for every $A\in\alpha$ there exists $B\in\beta$ with $\mu(A\,\triangle\,B)<\epsilon$. Similarly, the notation $\alpha\overset{\epsilon}{\subset}\mathcal{F}$ is used for a partition α and an algebra $\mathcal{F}\subset\mathcal{X}$.

20.3. LEMMA (Ornstein's fundamental lemma). *Let $\mathbf{X}$ and $\mathbf{Y}$ be ergodic non-periodic systems, and let $\lambda\in J_e(\mu,\nu)$.*

1. *If $(\mathbf{X},\alpha)$ and $(\mathbf{Y},\beta)$ are processes with $h_\mu(\alpha)>h_\nu(\beta)$ and $\epsilon>0$, then there exists a partition γ of Y such that for the symbolic representations $\rho_1 = \rho(X\times Y,\lambda,\alpha\times\beta)$ and $\rho_2 = \rho(Y,\nu,\beta\vee\gamma)$,*
   (a) *$d_{\text{meas}}(\rho_1,\rho_2)<\epsilon$, and*
   (b) *$\beta\overset{\epsilon}{\subset}\gamma^T$.*
2. *If $(\mathbf{X},\alpha)$ and $(\mathbf{Y},\beta)$ are processes with $h_\mu(\alpha)>h_\nu$ and $\epsilon>0$, then there exists an ergodic non-periodic system $\mathbf{Z}=(Z,\mathcal{Z},\eta)$, a partition ζ of Z and an ergodic joining $\hat\lambda\in J_e(\eta,\nu)$ such that, for the symbolic representations $\rho_1=\rho(X\times Y,\lambda,\alpha\times\beta)$ and $\rho_2=\rho(Z\times Y,\hat\lambda,\zeta\times\beta)$, the following properties hold*
   (a) $h_\eta(\zeta)>h_\nu$

(b) $d_{\text{meas}}(\rho_1, \rho_2) < \epsilon$,
(c) $\zeta \times Y \overset{\epsilon}{\subset} Z \times \mathcal{Y} \quad (\text{mod } \hat{\lambda})$, *and*
(d) $Z \times \beta \overset{\epsilon}{\subset} \mathcal{Z} \times Y \quad (\text{mod } \hat{\lambda})$.

In fact we have

$$\mathbf{Z} = (\Omega(\ell), \eta, S) \quad \textit{and} \quad \zeta = P_0,$$

where $\ell = \text{card}\, \alpha$.

PROOF. 1. Our first step is to apply Lemma 20.2 and replace α by a partition α' such that $d_{\text{part}}(\alpha', \alpha) < \delta$ and such that for some positive integer K

$$(*) \qquad \mu(\{x \in X : \alpha'(T^j x) = 1 \;\forall\, 0 \le j < K\}) = 0.$$

If δ is sufficiently small we get by Lemma 15.8

$$d_{\text{meas}}(\rho(X \times Y, \lambda, \alpha \times \beta), \rho(X \times Y, \lambda, \alpha' \times \beta)) < \epsilon$$

and by Lemma 15.9, that $h_\mu(\alpha') > h_\nu(\beta)$. Thus with no loss of generality we now assume that $(*)$ holds for α and in particular the string $211\ldots112$ will never occur in (N, α)-names.

Let N be a number provided by Lemma 20.1 and as in this lemma, construct a tower in Y of height $N+K+2$, $\mathfrak{c} = \{C, TC, \ldots, T^{N+K+1}C\}$, with $\nu(|\mathfrak{c}|) > 1 - \epsilon$. As in the proof of Lemma 20.1 let $B = X \times C$ and let $\mathfrak{t} = \{B, (T \times T)B, \ldots, (T \times T)^{N+K+1}B\}$ be the corresponding tower in $X \times Y$. Ignoring for the time being the top $K+2$ levels we now refine $\mathfrak{t}$ according to the various (N, β) names. Thus $B_{\mathbf{b}}$ for an (N, β) name $\mathbf{b}$ consists of those points $(x, y) \in B$ such that $\mathbf{b}_N(y) = \mathbf{b}$.

Denoting by I and J the collections of (N, ϵ)-generic $(\mathbf{X}, \alpha)$ and $(\mathbf{Y}, \beta)$ names respectively, we now have, by Lemma 20.1, a subset $J_0 \subset J$ and a 1-1 function $\phi : J_0 \to I$ such that

1. for every $\mathbf{b} \in J_0$ the double name $(\phi(\mathbf{b}), \mathbf{b})$ is (N, ϵ)-generic for $(X \times Y, \alpha \times \beta, \lambda)$,
2. $\nu(\bigcup_{\mathbf{b} \in J_0} C_{\mathbf{b}}) > (1 - \epsilon)\nu(C)$.

We are now ready to define the partition γ. We do this by copying over the tower $\mathfrak{c}$ as follows: for $\mathbf{b} \in J_0$ we copy the name $\phi(\mathbf{b})$ over the first N levels of the column over $C_{\mathbf{b}}$; on the last $K+2$ levels of each column we copy the string $211\cdots112$. The remaining columns and whatever is left of Y we copy arbitrarily. This defines a partition γ of Y which we claim satisfies our requirements.

To verify 1.(a) observe that, in fact, by copying the γ name $\phi(\mathbf{b})211\cdots112$ on the column over $C_{\mathbf{b}}$ we obtain also the partition $\beta \vee \gamma$ of Y. Since $(\phi(\mathbf{b}), \mathbf{b})$ is (N, ϵ)-generic for $(X \times Y, \alpha \times \beta, \lambda)$ we conclude by 15.15 that $d_{\text{meas}}(\rho_1, \rho_2) < \epsilon$ as required.

To verify 1.(b) we shall show that there exists a partition β', measurable with respect to γ_{-N+1}^{N+K+1} (hence of course γ^T-measurable) which agrees with the partition β on the subset

$$A = \bigcup_{\mathbf{b} \in J_0} \bigcup_{j=0}^{N-1} T^j C_{\mathbf{b}}.$$

Now for $y \in T^j(C_{\mathbf{b}}) \subset A$, in the γ-name

$$y_0^{N+K+1} = \gamma(y)\gamma(Ty)\cdots\gamma(T^{N+K+1}y),$$

the first occurrence of the string $211\cdots 112$ will be precisely at

$$\gamma(T^{N-j}y)\cdots\gamma(T^{N+K+1-j}y)$$

and the γ name

$$\mathbf{c} = y_{-j}^{N-1-j} = \gamma(T^{-j}y)\gamma(T^{1-j}y)\cdots\gamma(T^{N-1-j}y)$$

will satisfy $\mathbf{c} = \phi(\mathbf{b})$. Thus, denoting by $\mathbf{c}_N(z)$ the (N,γ)-name of a point $z \in Y$, we now define the partition β' on A by

$$\beta'(y) = [\phi^{-1}(\mathbf{c}_N(T^{-j}y))]_j, \qquad \text{for } y \in T^j(C_{\mathbf{b}}),\ \mathbf{b} \in J_0$$

and arbitrarily on $Y \setminus A$. Clearly then β' is γ_{-N+1}^{N+K+1}-measurable and β' and β agree on A. Since, by a simple calculation, $\nu(A) > \frac{N}{N+K+2}(1-2\epsilon)$, it follows that for sufficiently large N $d_{\text{part}}(\beta',\beta) < 2\epsilon$, hence $\beta \overset{2\epsilon}{\subset} \gamma^T$. This concludes the proof of part 1 of the lemma.

2. The condition $h_\mu(\alpha) > h_\nu$ (rather than $h_\mu(\alpha) > h_\nu(\beta)$), provides some flexibility in the choice of the partition β. In fact we note that when the conditions 2.(b) and (d) are satisfied for a refinement of β it is a fortiori true for β. We can therefore assume that the partition β satisfies

$$0 \le h_\nu - h_\nu(\beta) < \theta\epsilon/2,$$

with $0 < \theta < \min(\log(\ell) - h_\nu, 1)$. Now apply part 1 of the lemma to obtain a partition γ of Y such that

$$d_{\text{meas}}(\rho(X \times Y, \lambda, \alpha \times \beta), \rho(Y, \nu, \gamma \vee \beta)) < \epsilon_1, \quad \text{and} \tag{1.2}$$

$$\beta \overset{\epsilon_1}{\subset} \gamma^T, \tag{1.3}$$

where $\epsilon_1 < \epsilon/2$.

As $\beta \overset{\epsilon_1}{\subset} \gamma^T$, there exists an N so that $\beta \overset{\epsilon_1}{\subset} \gamma_{-N}^N$. Next set $\rho = \rho(\mathbf{Y},\gamma)$ and apply Theorem 15.22 to find an ergodic $\eta \in M_S(\Omega(\ell))$ such that

$$h_\eta \ge h_\rho + \epsilon(\log(\ell) - h_\rho), \quad \text{and} \tag{1.4}$$

$$\bar{d}^N(\rho,\eta) < \epsilon. \tag{1.5}$$

We are now ready to define the process $(\mathbf{Z},\zeta)$ and the joining $\hat{\lambda}$. We set

$$\mathbf{Z} = (\Omega(\ell), \eta, S) \quad \text{and} \quad \zeta = P_0.$$

The joining $\hat{\lambda}$ will be any ergodic joining that achieves the $\bar{d}^N$ distance $\bar{d}^N(\rho,\eta)$. We proceed to check that these objects satisfy the requirements of the lemma.

Regarding 2.(a), we have $\log(\ell) \ge h_\mu(\alpha) > h_\nu \ge h_\rho = h_\nu(\gamma)$ and we observe that Lemma 15.9.5 and (1.3) imply that for sufficiently small ϵ_1, $|h_\nu(\gamma) - h_\nu(\beta)| < \theta\epsilon/2$, hence

$$\begin{aligned} h_\eta(\zeta) = h_\eta &\ge h_\rho + \epsilon(\log(\ell) - h_\rho) \\ &\ge h_\rho + \epsilon(\log(\ell) - h_\nu) > h_\rho + \epsilon\theta \\ &\ge h_\rho + h_\nu - h_\nu(\beta) + \epsilon\theta/2 > h_\nu. \end{aligned}$$

To get 2.(b) define a joining $\tilde{\lambda} \in J(\hat{\lambda},\nu)$, a measure on $(Z \times Y) \times Y$, by

$$\tilde{\lambda}((D \times B) \times C) = \hat{\lambda}(D \times (B \cap C)), \quad D \in \mathcal{Z}, B, C \in \mathcal{Y}.$$

A calculation then shows that

$$\tilde{\lambda}((\zeta \times \beta \times Y)_{-N}^{N} \triangle (Z \times Y \times (\gamma \vee \beta)_{-N}^{N})) = \hat{\lambda}((\zeta \times \beta)_{-N}^{N} \triangle (Z \times (\gamma \vee \beta))_{-N}^{N}).$$

Now Lemma 15.8.3 and (1.5) show that

$$\begin{aligned}\hat{\lambda}((\zeta \times \beta)_{-N}^{N} \triangle (Z \times (\gamma \vee \beta))_{-N}^{N}) &= \hat{\lambda}((\zeta \times Y)_{-N}^{N} \triangle (Z \times \gamma)_{-N}^{N}) \\ &= d_{\text{part}}^{\hat{\lambda}}((\zeta \times Y)_{-N}^{N}, (Z \times \gamma)_{-N}^{N}) \\ &= \bar{d}^{N}(\rho, \eta) < \epsilon/2.\end{aligned}$$

Thus

$$\bar{d}^{N}(\rho(Z \times Y, \hat{\lambda}, \zeta \times \beta), \rho(Y, \nu, \gamma \vee \beta) \leq \tilde{\lambda}((\zeta \times \beta \times Y)_{-N}^{N} \triangle (Z \times Y \times (\gamma \vee \beta))_{-N}^{N})) < \epsilon/2.$$

Now Proposition 15.21 implies

$$d_{\text{meas}}(\rho(Z \times Y, \hat{\lambda}, \zeta \times \beta), \rho(Y, \nu, \gamma \vee \beta)) < \epsilon/2,$$

and (1.2) implies

$$d_{\text{meas}}(\rho(X \times Y, \lambda, \alpha \times \beta), \rho(Z \times Y, \hat{\lambda}, \zeta \times \beta)) < \epsilon.$$

This proves 2.(b).

For 2.(c) we only have to observe that by (1.5)

$$\hat{\lambda}(P_0 \times Y \triangle Z \times \gamma) \leq \bar{d}^{N}(\rho, \eta) < \epsilon,$$

hence $\zeta \times Y \overset{\epsilon}{\subset} Z \times \mathcal{Y} \pmod{\hat{\lambda}}$.

It only remains to check 2.(d). Now $\beta \overset{\epsilon_1}{\subset} \gamma_{-N}^{N}$ means that there exists a partition γ_0 which is γ_{-N}^{N}-measurable and satisfies

$$\nu(\beta \triangle \gamma_0) < \epsilon_1. \tag{1.6}$$

Since by (1.5)

$$\hat{\lambda}(P_{-N}^{N} \times Y \triangle Z \times \gamma_{-N}^{N}) = \bar{d}^{N}(\rho, \eta) < \epsilon,$$

we can — by imitating the way γ_0 is built up from elements of γ_{-N}^{N} — define a partition ζ_0 of Z which is P_{-N}^{N}-measurable. Clearly then $\hat{\lambda}(\zeta_0 \times Y \triangle Z \times \gamma_0) < \epsilon$, and with (1.6) we get

$$\hat{\lambda}(\zeta_0 \times Y \triangle Z \times \beta) < 2\epsilon,$$

whence 2.(d). □

2. Krieger's finite generator theorem

20.4. THEOREM (Krieger's finite generator theorem). *If* $\mathbf{Y}$ *is an ergodic system with* $h(\mathbf{Y}) < \log(\ell)$ *then there exists a faithful symbolic representation* $\phi : \mathbf{Y} \to \Omega(\ell)$.

PROOF. Since the periodic case is trivial we shall assume that $\mathbf{Y}$ is not periodic. Fix a Cantor topology on Y with a sequence $\beta_n = \{B_1^n, B_2^n, \ldots, B_{k_n}^n\}$ of nesting partitions consisting of clopen sets that together form a basis for the topology of Y and with $T : Y \to Y$ a homeomorphism (Theorem 2.15). Let $\Omega = \Omega(\ell)$, then the product space $\Omega \times Y$ is a compact metric space with $\cup_n P_n \times \beta_n$ a basis for its topology consisting of clopen sets. Set

$$\mathfrak{J} = \text{cls}\,\{\lambda \in M_{S \times T}(\Omega \times Y) : \lambda \in J_e(\pi_\Omega(\lambda), \nu) \text{ and } h(\pi_\Omega(\lambda)) > h_\nu\}.$$

The topology on $M_{S\times T}(\Omega\times Y)$ is of course the weak* topology and is given by the metric

$$d_{\text{meas}}(\lambda,\lambda')=\sum_{n=1}^{\infty}2^{-n}d_n(\lambda,\lambda'), \tag{2.1}$$

where

$$d_n(\lambda,\lambda')=\frac{1}{2}\sum_{A\times B\in P_n\times\beta_n}|\lambda(A\times B)-\lambda'(A\times B)|. \tag{2.2}$$

$\mathfrak{J}$ as a closed subset of $M_{S\times T}(\Omega\times Y)$ is a compact metric space with respect to the weak* topology. It is non-empty as $\eta_0\times\mu\in\mathfrak{J}$ (where $\eta_0\in M_S(\Omega(\ell))$ is the Bernoulli measure $\eta_0=(\ell^{-1},\ell^{-1},\dots,\ell^{-1})^{\mathbb{Z}}$ with $h_{\eta_0}=\log(\ell)>h_\nu$). Also note that by Lemma 15.1, $h(\pi_\Omega(\lambda))\geq h_\nu$ for every $\lambda\in\mathfrak{J}$.

The idea of the proof is to show that the set of $\lambda\in\mathfrak{J}$ with

$$\mathcal{B}\times Y=\Omega\times\mathcal{Y}\qquad(\text{mod }\lambda)$$

forms a dense G_δ subset of $\mathfrak{J}$ which, in view of Theorem 6.5, will establish our theorem. For that purpose it is enough, by the Baire category theorem, to show that for each n the set V_n of all $\lambda\in\mathfrak{J}$ with

$$P_0\times Y\overset{1/n}{\subset}\Omega\times\mathcal{Y}\quad(\text{mod }\lambda)\quad\text{and}$$

$$\Omega\times\beta_n\overset{1/n}{\subset}\mathcal{B}\times Y\quad(\text{mod }\lambda),$$

is an open and dense subset of $\mathfrak{J}$. Note that $V_m\subset V_n$ for $m\geq n$.

The condition $\lambda\in V_n$ can be expressed by asserting that for some N there are β_N-measurable partition P_0' and P_{-N}^N-measurable partition β_n' with

$$\lambda(P_0\times Y\,\triangle\,\Omega\times P_0')<1/n\quad\text{and}$$

$$\lambda(\Omega\times\beta_n\,\triangle\,\beta_n'\times X)<1/n.$$

Since all these sets are clopen this clearly implies that V_n is weak* open.

In order to show that V_n is dense pick any ergodic $\lambda\in\mathfrak{J}$ with $h(\pi_\Omega(\lambda))>h_\nu$ (such measures are dense in $\mathfrak{J}$) and $0<\epsilon<1/n$. In order to get a $\hat\lambda\in V_n$ with $d_{\text{meas}}(\lambda,\hat\lambda)<\epsilon$ it is clearly enough to find $\hat\lambda\in V_n$ that satisfies $d_{\text{meas}}(\rho_1,\rho_2)<\epsilon/2$, where $\rho_1=\rho(\Omega\times Y,\lambda,P_0\times\beta_m)$ and $\rho_2=\rho(\Omega\times Y,\hat\lambda,P_0\times\beta_m)$ for a sufficiently large m. Fix such an $m\geq n$.

Now apply Lemma 20.3.2 (with $(\Omega,\mathcal{B},\pi_\Omega,S,P_0)$ as $(\mathbf{X},\alpha)$) to get an ergodic non-periodic measure $\eta\in M_S(\Omega(\ell))$ and an ergodic joining $\hat\lambda\in J_e(\eta,\nu)$ such that, for the symbolic representations $\rho_1=\rho(\Omega\times Y,\lambda,P_0\times\beta_m)$ and $\rho_2=\rho(\Omega\times Y,\hat\lambda,P_0\times\beta_m)$, the following properties hold

1. $h_\eta>h_\nu$,
2. $d_{\text{meas}}(\rho_1,\rho_2)<\epsilon/2$,
3. $P_0\times Y\overset{\epsilon}{\subset}\Omega\times\mathcal{Y}\quad(\text{mod }\hat\lambda)$,
4. $\Omega\times\beta_m\overset{\epsilon}{\subset}\mathcal{B}\times Y\quad(\text{mod }\hat\lambda)$.

By property 1. $\hat\lambda\in\mathfrak{J}$, by properties 3 and 4 it is in $V_m\subset V_n$, and by our previous observation property 2 implies $d_{\text{meas}}(\lambda,\hat\lambda)<\epsilon$. This completes the proof that V_n is dense and therefore also the proof of the theorem. □

3. Finitely determined processes

If a sequence of processes $(\mathbf{X}_n, \alpha_n)$ converges in $\bar{d}$ to a process $(\mathbf{X}, \alpha)$ then it also tends to it in d_{meas} and in entropy. We call those processes for which the converse is true finitely determined.

20.5. DEFINITION. An ergodic process $(\mathbf{X}, \alpha)$ is called *finitely determined* if for every $\epsilon > 0$ there exists a $\delta > 0$ such that any ergodic $(\mathbf{Y}, \beta)$ satisfying

$$\text{(i)} \qquad d_{\text{meas}}(\rho(\mathbf{X}, \alpha), \rho(\mathbf{Y}, \beta)) < \delta \quad \text{and}$$

$$\text{(ii)} \qquad |h_\mu(\alpha) - h_\nu(\beta)| < \delta,$$

also satisfies

$$\text{(iii)} \qquad \bar{d}(\rho(\mathbf{X}, \alpha), \rho(\mathbf{Y}, \beta)) < \epsilon.$$

As we shall see in the next section Bernoulli processes are finitely determined.

20.6. THEOREM. *Let $(\mathbf{X}, \alpha)$ be a finitely determined process, $(\mathbf{Y}, \beta)$ an ergodic process, where α and β are generators. Suppose further that $h_\mu(\alpha) = h_\mu = h_\nu = h_\nu(\beta)$. Let $\mathfrak{J}$ be the weak* closure of $J_e(\mu, \nu)$ in $J(\mu, \nu)$. The set of graph joinings $\lambda \in \mathfrak{J}$, i.e. (Theorem 6.5) those joinings $\lambda \in \mathfrak{J}$ for which*

$$\alpha \times Y \subset X \times \mathcal{Y} \qquad (\text{mod } \lambda),$$

forms a dense G_δ subset of $\mathfrak{J}$.

PROOF. As in the proof of Krieger's finite generator theorem (Theorem 20.4) it is enough, by the Baire category theorem, to show that for each n the set V_n of all $\lambda \in \mathfrak{J}$ with

$$\alpha \times Y \overset{1/n}{\subset} X \times \mathcal{Y} \quad (\text{mod } \lambda),$$

is an open and dense subset of $\mathfrak{J}$. The proof that V_n is open is very similar to that in Krieger's theorem. To show that it is dense it suffices to show that given $\lambda \in \mathfrak{J}$ and $\epsilon > 0$, there exists $\lambda' \in V_n$ with $d_{\text{meas}}(\lambda, \lambda') < \epsilon$.

By Theorem 15.20, there exists an $\epsilon_1 > 0$ such that for $\rho, \rho' \in M_S(\Omega(\ell \times \ell))$,

$$\bar{d}(\rho, \rho') < \epsilon_1 \quad \Rightarrow \quad d_{\text{meas}}(\rho, \rho') < \epsilon/4. \tag{3.1}$$

Let $0 < \delta$ correspond to this ϵ_1 in the definition of a finitely determined process. We also require $\delta < \min(\epsilon/4, \epsilon_1)$ and $\epsilon_1 < 1/(2n)$.

We start by applying Theorem 15.17 to the process $(\mathbf{Y}, \beta)$ and δ as above. This yields a partition γ of Y with the properties

1. $d^{\nu}_{\text{part}}(\beta, \gamma) \leq \delta$ and,
2. $h_\nu(\beta) - \delta < h_\nu(\gamma) < h_\nu(\beta)$.

Now consider λ as a joining of the processes $(\mathbf{X}, \alpha)$ and $(\mathbf{Y}, \gamma)$ and the last inequality enables us to apply Lemma 20.3 to obtain an ergodic non-periodic system $\mathbf{Z} = (Z, \mathcal{Z}, \eta)$, a partition ζ of Z and an ergodic joining $\lambda_1 \in J_e(\eta, \nu)$ such that, for the symbolic representations $\rho_0 = \rho(X \times Y, \lambda, \alpha \times \gamma)$ and $\rho_1 = \rho(Z \times Y, \lambda_1, \zeta \times \gamma)$, the following properties hold

(i) $$h_\eta(\zeta) > h_\nu$$

(ii) $$d_{\text{meas}}(\rho_0, \rho_1) < \delta$$

(iii) $$\zeta \times Y \overset{\delta}{\subset} Z \times \gamma^T \quad (\text{mod } \lambda_1) \quad \text{and}$$

(iv) $$Z \times \gamma \overset{\delta}{\subset} \mathcal{Z} \times Y \quad (\text{mod } \lambda_1).$$

These properties imply

(v) $$|h_\mu(\alpha) - h_\eta(\zeta)| < \delta, \quad \text{and}$$

(vi) $$d_{\text{meas}}(\rho(\mathbf{X}, \alpha), \rho(\mathbf{Z}, \zeta)) < \delta,$$

and since $(\mathbf{X}, \alpha)$ is finitely determined, our choice of δ implies

$$\bar{d}(\rho(\mathbf{X}, \alpha), \rho(\mathbf{Z}, \zeta)) < \epsilon_1.$$

We now let $\hat{\lambda} \in J_e(\mu, \eta)$ be an ergodic joining which achieves the $\bar{d}$ distance

$$\hat{\lambda}((\alpha \times Z) \triangle (X \times \zeta)) = \bar{d}(\rho(\mathbf{X}, \alpha), \rho(\mathbf{Z}, \zeta)) < \epsilon_1. \tag{3.2}$$

Next construct $\hat{\lambda} \underset{Z}{\times} \lambda_1$, the relatively independent joining of $\hat{\lambda}$ and λ_1 over their common factor $\mathbf{Z}$. Choose Λ to be a typical ergodic component of this relatively independent product measure (typical means that it has the right marginals, of course almost every ergodic component has this property). We let $\lambda' \in \mathfrak{J}$ be the projection of Λ onto the $X \times Y$ component of $X \times Y \times Z$.

Denoting

$$\begin{aligned}
\rho &= \rho(X \times Y, \lambda, \alpha \times \beta)\\
\rho_0 &= \rho(X \times Y, \lambda, \alpha \times \gamma)\\
\rho_1 &= \rho(Z \times Y, \lambda_1, \zeta \times \gamma)\\
\rho_2 &= \rho(X \times Y, \lambda', \alpha \times \gamma)\\
\rho' &= \rho(X \times Y, \lambda', \alpha \times \beta),
\end{aligned}$$

we have

$$\begin{aligned}
d_{\text{meas}}(\rho, \rho') \leq & d_{\text{meas}}(\rho, \rho_0) + d_{\text{meas}}(\rho_0, \rho_1)\\
& + d_{\text{meas}}(\rho_1, \rho_2) + d_{\text{meas}}(\rho_2, \rho').
\end{aligned} \tag{3.3}$$

By (ii) we have $d_{\text{meas}}(\rho_0, \rho_1) < \delta < \epsilon/4$. The inequality $d^\nu_{\text{part}}(\beta, \gamma) \leq \delta$ implies (Theorem 15.20.5) $\bar{d}(\rho, \rho_0) < \delta < \epsilon_1$ and $\bar{d}(\rho_2, \rho') < \delta < \epsilon_1$, hence, by the choice of ϵ_1, (3.1), $d_{\text{meas}}(\rho, \rho_0) < \epsilon/4$ and $d_{\text{meas}}(\rho_2, \rho') < \epsilon/4$. It only remains to bound the third summand in (3.3). For that purpose note that Λ is a joining of the processes $(X \times Y, \lambda', \alpha \times \gamma)$ and $(X \times Y, \lambda', \alpha \times \beta)$. Therefore a bound on

$$\Lambda((\alpha \times \gamma \times Z) \triangle (X \times \gamma \times \zeta)),$$

will give a bound on $\bar{d}(\rho_1, \rho_2)$.

Now (with some disregard to the order on our triple product space)

$$\begin{aligned}\Lambda\big((\alpha \times \gamma \times Z) \triangle (X \times \gamma \times \zeta)\big) &= \sum_{A\times C\in\alpha\times\zeta}\ \sum_{B\in\gamma} \Lambda\big((A \times B \times Z) \triangle (X \times B \times C)\big)\\ &= \sum_{A\times C\in\alpha\times\zeta}\ \sum_{B\in\gamma} \Lambda\big(((A \times Z) \triangle (X \times C)) \times B\big)\\ &= \sum_{A\times C\in\alpha\times\zeta} \Lambda\big(((A \times Z) \triangle (X \times C)) \times Y\big)\\ &= \hat{\lambda}\big((\alpha \times Z) \triangle (X \times \zeta)\big) < \epsilon_1.\end{aligned}$$

Thus $\bar{d}(\rho_1, \rho_2) < \epsilon_1$, whence $d_{\mathrm{meas}}(\rho_1, \rho_2) < \epsilon/4$. Put together these estimations yield

$$d_{\mathrm{meas}}(\rho, \rho') < \epsilon.$$

Finally we show that $\lambda' \in V_n$. To this end we express the relations (*iii*) and (3.2) in terms of the joining Λ. First observe that the relation (*iii*) implies the existence of a γ^T-measurable partition γ' of Y such that (with a change of order) $\lambda_1((Y \times \zeta) \triangle (\gamma' \times Z)) < \delta$. Now

$$\begin{aligned}\Lambda\big((X \times Y \times \zeta) \triangle (\alpha \times Y \times Z)\big) &= \hat{\lambda}\big((X \times \zeta) \triangle (\alpha \times Z)\big) < \epsilon_1\\ \Lambda\big((X \times Y \times \zeta) \triangle (X \times \gamma' \times Z)\big) &= \lambda_1\big((Y \times \zeta) \triangle (\gamma' \times Z)\big) < \delta,\end{aligned}$$

hence

$$\begin{aligned}\Lambda\big((\alpha \times Y \times Z) \triangle (X \times \gamma' \times Z)\big) &= \lambda'\big((\alpha \times Y) \triangle (X \times \gamma')\big)\\ &< \epsilon_1 + \delta < 1/(2n) + 1/(2n) = 1/n,\end{aligned}$$

which clearly implies

$$\alpha \times Y \overset{1/n}{\subset} X \times \mathcal{Y} \pmod{\lambda}.$$

This completes the proof that V_n is dense in $\mathfrak{J}$ and hence also the proof of the theorem. □

20.7. COROLLARY. *Two finitely determined processes* $(\mathbf{X}, \alpha)$ *and* $(\mathbf{Y}, \beta)$ *are isomorphic iff* $h_\nu(\beta) = h_\mu(\alpha)$.

PROOF. Of course the equality of entropies is a necessary condition. If $(\mathbf{X}, \alpha)$ and $(\mathbf{Y}, \beta)$ have equal entropies then by Theorem 20.6, the sets of $\lambda \in \mathfrak{J}$ satisfying

$$\alpha \times Y \subset X \times \mathcal{Y} \qquad \pmod{\lambda}$$

and

$$X \times \beta \subset \mathcal{X} \times Y \qquad \pmod{\lambda}$$

respectively, are both dense G_δ subsets of $\mathfrak{J}$. Therefore their intersection is non-empty and any element in this intersection is an isomorphism of the two systems. □

4. Bernoulli processes are finitely determined

20.8. DEFINITION. 1. Let $\mathbf{X}$ be a dynamical system, $\epsilon > 0$. Two partitions $\alpha = \{A_1, A_2, \ldots, A_\ell\}$ and $\beta = \{B_1, B_2, \ldots, B_k\}$ are called *ϵ-independent* if

$$\sum_{i=1}^{\ell}\sum_{j=1}^{k} |\mu(A_i \cap B_j) - \mu(A_i)\mu(B_j)| < \epsilon.$$

2. A processes $(\mathbf{Y}, \beta), \beta \in \mathcal{P}_\ell$ is *ϵ-independent* if for all $n \geq 0$ the partitions $T^{-n-1}\beta$ and $\beta_0^n = \beta \vee T^{-1}\beta \vee \cdots \vee T^{-n}\beta$ are ϵ-independent; i.e.

$$\sum_{y_0^n \in \mathcal{L}^n} \sum_{y_{n+1} \in \mathcal{L}} |\nu(y_0^{n+1}) - \nu(y_0^n)\nu(y_{n+1})| \leq \epsilon$$

(with $\mathcal{L} = \{1, \ldots, \ell\}$).

We shall need the following inequality whose proof we leave as an exercise.

20.9. LEMMA (Pinsker's inequality). *Let $p = (p_1, p_2, \ldots, p_\ell)$ be a probability vector and $q = (q_1, q_2, \ldots, q_\ell)$ a sub-probability vector; i.e. $\forall i,\ q_i > 0$ and $\sum_{i=1}^{\ell} q_i \leq 1$, then with $c = \frac{1}{2}\log 2$,*

$$\sum_{i=1}^{\ell} p_i \log(\frac{p_i}{q_i}) \geq c\left(\sum_{i=1}^{\ell} |p_i - q_i|\right)^2.$$

20.10. LEMMA. *There exists a constant $c > 0$ such that if the finite partitions $\alpha, \beta \in \mathcal{P}$, satisfy*

$$H(\alpha) - H(\alpha|\beta) < c\epsilon^2,$$

then they are ϵ-independent.

PROOF. By Pinsker's inequality, Lemma 20.9,

$$\begin{aligned} H(\alpha) - H(\alpha|\beta) &= -\sum_{A \in \alpha} \mu(A)\log(\mu(A)) + \sum_{B \in \beta}\sum_{A \in \alpha} \mu(A \cap B)\log\frac{\mu(A \cap B)}{\mu(B)} \\ &= \sum_{B \in \beta}\sum_{A \in \alpha} \mu(A \cap B)\log\frac{\mu(A \cap B)}{\mu(A)\mu(B)} \\ &\geq c\Big(\sum_{B \in \beta}\sum_{A \in \alpha} |\mu(A \cap B) - \mu(A)\mu(B)|\Big)^2. \end{aligned}$$

□

Given a partition α of $\mathbf{X}$ we denote its probability distribution vector by

$$\operatorname{dist}(\alpha) = (\mu(A_1), \mu(A_2), \ldots, \mu(A_\ell)).$$

20.11. LEMMA. *Let $(\mathbf{X}, \alpha)$ and $(\mathbf{Y}, \beta)$ be processes where α is independent. Given $\epsilon > 0$ there exists a $\delta > 0$ such that if*

(i) $$\|\operatorname{dist}(\alpha) - \operatorname{dist}(\beta)\|_1 < \delta \quad \text{and}$$

(ii) $$|h_\mu(\alpha) - h_\nu(\beta)| < \delta,$$

then $(\mathbf{Y}, \beta)$ *is* ϵ*-independent.*

PROOF. Since $(\mathbf{X}, \alpha)$ is an independent process, we have $h_\mu(\alpha) = H(\mathrm{dist}(\alpha))$. We also have

$$H(\beta|\beta_{-n}^{-1}) \searrow h_\nu(\beta).$$

Now for a sufficiently small δ condition (i) implies

$$|H(\alpha) - H(\beta)| < c\epsilon^2/2$$

and if we also take $\delta < c\epsilon^2/2$ we then have

$$\begin{aligned} H(\beta) &\geq H(\beta|\beta_{-n}^{-1}) \searrow h_\nu(\beta) \\ &\geq h_\mu(\alpha) - \delta = H(\alpha) - \delta \\ &\geq H(\beta) - c\epsilon^2/2 - \delta \geq H(\beta) - c\epsilon^2. \end{aligned}$$

Thus

$$0 \leq H(\beta) - H(\beta|\beta_{-n}^{-1}) \leq c\epsilon^2$$

and the ϵ-independence of $(\mathbf{Y}, \beta)$ follows from Lemma 20.10. □

20.12. THEOREM. *Bernoulli processes are finitely determined.*

PROOF. We are given a Bernoulli measure $\mu \in M_S(\Omega(\ell))$ and $\epsilon > 0$. Let P_0 be the standard partition of $\Omega(\ell)$ and for $\nu \in M_S(\Omega(\ell))$ let $(\nu(1), \nu(2), \dots, \nu(\ell)) = \mathrm{dist}_\nu(P_0)$. By Lemma 20.11 there exists a $0 < \delta < \epsilon$ such that if $\nu \in M_S(\Omega(\ell))$ satisfies

(i) $$d_0(\mu, \nu) = \sum_{j=1}^{\ell} |\mu(j) - \nu(j)| < \delta \quad \text{and}$$

(ii) $$|h_\mu - h_\nu| < \delta,$$

then ν is ϵ-independent. We shall show that, in fact any such ν already satisfies

(iii) $$\bar{d}(\mu, \nu) < \epsilon,$$

thus proving that μ is finitely determined.

In order to get (iii) we use the characterization of the $\bar{d}$ distance given in Theorem 15.25, namely

$$\bar{d}(\mu, \nu) = \lim_{n \to \infty} \bar{d}(\mu_n, \nu_n),$$

where μ_n and ν_n are the projections of μ and ν onto the finite space $\mathfrak{L}^{2n+1}$ corresponding to the coordinates $-n, -n+1, \dots, 0, 1, \dots, n$. It will, however, be more convenient to work with the projections onto the coordinates $0, 1, \dots, n$. Accordingly, in this proof μ_n and ν_n will denote the projections onto the coordinates $0, 1, \dots, n$. This we may well do by the stationarity of the measures μ and ν.

We use induction to show that for every $n \geq 0$ there exists a joining $\lambda_n \in J(\mu_n, \nu_n)$ with

$$\mathbb{E}_{\lambda_n}(d_{\mathrm{ham}}) = \int d_{\mathrm{ham}}(x_0^n, y_0^n)\, d\lambda(x_1^n, y_1^n) < \epsilon.$$

Now for $n = 0$ we have, by Lemma 15.26.2,

$$\bar{d}(\mu_0, \nu_0) = d_0(\mu_0, \nu_0) = \frac{1}{2} \sum_{j \in \mathfrak{L}} |\mu(j) - \nu(j)| < \delta < \epsilon.$$

By Lemma 15.24.1. there exists $\lambda_0 \in J(\mu_0, \nu_0)$ with $\int d_{\text{ham}}(x_0, y_0)\, d\lambda_0(x_0, y_0) = \bar{d}(\mu_0, \nu_0)$ so that $\int d_{\text{ham}}(x_0, y_0)\, d\lambda_0(x_0, y_0) < \epsilon$. Assume that for some $\lambda_n \in J(\mu_n, \nu_n)$ we have $\int d_{\text{ham}}(x_0^n, y_0^n)\, d\lambda_n(x_0^n, y_0^n) < \epsilon$. We shall construct $\lambda_{n+1} \in J(\mu_{n+1}, \nu_{n+1})$ with

$$\int d_{\text{ham}}(x_0^{n+1}, y_0^{n+1})\, d\lambda_{n+1}(x_0^{n+1}, y_0^{n+1}) < \epsilon.$$

Let

$$\mu_{n+1} = \int \delta_{x_0^n} \times \mu_{x_0^n}\, d\mu_n(x_0^n) = \sum_{x_0^n \in \mathfrak{L}^n} \mu_n(x_0^n)\delta_{x_0^n} \times \mu_{x_0^n}, \quad \text{and}$$

$$\nu_{n+1} = \int \delta_{x_0^n} \times \nu_{x_0^n}\, d\nu_n(x_0^n) = \sum_{x_0^n \in \mathfrak{L}^n} \nu_n(x_0^n)\delta_{x_0^n} \times \nu_{x_0^n}.$$

be the disintegrations of μ_{n+1} and ν_{n+1} over μ_n and ν_n respectively. Since μ, being Bernoulli, is the product measure $\mu_0^{\mathbb{Z}}$ we have $\mu_{x_0^n} \equiv \mu_0$. Let $\lambda_{x_0^n, y_0^n} \in J(\mu_{x_0^n}, \nu_{y_0^n})$ achieve the $\bar{d}$ distance

$$\bar{d}(\mu_{x_0^n}, \nu_{y_0^n}) = \int d_{\text{ham}}(x_{n+1}, y_{n+1})\, d\lambda_{x_0^n, y_0^n}(x_{n+1}, y_{n+1}),$$

and set

$$\lambda_{n+1} = \int \delta_{x_0^n} \times \lambda_{x_0^n, y_0^n}\, d\lambda_n(x_0^n, y_0^n) = \sum_{(x_0^n, y_0^n) \in (\mathfrak{L} \times \mathfrak{L})^n} \lambda_n(x_0^n, y_0^n)\delta_{(x_0^n, y_0^n)} \times \lambda_{x_0^n, y_0^n}.$$

It is easy to check that $\lambda_{n+1} \in J(\mu_{n+1}, \nu_{n+1})$.

Now the definition of the function $d_{\text{ham}}(\cdot, \cdot)$ as an average yields the formula

$$(n+1)d_{\text{ham}}(x_0^{n+1}, y_0^{n+1}) = n d_{\text{ham}}(x_0^n, y_0^n) + d_{\text{ham}}(x_{n+1}, y_{n+1}),$$

whence

$$\begin{aligned} &(n+1)\mathbb{E}_{\lambda_{n+1}}(d_{\text{ham}}) \\ &= n\mathbb{E}_{\lambda}(d_{\text{ham}}) \\ &\quad + \sum_{(x_0^{n+1}, y_0^{n+1}) \in (\mathfrak{L} \times \mathfrak{L})^{n+1}} \lambda_n(x_0^n, y_0^n) d_{\text{ham}}(x_{n+1}, y_{n+1}) \lambda_{x_0^n, y_0^n}(x_{n+1}, y_{n+1}). \end{aligned} \tag{4.1}$$

The first summand is by assumption $\leq n\epsilon$. The second equals

(4.2)

$$\begin{aligned} \sum_{(x_0^n, y_0^n) \in (\mathfrak{L} \times \mathfrak{L})^n} \lambda_n(x_0^n, y_0^n) \sum_{(x_{n+1}, y_{n+1}) \in \mathfrak{L} \times \mathfrak{L}} d_{\text{ham}}(x_{n+1}, y_{n+1}) \lambda_{x_0^n, y_0^n}(x_{n+1}, y_{n+1}) \\ = \sum_{(x_0^n, y_0^n) \in (\mathfrak{L} \times \mathfrak{L})^n} \lambda_n(x_0^n, y_0^n) \bar{d}(\mu_{x_0^n}, \nu_{y_0^n}), \end{aligned}$$

since we assumed that $\lambda_{x_0^n, y_0^n}$ achieves $\bar{d}(\mu_{x_0^n}, \nu_{y_0^n})$.

By the triangle inequality

$$\bar{d}(\mu_{x_0^n}, \nu_{y_0^n}) \leq \bar{d}(\mu_{x_0^n}, \mu_{n+1}) + \bar{d}(\mu_{n+1}, \nu_{n+1}) + \bar{d}(\nu_{n+1}, \nu_{y_0^n}). \tag{4.3}$$

The first term on the right hand side of (4.3) is 0, as $\mu_{x_0^n} = \mu_{n+1} = \mu_0$. By stationarity the second term equals $\bar{d}(\mu_0, \nu_0)$, which we have seen is $\leq \epsilon/2$.

Considering the last term we observe that again, by Lemma 15.26.3,

$$\bar{d}(\nu_{n+1}, \nu_{y_0^n}) = d_0(\nu_{n+1}, \nu_{y_0^n}) = \frac{1}{2}\sum_{j\in\mathfrak{L}} |\nu_{n+1}(j) - \nu_{y_0^n}(j)|,$$

therefore, integrating this term according to λ_n as prescribed in (4.2), we get

$$\begin{aligned}
&\sum_{(x_0^n, y_0^n)\in(\mathfrak{L}\times\mathfrak{L})^n} \lambda_n(x_0^n, y_0^n)\bar{d}(\nu_{n+1}, \nu_{y_0^n}) \\
=&\frac{1}{2}\sum_{y_0^n\in\mathfrak{L}^n} \nu_n(y_0^n)\sum_{j\in\mathfrak{L}} |\nu_{n+1}(j) - \nu_{y_0^n}(j)| \\
=&\frac{1}{2}\sum_{y_0^n\in\mathfrak{L}^n}\sum_{j\in\mathfrak{L}} |\nu_n(y_0^n)\nu_{n+1}(j) - \nu_n(y_0^n)\nu_{y_0^n}(j)| \\
=&\frac{1}{2}\sum_{y_0^n\in\mathfrak{L}^n}\sum_{j\in\mathfrak{L}} |\nu_n(y_0^n)\nu_{n+1}(j) - \nu_{n+1}(y_0^{n+1})|.
\end{aligned}$$

The ϵ-independence of the measure ν asserts that the last line is $\leq \epsilon/2$. Thus substituting (4.3) in (4.2) and (4.1) and taking into account the bounds we derived we get from (4.1)

$$(n+1)\mathbb{E}_{\lambda_{n+1}}(d_{ham}) \leq n\epsilon + \epsilon/2 + \epsilon/2 = (n+1)\epsilon.$$

This completes the induction and also concludes the proof of the theorem since we now have for all $n \geq 0$

$$\bar{d}(\mu_n, \nu_n) \leq \mathbb{E}_{\lambda_n}(d_{ham}) \leq \epsilon,$$

and therefore also

$$\bar{d}(\mu, \nu) = \lim_{n\to\infty} \bar{d}(\mu_n, \nu_n) \leq \epsilon.$$

□

5. Sinai's factor theorem and Ornstein's isomorphism theorem

20.13. THEOREM (Sinai's factor theorem). *If $\mathbf{X}$ is an ergodic system with entropy $h = h_\mu > 0$ then for every Bernoulli system $\mathbf{Y}$ with entropy $h' \leq h$ there exists a factor map $\pi : \mathbf{X} \to \mathbf{Y}$.*

PROOF. Let $\mathbf{Y}$ be a Bernoulli system with entropy h'; by Theorem 20.12 $\mathbf{Y}$ is finitely determined. By Theorem 15.11 $\mathbf{X}$ admits a factor $\mathbf{X}'$ with entropy h'. Finally by Theorem 20.6 $\mathbf{X}'$ has a factor $\mathbf{X}''$ isomorphic to $\mathbf{Y}$ and the map $\pi : \mathbf{X} \to \mathbf{X}''$ is the required factor map. □

20.14. THEOREM (Ornstein's isomorphism theorem). *Two Bernoulli systems with the same entropy are isomorphic.*

PROOF. This is a direct corollary of the fact that every Bernoulli system is finitely determined (Theorem 20.12) and Corollary 20.7. □

Here is a simple application:

20.15. THEOREM. *Let $\mathbf{X}$ be a Bernoulli system and $n \geq 2$ an integer, then T has an n'th root; i.e. there exists an automorphism $S \in \mathrm{Aut}(\mathbf{X})$ with $S^n = T$.*

PROOF. Denote $h = h_\mu(X, \mathcal{X}, T)$ and let $\mathbf{Y} = (Y, \mathcal{Y}, \nu, R)$ be a Bernoulli system with $h_\nu(Y, \mathcal{Y}, R) = h/n$. It is easy to see that the system $(Y, \mathcal{Y}, \nu, R^n)$ is again a Bernoulli system and $h_\nu(Y, \mathcal{Y}, R^n) = h$. By Ornstein's theorem the two systems $(Y, \mathcal{Y}, \nu, R^n)$ and $(X, \mathcal{X}, \mu, T)$ are isomorphic and the automorphism $S \in \mathrm{Aut}(\mathbf{X})$ which corresponds to R via this isomorphism is the sought for n'th root of T. □

6. Notes

Sources: mainly Rudolph's book [**220**] and, for Sections 3 and 4, also Shields's book [**235**].

APPENDIX A

Prerequisite Background and Theorems

A.1. THEOREM (The subadditivity lemma). *Let $\{a_n\}_{n=1}^{\infty}$ be a sequence of non-negative numbers which is subadditive; i.e. for every m and n, $a_{m+n} \leq a_m + a_n$, then the limit $\lim_{n\to\infty} a_n/n$ exists and equals $\inf_n a_n/n$.*

(See e.g. [**257**], Theorem 4.9.)

Call a commutative Banach algebra B over $\mathbb{R}$, with unit and satisfying $||f^2|| = ||f||^2, \forall f \in B$, a real commutative C^* algebra.

A.2. THEOREM. 1. *There is a 1-1 correspondence between real commutative C^* algebras B and compact Hausdorff spaces X. Under this correspondence B is isometrically isomorphic with $C(X)$ (the Banach algebra of real valued continuous functions on X with sup norm: $||f|| = \sup_{x\in X}|f(x)|$). The space X is the* Gelfand space *or the* spectrum *of B; i.e. the set of all continuous multiplicative homomorphisms of the algebra B into $\mathbb{R}$. The topology on X can be described as the weak* topology induced on X when it is considered as a subset of the dual Banach space B^*. The algebra B is separable iff X is metrizable.*

2. *There is a 1-1 correspondence between closed subalgebras containing the unit $1 \in D \subset B$, and continuous onto maps $\phi : X \to Y$ from the compact space X, the spectrum of B, onto the compact space Y, the spectrum of D, where for a functional $x \in X$, $\phi(x) \in Y$ is the restriction of x to the subalgebra D.*
3. *There is a 1-1 correspondence between self homeomorphisms T of the compact space X and automorphisms of $C(X) \cong B$ (also denoted by T) given by: $Tf(x) = f(T^{-1}x)$, $(x \in X, f \in C(X))$.*

(See e.g. [**53**], Section IV.5.)

A.3. THEOREM (The Stone-Čech compactification). *Let X be a completely regular (Hausdorff) topological space and let βX denote the Gelfand space of the real commutative C^* algebra $C_b(X)$ consisting of the bounded continuous real valued functions on X with supremum norm.*

1. *The map $j : X \to \beta X$ sending the point $x \in X$ to the multiplicative homomorphism $j(x) \in \beta X$ defined by $j(x)(F) = F(x), F \in C_b(X)$, is a homeomorphism of X onto a dense subset of βX.*
2. *The compactification $j : X \to \beta X$ (which is called the* Stone-Čech *compactification of X) has the following universal property. Every continuous map $f : X \to Z$ into a compact Hausdorff space Z has a unique extension to a continuous function $\bar{f} : \beta X \to Z$.*

3. *The Stone-Čech compactification of X is unique in the following sense. If $i : X \to Z$ is a homeomorphism of X into a compact Hausdorff space Z with dense image then there exists a homeomorphism onto $\phi : \beta X \to Z$ which makes the diagram*

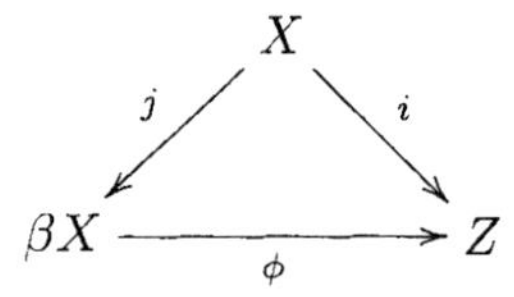

commutative.

(See e.g. [**53**], Section IV.6, and [**62**], Sections 3.5 and 3.6.)

Let X be a separable metrizable space. Let $C_b(X)$ (or just $C(X)$ when X is compact) be the algebra of bounded continuous real valued functions on X, and $M(X)$ the set of probability Borel measures on X. We equip $M(X)$ with the smallest topology which makes all maps $\mu \mapsto \mu(f)$, $f \in C_b(X)$, of $M(X)$ into $\mathbb{R}$, continuous. Let d be a compatible metric on X such that the corresponding completion $(\hat{X}, \hat{d})$ is compact (there always is such a metric). X is a dense subset of the compact metric space $\hat{X}$. It is a G_δ subset of $\hat{X}$ iff X is Polish. Denote by $U_d(X)$ the space of d-uniformly continuous, real valued functions on X. (Thus, $U_d(X)$ is isometrically isomorphic with the Banach algebra $C(\hat{X})$). Let $\{f_n\}_{n\geq 1}$ be a dense sequence in $U_d(X)$ and let:

$$D(\mu, \nu) = \sum_{n=1}^{\infty} 2^{-n} \frac{|\mu(f_n) - \nu(f_n)|}{||f_n||_\infty}.$$

By the Riesz representation theorem we can identify $M(\hat{X})$ with the convex weak* compact subset of $C(\hat{X})^*$, the dual Banach space, consisting of all positive functionals μ with $\mu(1) = 1$. It is easy to see that the map $\mu \mapsto \hat{\mu}$, defined by $\hat{\mu}(A) = \mu(A \cap X)$ is an embedding of $M(X)$ into $M(\hat{X})$ with image $\{\hat{\mu} \in M(\hat{X}) : \mu(X) = 1\}$ and that this image is a dense subset of $M(\hat{X})$.

A.4. THEOREM. *Let X be a separable metrizable space, then*

1. *$M(X)$ is a separable metrizable space and D is a compatible metric.*
2. *X is compact iff $M(X)$ is compact.*
3. *X is Polish iff $M(X)$ is Polish.*
4. *The Borel σ-algebra $\mathcal{B}(M(X))$ is the smallest σ-algebra with respect to which the maps $\mu \mapsto \mu(A)$, $A \in \mathcal{B}(X)$ are measurable and also the smallest σ-algebra with respect to which the maps $\mu \mapsto \mu(f)$, f a bounded Borel function on X, are measurable. In particular the Borel structure on $M(X)$ for a standard Borel space X does not depend on a particular Polish topology.*

(See e.g. [**152**].)

A.5. THEOREM. *Let X be a separable metrizable space, $\mu_n, \mu \in M(X)$, then the following are equivalent:*

1. $\lim_{n\to\infty} \mu_n = \mu$.
2. *For every $f \in C_b(X)$, $\lim_{n\to\infty} \int f\, d\mu_n = \int f\, d\mu$ (or if this holds for every f in a dense subset of $U_d(X)$).*

3. $\limsup_n \mu_n(F) \le \mu(F)$ *for every closed subset* F *of* X.
4. $\liminf_n \mu_n(U) \ge \mu(U)$ *for every open subset* U *of* X.
5. $\lim_{n\to\infty} \mu_n(A) = \mu(A)$ *for every Borel subset* A *of* X *with* $\mu(\partial A) = 0$.

(See [**152**], Theorem 17.20.)

A.6. THEOREM (The Kolmogorov consistency theorem). *Let* $\{X_n : n \in \mathbb{N}\}$ *be a sequence of standard Borel spaces and suppose that for every finite subset* $F \subset \mathbb{N}$ *there is a probability Borel measure* μ_F *defined on* $X_F = \prod_{j\in F} X_j$ *in such a way that whenever* $F_1 \subset F_2$ *then the image of* μ_{F_2} *under the projection map* $\pi_{F_2,F_1} : X_{F_2} \to X_{F_1}$ *is* μ_{F_1}, *then there is a unique probability Borel measure* μ *on the standard Borel space* $X = \prod_{j\in\mathbb{N}} X_j$ *with* $\pi_F(\mu) = \mu_F$ *for every finite* $F \subset \mathbb{N}$, *where* $\pi_F : X \to X_F$ *is the projection map.*

(See e.g. [**186**], page 82.)

A.7. THEOREM (Measure disintegration). *Let* X, Y *be standard Borel spaces, and* $\pi : X \to Y$ *a Borel map. Let* $\mu \in M(X)$ *and let* $\nu = \pi(\mu) = \mu \circ \pi^{-1}$ *be its image in* $M(Y)$, *then there is a Borel map* $y \mapsto \mu_y$, *from* Y *into* $M(X)$ *such that:*

1. *For* ν *almost every* $y \in Y$, $\mu_y(\pi^{-1}(y)) = 1$.
2. $$\mu = \int_Y \mu_y \, d\nu(y).$$
   (The latter means that $\mu(A) = \int \mu_y(A)\, d\nu(y)$ *for every Borel subset* A *of* X, *or equivalently that* $\int f\, d\mu = \int\int f(x)\, d\mu_y(x)\, d\nu(y)$, *for every bounded Borel function* $f : X \to \mathbb{R}$.*)*

Moreover such a map is unique in the following sense: If $y \mapsto \mu'_y$ *is another such map, then* $\mu_y = \mu'_y$, ν*-a.e.*

(See e.g. [**35**] and [**77**], Theorem 5.8.)

Let $\mathbf{X} = (X, \mathcal{X}, \mu)$ and $\mathbf{Y} = (Y, \mathcal{Y}, \nu)$ be separable probability measure spaces and let $\phi : \mathbf{X} \to \mathbf{Y}$ be a measurable measure-preserving map. This map defines an embedding of $L^1(\mathbf{Y})$ into $L^1(\mathbf{X})$ and we shall thus consider $L^1(\mathbf{Y})$ as a closed subspace of $L^1(\mathbf{X})$.

A.8. THEOREM (Conditional expectation). *The conditional expectation operator* $\mathbb{E}^{\mathbf{Y}} : L^1(\mathbf{X}) \to L^1(\mathbf{Y})$ *is characterized by the following properties:*

1. $\mathbb{E}^{\mathbf{Y}}$ *is a linear operator.*
2. $\mathbb{E}^{\mathbf{Y}} f \ge 0$ *for* $f \ge 0$.
3. $\mathbb{E}^{\mathbf{Y}} g = g$ *for* $g \in L^1(\mathbf{Y})$.
4. $\mathbb{E}^{\mathbf{Y}}(gf) = g\mathbb{E}^{\mathbf{Y}} f$ *for* $g \in L^1(\mathbf{Y})$ *and* $f \in L^1(\mathbf{X})$.
5. $\forall\, f \in L^1(\mathbf{X})$, $\int \mathbb{E}^{\mathbf{Y}} f\, d\nu = \int f\, d\mu$.

In addition we have the following properties:

6. *Let* ψ *be a convex nonnegative function of a real variable. If* f *and* $\psi \circ f$ *are in* $L^1(\mathbf{X})$ *then*
   $$\psi(\mathbb{E}^{\mathbf{Y}} f) \le \mathbb{E}^{\mathbf{Y}}(\psi \circ f).$$
7. *The conditional expectation operator maps each* $L^p(\mathbf{X})$ *to* $L^p(\mathbf{Y})$, $1 \le p \le \infty$, *with* $\|\mathbb{E}^{\mathbf{Y}} f\|_p \le \|f\|_p$.

8. *On* $L^2(\mathbf{X})$, $\mathbb{E}^{\mathbf{Y}}$ *is the orthogonal projection of* $L^2(\mathbf{X})$ *onto* $L^2(\mathbf{Y})$.
9. *When* X *and* Y *are standard Borel spaces then — letting* $\mu = \int \mu_y \, d\nu(y)$ *be the disintegration of* μ *over* ν *— for every* $f \in L^1(\mathbf{X})$, f *is in* $L^1(\mu_y)$ *for* ν*-a.e.* y *and*

$$\mathbb{E}^{\mathbf{Y}} f(y) = \int f \, d\mu_y.$$

(See e.g. [**186**].)

A.9. THEOREM. *Suppose that* $(X, \mathcal{X}, \mu)$ *is a standard Lebesgue space and* $A \subset L^\infty(X, \mathcal{X}, \mu)$ *a conjugation invariant subalgebra. Let* $\mathcal{B}$ *be the smallest* σ*-algebra of Borel sets in* X *with respect to which all functions in* A *are measurable. Then*

$$\operatorname{cls} A = L^2(X, \mathcal{B}) = \{f \in L^2(X, \mathcal{X}, \mu) : f \textit{ is } \mathcal{B}\textit{-measurable} \}.$$

(See [**267**], Theorem 1.2.)

A.10. THEOREM. *Let* $(X, \mathcal{X}, \mu)$ *and* $(Y, \mathcal{Y}, \nu)$ *be probability spaces and let* $\Phi : (\mathcal{Y}, \nu) \to (\mathcal{X}, \mu)$ *be a homomorphism of measure algebras, then the map* $V\mathbf{1}_B = \mathbf{1}_{\Phi(B)}$ *has a unique extension to a linear operator* $V : L^2(\nu) \to L^2(\mu)$ *satisfying*

1. V *is an isometry:* $(Vf, Vg) = (f, g)$ *(hence* $||Vf|| = ||f||$*).*
2. V *sends bounded functions to bounded functions.*
3. $V(fg) = V(f)V(g)$ *for bounded functions.*

If moreover Φ *is an isomorphism (i.e. onto), then* V *is a unitary isomorphism of* $L^2(\nu)$ *onto* $L^2(\mu)$*. Conversely, if* $V : L^2(\nu) \to L^2(\mu)$ *is a linear operator satisfying the conditions* $1-3$ *above, then for every* $B \in \mathcal{Y}$ *there is unique (mod 0)* $A \in \mathcal{X}$ *such that* $V\mathbf{1}_B = \mathbf{1}_A$ *and the map* $\Phi(B) = A$ *is a homomorphism of measure algebras; an isomorphism when* V *is a unitary isomorphism.*

(See [**257**], page 56.)

A.11. THEOREM. *Let* $(\mathbb{T}, \lambda)$ *be the circle equipped with Lebesgue's measure and* $V : L^2(\lambda) \to L^2(\lambda)$ *a unitary operator which is also positive (i.e.* $Vf \geq 0$ *for* $f \geq 0$*). Then* V *has the form:* $Vf(t) = \sqrt{\frac{dT\lambda}{d\lambda}(t)} f(Tt)$*, where* $T : \mathbb{T} \to \mathbb{T}$ *is an invertible measurable map.*

For a proof refer to the proof of Lamperti's Theorem in [**215**], Chapter 15, Theorem 24, together with the following observation: Royden proves a version of Theorem A.11 for *any* surjective isometry of $L^p(\lambda)$ which applies to all $1 \leq p < \infty$ but $p = 2$. Of course this more general statement is false for a general unitary operator $V : L^2(\lambda) \to L^2(\lambda)$. However, when in addition one assumes that V is positive, then one gets that for two disjoint sets $A, B \subset X$ the non-negative functions $f = V\mathbf{1}_A$ and $g = V\mathbf{1}_B$ satisfy

$$\int_{\mathbb{T}} f \cdot g \, d\lambda = \langle V\mathbf{1}_A, V\mathbf{1}_B \rangle = \langle \mathbf{1}_A, \mathbf{1}_B \rangle = \int_{\mathbb{T}} \mathbf{1}_A \cdot \mathbf{1}_B \, d\lambda = 0$$

and it follows that $(V\mathbf{1}_A) \cdot (V\mathbf{1}_B) = 0$ a.e. With this information one can complete the proof of Theorem A.11 in exactly the same way it is done in Royden's book.

A.12. THEOREM (Eberlein-Šmulian's theorem). *Let* A *be a subset of a Banach space* X*, then for the weak topology on* X *the following are equivalent*

1. *A is conditionally compact; i.e.* weak-cls A *is weakly compact.*
2. *A is conditionally sequentially compact; i.e. every sequence of elements of A has a weakly convergent subsequence.*
3. *A is conditionally countably compact; i.e. every sequence of elements of A has a cluster point in X.*

(See [**53**], [**47**], or [**158**].)

For the next theorem recall the following definitions. Let E be a locally convex topological vector space and $P \subset E$ a closed convex cone such that $P - P = E$. A compact convex $Q \subset P$ is a *base* for the cone P if for every $y \in P$ there is a unique $x \in X$ and $\alpha \geq 0$ such that $y = \alpha x$. When P admits a base then clearly $P \cap (-P) = \{0\}$ and it follows that the partial order defined on E by $x \geq y \Leftrightarrow x - y \in P$ satisfies: $x \geq y$ and $y \geq x$ implies $x = y$. The pair (E, P) is a *lattice* if for every $x, y \in X$ there exists a *greatest lower bound*, i.e. a lower bound z such that for every lower bound w we have $w \leq z$ (we write $z = x \wedge y$). Finally the convex compact subset $Q \subset E$ is a *simplex* if it is the base of a cone P as above for which (E, P) is a lattice; it is a *Bauer simplex* if the set ext (Q) of extreme points of Q is closed.

A.13. THEOREM (Choquet's theorem). *Let Q be a metrizable compact convex subset of a locally convex topological vector space E.*

1. *The subset* ext (Q) *of extreme points of Q is a G_δ subset of Q.*
2. *For every $x_0 \in Q$ there exists a probability measure κ on Q such that* $\kappa(\text{ext}\,(Q)) = 1$ *and such that*
$$\int_{\text{ext}\,(Q)} f(x)\, d\kappa(x) = f(x_0)$$
for every bounded upper semi-continuous affine function $f : Q \to \mathbb{R}$. (Such a measure κ is said to represent *the point x_0.)*
3. *The point x_0 is in* ext (Q) *iff the point mass $\kappa = \delta_{x_0}$ is the only probability measure on* ext (Q) *that represents x_0.*
4. *Q is a simplex iff for every $x_0 \in Q$ there is just one representing measure.*
5. *Q is a Bauer simplex iff it is affinelly homeomorphic to the simplex $M(X)$, where $X =$* ext (Q) *and $M(X)$ is the space of Borel probability measures on X with the weak* topology.*

(See [**206**] for details.)

A topological group G is called *monothetic* if there exists an element $g_0 \in G$ (called a *topological generator*) such that the set $\{g_0^n : n \in \mathbb{Z}\}$ is dense in G. It is easy to check that every monothetic group is abelian and that, when G is second countable, the set of topological generators is a dense G_δ subset of G.

A.14. THEOREM. *Let K be a compact, abelian, second countable, topological group. Then K is monothetic iff its dual group $\hat{K}$ is isomorphic (as an abstract group) to a countable subgroup of the circle $S^1 = \{z \in \mathbb{C} : |z| = 1\}$. When K is monothetic every such isomorphism is implemented by a choice of a topological generator $a \in K$ and then by mapping $\chi \in \hat{K}$ into $\chi(a) \in S^1$.*

(See [**119**], page 390.)

A.15. THEOREM (Peter-Weyl's theorem). *Let K be a compact, second countable, topological group with normalized Haar measure m. Let Σ denote the countable set of all equivalence classes (with respect to the relation of unitary equivalence) of irreducible unitary representations of G on (necessarily) finite dimensional Hilbert spaces. For every $\sigma \in \Sigma$ let $U^{(\sigma)} : G \to \mathfrak{U}(\mathfrak{H}_\sigma)$ be a representative element of σ with dim-$\mathfrak{H}_\sigma = d_\sigma$ and $\{\zeta_1, \zeta_2, \dots, \zeta_{d_\sigma}\}$ a fixed orthonormal basis. Let*

$$u_{jk}^{(\sigma)}(x) = \langle U_x^{(\sigma)} \zeta_k, \zeta_j \rangle,$$

be the unitary matrix of the operator $U_x^{(\sigma)}$, $(x \in G)$ in this basis. The set of functions

$$d_\sigma^{1/2} u_{jk}^{(\sigma)}, \quad (\sigma \in \Sigma, \ 1 \le j, k \le d_\sigma),$$

is an orthonormal basis for $L^2(G, m)$.

(See [**120**], page 24.)

Bibliography

[1] J. Aaronson, *An introduction to infinite ergodic theory*, Math. Surveys and Monographs **50**, Amer. Math. Soc. 1997.

[2] L. M. Abramov, *Metric automorphisms with quasi-discrete spectrum*, Izv. Akad. Nauk, SSR Ser. Mat. **26**, (1962), 513-530. Amer. Math. Soc. Transl. **39**, (1964), 37-56.

[3] S. Adams, *Another proof of Moore's ergodicity theorem for $SL(2,\mathbb{R})$*, Topological dynamics and applications, Contemporary Mathematics **215**, a volume in honor of R. Ellis, 1988, pp. 183-187.

[4] R. L. Adler, A. G. Konheim and M. H. McAndrew, *Topological entropy*, Trans. Amer. Math. Soc. **114**, (1965), 309-319.

[5] R. L. Adler and B. Weiss, *Entropy, a complete metric invariant for automorphisms of the torus*, Proc. Nat. Acad. Sci. USA **57**, (1967), 1573-1576.

[6] ———, *Similarity of automorphisms of the torus*, Memoirs of the Amer. Math. Soc. **98**, 1970.

[7] E. Akin, *The general topology of dynamical systems*, Amer. Math. Soc. , Providence, RI, 1993.

[8] ———, *Recurrence in topological dynamics: Furstenberg families and Ellis actions*, Plenum Press, New York, 1997.

[9] E. Akin, J. Auslander, and K. Berg, *When is a transitive map chaotic* , Convergence in Ergodic Theory and Probability, Walter de Gruyter & Co. 1996, pp. 25-40.

[10] ———, *Almost equicontinuity and the enveloping semigroup*, Topological dynamics and applications, Contemporary Mathematics **215**, a volume in honor of R. Ellis, 1998, pp. 75-81.

[11] J. Auslander, *Minimal Flows and their Extensions*, Mathematics Studies 153, Notas de Matemática, 1988.

[12] L. Auslander, L. Green and F. Hahn *Flows on homogeneous spaces*, Annals of mathematics studies 53, Princeton University Press, Princeton, New Jersey, 1963.

[13] J. Banks, J. Brooks, G. Cairns, G. Davis and P. Stacey, *On Devaney's Definition of Chaos*, Amer. Math. Monthlly **99**, (1992),332-334.

[14] J. R. Baxter, *A class of ergodic transformations having simple spectrum*, Proc. Amer. Math. Soc. **27**, (1971), 275-279.

[15] M. B. Bekka and M. Mayer, *Ergodic theory and topological dynamics of group actions on homogeneous spaces*, London Math. Soc. Lecture Note Series **269**, Cambridge University Press, 2000.

[16] M. B. Bekka and A. Valette, *Kazhdan's property (T) and amenable representations*, Math. Zeitschrift **212**, (1993), 293-299.

[17] F. Beleznay and M. Foreman, *The collection of distal flows is not Borel*, Amer. J. Math. **117**, (1995), 203-239.

[18] K. R. Berg, *Entropy of torus automorphisms*, Topological dynamics, an international symposium, J. Auslander and W.A. Gottschalk, eds., W. A. Benjamin, New York, 1968.

[19] V. Bergelson, H. Furstenberg, N. Hindman and Y. Katznelson, *An algebraic proof of van der Waerden's theorem*, L'Enseignement Math. **35**, (1989), 209-215.

[20] V. Bergelson and A. Leibman, *Polynomial extensions of van der Waerden's and Szemerédi's theorems*, Journal of Amer. Math. Soc. **9**, (1996), 725-753

[21] V. Bergelson and J. Rosenblatt, *Mixing actions of groups*, Ill. J. of Math. **32**, (1988), 65-80.

[22] P. Billingsley, *Ergodic theory and information*, Wiley, New York, 1965.

[23] G. D. Birkhoff, *Recurrence theorem on strongly transitive systems*, Proc. of the Academy of Science USA **17**, (1931), 650-655.

[24] ———, *Proof of the ergodic theorem*, Proc. of the Academy of Science USA **17**, (1931), 656-660.

[25] F. Blanchard, *Fully positive topological entropy and topological mixing*, Contemp. Math. Symbolic Dynamics and Applications **135**, Amer. Math. Soc. , Providence, 1992, pp. 95-105.

[26] ———, *A disjointness theorem involving topological entropy*, Bull. de la Soc. Math. de France **121**, (1993), 465-478.

[27] F. Blanchard, E. Glasner and B. Host, *A variation on the variational principle and applications to entropy pairs*, Ergod. Th. Dynam. Sys. **17**, (1997), 29-43.

[28] F. Blanchard, B. Host, A. Maass, S. Martínez and D. Rudolph *Entropy pairs for a measure*, Ergod. Th. and Dynam. Sys. **15**, (1995), 621-632.

[29] F. Blanchard, B. Host and A. Maass *Topological complexity*, Ergod. Th. Dynam. Sys. **20**, (2000), 641-662.

[30] F. Blanchard and Y. Lacroix *Zero-entropy factors of topological flows*, Proc. Amer. Math. Soc. **119**, (1993), 985-992.

[31] J. Bourgain, *On the spectral type of Ornstein's class one transformations*, Israel J. Math. **84**, (1993), 53-63.

[32] R. Bowen, *Entropy for group endomorphisms and homogeneous spaces*, Trans. Amer. Math. Soc. **153**, (1971), 401-414.

[33] ———, *On axiom A diffeomorphisms*, Regional Conference Series in Mathematics, Amer. Math. Soc. **35**, 1978.

[34] M. Boyle *Algebraic aspects of symbolic dynamics*, Topics in symbolic dynamics and applications, Editors: F. Blanchard, A. Maass and A. Nogueira, LMS Lecture note Series **279**, Cambridge University Press, 2000, 57-88.

[35] L. Breiman, *Probability*, Addison-Wesley, Reading, Massachusetts, 1968.

[36] I. U. Bronstein, *Extensions of minimal transformation groups*, Sijthoff & Noordhoff, 1979.

[37] R.B. Burckel, *Weakly almost periodic functions on semigroups*, Notes on mathematics and its applications, Gordon and Breech, New York, 1970.

[38] R. Burton, M. Keane and J. Serafin, *Residuality of dynamical morphisms*, Colloq. Math. **84/85**, Dedicated to the memory of Anzelm Iwanik, part 2, (2000), 307-317.

[39] R. Burton and A. Rothstein, *Isomorphism theorems in ergodic theory*, Technical report, Oregon State University, 1977.

[40] R. V. Chacón, *Weakly mixing transformations which are not strongly mixing*, Proc. Amer. Math. Soc. **22**, (1969), 559-562.

[41] P.-A. Cherix, M. Cowling, P. Jolissaint, P. Julg and A. Valette, *Groups with the Haagerup property (Gromov's a-T-menability)*, Progress in Mathematics, **197**, Birkhäuser Verlag, Basel, 2001.

[42] A.Connes and B. Weiss, *Property T and almost invariant sequences*, Israel J. Math. **37**, (1980), 209-210.

[43] I. P. Cornfeld, S. V. Fomin and Ya. G. Sinai, *Ergodic theory*, Springer-Verlag, New York, 1982.

[44] A. I. Danilenko, *Entropy theory from orbital point of view*, Monatsh. Math. **134**, (2001), 121-141.

[45] A. I. Danilenko and K. K. Park, *Generators and bernoullian factors for amenable actions and cocycles on their orbits*, to appear, Ergod. Th. Dynam. Sys.

[46] M. Denker, C. Grillenberger and K. Sigmund, *Ergodic theory on compact spaces*, Lecture Notes in Math. **527**, Springer-Verlag, 1976.

[47] J. Diestel, *Sequences and series in Banach spaces*, Graduate texts in mathematics **92**, Springer-Verlag, 1984.

[48] J. L. Doob, *Stochastic processes*, Wiley, New York, 1953.

[49] A. H. Dooley and V.Ya. Golodets, *The spectrum of completely positive entropy actions of countable amenable groups*, preprint, (2001).

[50] T. Downarowicz, *The Choquet simplex of invariant measures for minimal flows*, Israel J. of Math. **74**, (1991), 241-256.

[51] ———, *Weakly almost periodic flows and hidden eigenvalues*, Topological dynamics and applications, Contemporary Mathematics **215**, a volume in honor of R. Ellis, 1998, pp. 101-120.

[52] L. Dubins, *Bernstein-like polynomial approximation in higher dimensions*, Pacific J. of Math. **109**, (1983), 305-311.

[53] N. Dunford and J. Schwartz, *Linear operators*, Part I, Interscience, New York, third printing 1966.

[54] ———, *Linear operators*, Part II, Interscience, New York, third printing 1967.

[55] W. F. Eberlein, *abstract ergodic theorems and weak almost periodic functions*, Trans. Amer. Math. Soc. **67**, (1949), 217-240.

[56] R. Ellis, *Locally compact transformation groups*, Duke Math. J. **24**, (1957), 119-126.

[57] ———, *Distal transformation groups*, Pacific J. Math. **8**, (1957), 401-405.

[58] ———, *Lectures on Topological Dynamics*, W. A. Benjamin, Inc. , New York, 1969.

[59] R. Ellis and W. H. Gottschalk, *Homomorphisms of transformation groups*, Trans. Amer. Math. Soc. **94**, (1960), 258-271.

[60] R. Ellis and M. Nerurkar, *Weakly almost periodic flows*, Trans. Amer. Math. Soc. **313**, (1989), 103-119.

[61] R. Ellis and W. Perrizo, *Unique ergodicity of flows on homogeneous spaces*, Israel J. of Math. **29**, (1978), 276-284.

[62] R. Engelking, *General topology*, revised and completed edition, Heldermann Verlag, Berlin, 1989.

[63] J. Feldman, *Representations of invariant measures*, (dittoed notes, 17 pp.), 1963.

[64] J. M. G. Fell, *An extension of Mackey's method to Banach* algebraic bundles*, Memoires of the Amer. Math. Soc. **90**, 1969.

[65] W. Feller, *An introduction to probability theory and its applications, Vol. I* , second eddition, Wiley, New York, 1957.

[66] S. Ferenczi, *Systems of finite rank*, Colloquium. Math. **73**, (1997), 35-65.

[67] A. Fieldsteel, *Stability of the weak Pinsker property for flows*, Ergod. Th. Dynam. Sys. **4**, (1984), 381-390.

[68] M. Foreman and B. Weiss *An anti-classification theorem for ergodic measure-preserving transformations*, in preparation.

[69] M. Frank, *Ergodic theory in the 1930's: a study in international mathematical activity.* Mathematics Unbound: The evolution of an international mathematical community, 1800-1945, Eds.: K. Parshall & A. Rice

[70] N. Friedman, *Introduction to ergodic theory*, van Nostrand Reinhold, New York, 1970.

[71] H. Furstenberg, *Stationary processes and prediction theory*, Princeton University Press, Princeton, New Jersey, 1960.

[72] ———, *Strict ergodicity and transformation of the torus*, American J. of Math. **83**, (1961), 573-601.

[73] ———, *The structure of distal flows*, American J. of Math. **85**, (1963), 477-515.

[74] ———, *Disjointness in ergodic theory, minimal sets, and a problem in Diophantine approximation*, Math. System Theory **1**, (1967), 1-49.

[75] ———, *The unique ergodicity of the horocycle flow*, Recent advances in Topological dynamics (A. Beck, ed.) Springer-Verlag, Berlin, 1972, pp. 95-115.

[76] ———, *Ergodic behavior of diagonal measures and a theorem of Szemerédi on arithmetic progressions*, J. d'Analyse Math. **31**, (1977), 204-256.

[77] ———, *Recurrence in ergodic theory and combinatorial number theory*, Princeton university press, Princeton, N.J., 1981.

[78] H. Furstenberg and E. Glasner, *On the existence of isometric extensions*, Amer. J. of Math. **100**, (1978), 1185-1212.

[79] ———, *m-systems*, preprint, (1998).

[80] H. Furstenberg and Y. Katznelson, *A density version of the Hales-Jewett theorem*, J. d'Analyse Math. **57**, (1991), 64-119.

[81] H. Furstenberg and B. Weiss, *Topological dynamics and combinatorial number theory*, J. d'Analyse Math. **34**, (1978), 61-85.

[82] ———, *The finite multipliers of infinite transformation*, Spriger Verlag Lecture Notes in Math. **688**, (1978), 127-132.

[83] F. R. Gantmacher, *Applications of the theory of matrices*, Interscience, New York, 1959.

[84] A. M. Garsia, *A simple proof of E. Hopf's maximal ergodic theorem*, J. Math. Mech. **14**, (1956), 381-382.
[85] E. Glasner, *Proximal flows*, Lecture Notes in Math. **517**, Springer-Verlag, 1976.
[86] ———, *Quasifactors in ergodic theory*, Israel J. of Math. **45**, (1983), 198-208.
[87] ———, *A simple characterization of the set of μ-entropy pairs and applications*, Israel J. of Math. **102**, (1997), 13-27.
[88] ———, *Topological ergodic decompositions and applications to products of powers of minimal transformations*, J. d'Analyse Math. **64**, (1994), 241-262.
[89] ———, *Structure theory as a tool in topological dynamics*, Descriptive set theory and dynamical systems, LMS Lecture note Series **277**, Cambridge University Press, 2000, 173-209.
[90] ———, *Quasifactors of minimal systems*, Topol. Meth. in Nonlinear Anal. **16**, (2000), 351-370.
[91] E. Glasner, B. Host and D. Rudolph, *Simple systems and their higher order self-joinings*, Israel J. of Math. **78**, (1992), 131-142.
[92] E. Glasner and D. Maon, *Rigidity in topological dynamics*, Ergod. Th. Dynam. Sys. **9**, (1989), 309-320.
[93] E. Glasner J.-P. Thouvenot and B. Weiss, *Entropy theory without a past*, Ergodic theory and Dynam. Sys. **20**, (2000), 1355-1370.
[94] E. Glasner and B. Weiss, *Minimal transformations with no common factor need not be disjoint*, Israel J. of Math. **45**, (1983), 1-8.
[95] ———, *Processes disjoint from weak mixing*, Trans. Amer. Math. Soc. **316**, (1989), 689-703.
[96] ———, *On the construction of minimal skew-products*, Israel J. of Math. , **34**, (1979), 321-336.
[97] ———, *Sensitive dependence on initial conditions*, Nonlinearity **6**, (1993), 1067-1075.
[98] ———, *Strictly ergodic, uniform positive entropy models*, Bull. Soc. Math. France **122**, (1994), 399-412.
[99] ———, *A simple weakly mixing transformation with nonunique prime factor*, Amer. J. Math. **116**, (1994), 361-375.
[100] ———, *Quasifactors of zero-entropy systems*, J. of Amer. Math. Soc. **8**, (1995), 665-686.
[101] ———, *Topological entropy of extensions*, Proceedings of the 1993 Alexandria conferece, Ergodic theory and its connections with harmonic analysis, Editors: K. E. Petersen and I. A. Salama, LMS Lecture note Series **205**, Cambridge University Press, 1995, 299-307.
[102] ———, *Kazhdan's property T and the geometry of the collection of invariant measures*, GAFA, Geom. funct. anal. **7**, (1997), 917-935.
[103] ———, *Quasifactors of ergodic systems with positive entropy*, to appear, Israel J. of Math.
[104] T. N. T. Goodman, *Relating topological entropy with measure theoretic entropy*, Bull. London. Math. Soc. **3**, (1971), 176-180.
[105] G. R. Goodson, *A survey of recent results in the spectral theory of ergodic dynamical systems*, J. Dynam. Control Systems **5**, (1999), 173-226.
[106] W. H. Gottschalk and G. A. Hedlund, *Topological Dynamics*, AMS Colloquium Publications, Vol. 36, 1955.
[107] F. P. Greenleaf, *Invariant means on topological groups*, Van Nostrand, New York, 1969
[108] U. Grenander, *Stochastic processes and statistical inference*, Archiv for Matematik **1**, (1950), 195-277.
[109] A. Grothendieck, *Critères de compacité dans les espaces fonctionnels généraux*, Amer. J. of Math. **74**, (1952), 168-186.
[110] F. Hahn and Y. Katznelson, *On the entropy of uniquely ergodic transformations*, Trans. Amer. Math. Soc. **126**, (1967), 335-360.
[111] F. Hahn and W. Parry, *Some characteristic properties of dynamical systems with quasi-discrete spectrum*, Math. Systems Theory **2**, (1968), 179-190.
[112] P. R. Halmos, *Measure theory*, Van Nostrand, Princeton, 1950.
[113] ———, *Lectures on ergodic theory*, Chelsea Publishing Company, New York, 1956.
[114] J. Hansel and J. P. Raoult, *Ergodicity, uniformity and unique ergodicity*, Indiana. Univ. Math. J. **23**, (1974), 221-237.

[115] P. de la Harpe, *Topics in geometric group theory*, Chicago Lectures in Mathematics, University of Chicago Press, Chicago, 2000.

[116] P. de la Harpe et A. Valette, *La propriété (T) de Kazhdan pour les group localement compacts*, Astérisque, 1989.

[117] G. A. Hedlund, *Fuchsian groups and transitive horocycle*, Duke Jour. of Math. **2**, (1936), 530-542.

[118] ———, *Dynamics of geodesic flows*, Bull. Am. Math. Soc. **45**, (1939), 241-260.

[119] E. Hewitt and K. Ross, *Abstract harmonic analysis, Volume I*, Springer-Verlag, Berlin, 1963.

[120] ———, *Abstract harmonic analysis, Volume II*, Springer-Verlag, Berlin, 1970.

[121] E. Hewitt and L.J. Savage, *Symmetric measures on Cartesian products*, Trans. Amer. Math. Soc. **80**, (1955), 470-501.

[122] G. Hjorth, *On invariants for measure preserving transformations*, Fund. Math. **169**, (2001), 51-84.

[123] B. Host, *Mixing of all orders and independent joinings of systems with singular spectrum*, Israel J. Math. **76**, (1991), 289-298.

[124] ———, *Substitution subshifts and Bratteli diagrams*, Topics in symbolic dynamics and applications, Editors: F. Blanchard, A. Maass and A. Nogueira, LMS Lecture note Series **279**, Cambridge University Press, 2000, 35-55.

[125] R. Howe and C.C. Moore, *Asymptotic properties of unitary representations*, J. Funct. Anal. **32**, (1979), 299-314.

[126] K. Itô, *On the ergodicity of a certain stationary process*, Proc. Imper. Acad. Tokyo **20**, (1944), 54-55.

[127] K. Jacobs, *Lecture notes on ergodic theory*, Aarhus University, 1962.

[128] K. Jacobs and M. Keane, *0-1-sequences of Toeplitz type*, Z. Wahrsch. **13**, (1969), 123-131.

[129] R. I. Jewett, *The prevalence of uniquely ergodic systems*, J. Math. Mech. **19**, (1970), 717-729.

[130] V. F. R. Jones and K. Schmidt, *Asymptotically invariant sequences and approximate finiteness*, Amer. J. of Math. **109**, (1987), 91-114.

[131] A. del Junco, *Transformations with discrete spectrum are stacking transformations*, Canad. J. of Math. **24**, (1976), 836-839.

[132] ———, *A simple map with no prime factors*, Israel. J. of Math. **104**, (1998), 301-320.

[133] A. del Junco and M. Keane, *On generic points in the cartesian square of Chacón's transformation*, Ergodic theory and Dynam. Sys. **5**, (1985), 59-69.

[134] A. del Junco and M. Lemańczyk, *Generic spectral properties of measure-preserving maps and applications*, Proc. Amer. Math. Soc. **115**, (1992), 725-736.

[135] ———, *Simple systems are disjoint with Gaussian systems*, Studia Math. **133**, (1999), 249-256.

[136] A. del Junco, M. Lemańczyk and M. K. Mentzen, *Semisimplicity, joinings and group extensions*, Studia Math. **112**, (1995), 141-164.

[137] A. del Junco, M. Rahe and L. Swanson *Chacón's automorphism has minimal self-joinings*, J. d'Analyse Math. **37**, (1980), 276-284.

[138] A. del Junco and D.J. Rudolph, *On ergodic actions whose self-joinings are graphs*, Ergod. Th. Dynam. Sys. **7**, (1987), 531-557.

[139] ———, *A rank-one rigid simple prime map*, Ergod. Th. Dynam. Sys. **7**, (1987), 229-247.

[140] S. Kalikow, *T, T^{-1} transformation is not loosely Bernoulli*, Annals of Math. **115**, (1982), 393-409.

[141] ———, *Two fold mixing implies three fold mixing for rank one transformations*, Ergod. Th. Dynam. Sys. **4**, (1984), 237-259.

[142] R. Kallman, *Certain quatint spaces are countably separated, III*, J. Funct. Anal. **22**, (1976), 225-241.

[143] J. W. Kammeyer and D. Rudolph, *Restricted orbit equivalence for actions of discrete amenable groups*, preprint, (181 pages, 1995).

[144] A. Katok, *Smooth non-Bernoulli K-automorphisms*, Invent. Math. **61**, (1980), 291-300.

[145] A. Katok (assisted by E. A. Robinson), *Constructions in Ergodic Theory*, unpublished notes, 1986.

[146] A. Katok and B. Hasselblatt, *Modern theory of dynamical systems*, Encyclopedia of mathematics and its applications, Vol. **54**, Cambridge University Press, 1995.

[147] Y. Katznelson and B. weiss, *When all points are recurrent/generic* , Ergodic theory and dynamical systems I, Proceedings, Special year, Maryland 1979-80, Birkhäuser, Boston, 1981.

[148] D. A. Kazhdan, *Connection of the dual space of a group with the structure of its closed subgroups*, Funct. Anal. and its Appl. **1**, (1967), 63-65.

[149] M. Keane, *Ergodic theory and subshifts of finite type*, Ergodic theory, symbolic dynamics and hyperbolic spaces, Editors: T. Bedford, M. Keane and C. Series, Oxford University Press, 1991, 35-70.

[150] M. Keane and J. Serafin, *On the countable generator theorem*, Fund. Math. **157**, (1998), 255-259.

[151] M. Keane and M. Smorodinsky, *Bernoulli schemes of the same entropy are finitarily isomorphic*, Ann. Math. **109**, (1979), 397-406.

[152] A. S. Kechris, *Classical descriptive set theory*, Springer-Verlag, Graduate texts in mathematics **156**, 1991.

[153] H. B. Keynes and J. B. Robertson, *Eigenvalue theorems in topological transformation groups*, Trans. Amer. Math. Soc. **139**, (1969), 359-369.

[154] J. King, *The commutant is the weak closure of the powers for rank one transformations*, Ergod. Th. Dynam. Sys. **6**, (1986), 363-384.

[155] ______, *Ergodic properties where order 4 implies infinite order*, Israel J. of Math. **80**, (1992), 65-86.

[156] B. Kitchens, *Symbolic dynamics. One-sided, two-sided and countable Markov shifts*, Springer-Verlag, 1998.

[157] A. N. Kolmogorov, *A new metric invariant of transitive dynamical systems and Lebesgue space automorphisms*, Dokl. Acad. Sci. SSSR **119**, (1958), 861-864.

[158] S. Kremp, *An elementary proof of the Eberlein-Šmulian's theorem and the double limit criterion*, Arc. Math. **47**, (1986), 66-69.

[159] U. Krengel, *Ergodic theorems*, W. de Gruyter, Berlin, 1985.

[160] W. Krieger, *On entropy and generators of measure-preserving transformations*, Trans. Amer. Math. Soc. **149**, (1970), 453-464.

[161] ______, *On unique ergodicity*, Proc. Sixth Berkeley Simposium on Math. Stat. and Prob., (1970), 327-346.

[162] F. Ledrappier, *Un champ markovien peut etre d'entropie nulle et meëlangeant*, C. R. Acad. Sc. Paris, Ser. A **287**, (1987), 561-562.

[163] K. de Leeuw and I. Glicksberg, *Applications of almost periodic compactifications*, Acta Math. **105**, (1961), 63-97.

[164] E. Lehrer, *Topological mixing and uniquely ergodic systems*, Israel J. of Math. **57**, (1987), 239-255.

[165] M. Lemańczyk, *Introduction to ergodic theory from the point of view of spectral theory*, Lecture notes, Edited by Geon Ho Choe, Korea Advanced Institute of Science and Technology, Mathematics Research center, Korea, 1995.

[166] M. Lemańczyk and E. Lesigne *Ergodicity of Rohlin cocycles*, J. d'Analyse Math. **85**, (2001), 43-86.

[167] M. Lemańczyk, F. Parreau and J.-P. Thouvenot, *Gaussian automorphisms whose ergodic self-joinings are Gaussian*, Fund. Math. **164**, (2000), 253–293.

[168] M. Lemańczyk, J.-P. Thouvenot and B. Weiss, *Relative discrete spectrum and joinings*, Monatsh. Math. **137**, (2000), 57-75.

[169] D. Lind and B. Marcus, *An introduction to symbolic dynamics and coding*, Cambridge University Press, 1995.

[170] E. Lindenstrauss, *Lowering topological entropy*, J. d'Analyse Math. **67**, (1995), 231-267.

[171] ______, *Measurable distal and topological distal systems*, Ergod. Th. Dynam. Sys. **19**, (1999), 1063-1076.

[172] ______, *Pointwise theorems for amenable groups*, Invent. Math. **146**, (2001), 259-295.

[173] J. Lindenstrauss, G. H. Olsen and Y. Sternfeld, *The Poulsen simplex*, Ann. Inst. Fourier (Grenoble) **28**, (1978), 91-114.

[174] G. W. Mackey, *Borel structures in groups and their duals*, Trans. Amer. Math. Soc. **85**, (1957), 134-165.

[175] ______, *Ergodic theory and virtual groups*, Math. Ann. **166**, (1966), 187-297.

[176] ______, *Point realizations of transformation groups*, Illinois J. Math. **6**, (1962), 327-335.

[177] V. Mandrekar and M. Nadkarni, *On ergodic quasi-invariant measures on the circle group*, J. Funct. Anal. **3**, (1969), 157-163.

[178] B. Marcus, *Unique ergodicity of some flows related to axiom A diffeomorphiss*, Israel. J. of Math. **21**, (1975), 111-132.

[179] ———, *The horocycle flow is mixing of all degrees*, Invent. Math. **46**, (1978), 201-209.

[180] N. G. Markley and M. E. Paul, *Almost automorphic minimal sets without unique ergodicity*, Israel. J. of Math. **34**, (1979), 259-272.

[181] D. C. McMahon, *Relativized weak disjointness and relative invariant measures*, Trans. Amer. Math. Soc. **236**, (1978), 225-137.

[182] M. Misiurewicz, *A short proof of the variational principle for a* $\mathbb{Z}_+^n$ *action on a compact space*, Int. Conf. Dyn. Systems in Math. Physics, Société Mathématique de France, Astérisque **40** (1976), 147-158.

[183] S. Mozes, *Mixing of all orders of Lie group actions*, Invent. Math. **107** (1992), 235-241. Erratum, Invent. Math. **119** (1995), 399.

[184] M. G. Nadkarni, *Spectral theory of dynamical systems*, Birkhauser advanced texts, 1998.

[185] I. Namioka, *Separate continuity and joint continuity*, Pacific J. of Math. **51** (1974), 515-531.

[186] J. Neveu, *Mathemaical foundations of the calculus of probability*, Holden-Day, Inc., San Francisco, 1965.

[187] ———, *Processus aléatoires gaussiens*, Presses de l'Université de Montréal, 1971.

[188] P. J. Nicholls, *The ergodic theory of discrete groups*, London Math. Soc. Lecture Note Series **143**, Cambridge University Press, 1989.

[189] D. S. Ornstein, *Bernoulli shifts with the same entropy are isomorphic*, Advances in Math. **4**, (1970), 337-352.

[190] ———, *On the root problem in ergodic theory*, Proc. sixth Berkeley symposium Math. Statist. Probab. Univ. of California Press, 1970, 347-356.

[191] ———, *A K-automorphism with no square root and Pinsker's conjecture*, Advances in Math. **10**, (1973), 89-102.

[192] ———, *Ergodic theory, randomness and dynamical systems*, Yale Mathematical Monographs **5**, Yale University Press, New Haven, CT, 1974.

[193] D.S. Ornstein and P. Shields, *An uncountable family of K-auotomorphisms*, Advances in Math. **10**, (1973), 63-88.

[194] D.S. Ornstein and B. Weiss, *Geodesic flows are Bernoullian*, Israel. J. of Math. **14**, (1973), 184-198.

[195] ———, *Ergodic theory of amenable group actions I. The Rohlin lemma*, Bull. Amer. Math. Soc. **2**, (1980), 161-164.

[196] ———, *The Sannon-McMillan-Breiman theorem for amenable groups*, Israel. J. of Math. **44**, (1983), 53-60.

[197] ———, *Entropy and isomorphism theorems for actions of amenable groups*, J. d'Analyse Math. **48**, (1987), 1-141.

[198] ———, *On the Bernoulli nature of systems with some hyperbolic structure*, Ergod. Th. Dynam. Sys. **18**, (1998), 441-456.

[199] W. Parry, *Intrinsic Markov chains*, Trans. Amer. Math. Soc. **112**, (1964), 55-66.

[200] ———, *Zero entropy of distal and related transformations*, in *Topological dynamics*, J. Auslader and W. Gottschalk, eds., Benjamin, New York, 1967.

[201] ———, *Entropy and generators in ergodic theory*, W. A. Benjamine, INC, New York, 1969.

[202] ———, *Dynamical systems on nilmanifolds*, London Math. Soc. **2**, (1970), 37-40.

[203] ———, *Topics in ergodic theory*, Cambridge University Press, Cambridge, 1981.

[204] K. Petersen, *Disjointness and weak mixing of minimal sets*, Proc. Amer. Math. Soc. **24**, (1970), 278-280.

[205] ———, *Ergodic theory*, Cambridge University Press, Cambridge, 1983.

[206] R. R. Phelps, *Lectures on Choquet's theorem*, Van Nostrand mathematical studies **7**, new York, 1966.

[207] S. Polit, *Weakly isomorphic maps need not be isomorphic*, Ph.D. Dissertation, Stanford university, 1974.

[208] M. Queffelec, *Substitution dynamical systems*, Lecture Notes in Math. **1294**, Springer-Verlag, 1987.

[209] A. Ramsay, *Virtual groups and group actions*, Advances in Math. **6**, (1971), 253-322.

[210] M. Ratner, *Horocycle flows, joinings and rigidity of products*, Anna. of Math. **118**, (1983), 277-313.

[211] V. A. Rohlin, *Exact endomorphisms of Lebesgue spaces*, Izv. Acad. Si. USSR, Ser. Mat. **25**, (1961), 499-530.

[212] ———, *Lectures on the entropy theory of measure-preserving transformations*, Russian Math. Surveys **22**, (1967), no. 5, 1-52.

[213] V. A. Rohlin and Ja. G. Sinai, *Construction and properties of invariant measurable partitions*, Dokl. Acad. Nauk. SSSR. **141**, (1961), 1038-1041. Sov. Math. **2** (6) (1961), 1611-1614.

[214] A. Rosental, *Strictly ergodic models and amenable group actions*, preprint, (1988).

[215] H. L. Royden, *Real Analysis*, third edition, Macmillan Publishing Co., New York, 1988.

[216] W. Rudin, *Fourier analysis on groups*, Interscience Publishers, New York, 1960.

[217] D. J. Rudolph, *An example of a measure-preserving transformation with minimal self-joinings and applications*, J. d'Analyse Math. **35**, (1979), 97-122.

[218] ———, *k-fold mixing lifts to weakly mixing isometric extensions*, Ergod. Th. Dynam. Sys. **5**, (1985), 445-447.

[219] ———, *$\mathbb{Z}^n$ and $\mathbb{R}^n$ cocycle extensions and complementary algebras*, Ergod. Th. Dynam. Sys. **6**, (1986), 583-599.

[220] ———, *Fundamentals of Measurable Dynamics*, Clarendon Press, Oxford, 1990.

[221] D. J. Rudolph and B. Weiss, *Entropy and mixing for amenable group actions*, Ann. of Math. **151**, (2000), 1119-1150.

[222] V.V. Ryzhikov, *Joinings and multiple mixing of finite rank actions*, Funktsional. Anali. i Prilozhen. **27**, (1993), 63-78; translation in Funct. Anal. Appl. **27**, (1993), 128-140.

[223] ———, *Joinings, intertwining operators, factors and mixing properties of dynamical systems*, Russian Acad. Izv. Math. **42**, (1994), 91-114.

[224] ———, *Around simple dynamical systems. Induced joinings and multiple mixing*, J. Dynam. Control Systems **3**, (1997), 111-127.

[225] J. Serafin *Finitary codes and isomorphisms*, Ph.D. Thesis, Technische Universiteit Delft, 1996.

[226] K. Schmidt, *Cocycles of ergodic transformation groups*, Lecture Notes in Math. Vol. **1**, MacMillan Co. of India, 1977.

[227] ———, *Asymptotically invariant sequences and an action of $SL(2,\mathbb{R})$ on the 2-sphere*, Israel J. of Math. **37**, (1980), 193-208.

[228] ———, *Amenability, Kazhdan's property T, strong ergodicity and invariant means for ergodic group actions*, Ergod. Th. and Dynam. Sys. **1**, (1981), 223-236.

[229] ———, *Asymptotic properties of unitary representations and mixing*, Proc. London Math. Soc. **48**, (1984), 445-460.

[230] ———, *Algebraic ideas in ergodic theory*, AMS-CBMS Reg. Conf. **76**, Providence, 1990.

[231] ———, *Dynamical systems of algebraic origin*, Birkhäuser, 1995.

[232] ———, *Mildly mixing actions of locally compact groups*, Proc. London Math. Soc. **45**, (1982), 506-518.

[233] L. Shapiro, *Proximality in minimal transformation groups*, Proc. Amer. Math. Soc. **26**, (1970), 521-525.

[234] P. Shields, *The theory of Benoulli shifts*, University of Chicago Press, 1973.

[235] ———, *The ergodic theory of discrete sample paths*, Graduate studies in mathematics **13**, Amer. Math. Soc. 1996.

[236] M. Shub and B. Weiss, *Can one always lower topological entropy?*, Ergod. Th. and Dynam. Sys. **11** (1991), 535-546.

[237] Ja. G. Sinai, *A weak isomorphism of transformations with an invariant measure*, Dokl. Acad. Nauk. S.S.S.R. **147**, (1962), 797-800.

[238] M. Smorodinsky, *Ergodic theory, entropy*, Lecture Notes in Math. **214**, Springer-Verlag, New Yourk, 1971.

[239] ———, *Information, entropy and Bernoulli systems*, Development of mathematics 1950–2000, 993-1012, Birkhuser, Basel, 2000.

[240] M. Smorodinsky and J.-P. Thouvenot, *Bernoulli factors that span a transformation*, Israel J. of Math. **32**, (1979), 39-43.

[241] A. M. Stepin, *Spectral properties of generic dynamical systems*, Math. U.S.S.R. Izv. **50**, (1986), 159-192.

[242] E. Szemerédi's, *On sets of integers containing no k elements in arithmetic progression*, Acta Arith. **27**, (1975), 199-245.

[243] A. Tempelman, *Ergodic theorems for group actions*, Kluwer Acad. Publ. , Dordrecht, 1992.

[244] J.-P. Thouvenot, *Quelques propriétés des systèmes dynamiques qui se décomposent en un produit de deux systmes dont l'un est un schéma de Bernoulli*, Israel J. of Math. **21**, (1975), 177-207.

[245] ———, *On the stability of the weak Pinsker property*, Israel J. of Math. **27**, (1977), 150-162.

[246] ———, *Some properties and applications of joinings in ergodic theory*, Proceedings of the 1993 Alexandria conferece, Ergodic theory and its connections with harmonic analysis, Editors: K. E. Petersen and I. A. Salama, LMS Lecture note Series **205**, Cambridge University Press, 1995, 207-238.

[247] V. S. Varadarajan, *Groups of automorphisms of Borel spaces*, Trans. Amer. Math. Soc. **109**, (1963), 191-220.

[248] ———, *Geometry of quantum theory*, Vol II, Van Nostrand, Princeton, 1970.

[249] ———, *Lie groups, Lie algebras, and their representations*, Graduate texts in mathematics **102**, Springer-Verlag, 1984.

[250] W. A. Veech, *The equicontinuous structure relation for minimal Abelian transformation groups*, Amer. J. of Math. **90**, (1968), 723-732.

[251] ———, *A fixed point theorem-free approach to weak almost periodicity*, Trans. Amer. Math. Soc. **177**, (1973), 353-362.

[252] ———, *Topological dynamics*, Bull. Amer. Math. Soc. **83**, (1977), 775-830.

[253] ———, *Unique ergodicity of horospherical flows*, Amer. J. of Math. **99**, (1977), 827-859.

[254] ———, *A criterion for a process to be prime*, Monatsh. Math. **94**, (1982), 335-341.

[255] J. von Neumann, *Proof of the quasi-ergodic hypothesis*, Proc. Nat. Acad. Sci. USA **18**, (1932), 70-82.

[256] J. de Vries, *Elements of Topological Dynamics*, Kluwer Academic Publishers, 1993.

[257] P. Walters, *An introduction to ergodic theory*, Springer-Verlag, New York, 1982.

[258] A. Weil, *L'intégration dans les groupes topologiques et ses applications*, Actualités Sci. et Ind. 869, 1145. Paris: Hermann & Cie. 1941 and 1951.

[259] B. Weiss, *Topological transitivity and ergodic measures*, Math. Systems Theory **5**, (1971), 71-75.

[260] ———, *Ergodic theory*, Unpublished lecture notes, 1972.

[261] ———, *Strictly ergodic models for dynamical systems*, Bull. of the Amer. Math. Soc. **13**, (1985), 143-146.

[262] ———, *Countable generators in dynamics — Universal minimal models*, Contemporary Mathematics **94**, (1989), 321-326.

[263] ———, *Multiple recurrence and doubly minimal systems*, Contemporary Mathematics **215**, (1998), 189-196.

[264] ———, *Single orbit dynamics*, CBMS, Regional Conference Series in Math. **95**, Amer. Math. Soc. Providence RI, 2000.

[265] H. Weyl, *Über die Gleichverteilung von Zahlen mod Eins*, Math. Ann. **77**, (1916), 313-352.

[266] S. Williams, *Toeplitz minimal flows which are not uniquely ergodic*, Z. Wahrsch. **67**, (1984), 95-107.

[267] R. J. Zimmer, *Extensions of ergodic group actions*, Illinois J. Math. **20**, (1976), 373-409.

[268] ———, *Ergodic actions with generalized discrete spectrum*, Illinois J. Math. **20**, (1976), 555-588.

[269] ———, *Ergodic theory and semisimple groups*, Birkhäuser, 1984.

Index of Symbols

$\mathcal{A}(K)$, $H(\mathcal{Y})$ (Galois correspondence for group extensions), **57**, **133**, 135
$\langle A \rangle$ (icer generated by A), **341**
$\mathbb{A}_n$, $\mathbb{A}_F$ (ergodic sums), **79**, **83**
$AP(\Gamma)$ (algebra of almost periodic functions), **38**
$\operatorname{Aut}(X, \Gamma)$ (group of automorphisms of a topological dynamical system), **13**
$\operatorname{Aut}(\mathbf{X})$ (group of automorphisms of a measure dynamical system), **56**
α^-, α^T, **258**
$\alpha \overset{\epsilon}{\subset} \beta$, **349**
$B(p) = B(p_1, \ldots, p_\ell)$ (Bernoulli shifts), **82**
$\mathbf{B}(\mathbf{X})$, **67**
BV, BV^-, **235**
$B(\Gamma)$, B_π (spaces of matrix coefficients), **38**
βX, $\beta\Gamma$ (Stone-Čech compactification), **30**, **363**
$\mathfrak{c}$, $|\mathfrak{c}|$ (a column and its carrier), **271**
$c(K)$, $c_\mu(K)$ (entropy capacity), **316**
$c_0(\Gamma)$ (space of functions vanishing at infinity), **38**
$\mathcal{C}_0$, **178**
$\mathcal{C}_1$, **186**
$\Delta_X = \{(x, x) : x \in X\}$ (diagonal), **13**
$\bar{d}$ (metric), **283**
$\bar{d}^N$ (metric), **285**
d_{ent}, **275**
d_{ham}, **271**
d_{meas}, **270**, 274, 278, 285, 349, 354
d_{part}, **274**, 275
$\operatorname{dist}(\alpha)$, **357**
$\mathbb{E}$, $\mathbb{E}^{\mathcal{F}}$ (expectation, conditional expectation with respect to the σ-algebra $\mathcal{F}$), **365**
$\mathbf{E}(\mathbf{X})$, **67**
$E(X, \Gamma)$, $E(X, T)$ (enveloping semigroup), **28**
E_μ (μ-entropy pairs), **332**
$E_{(X,T)}$, E_X (topological entropy pairs), **332**
$\mathcal{E}(\mathbf{X}/\mathbf{Y})$ (closure of $\mathbf{Y}$-eigenfunctions), **181**
$EQ(X)$ (set of equicontinuity points), **34**
$\operatorname{ext}(Q)$, $\operatorname{ext}(K)$ (set of extreme points), **367**, 97, 168, 231
$\mathbf{G}(\mathbf{X})$, **67**
$\operatorname{gr}(\mu, \phi)$ (graph joining), **126**
$\dot{\mathfrak{G}}$ (Hilbert-Schmidt bundle), **185**
$H(\mathcal{U})$ (topological entropy of the cover $\mathcal{U}$), **249**
$h(\mathcal{U})$ (topological T-entropy of the cover $\mathcal{U}$), **249**
$h_{\text{top}}(X, T)$ (topological entropy of the system (X, T)), **249**
$H(\alpha)$ (entropy of the partition α), **254**
$H(\alpha|\mathcal{F})$ (conditional entropy of the partition α with respect to the σ-algebra $\mathcal{F}$), **254**
$h_\mu(\alpha) = h_\mu(\alpha, T)$ (T, μ-entropy of the partition α), **258**
$h_\mu(\alpha|\mathcal{F})$ (conditional T, μ-entropy of the partition α with respect to the σ-algebra $\mathcal{F}$), **258**
$h(\mathbf{X})$ (entropy of the dynamical system $\mathbf{X}$), **258**
$h(\mathbf{X}|\mathbf{Y})$ (conditional entropy of the dynamical system $\mathbf{X}$ relative to the factor $\mathbf{Y}$), **258**
$H(\mathcal{Y})$, (Galois correspondence for group extensions), **133**, 135
$\mathfrak{H}$ (Hilbert space), **62**
$\dot{\mathfrak{H}}$ (Hilbert bundle), **178**
$I(\alpha)$, $I^{\mathcal{F}}(\alpha)$ ((conditional) information function), **254**
$\mathbf{I}(\mathbf{X})$, **67**
$\mathfrak{I}$, $\operatorname{Inv}(L^2(\mathbf{X}))$, **63**, **67**, **181**
$\operatorname{Iso}(\mathfrak{H})$ (group of isometries of a Hilbert space), **235**
$J(\mathbf{X})$, $J_e(\mathbf{X})$, $J(\mathbf{X}, \mathbf{Y})$, $J_e(\mathbf{X}, \mathbf{Y})$ (set of (ergodic) joinings), **125**
$J^\rho(\mathbf{X})$, $J^{\nu \times \nu}(\mathbf{X})$, **127**
$J^{\mathbf{Z}}(\mathbf{X}, \mathbf{Y})$ (joinings over a common factor), **127**
$\hat{\jmath}^{(k)}(\mathbf{X})$, **148**
κ', $\kappa^{(k)}$, $\bar{\kappa}^{(k)}$, $\kappa^{(\infty)}$, $\bar{\kappa}^{(\infty)}$ (joinings associated with a quasifactor), **164**
$\mathfrak{L} = \{1, 2, \ldots, \ell\}$ (an alphabet), **16**, 269, 310
$\mathcal{L}(\mathfrak{H})$ (The Banach algebra of bounded linear operators on a Hilbert space), **62**

λ^* (adjoint joining), **129**
$\lambda(\alpha \triangle \beta)$, **274**
$\mathfrak{m}$ (invariant mean on WAP functions), **43**
$M_\Gamma(X)$, $M_T(X)$ (set of Γ-invariant probability measures), **96**
$M_\Gamma^{\mathrm{erg}}(X)$, $M_T^{\mathrm{erg}}(X)$ (set of ergodic Γ-invariant probability measures), **97**, **147**
$M_{\max}(X)$ (set of measures of maximal entropy), **310**
$\mathbf{M(X)}$, $\mathbf{M(X, Y)}$ (semigroup (set) of commuting Markov operators), **132**
$\mu(\alpha \triangle \beta)$ $(= d^\mu_{\mathrm{part}}(\alpha, \beta))$, **274**
$N(A, B) = \{\gamma \in \Gamma : \gamma A \cap B \neq \emptyset\}$, **13**
$N(x, B) = \{\gamma \in \Gamma : \gamma x \in B\}$, **13**
$N(\mathcal{U})$ (minimal cardinality of a subcover of the cover $\mathcal{U}$), **249**
$\mathcal{O}_\Gamma x$ (Γ-orbit of x), **13**
$\bar{\mathcal{O}}_\Gamma x$ (Γ-orbit closure of x), **13**
$P = P(X, \Gamma), P_\pi, P^{(n)}$ (proximal relations), **22**
$P_{\mathfrak{C}}, P_{\mathfrak{I}}$ (orthogonal projections), **67**
$\Pi(\mathbf{X})$ (Pinsker algebra), **319**
$Q(\mu)$, $Q(\mathbf{X})$ (set of quasifactors), **160**
$Q_J(\mu)$, $Q_J(\mathbf{X})$, $Q_J^e(\mathbf{X})$ (set of (ergodic) joining quasifactors), **171**
$Q(\pi), Q_J(\pi)$, etc. (quasifactors over a factor), **160**, **171**
$Q = Q(X, \Gamma), Q_\pi$ (regionally proximal relations), **22**
$QG(x)$ (set of measures for which x is quasi-generic), **98**
$R_\pi = \{(x, x') : \pi(x) = \pi(x')\}$ (relation determined by a homomorphism), **15**
$S(\mu)$ (property), **333**
$\mathrm{s}(\mu)$, $\mathrm{s}^2(\mu)$ (support), **336**
SFT (subshift of finite type), **17**
$sp_n(\epsilon)$, $sr_n(\epsilon)$, $sp(\epsilon, T)$, $sr(\epsilon, T)$, $sp(T, d)$, $sr(T, d)$ (Bowen's entropy), **251**
sym (symmetrization map), **58**
$\mathfrak{S}$ (group of permutations), **168**
σ_x (spectral measure), **116**
$\mathfrak{t}$, $|\mathfrak{t}|$ (a tower and its carrier), **271**
T (Kazhdan's property), **231**
T_w (Weak Kazhdan property), **243**
$\tau \circ \lambda$ (composition of joinings), **130**
u (minimal idempotent), **30**
u (minimal idempotent in $\mathbf{E}$), **63**, **67**
U_γ, U_T (Koopman representation), **67**
$\mathcal{U}(\mathfrak{H})$ (The group of unitary operators), **38**
$W_0(\Gamma)$ (space of flight functions), **44**
$WAP(\Gamma)$ (algebra of weakly almost periodic functions), **38**
$\mathbf{X} = (X, \mathcal{X}, \mu, \Gamma)$, $\mathbf{X} = (X, \mathcal{X}, \mu, T)$ (measure-preserving dynamical systems), **53**, **55**
$\mathbf{X}^{\langle r \rangle}$ (r-symmetric product system), **58**
(X, Γ), $(X, \mathbb{Z})$, (X, T) (topological dynamical systems), **13**
(X^n, Γ) (n-fold product system), **13**
X_d, X_{eq} (maximal distal (equicontinuous) factor), **18**
X_P, $X_P(\mu)$ (topological Pinsker factors), **341**
$\mathbf{Y} = (Y, \mathcal{Y}, \nu, \Gamma)$, $\mathbf{Y} = (Y, \mathcal{Y}, \nu, T)$ (measure-preserving dynamical systems), **53**, **55**
$\mathbf{Y} \underset{\alpha}{\times} (U, \rho)$ (skew-product with cocycle α), **57**
$\mathbf{Z} = (Z, \mathcal{Z}, \eta, \Gamma)$, $\mathbf{Z} = (Z, \mathcal{Z}, \eta, T)$ (measure-preserving dynamical systems), **53**, **55**
$Z(x)$ (cyclic subspace generated by x), **116**
$\mathbf{1}_\Gamma \prec \pi, \mathbf{1}_\Gamma \subset \pi$, **235**

Index of Terms

Titles in This Series

99 **Philip S. Hirschhorn,** Model categories and their localizations, 2003
98 **Victor Guillemin, Viktor Ginzburg, and Yael Karshon,** Moment maps, cobordisms, and Hamiltonian group actions, 2002
97 **V. A. Vassiliev,** Applied Picard-Lefschetz theory, 2002
96 **Martin Markl, Steve Shnider, and Jim Stasheff,** Operads in algebra, topology and physics, 2002
95 **Seiichi Kamada,** Braid and knot theory in dimension four, 2002
94 **Mara D. Neusel and Larry Smith,** Invariant theory of finite groups, 2002
93 **Nikolai K. Nikolski,** Operators, functions, and systems: An easy reading. Volume 2: Model operators and systems, 2002
92 **Nikolai K. Nikolski,** Operators, functions, and systems: An easy reading. Volume 1: Hardy, Hankel, and Toeplitz, 2002
91 **Richard Montgomery,** A tour of subriemannian geometries, their geodesics and applications, 2002
90 **Christian Gérard and Izabella Łaba,** Multiparticle quantum scattering in constant magnetic fields, 2002
89 **Michel Ledoux,** The concentration of measure phenomenon, 2001
88 **Edward Frenkel and David Ben-Zvi,** Vertex algebras and algebraic curves, 2001
87 **Bruno Poizat,** Stable groups, 2001
86 **Stanley N. Burris,** Number theoretic density and logical limit laws, 2001
85 **V. A. Kozlov, V. G. Maz′ya, and J. Rossmann,** Spectral problems associated with corner singularities of solutions to elliptic equations, 2001
84 **László Fuchs and Luigi Salce,** Modules over non-Noetherian domains, 2001
83 **Sigurdur Helgason,** Groups and geometric analysis: Integral geometry, invariant differential operators, and spherical functions, 2000
82 **Goro Shimura,** Arithmeticity in the theory of automorphic forms, 2000
81 **Michael E. Taylor,** Tools for PDE: Pseudodifferential operators, paradifferential operators, and layer potentials, 2000
80 **Lindsay N. Childs,** Taming wild extensions: Hopf algebras and local Galois module theory, 2000
79 **Joseph A. Cima and William T. Ross,** The backward shift on the Hardy space, 2000
78 **Boris A. Kupershmidt,** KP or mKP: Noncommutative mathematics of Lagrangian, Hamiltonian, and integrable systems, 2000
77 **Fumio Hiai and Dénes Petz,** The semicircle law, free random variables and entropy, 2000
76 **Frederick P. Gardiner and Nikola Lakic,** Quasiconformal Teichmüller theory, 2000
75 **Greg Hjorth,** Classification and orbit equivalence relations, 2000
74 **Daniel W. Stroock,** An introduction to the analysis of paths on a Riemannian manifold, 2000
73 **John Locker,** Spectral theory of non-self-adjoint two-point differential operators, 2000
72 **Gerald Teschl,** Jacobi operators and completely integrable nonlinear lattices, 1999
71 **Lajos Pukánszky,** Characters of connected Lie groups, 1999
70 **Carmen Chicone and Yuri Latushkin,** Evolution semigroups in dynamical systems and differential equations, 1999
69 **C. T. C. Wall (A. A. Ranicki, Editor),** Surgery on compact manifolds, second edition, 1999
68 **David A. Cox and Sheldon Katz,** Mirror symmetry and algebraic geometry, 1999
67 **A. Borel and N. Wallach,** Continuous cohomology, discrete subgroups, and representations of reductive groups, second edition, 2000

TITLES IN THIS SERIES

40.1 **Daniel Gorenstein, Richard Lyons, and Ronald Solomon,** The classification of the finite simple groups, number 1, 1994

39 **Sigurdur Helgason,** Geometric analysis on symmetric spaces, 1994

38 **Guy David and Stephen Semmes,** Analysis of and on uniformly rectifiable sets, 1993

37 **Leonard Lewin, Editor,** Structural properties of polylogarithms, 1991

36 **John B. Conway,** The theory of subnormal operators, 1991

35 **Shreeram S. Abhyankar,** Algebraic geometry for scientists and engineers, 1990

34 **Victor Isakov,** Inverse source problems, 1990

33 **Vladimir G. Berkovich,** Spectral theory and analytic geometry over non-Archimedean fields, 1990

32 **Howard Jacobowitz,** An introduction to CR structures, 1990

31 **Paul J. Sally, Jr. and David A. Vogan, Jr., Editors,** Representation theory and harmonic analysis on semisimple Lie groups, 1989

30 **Thomas W. Cusick and Mary E. Flahive,** The Markoff and Lagrange spectra, 1989

29 **Alan L. T. Paterson,** Amenability, 1988

28 **Richard Beals, Percy Deift, and Carlos Tomei,** Direct and inverse scattering on the line, 1988

27 **Nathan J. Fine,** Basic hypergeometric series and applications, 1988

26 **Hari Bercovici,** Operator theory and arithmetic in H^∞, 1988

25 **Jack K. Hale,** Asymptotic behavior of dissipative systems, 1988

24 **Lance W. Small, Editor,** Noetherian rings and their applications, 1987

23 **E. H. Rothe,** Introduction to various aspects of degree theory in Banach spaces, 1986

22 **Michael E. Taylor,** Noncommutative harmonic analysis, 1986

21 **Albert Baernstein, David Drasin, Peter Duren, and Albert Marden, Editors,** The Bieberbach conjecture: Proceedings of the symposium on the occasion of the proof, 1986

20 **Kenneth R. Goodearl,** Partially ordered abelian groups with interpolation, 1986

19 **Gregory V. Chudnovsky,** Contributions to the theory of transcendental numbers, 1984

18 **Frank B. Knight,** Essentials of Brownian motion and diffusion, 1981

17 **Le Baron O. Ferguson,** Approximation by polynomials with integral coefficients, 1980

16 **O. Timothy O'Meara,** Symplectic groups, 1978

15 **J. Diestel and J. J. Uhl, Jr.,** Vector measures, 1977

14 **V. Guillemin and S. Sternberg,** Geometric asymptotics, 1977

13 **C. Pearcy, Editor,** Topics in operator theory, 1974

12 **J. R. Isbell,** Uniform spaces, 1964

11 **J. Cronin,** Fixed points and topological degree in nonlinear analysis, 1964

10 **R. Ayoub,** An introduction to the analytic theory of numbers, 1963

9 **Arthur Sard,** Linear approximation, 1963

8 **J. Lehner,** Discontinuous groups and automorphic functions, 1964

7.2 **A. H. Clifford and G. B. Preston,** The algebraic theory of semigroups, Volume II, 1961

7.1 **A. H. Clifford and G. B. Preston,** The algebraic theory of semigroups, Volume I, 1961

6 **C. C. Chevalley,** Introduction to the theory of algebraic functions of one variable, 1951

5 **S. Bergman,** The kernel function and conformal mapping, 1950

4 **O. F. G. Schilling,** The theory of valuations, 1950

3 **M. Marden,** Geometry of polynomials, 1949

2 **N. Jacobson,** The theory of rings, 1943

1 **J. A. Shohat and J. D. Tamarkin,** The problem of moments, 1943